LONDON MATHEMATICAL SOCIETY LECTURE NOTE SERIES

Managing Editor: Professor N. J. Hitchin, Mathematical Institute,
University of Oxford, 24–29 St Giles, Oxford OX1 3LB, United Kingdom

The titles below are available from booksellers, or from Cambridge University Press
at www.cambridge.org/mathematics

London Mathematical Society Lecture Note Series: 351

Finite von Neumann Algebras and Masas

ALLAN M. SINCLAIR
University of Edinburgh

ROGER R. SMITH
Texas A&M University

CAMBRIDGE
UNIVERSITY PRESS

CAMBRIDGE
UNIVERSITY PRESS

Shaftesbury Road, Cambridge CB2 8EA, United Kingdom

One Liberty Plaza, 20th Floor, New York, NY 10006, USA

477 Williamstown Road, Port Melbourne, VIC 3207, Australia

314–321, 3rd Floor, Plot 3, Splendor Forum, Jasola District Centre, New Delhi – 110025, India

103 Penang Road, #05–06/07, Visioncrest Commercial, Singapore 238467

Cambridge University Press is part of Cambridge University Press & Assessment, a department of the University of Cambridge.

We share the University's mission to contribute to society through the pursuit of education, learning and research at the highest international levels of excellence.

www.cambridge.org
Information on this title: www.cambridge.org/9780521719193

First published 2008

A catalogue record for this publication is available from the British Library

Library of Congress Cataloging-in-Publication data
Sinclair, Allan M.
Finite von Neumann algebras and masas / Allan M. Sinclair, Roger R. Smith.
p. cm.
Includes index.
ISBN 978-0-521-71919-3 (pbk.)
1. Von Neumann algebras. I. Smith, Roger R. II. Title.
QA326.S565 2008
512´.556 – dc22 2007050518

ISBN 978-0-521-71919-3 Paperback

London Mathematical Society Lecture Note Series: 351

Finite von Neumann Algebras and Masas

ALLAN M. SINCLAIR
University of Edinburgh

ROGER R. SMITH
Texas A&M University

CAMBRIDGE
UNIVERSITY PRESS

CAMBRIDGE
UNIVERSITY PRESS

Shaftesbury Road, Cambridge CB2 8EA, United Kingdom

One Liberty Plaza, 20th Floor, New York, NY 10006, USA

477 Williamstown Road, Port Melbourne, VIC 3207, Australia

314–321, 3rd Floor, Plot 3, Splendor Forum, Jasola District Centre, New Delhi – 110025, India

103 Penang Road, #05–06/07, Visioncrest Commercial, Singapore 238467

Cambridge University Press is part of Cambridge University Press & Assessment,
a department of the University of Cambridge.

We share the University's mission to contribute to society through the pursuit of
education, learning and research at the highest international levels of excellence.

www.cambridge.org
Information on this title: www.cambridge.org/9780521719193

First published 2008

A catalogue record for this publication is available from the British Library

Library of Congress Cataloging-in-Publication data
Sinclair, Allan M.
Finite von Neumann algebras and masas / Allan M. Sinclair, Roger R. Smith.
p. cm.
Includes index.
ISBN 978-0-521-71919-3 (pbk.)
1. Von Neumann algebras. I. Smith, Roger R. II. Title.
QA326.S565 2008
512´.556 – dc22 2007050518

ISBN 978-0-521-71919-3 Paperback

Contents

Preface

The authors wish to thank all the mathematicians who have contributed to the ideas that are presented in these notes. The informal discussions, seminars, papers and reprints of our friends and colleagues have shaped our approach to these notes. We are indebted to our co-authors, Erik Christensen, Ken Dykema, Florin Pop, Sorin Popa, Guyan Robertson, Stuart White and Alan Wiggins, who have directly, and indirectly, influenced us.

We wish to thank Jan Cameron, Kunal Mukherjee, Gabriel Tucci and Alan Wiggins who, while research students, read parts of the notes and made many constructive comments on them. Particular thanks go to Stuart White who read the entire manuscript, made many suggestions for improvements, and saved us from numerous errors.

Our wives Pat and Ginny have been extremely patient and understanding during the long periods that we have been immersed in this project. We have received much support and advice from Roger Astley at Cambridge University Press. Our thanks go to Robin Campbell who with great expertise turned our manuscripts and rough LaTeX files into this book.

We take this opportunity to record our sincere gratitude to the National Science Foundation. Grants to the second author have enabled us to meet on numerous occasions in the last few years, allowing us to bring this book to completion.

Chapter 1

General introduction

These notes are an introduction to some of the theory of finite von Neumann algebras and their von Neumann subalgebras, with the emphasis on maximal abelian self-adjoint subalgebras (usually abbreviated *masas*). Assuming basic von Neumann algebra theory, the notes are fairly detailed in covering the basic construction, perturbations of von Neumann subalgebras, general results on masas and detailed ones on singular masas in II_1 factors. Due to the large volume of research on finite von Neumann algebras and their masas the authors have been forced to be selective of the topics included. Nevertheless, a substantial body of recent research has been covered.

Each chapter of the book has its own introduction, so the overview of the contents below will be quite brief. We have also included a discussion of a few important results which have been omitted from the body of the text. In each case, we felt that the amount of background required for a reasonably self-contained account was simply too much for a book of this kind.

We have tried to make the material accessible to graduate students who have some familiarity with von Neumann algebras at the level of a first course in the subject. The early chapters review some of this, but are best read by the beginner with one of the standard texts, [104, 105, 187], to hand to fill in any gaps.

1.1 Synopsis

The book falls naturally into five parts. The first of these comprises Chapters 2, 3, 4, 5, 6 and 8 in which we lay out some of the foundations of the subject. The papers of Murray and von Neumann [116, 117, 202, 118] introduced the subject of von Neumann algebras (then called *Rings of Operators*) and are still influential today. The finite algebras are, roughly speaking, those that admit a faithful normal tracial state, and are closest in spirit to the matrix algebras, which are particular examples. Murray and von Neumann paid particular attention to the finite algebras, and established the close connection to the theory of discrete groups that persists today. They also introduced the crossed product construction from a group acting by automorphisms of a von Neumann algebra, notably an abelian one. In this way masas appear naturally in von Neumann algebras. In principle, masas can always be constructed by using Zorn's lemma, but this is rarely enlightening. The important examples always make their appearance from some auxiliary algebraic structure, for example maximal abelian subgroups of groups.

Chapter 2 classifies the masas in $B(H)$, the algebra of bounded operators on a separable Hilbert space. Although our focus is on II_1 factors, these cannot be studied in isolation, and various constructions will produce algebras of types I and II_∞ (but no type III factors will appear in these notes).

Chapter 3 gives an overview of the basic theory of finite von Neumann algebras and of the standard constructions of tensor products and crossed products. Since there is such a close connection to discrete groups, examples of masas arising from groups are presented, and the precursor to the conditional expectations are discussed at the level of group algebras where they are very easy to understand. An important characterisation of diffuse abelian algebras is given, and the chapter ends with a brief discussion of hyperfiniteness. The fundamental work of Connes, [36], on this topic is summarised without proofs.

The following chapter is devoted to the basic construction. This is an algebra $\langle N, e_B \rangle$ which arises from a von Neumann subalgebra B of a finite factor N. It is of fundamental importance in the theory of subfactors and also in perturbation theory. A detailed exposition of its properties is given, including the construction of its canonical semifinite trace (see also Appendix C for a different approach to this construction). Some simple examples are included.

Chapters 5, 6 and 8 deal with various technical issues. The first of these concerns the basic operators of von Neumann algebra theory—the unitaries, projections and partial isometries. Various approximation results are proved, and several important $\|\cdot\|_2$-norm estimates are given, all to be used subsequently. The next chapter continues in this spirit, and discusses various technical issues concerning normalising unitaries as well as orthogonality in von Neumann algebras. The background material is rounded out in Chapter 8 by presenting some estimates for operators in type I_∞ von Neumann algebras. We have avoided any discussion of direct integrals in these notes, but the material of Chapter 8 is essentially this topic in an embryonic form.

Chapter 7 introduces the Pukánszky invariant of a masa. At one level, all masas in separable II_1 factors are the same since all are isomorphic to $L^\infty[0,1]$. However, this ignores the relationship between a masa A and its containing factor N, and the invariant $\text{Puk}(A)$ addresses this. The masa A and its reflection JAJ in the commutant combine to generate an abelian algebra whose commutant restricted to $L^2(N) \ominus L^2(A)$ decomposes as a direct sum of type I_n von Neumann algebras for n in the range $\{1, 2, \ldots, \infty\}$. Those integers that appear then constitute $\text{Puk}(A)$. There is a discussion of this invariant in the context of group factors where everything can be related to the group structure, allowing many examples to be presented. An alternative approach, based on important work of Tauer, [190], from the 1960s, has enabled White, [207], to show that all possible values of Puk(A) can be realised in certain factors including the hyperfinite one. Space considerations have forced us to omit this work.

The third group of chapters concerns perturbations of masas. This topic is split between Chapters 9 and 10, the first dealing with basic theory and second with extensions to general subalgebras. The problem is to consider two subalgebras A and B of a factor N which are close in an appropriate sense and then to look for a partial isometry w such that $wAw^* \subseteq B$. This creates a spatial isomorphism between compressions of the two algebras. This theory has played a decisive role in the resolution of some old questions in von Neumann algebra theory. We expand on this below.

Chapters 11–16 present various special aspects of masas. The focus of Chapter 11 is the theory of singular masas and we include a discussion of the Laplacian masa in free group factors. Chapter 12 is devoted to the construction of singular and semiregular masas in all finite factors. Chapter 13 explores the topic of Cartan masas, which is closely connected to the theory of hyperfinite subfactors, and there is also a discussion of property Γ and its relationship to masas. Maximally injective masas and subfactors are presented in Chapter 14, and the subsequent chapter looks at non-separable factors which can arise from ultrapowers. The last chapter presents some recent work of Shen [171] on singly generated algebras, a subject which relies heavily on the theory of masas.

The book concludes with three appendices. The first develops the theory of ultrapowers and includes some further material on property Γ. The second discusses the basic theory of unbounded operators. These types of operators appear in the perturbations of Chapters 9 and 10, so this appendix covers just that part of the theory which is used in these applications. The final appendix gives a second approach to the existence of the trace in the basic construction, first presented in Chapter 4.

1.2 Further results

There are three major results about masas which we have omitted from this book, due to the amount of background material that would have been needed to give a reasonably self-contained account of them.

The first of these is a uniqueness result for Cartan masas in the hyperfinite factor R. No automorphism of R can satisfy $\phi(A) = B$ when A and B are masas with distinct Pukánszky invariants, and such pairs do occur. However, the result of Connes, Feldman and Weiss, [40], for Cartan masas is as follows:

Theorem 1.2.1. *Let A and B be Cartan masas in the hyperfinite* II_1 *factor R. Then there exists an automorphism ϕ of R such that $\phi(A) = B$.*

The original proof in [40] is in the context of ergodic theory. A more operator-theoretic proof is presented in [141].

The second theorem answers an old question concerning the existence of Cartan masas in finite factors. The theory of free probability, [200], was devel-

oped as part of an investigation of the free group factors. In [199], this was used
by Voiculescu to obtain

Theorem 1.2.2. *There do not exist Cartan masas in the free group factors*
$L(\mathbb{F}_n)$, $2 \leq n \leq \infty$.

This settled the existence question for the standard types of masas since it
had been shown earlier by Popa, [139, 140], that singular and semiregular masas
exist in all separable II_1 factors (see Chapter 12).

The third problem that we wish to mention concerns the fundamental group,
which is a multiplicative subgroup of $\mathbb{R}^+$ associated to any II_1 factor. The
question is whether any subgroup can be the fundamental group of a II_1 factor.
There had been some partial results that are discussed in Section 16.4, but the
definitive answer was obtained by Popa [149]:

Theorem 1.2.3. *If G is a subgroup of $\mathbb{R}^+$, then there is a II_1 factor N whose
fundamental group in G. If G is countable, then N may be taken to be separable.*

The proof depends on Gabariau's work in ergodic theory [70], as well as the
perturbation results on masas discussed in Chapters 9 and 10.

Many mathematicians have made important contributions to the theory of
masas, but amongst these two names stand out: Jacques Dixmier and Sorin
Popa. This early work on masas in the 1950s is due to Dixmier, who did much
to establish this topic as a viable field of study. Many of the later developments,
in the periods 1980–1985 and from 2000 to the present, are due to Popa and
several of our chapters are drawn substantially from his work. Without the
fundamental contributions of these two researchers, this book could not have
been written.

Chapter 2

Masas in $B(H)$

2.1 Introduction

Our main objective in this chapter is to describe the maximal abelian self-adjoint subalgebras (masas) of $B(H)$ in Section 2.3. To avoid technicalities concerning cardinalities, we will only discuss the case when H is separable. There are two basic types of masas, discrete and diffuse. For the first, fix an orthonormal basis $\{\xi_n\}_{n=1}^{\infty}$ for H and let p_n be the rank one projection onto $\mathbb{C}\xi_n$, $n \geq 1$. Then A, the von Neumann algebra generated by these projections, is a masa, and has many minimal projections. For the second type, let $L^{\infty}[0,1]$ act on $L^2[0,1]$ as multiplication operators. This is a masa, established below, but in contrast to the first type, it has no minimal projections. Up to unitary equivalence, each masa in $B(H)$ will be a direct sum of the two types.

In Section 2.4, we discuss masas in type I_n von Neumann algebras, where $n \in \mathbb{N}$ is arbitrary. These algebras have the form $A \otimes \mathbb{M}_n$ for some abelian von Neumann algebra A, where $\mathbb{M}_n$ denotes the algebra of scalar $n \times n$ matrices. Each of these algebras contains an obvious diagonal masa $A \otimes \mathbb{D}_n$, where $\mathbb{D}_n$ is the algebra of diagonal $n \times n$ matrices, and Theorem 2.4.3 establishes that this is the only masa up to unitary equivalence. We then conclude the chapter by introducing abelian projections and proving the useful result that they occur in all masas in finite type I von Neumann algebras.

Before we embark on the study of masas, we recall in Section 2.2 some of the standard theorems of von Neumann algebra theory. The ones chosen are those which will be used many times in the succeeding pages of this book.

2.2 Standard theorems

There are many important theorems in operator algebras, but the four that we recall here are of particular relevance to the topics of these notes. They can be found in all of the textbooks in this field, and are restated here for the reader's convenience (see, for example, the books by Dixmier [48, 49], Kadison and Ringrose [104, 105], Pedersen [131], Sakai [167] and Takesaki [187]). All are valid without restriction on the cardinality of the Hilbert space.

If a von Neumann algebra N is represented on a Hilbert space H, then a net (x_α) from N converges strongly to $x \in N$ if $\lim_\alpha \|x_\alpha \eta - x\eta\|_2 = 0$ for all vectors $\eta \in H$. If, in addition, $\lim_\alpha x_\alpha^* = x^*$ strongly, then we say that $x_\alpha \to x$ *-strongly. These notions are distinct: if v denotes the adjoint of the unilateral shift operator on $\ell^2(\mathbb{N})$, then $\lim_{n\to\infty} v^n = 0$ strongly, while this is not so for powers of v^*. Basic functional analysis shows that weak and strong closures of convex sets coincide, giving a choice of how to state these results.

The first one, due to von Neumann, [201], describes the strong closure of a *-algebra of operators in purely algebraic terms.

Theorem 2.2.1 (The double commutant theorem). *Let A be a *-algebra of operators on a Hilbert space H and suppose that 1 is in the strong (or weak) closure of A. Then the strong and weak closures of A are both equal to the double commutant A''. In particular, A is a von Neumann algebra if and only if $A = A''$.*

The next theorem, due to Dixmier, [46], is a type of averaging result, and such techniques will appear frequently in these notes, taking various forms. We let Z denote the centre of a von Neumann algebra N, and the closure of the convex set below is taken in the norm topology. Another way of describing the theorem is to say that appropriately chosen convex combinations of unitary conjugates of any fixed element approach the centre arbitrarily closely. The simplicity of finite factors is one consequence of this result (see Theorem A.3.2).

Theorem 2.2.2 (The Dixmier approximation theorem). *Let N be a von Neumann algebra with unitary group $\mathcal{U}(N)$. For each $x \in N$,*

$$Z \cap \overline{\mathrm{conv}}\{uxu^* : u \in \mathcal{U}(N)\} \neq \emptyset.$$

The third theorem is due to Kaplansky [106]. The most important feature for us is the norm estimate of the first part, since it is quite possible to have nets which are unbounded in norm, but nevertheless converge strongly. Our statement of the result combines the versions of [105, 187].

Theorem 2.2.3 (The Kaplansky density theorem). *Let $N \subseteq B(H)$ be a von Neumann algebra and let A be a strongly dense *-subalgebra, not assumed to be unital.*

(i) *If $x \in N$, then there exists a net (x_α) from A converging *-strongly to x and satisfying $\|x_\alpha\| \leq \|x\|$ for all α.*

(ii) *If $x \in N$ is self-adjoint then the net in (i) may be chosen with the additional property that each x_α is self-adjoint.*

(iii) *If $u \in N$ is a unitary and A is a unital C^*-algebra, then there is a net (u_α) of unitaries from A converging *-strongly to u.*

The last of these theorems is due to Tomita and may be found in [105, Theorem 11.2.16]. It is a consequence of another of Tomita's theorems, [105, Theorem 9.2.9], that has played a fundamental role in the modern development of certain aspects of the theory. The simplicity of its statement is in contrast to the difficulty of its proof.

Theorem 2.2.4 (Tomita's commutant theorem). *Let $M \subseteq B(H)$ and $N \subseteq B(K)$ be von Neumann algebras. Then*

$$(M \overline{\otimes} N)' = M' \overline{\otimes} N'.$$

Early in the development of the subject, von Neumann introduced direct integrals [203] which allowed him to decompose a von Neumann algebra on a separable Hilbert space into a direct integral over the centre of factors (those algebras with trivial centre). This focused attention on factors, the prevailing view being that direct integral theory would extend results for factors to separably acting von Neumann algebras, which is largely correct. Murray and von Neumann introduced the type classification of factors in their seminal series of papers [116, 117, 118, 202]. There are algebras of types I, II_1, II_∞ and III. The reader will find the original definitions in [105], which also contains theorems making these equivalent to the following descriptions. Type I breaks down further into I_n, $n \geq 1$, and type I_∞, and these correspond respectively to the algebras $\mathbb{M}_n$ of $n \times n$ matrices over $\mathbb{C}$, and to the algebras $B(H)$ of bounded operators on infinite dimensional Hilbert spaces H, one for each cardinality of H. The II_1 factors are those which are infinite dimensional and admit a finite trace, and are often called finite factors. The II_∞ factors are those which arise from the tensor product of a type II_1 with a type I_∞, while any factor which does not fall into the classes already described is called type III. The original definitions also cover algebras with nontrivial centre and the direct integral theory works well here: a separably acting von Neumann algebra is of type α precisely when the factors in its direct integral decomposition are all of type α ($\alpha \in \{\mathrm{I}, \mathrm{II}_1, \mathrm{II}_\infty, \mathrm{III}\}$). The type III algebras play no further role in these notes, which mainly concern those of type II_1. However, various constructions lead to algebras of types I and II_∞. As we will see, the trace is fundamental for finite algebras, and we lose this when we move to the other two types. Fortunately, both have densely defined semifinite traces, and considerable use will be made of this.

All von Neumann algebras have identity elements, which we usually denote by 1. We will often need to consider containments $B \subseteq N$ of von Neumann algebras, and we adopt the following convention. We always suppose that the identities of N and B are equal unless **either** it is explicitly stated to the contrary, **or** it is clearly not the case from the context (e.g. $B = pNp$ for some projection $p \in N$).

2.3 Masas

The material in this section owes much to the presentation of these results to be found in [104, Chapter 5].

Recall that a vector ξ is *cyclic* for a von Neumann algebra $N \subseteq B(H)$ if the subspace $N\xi$ is norm dense in H. We say that ξ is *separating* for N if $x\xi = 0$ implies that $x = 0$ when $x \in N$. The first lemma gives a useful relationship between cyclic and separating vectors.

Lemma 2.3.1. *Let $N \subseteq B(H)$ be a von Neumann algebra. Then $\xi \in H$ is cyclic for N if and only if ξ is separating for N'.*

Proof. Suppose that ξ is cyclic for N, and let $x' \in N'$ be such that $x'\xi = 0$. Then

$$x'x\xi = xx'\xi = 0, \qquad x \in N, \tag{2.3.1}$$

and so $x' = 0$ since $N\xi$ is dense in H. Thus ξ is separating for N'.

Conversely, suppose that ξ is separating for N'. Let $p \in N'$ be the projection onto the norm closed span of $N\xi$. Then $p\xi = \xi$, so $(1 - p)\xi = 0$. Since ξ is separating, $p = 1$, which says that ξ is cyclic for N. $\qquad\square$

Both examples of discrete and diffuse masas above have cyclic vectors. In the first case $\xi = \sum_{n=1}^{\infty} \xi_n/2^n$ is cyclic, while in the second case the constant function 1 is cyclic for $L^\infty[0,1]$. The situation would change if we allowed non-separable Hilbert spaces. If S is an uncountable set then $\ell^\infty(S)$ is a masa when acting by multiplication on $\ell^2(S)$. Any vector in $\ell^2(S)$ has only a countable number of non-zero entries, and thus no cyclic vector can exist. The obstructions, of course, are the cardinality S and the resulting dimension of $\ell^2(S)$.

Lemma 2.3.2. *Let $A \subseteq B(H)$ be a masa, where H is a separable Hilbert space. Then there is a vector $\xi \in H$ which is both cyclic and separating for A.*

Proof. By Zorn's lemma, choose a maximal set $\{\xi_n\}_{n=1}^{\infty}$ of non-zero vectors such that the subspaces $\overline{A\xi_n}$, $n \geq 1$, are pairwise orthogonal. The separability of H allows us to enumerate this set. If $\eta \neq 0$ was orthogonal to all of these subspaces then $\overline{A\eta}$ would be orthogonal to each $\overline{A\xi_n}$, contradicting maximality. Thus $H = \bigoplus_{n=1}^{\infty} \overline{A\xi_n}$. Let p_n be the projection onto $\overline{A\xi_n}$, $n \geq 1$. Then $p_n \in A' = A$. Let $\xi = \sum_{n=1}^{\infty} \xi_n/(2^n \|\xi_n\|_2) \in H$. Then $\xi_n = (2^n \|\xi_n\|_2 p_n)(\xi)$, so $\xi_n \in \overline{A\xi}$, $n \geq 1$. It then follows that $\overline{A\xi_n} \subseteq \overline{A\xi}$, $n \geq 1$, and so ξ is cyclic for A. By Lemma 2.3.1, ξ is separating for A', which equals A. $\qquad\square$

We now establish a converse to this result, by showing that any abelian von Neumann algebra on a separable Hilbert space which has a cyclic vector must be a masa. This then proves that the example $L^\infty[0,1]$ above is indeed a masa on $L^2[0,1]$. We will require a preliminary lemma concerning tracial vectors. A vector $\xi \in H$ is said to be *tracial* for a von Neumann algebra $N \subseteq B(H)$ if

$$\langle xy\xi, \xi \rangle = \langle yx\xi, \xi \rangle, \qquad x, y \in N. \tag{2.3.2}$$

Note that if N is abelian then any vector is tracial for N.

Lemma 2.3.3. *Let $N \subseteq B(H)$ be a von Neumann algebra with a tracial cyclic and separating vector ξ. Then there is a conjugate linear isometry $J \colon H \to H$ with the following properties:*

(i) *$t \mapsto Jt^*J$, $t \in B(H)$, defines an anti-isomorphism $\pi \colon B(H) \to B(H)$;*

(ii) *π maps N onto N' and N' onto N.*

Proof. (i) For each $x \in N$,

$$\|x\xi\|_2^2 = \langle x\xi, x\xi \rangle = \langle x^*x\xi, \xi \rangle = \langle xx^*\xi, \xi \rangle = \|x^*\xi\|_2^2, \qquad (2.3.3)$$

so we may define a conjugate linear isometry $J \colon N\xi \to N\xi$ by $J(x\xi) = x^*\xi$, $x \in N$. Since ξ is cyclic for N, this map extends to H, also denoted J. If we define $\pi \colon B(H) \to B(H)$ by $\pi(t) = Jt^*J$ then, for $s, t \in B(H)$,

$$\pi(st) = Jt^*s^*J = Jt^*JJs^*J = \pi(t)\pi(s), \qquad (2.3.4)$$

using $J^2 = I$. This proves (i).

(ii) Now consider $x, y, z \in N$. We have

$$\begin{aligned}
(x\pi(y) - \pi(y)x)z\xi &= xJy^*Jz\xi - Jy^*Jxz\xi \\
&= xJy^*z^*\xi - Jy^*z^*x^*\xi \\
&= xzy\xi - xzy\xi = 0. \qquad (2.3.5)
\end{aligned}$$

Letting z vary over N, we conclude that $\pi(y)$ commutes with all $x \in N$, so $\pi(y) \in N'$. Thus π maps N to N'.

Let $y \in N'$ be self-adjoint and choose a sequence $\{x_n\}_{n=1}^\infty \in N$ such that $y\xi = \lim\limits_{n\to\infty} x_n\xi$, possible because ξ is cyclic for N. Then $\{x_n^*\xi\}_{n=1}^\infty$ is Cauchy, since $x_n^*\xi = Jx_n\xi$, and so this sequence converges to $\eta \in H$. For $z \in N$,

$$\begin{aligned}
\langle y\xi - \eta, z\xi \rangle &= \lim_{n\to\infty}\left(\langle y\xi, z\xi \rangle - \langle x_n^*\xi, z\xi \rangle\right) \\
&= \lim_{n\to\infty}\left(\langle z^*\xi, y\xi \rangle - \langle z^*x_n^*\xi, \xi \rangle\right) \\
&= \lim_{n\to\infty}\left(\langle z^*\xi, y\xi \rangle - \langle x_n^*z^*\xi, \xi \rangle\right) \\
&= \lim_{n\to\infty}\left(\langle z^*\xi, y\xi \rangle - \langle z^*\xi, x_n\xi \rangle\right) = 0, \qquad (2.3.6)
\end{aligned}$$

where we have used the tracial property of ξ and the fact that y is self-adjoint and commutes with z^*. Letting z vary, we conclude that $y\xi = \eta$, and so $y\xi$ is the limit of the sequence $\{((x_n + x_n^*)/2)\xi\}_{n=1}^\infty$. Replacing x_n by $(x_n + x_n^*)/2$, we may assume that x_n is self-adjoint. Then

$$Jy\xi = \lim_{n\to\infty} Jx_n\xi = \lim_{n\to\infty} x_n\xi = y\xi. \qquad (2.3.7)$$

A general element $y \in N'$ may be written as $y = y_1 + iy_2$ with y_1 and y_2 self-adjoint in N'. From (2.3.7), we obtain $Jy\xi = y^*\xi$. By Lemma 2.3.1, ξ is also cyclic and separating for N', so we may repeat the argument of (2.3.5), this time with $x, y, z \in N'$, to conclude that π maps N' to N. Since $\pi^2 = I$, it follows easily that π defines an anti-isomorphism of N onto N', proving (ii). $\square$

Theorem 2.3.4. Let $A \subseteq B(H)$ be an abelian von Neumann algebra on a separable Hilbert space H. Then A is a masa if and only if A has a cyclic vector.

Proof. We have already shown in Lemma 2.3.2 that if A is a masa then it has a cyclic vector. To show the converse, suppose that A has a cyclic vector ξ. Since A is abelian, we have that $A \subseteq A'$, and ξ is also cyclic for A'. By Lemma 2.3.1, ξ is separating for A, and it is of course tracial. By Lemma 2.3.3, there is a surjective anti-isomorphism $\pi \colon A \to A'$, showing that A' is also abelian. Thus $A' \subseteq (A')' = A$, proving that $A = A'$. Then A is a masa. □

Our next objective is to give a complete description of masas in $B(H)$. We have already met two types: $L^\infty[0,1]$ acting on $L^2[0,1]$ and the masas of diagonal operators relative to given orthonormal bases. A general masa will be a direct sum of the two types (where either type may be missing), and the diagonal masas are characterised by the cardinality of the orthonormal bases. We call a masa *diffuse* (or *continuous*) if it has no minimal non-zero projections. If it is generated by its minimal projections then we refer to it as *discrete*. The following lemma is valid for all separably acting von Neumann algebras and so we prove it in full generality, although we only apply it to masas. It will be useful in describing diffuse masas.

Lemma 2.3.5. Let H be a separable Hilbert space and let $N \subseteq B(H)$ be a von Neumann algebra. Then there is a sequence $\{p_n\}_{n=1}^\infty$ of projections in N which generates this von Neumann algebra.

Proof. The spectral theorem allows us to approximate a given self-adjoint operator by a finite linear combination of its spectral projections, and each operator is in the span of two self-adjoint ones. Thus we may reduce to showing that N is generated by a countable set of elements.

Fix a countable dense set of vectors $\{\xi_i\}_{i=1}^\infty$ in H. Let n be a fixed integer, and let

$$S_n = \{(x\xi_1, \ldots, x\xi_n)^T \colon x \in N, \ \|x\| \leq 1\} \subseteq H^n.$$

Then S_n is separable, and so there is a countable subset F_n of the unit ball of N such that $\{(x\xi_1, \ldots, x\xi_n)^T \colon x \in F_n\}$ is norm dense in S_n. A simple approximation argument then shows that $\bigcup_{n=1}^\infty F_n$ is a countable strongly dense subset of N, and thus generates the von Neumann algebra. □

Lemma 2.3.6. Let A be a diffuse masa in $B(H)$, where H is a separable Hilbert space. Then A is unitarily equivalent to the masa $L^\infty[0,1]$ in $B(L^2[0,1])$.

Proof. For each $\lambda \in [0,1]$, define the projection $f_\lambda \in L^\infty[0,1]$ to be $\chi_{[0,\lambda]}$. Since the constant function $1 \in L^2[0,1]$ is a cyclic vector for the von Neumann algebra B generated by these projections, Theorem 2.3.4 shows that B is a masa, and thus coincides with $L^\infty[0,1]$. We note that the set of f_λ's is totally ordered and that $f_\lambda f_\mu = f_{\min\{\lambda,\mu\}}$. We will construct a set of projections indexed by [0,1] inside A with similar properties, from which we will obtain the implementing unitary.

By Lemma 2.3.5, there is a sequence $\{g_n\}_{n=1}^{\infty}$ of projections in A which generates the masa. We now construct, inductively, finite totally ordered sets $S_1 \subseteq S_2 \ldots$ of projections in A with $g_n \in \mathrm{span}\{S_n\}$, guaranteeing that their union generates A. We let $S_1 = \{0, g_1, 1\}$. Suppose that $S_{n-1} = \{e_1 \leq e_2 \ldots \leq e_r\}$ has been constructed. Since $e_1 = 0$ and $e_r = 1$, we have that $g_n = \sum_{i=2}^{r} g_n(e_i - e_{i-1})$, and thus S_n may be constructed by inserting the projection $e_{i-1} + g_n(e_i - e_{i-1})$ between e_{i-1} and e_i, for $2 \leq i \leq r$. Then $g_n \in \mathrm{span}\{S_n\}$ because it may be expressed as $\sum_{i=2}^{r}((e_{i-1} + g_n(e_i - e_{i-1})) - e_{i-1})$. Now let S be the set of projections in A either of the form $\wedge\{p \in F\}$ or $\vee\{p \in F\}$, where F is an arbitrary subset of $\bigcup_{n=1}^{\infty} S_n$. By taking one point subsets, we have that S contains this union. We now show comparability of any two projections in S.

If $\wedge\{p \in F_1\}$ does not lie below $\wedge\{p \in F_2\}$, then there exists a non-zero vector $\eta \in H$ such that $\wedge\{p \in F_1\}\eta = \eta$ and $\wedge\{p \in F_2\}\eta = 0$, since these projections commute. Then there exists $p_0 \in F_2$ with $\|p_0\eta\|_2 < \|\eta\|_2$, so $p_0 \leq p$ for $p \in F_1$ since the opposite inequality cannot hold. It follows that $\wedge\{p \in F_2\}$ lies below $\wedge\{p \in F_1\}$, and any two projections of this type are comparable. The same is true for two suprema, by taking complements. Consider now $\wedge\{p \in F_1\}$ and $\vee\{p \in F_2\}$. If the first does not lie below the second then there is a non-zero vector η such that $\wedge\{p \in F_1\}\eta = \eta$ and $\vee\{p \in F_2\}\eta = 0$. Then $p \geq q$ for $p \in F_1$ and $q \in F_2$, showing that $\wedge\{p \in F_1\}$ dominates $\vee\{p \in F_2\}$. We conclude that S is totally ordered.

Since A is a masa, Lemma 2.3.2 gives a cyclic and separating vector ξ, which we may normalise so that $\|\xi\|_2 = 1$. Define $\phi\colon S \to [0,1]$ by

$$\phi(p) = \langle p\xi, \xi \rangle = \|p\xi\|_2^2, \qquad p \in S. \tag{2.3.8}$$

If $p_1, p_2 \in S$ then without loss of generality $p_1 \leq p_2$, so

$$\phi(p_2) - \phi(p_1) = \langle (p_2 - p_1)\xi, \xi \rangle = \|(p_2 - p_1)\xi\|_2^2. \tag{2.3.9}$$

Thus ϕ is order preserving and injective since ξ is separating. Any projection of the form $\vee\{p \in F\}$ is the supremum of an increasing sequence of projections as follows: enumerate F as $F = \{p_1, p_2, \ldots\}$ and let $\tilde{p}_i = \max\{p_1, \ldots, p_i\}$. Then $\vee\{p \in F\} = \vee\{\tilde{p}_i\colon i \geq 1\}$, showing that $\vee\{p \in F\}$ is the supremum of an increasing sequence of projections. A similar result holds for infima, and thus the range of ϕ is a closed subset of $[0,1]$ and is the closure of the range of ϕ when restricted to $\bigcup_{n=1}^{\infty} S_n$. If there is a gap (a, b) in the range of ϕ with projections $p_1, p_2 \in S$ satisfying $\phi(p_1) = a$, $\phi(p_2) = b$, then we may choose two orthogonal non-zero projections $q_1, q_2 \in A$ with $q_1 + q_2 = p_2 - p_1$, because A has no minimal projections. Let η_i, $i = 1, 2$, be unit vectors such that $q_i\eta_i = \eta_i$. When $p \in S$, either $p\eta_1 = p\eta_2 = 0$ for $p \leq p_1$, or $(1 - p)\eta_1 = (1 - p)\eta_2 = 0$ for $p \geq p_2$. It follows that $\|x\eta_1\|_2 = \|x\eta_2\|_2$ for any $x \in \mathrm{span}\{S\}$, and so q_1 is not in the strong closure which is A. This contradiction shows that the range of ϕ is $[0,1]$, and so the projections in S may be indexed by $[0,1]$ where p_λ is such that $\phi(p_\lambda) = \lambda$,

$\lambda \in [0, 1]$. Then

$$\left\| \sum_{i=1}^{k} \alpha_i p_{\lambda_i} \xi \right\|_2^2 = \sum_{i,j=1}^{k} \alpha_i \bar{\alpha}_j \langle p_{\lambda_i} p_{\lambda_j} \xi, \xi \rangle$$

$$= \sum_{i,j=1}^{k} \alpha_i \bar{\alpha}_j \min\{\lambda_i, \lambda_j\}$$

$$= \left\| \sum_{i=1}^{k} \alpha_i f_{\lambda_i} 1 \right\|_2^2, \qquad (2.3.10)$$

and so there is a unitary operator $u \colon H \to L^2[0,1]$ defined on a dense subspace by

$$u \left(\sum_{i=1}^{k} \alpha_i p_{\lambda_i} \xi \right) = \sum_{i=1}^{k} \alpha_i f_{\lambda_i} 1. \qquad (2.3.11)$$

Then

$$\left\langle u^* f_\lambda u \sum_{i=1}^{k} \alpha_i p_{\lambda_i} \xi, \sum_{j=1}^{k} \beta_j p_{\mu_j} \xi \right\rangle = \sum_{i,j=1}^{k} \alpha_i \bar{\beta}_j \langle f_\lambda f_{\lambda_i} f_{\mu_j} 1, 1 \rangle$$

$$= \sum_{i,j=1}^{k} \alpha_i \bar{\beta}_j \langle p_\lambda p_{\lambda_i} p_{\mu_j} \xi, \xi \rangle$$

$$= \left\langle p_\lambda \sum_{i=1}^{k} \alpha_i p_{\lambda_i} \xi, \sum_{j=1}^{k} \beta_j p_{\mu_j} \xi \right\rangle. \qquad (2.3.12)$$

We conclude that $u^* f_\lambda u = p_\lambda$, $\lambda \in [0, 1]$. Since these sets of projections have strongly dense spans in their respective masas, we see that u implements a spatial isomorphism between A and $L^\infty[0, 1]$. $\qquad \square$

This lemma constitutes most of the proof of the following theorem which is the main result of this section.

Theorem 2.3.7. *Let H be a separable Hilbert space and let A be a masa in $B(H)$. Then A is unitarily equivalent to a direct sum of a diagonal masa and $L^\infty[0, 1]$.*

Proof. Let $\{p_i\}_{i=1}^n$, $0 \le n \le \infty$, be a maximal set of orthogonal minimal projections in A, which is necessarily countable since H is separable. Let $p = \sum_{i=1}^n p_i$. Then $A(1-p)$ is diffuse and so unitarily equivalent to $L^\infty[0, 1]$ by Lemma 2.3.6, provided that $p \ne 1$. The masa $Ap \subseteq B(pH)$ is generated by its minimal projections, assuming $p \ne 0$. Each p_i has rank 1, otherwise it would be a sum of two non-zero projections commuting with A but not lying in A, an impossibility. It is then clear that Ap is a diagonal masa for a suitably chosen orthonormal basis for pH of cardinality n. $\qquad \square$

Remark 2.3.8. It seems difficult to simplify substantially the argument of Lemma 2.3.6, even in concrete cases. Consider, for example, the masa

$$L^\infty([0,1] \times [0,1]) \subseteq B(L^2([0,1] \times [0,1])).$$

2.4 Masas in type I_n algebras

In this section we consider type I_n von Neumann algebras. These have the form $A \otimes M_n$, where A is an abelian von Neumann algebra and M_n denotes the $n \times n$ matrices. The diagonal subalgebra $D_n \subseteq M_n$ leads to a masa $A \otimes D_n$ in $A \otimes M_n$, and our objective is to show that, up to unitary conjugation, this is the only masa in the latter algebra. More sophisticated results on diagonalising matrices over general von Neumann algebras may be found in [99, 100]. We will need two preliminary results, one on the structure of abelian von Neumann algebras A, and one on the equivalence of close projections. We return to the second of these in Lemma 5.2.10 where we also give some norm estimates. The first is standard in the theory of abelian von Neumann algebras: A is isomorphic to $C(\Omega)$, the algebra of continuous functions on a compact Hausdorff space Ω which is hyperstonean. We prove only that Ω is stonean, meaning that the closure of any open set in Ω is again open, since this is sufficient for our needs. The full result can be found in [104, Chapter 5]. Note that we place no restriction on the cardinality of A, so that these results are valid in full generality.

Lemma 2.4.1. *If A is an abelian von Neumann algebra with carrier space Ω, then Ω is stonean.*

Proof. Let $U \subseteq \Omega$ be a non-empty open set. The result is clear if the closure $\overline{U}$ is Ω, so we may assume that the complement $\overline{U}^c$ is non-empty. Let J be the ideal in $C(\Omega)$ of functions which vanish on U^c. Then the weak closure of J has the form Az for a projection $z \in C(\Omega)$ corresponding to a clopen set $\Omega_0 \subseteq \Omega$. For each $\omega \in U$, Urysohn's lemma gives a continuous function $f \in J$ vanishing on U^c and satisfying $f(\omega) \neq 0$, and so $z(\omega) \neq 0$. Thus $U \subseteq \Omega_0$. If $\omega_1 \in \overline{U}^c$, then applying Urysohn's lemma to $\{\omega_1\}$ and $\overline{U}$ gives a continuous function g so that $g(\omega_1) \neq 0$ and g vanishes on $\overline{U}$. Thus $fg = 0$ for all $f \in J$, and so $zg = 0$. Then $z(\omega_1) = 0$, so z vanishes on $\overline{U}^c$. Thus $U \subseteq \Omega_0 \subseteq \overline{U}$, showing that $\overline{U} = \overline{\Omega_0} = \Omega_0$. Thus $\overline{U}$ is open. $\square$

If A is an abelian von Neumann algebra with carrier space Ω, then $A \otimes M_n$ is identified with the algebra $C(\Omega, M_n)$ of continuous functions on Ω with values in M_n. The following lemma shows that two close projections in this algebra are unitarily equivalent, the proof being valid in any von Neumann algebra.

Lemma 2.4.2. *Let A be an abelian von Neumann algebra with carrier space Ω, and let $p, q \in A \otimes M_n$ be projections satisfying $\|p - q\| \leq k$ for some $k \in [0,1)$. Then there exists a unitary $u \in A \otimes M_n$ such that $u^*pu = q$.*

Proof. Let $x = pq - (1 - p)(1 - q)$. Two simple algebraic calculations show that $px = pq = xq$, and that $1 - x = (2p - 1)(p - q)$. Since $2p - 1$ is a unitary, the latter equation gives $\|1 - x\| \le k < 1$, and so x is invertible. The equation $x^*px = x^*xq$ shows that x^*x commutes with q, and the same is then true for $(x^*x)^{-1/2}$. Let $u = x(x^*x)^{-1/2}$, a unitary in $A \otimes \mathbb{M}_n$. Then

$$uq = x(x^*x)^{-1/2}q = xq(x^*x)^{-1/2} = px(x^*x)^{-1/2} = pu, \qquad (2.4.1)$$

and we reach the conclusion that $u^*pu = q$. $\square$

The following theorem gives the structure of masas in $A \otimes \mathbb{M}_n$ for any abelian von Neumann algebra A. If A happens to have a separable predual, then it is isomorphic to $L^\infty(\Omega, \mu)$ for some measure space, and we give a measure-theoretic proof of the following result for this case in Theorem 8.3.3.

Below we will use the *central support* (or *central carrier*) of an element x of a von Neumann algebra N. This is the smallest central projection z satisfying $zx = x$. In an abelian von Neumann algebra $A \cong C(\Omega)$ for a stonean space Ω, the central support of a function f is the characteristic function of the closure of the set $\{\omega \in \Omega \colon f(\omega) \ne 0\}$.

Theorem 2.4.3. *Let $n \in \mathbb{N}$, let A be an abelian von Neumann algebra, and let B be a masa in $A \otimes \mathbb{M}_n$. Then there exists a unitary $w \in A \otimes \mathbb{M}_n$ such that wBw^* is the diagonal masa $A \otimes \mathbb{D}_n$.*

Proof. We will prove this by induction on n. The case $n = 1$ is vacuously true, and so we will assume that the result is true in $A \otimes \mathbb{M}_k$ for all $1 \le k \le n - 1$. We will then complete the inductive step.

Identify A with $C(\Omega)$, where the carrier space Ω is stonean by Lemma 2.4.1. We then identify $A \otimes \mathbb{M}_n$ with $C(\Omega, \mathbb{M}_n)$, and we let Tr denote the unique trace on $\mathbb{M}_n$ such that $Tr(I_n) = n$. For a projection $p \in C(\Omega, \mathbb{M}_n)$, the function $\omega \mapsto Tr(p(\omega))$ is continuous and yields the rank of the projection $p(\omega) \in \mathbb{M}_n$. If we vary p over all projections in B and ω over Ω, then we may let r denote the smallest non-zero rank occurring amongst all the projections $p(\omega)$. Let $q \in B$ be a projection for which this value is taken at some point $\omega_0 \in \Omega$. Then

$$\Omega_0 = \{\omega \colon |Tr(q(\omega)) - Tr(q(\omega_0))| < 1\}$$

is an open set containing ω_0, and is also the set where $\operatorname{rank}(q(\omega)) = r$, so Ω_0 is a clopen set. The central projection $z \in A \otimes \mathbb{M}_n$ corresponding to the characteristic function of Ω_0 lies in B (since B is a masa), and $zp \ne 0$ has the property that the rank of $(zp)(\omega)$ is always 0 or r. Having found one projection with this property, Zorn's lemma allows us to find a family $(p_\lambda)_{\lambda \in \Lambda}$ of non-zero projections in B which is maximal with respect to the properties that their central supports z_λ are pairwise disjoint and that $\operatorname{rank}(p_\lambda(\omega))$ is always 0 or r for all $\omega \in \Omega$. Let $p \in B$ be the sum of these projections p_λ.

The z_λ's correspond to clopen sets $F_\lambda \subseteq \Omega$ which are pairwise disjoint. For each $\lambda \in \Lambda$, $z_\lambda p = p_\lambda$ and so $\operatorname{rank}(p(\omega)) = r$ for $\omega \in F_\lambda$. If F is the closure of $\cup_{\lambda \in \Lambda} F_\lambda$, then F is clopen and the corresponding central projection z satisfies

$p_\lambda(1-z) = 0$, for all $\lambda \in \Lambda$, and thus $p(1-z) = 0$. Consequently, z is the central support of p and rank $(p(\omega)) = r$ for $\omega \in F$, and is 0 on the complement. Cutting by z, we have a masa $Bz \subseteq Az \otimes M_n \cong C(F, M_n)$, and a projection $p \in Bz$ whose pointwise rank is always r. We now show by contradiction that $r = 1$, so suppose to the contrary that $r \geq 2$. The minimality of r implies that any projection $q \in Bz$ has the property that $q(\omega)$ is either orthogonal to, or contains $p(\omega)$ for each $\omega \in F$, otherwise the rank of $q(\omega)p(\omega)$ would lie strictly between 0 and r for some $\omega \in F$. Fix $\omega_0 \in F$ and choose a unit vector η_0 in the range of $p(\omega_0)$. Then let

$$U = \{\omega \in F \colon \|p(\omega)\eta_0\| > 1/2\}$$

be an open set with closure F_0, a clopen set on which $\|p(\omega)\eta_0\| \geq 1/2$. Then define a projection $p_0 \in C(F, M_n)$ by letting $p_0(\omega)$ be the rank one projection onto $\mathbb{C}p(\omega)\eta_0$ for $\omega \in F_0$, and 0 otherwise. Then $p_0 \leq p$, and by construction p_0 commutes with every projection in Bz, placing it in this algebra. If we extend the definition of p_0 to Ω by letting $p(\omega) = 0$ on the complement F^c of F, then $p_0 \in B' \cap (A \otimes M_n) = B$, and this contradicts the minimality of r. Thus $r = 1$.

We now show that $F = \Omega$. If this is not so, then the maximality of $(p_\lambda)_{\lambda \in \Lambda}$ implies that either the rank of $q(\omega)$ is 0 or at least 2 for all projections $q \in B$ and all $\omega \in F^c$. We may then repeat the whole argument to this point for the masa $B(1-z) \subseteq A(1-z) \otimes M_n$ to derive a contradiction. Thus $F = \Omega$, and we have a projection $p \in B$ such that rank $p(\omega) = 1$ for all $\omega \in \Omega$.

The set of rank 1 projections in M_n is compact and so we may fix a finite ε-net $\{P_1, \ldots, P_s\}$ for some $\varepsilon < 1/2$. For $1 \leq k \leq s$, define an open set

$$\Omega_k = \{\omega \colon \|p(\omega) - P_k\| < 1/2\}.$$

These sets cover Ω, as do their clopen closures $\overline{\Omega_k}$. Define disjoint clopen subsets G_k covering Ω by $G_1 = \overline{\Omega_1}$, $G_2 = \overline{\Omega_2} \setminus G_1$, ... Then we may define a continuous projection-valued function $f \in C(\Omega, M_n)$ by $f(\omega) = P_k$ for $\omega \in G_k$ (some G_k's could be empty and are then ignored). The choice of these projections ensures that $\|p - f\| \leq 1/2$, and so p and f are unitarily equivalent by a unitary $u \in A \otimes M_n$, by Lemma 2.4.2. Since each P_k is unitarily equivalent to the matrix unit $E_{11} \in M_n$, a further unitary conjugation moves p to the constant function with value E_{11}. Thus, up to unitary equivalence, B contains the projection $e \in A \otimes M_n$ whose value is always E_{11}. If we consider $B(1-e) \subseteq (1-e)(A \otimes M_n)(1-e) \cong A \otimes M_{n-1}$, then we may invoke the induction hypothesis to prove the result. $\square$

Remark 2.4.4. A finite type I von Neumann algebra is a direct sum of algebras $A_i \otimes M_{n_i}$ with n_i varying within $\mathbb{N}$ and each A_i an abelian von Neumann algebra. By applying Theorem 2.4.3 to each summand, it is easy to see that each masa in a finite type I von Neumann algebra is unitarily equivalent to a direct sum of diagonal masas. $\square$

Definition 2.4.5. A projection p in a von Neumann algebra N is called *abelian* if pNp is an abelian von Neumann algebra. $\square$

It is immediate from this definition that the set of abelian projections in N is closed under unitary conjugation. The principal example of an abelian projection is a rank one projection $p \in B(H)$, since $pB(H)p = \mathbb{C}p$. Indeed, non-zero abelian projections can only occur in type I summands of von Neumann algebras [105, Chapter 6].

Corollary 2.4.6. *Each masa B in a finite type* I *von Neumann algebra N contains an abelian projection p whose central support is 1.*

Proof. It suffices to consider the case where $N = A \otimes \mathbb{M}_n$ for an abelian von Neumann algebra $A \cong C(\Omega)$, since in general we may work independently in each type I_n summand. By Theorem 2.4.3, there is a unitary $u \in N$ such that uBu^* is the diagonal masa $A \otimes \mathbb{D}_n$. Let $Q \in \mathbb{D}_n$ be a fixed rank 1 projection, and let $q \in A \otimes \mathbb{D}_n$ be $1 \otimes Q$, the continuous function on Ω whose value is constantly Q. Then q is abelian in N with central support 1, and the same is also true for $p = u^*qu \in B$. $\qquad\square$

Chapter 3

Finite von Neumann algebras

3.1 Introduction

This chapter is concerned with the basic theory of finite von Neumann algebras, with an emphasis on those which arise from discrete groups. This is described in Section 3.2, with the objective of reaching quickly the examples of masas which we present in Section 3.3. These are based on the work of Dixmier, [47], who found algebraic conditions on an abelian subgroup H of a discrete group G which are sufficient to imply that $L(H)$ is a masa in the group von Neumann algebra $L(G)$, and to determine the type of this masa. We present various matrix groups which satisfy these conditions and thus obtain examples of singular, Cartan and semi-regular masas.

Section 3.4 discusses various other ways in which naturally occurring masas can arise. These are based on tensor products and crossed products, and we give a detailed construction of a crossed product with two natural Cartan masas. We also give a brief discussion of free products in which examples of semi-regular masas are easy to exhibit. In Section 3.5 we prove that all diffuse separable abelian von Neumann algebras with a faithful normal trace are ∗-isomorphic to $L^\infty[0, 1]$. In particular this applies to masas.

Section 3.6 of the chapter returns to the basic theory of general finite von Neumann algebras, and we construct the unique trace preserving conditional expectation onto a subalgebra, which is fundamental throughout these notes. We also introduce the very useful technique of using elements of minimal norm in closed convex subsets of Hilbert spaces, a topic that will be revisited several times subsequently. In Section 3.7, we return to Theorem 3.2.1 in order to give a proof, based on the results of Section 3.6. We conclude the chapter with a discussion of hyperfiniteness and various of its equivalent formulations, and we show that a discrete group G is amenable if and only in $L(G)$ is hyperfinite.

This chapter is not a comprehensive survey of finite algebras, and the topics have been selected with a view to later developments. We refer to the standard texts [104, 105, 167, 187, 188, 189] for a more substantial introduction to the subject.

3.2   Finite algebras

Let $N \subseteq B(H)$ be a von Neumann algebra with centre denoted Z. The comparison theory of projections in N was introduced by Murray and von Neumann, [116], who defined equivalence of projections p and q by requiring the existence of a partial isometry $v \in N$ such that $vv^* = p$ and $v^*v = q$. In $B(H)$, any two projections with infinite dimensional ranges are equivalent, and so it is possible to have equivalent projections p and q such that $p \leq q$ and $p \neq q$. The defining property of a finite von Neumann algebra is to exclude such occurrences: N is finite if the conclusion $p = q$ results from $p \leq q$ and p and q equivalent in N. The simplest examples of finite von Neumann algebras are as follows:

(1) Any abelian von Neumann algebra A. Here, projections are equivalent if and only if they are equal.

(2) The algebras $\mathbb{M}_n$ of $n \times n$ scalar matrices for any integer $n \geq 1$.

(3) $A \otimes \mathbb{M}_n$ where A is any abelian von Neumann algebra.

(4) $\bigoplus_{k=1}^{\infty} N_k$ where each N_k is finite.

(5) von Neumann algebras arising from discrete groups, discussed below.

(6) The weak closure of an increasing sequence of finite dimensional von Neumann algebras with a corresponding sequence of compatible normalised traces, defined below. In the case that these are matrix factors, the resulting algebra is the (unique) hyperfinite II_1 factor, a term that will be defined and discussed in Section 3.8.

A trace τ on a von Neumann algebra is a bounded normal positive linear functional in the predual N_* which has the property that $\tau(xy) = \tau(yx)$ for all $x, y \in N$. The trace is *normalised* if $\tau(1) = 1$, and *faithful* if $\tau(x^*x) = 0$ implies that $x = 0$. If a trace exists then, for any pair of equivalent projections p and q, we necessarily have $\tau(p) = \tau(q)$. When H is infinite dimensional, $B(H)$ contains an infinite set of equivalent orthogonal rank one projections, and so the only bounded trace on this algebra is 0, although it does have a densely defined unbounded trace. The situation is different for the first three examples above. If A is abelian then any $\theta \in A_*^+$ is a trace, while standard matrix theory gives a unique normalised trace τ_n on $\mathbb{M}_n$ defined by

$$\tau_n((a_{ij})) = n^{-1} \sum_{i=1}^{n} a_{ii}, \quad (a_{ij}) \in \mathbb{M}_n. \qquad (3.2.1)$$

These combine to give traces $\theta \otimes \tau_n$ on $A \otimes \mathbb{M}_n$.

One of the fundamental results for finite von Neumann algebras is the existence of a centre-valued trace. This is a normal positive bounded linear map $\mathbb{T}_Z \colon N \to Z$ with the properties

(i) $\mathbb{T}_Z(xy) = \mathbb{T}_Z(yx)$, $\quad x, y \in N$,

(ii) if $x \geq 0$ and $\mathbb{T}_Z(x) = 0$ then $x = 0$,

(iii) $\mathbb{T}_Z(z) = z$ for $z \in Z$,

(iv) $\mathbb{T}_Z(zx) = z\mathbb{T}_Z(x)$ for $x \in N$ and $z \in Z$.

The construction of the centre-valued trace may be found in [105, Chapter 8] where uniqueness is also shown. There is a simple converse to this result. If $\mathbb{T}_Z$ exists and p and q are equivalent projections with $p \leq q$ then $\mathbb{T}_Z(p) = \mathbb{T}_Z(q)$ by (i), so $p = q$ by (ii) since $q - p \geq 0$ and $\mathbb{T}_Z(q - p) = 0$. Consequently N is finite.

Once we have developed the theory of conditional expectations in Section 3.5, we will see in Lemma 3.6.5 and Remark 3.6.6 that this trace coincides with the expectation $\mathbb{E}_Z$ onto Z. Moreover, the proof of Lemma 3.6.5 gives a straightforward way to construct it.

The composition of any element of Z_*^+ with $\mathbb{T}_Z$ is a trace on N whenever N is finite. Varying the choice of linear functional gives a family $\{\tau_\alpha\}$ of traces on N that is faithful in the sense that if $x \geq 0$ and $\tau_\alpha(x) = 0$ for all α, then $\mathbb{T}_Z(x) = 0$ and thus $x = 0$ by (ii). In general this is all that can be said: $\ell^\infty(S)$ for an uncountable set has no single faithful normal trace since any element of the predual $\ell^1(S)$ has only a countable number of non-zero entries. However, if N is finite and acts on a separable Hilbert space (equivalent to the norm separability of N_*), then we may choose a countable orthonormal basis $\{\eta_n\}_{n=1}^\infty$ and define $\theta \in N_*$ by $\theta(x) = \sum_{n=1}^\infty 2^{-n}\langle x\eta_n, \eta_n \rangle$, $x \in N$. The composition $\theta \circ \mathbb{T}_Z$ is then a faithful trace on N. A von Neumann algebra with trivial centre is called a *factor*, so a finite factor N always has a faithful trace $\mathbb{T}_Z$ independently of the Hilbert space where it is represented. There are two types of finite factors. On the one hand we have the factors $\mathbb{M}_n$, $n \geq 1$, which are finite dimensional and are called type I_n; on the other we have the infinite dimensional ones which are called type II_1. These two classes are distinguished by the presence or absence of minimal non-zero projections.

The main examples of finite von Neumann algebras arise from discrete groups. Indeed, the question of Murray and von Neumann concerning the existence of a finite factor which is not a group factor was only settled (positively) in 1975 by Connes [35]. The method of obtaining a von Neumann algebra from a discrete group G is as follows. Form the Hilbert space $\ell^2(G)$, which has a natural orthonormal basis $\{\delta_h \colon h \in G\}$, and let G act on the left as permutation unitaries $\lambda_g(\delta_h) = \delta_{gh}$ for $g, h \in G$. This is the left regular representation and $L(G)$ is the von Neumann algebra generated by the operators $\{\lambda_g \colon g \in G\}$. The right regular representation is defined similarly by $\rho_g(\delta_h) = \delta_{hg^{-1}}$ for $g, h \in G$, and these operators generate the right group von Neumann algebra $R(G)$ which commutes with $L(G)$. The vector functional $\langle \cdot \delta_e, \delta_e \rangle$ is a trace τ on both algebras. The following calculation will demonstrate that τ is faithful on $L(G)$ (a similar one works for $R(G)$), showing that both are finite von Neumann algebras. If $x \in L(G)^+$ is such that $\tau(x) = 0$, then $x\delta_e = 0$. The equation

$$x\delta_h = x\rho_{h^{-1}}\delta_e = \rho_{h^{-1}}x\delta_e = 0, \quad h \in G, \tag{3.2.2}$$

shows that $x = 0$, establishing the faithfulness of τ.

The group G is said to be *Infinite Conjugacy Class* (I.C.C.) if each conjugacy class $\{hgh^{-1} : h \in G\}$ has infinite cardinality when $g \neq e$. The basic properties of $L(G)$ and $R(G)$ are summarised by the following theorem (see [105, Chapter 6]). We will delay the proof of this result until Section 3.7, since we will need the material of Section 3.6.

Theorem 3.2.1.   (i) $L(G)' = R(G)$ and $R(G)' = L(G)$;

(ii) $L(G)$ and $R(G)$ are factors if and only if G is I.C.C.

The general vector in $\ell^2(G)$ is $\sum_{h \in G} \alpha_h \delta_h$, where $\sum_{h \in G} |\alpha_h|^2 < \infty$, and so there is a conjugate linear isometry $J : \ell^2(G) \to \ell^2(G)$ defined by

$$J\left(\sum_{h \in G} \alpha_h \delta_h\right) = \sum_{h \in G} \overline{\alpha_h} \delta_{h^{-1}}. \tag{3.2.3}$$

Then

$$J\lambda_g J\delta_h = J\delta_{gh^{-1}} = \delta_{hg^{-1}} = \rho_g \delta_h, \quad h \in G, \tag{3.2.4}$$

showing that $J\lambda_g J = \rho_g$. Thus J implements a conjugate linear isomorphism between $L(G)$ and $R(G)$. This could have been achieved by omitting the conjugate in the definition of J, but in the present form it generalises to a finite von Neumann algebra with a faithful trace, a situation that we will discuss subsequently.

Suppose that a discrete group G is in its left regular representation on $\ell^2(G)$ and let H be a subgroup of G. The operators $\{\lambda_h : h \in H\}$ generate a von Neumann subalgebra M of $L(G)$, which has $\ell^2(H)$ as an invariant subspace. The projection p onto this subspace lies in M', so $x \mapsto xp$, $x \in M$, is a normal representation of M, the image of which is $L(H)$. The vector δ_e is separating for $L(G)$, and so we see that this representation of M is faithful, allowing us to identify M with $L(H)$. Subsequently we will regard $L(H)$ as embedded in $L(G)$ without further comment. It will be helpful later to have a criterion for determining whether an operator $x \in L(G)$ actually lies in $L(H)$, and that is the purpose of the next lemma. The map $\mathbb{E} : \mathbb{C}G \to \mathbb{C}H$ at the level of group algebras which we now introduce is the motivating example for the conditional expectations presented later in this chapter. For a finite sum $\sum \alpha_g g \in \mathbb{C}G$, let $\mathbb{E}(\sum \alpha_g \lambda_g) = \sum_{g \in H} \alpha_g \lambda_g$. The operators λ_g, for $g \in G \backslash H$, map $\ell^2(H)$ into $\ell^2(H)^\perp$ and thus the operator $x = \sum \alpha_g \lambda_g$ satisfies

$$\|x\eta\| = \|\mathbb{E}(x)\eta + (I - \mathbb{E})(x)\eta\| \geq \|\mathbb{E}(x)\eta\| \tag{3.2.5}$$

for vectors $\eta \in \ell^2(H)$, using the orthogonality of $\mathbb{E}(x)\eta$ and $(I - \mathbb{E})(x)\eta$. Since the norm of $\mathbb{E}(x)$ is determined by vectors in $\ell^2(H)$, we see that $\mathbb{E}$ is a contraction on $\mathbb{C}G \subseteq L(G)$. This map will extend to a normal contraction on $L(G)$ with range $L(H)$, but just having $\mathbb{E}$ defined on $\mathbb{C}G$ is sufficient for our immediate purposes.

Before we discuss group von Neumann algebras and their generalisation to crossed products, we need to bring to the reader's attention an error in the literature. Each $x \in L(G)$ can be viewed as an element of $\ell^2(G)$, and as such has a Fourier expansion $\sum_{h \in G} \alpha_h \delta_h$ where the sum converges in $\|\cdot\|_2$-norm. It is tempting, but incorrect, to think that this sum will converge either strongly or weakly to x. A more general series for elements of crossed products is stated to converge strongly in [187, Proposition V.7.6] and [131, Section 7.11.2], but such series converge neither strongly nor weakly in general, as an example by Mercer [115] shows. These remarks account for the rather circuitous proofs given from here to the end of Section 3.4. Having issued this warning, it should also be said that assuming strong convergence of such sums (while wrong) is a good guide to seeing what is true, and to constructing correct proofs.

Lemma 3.2.2. *Let $H \subseteq G$ be discrete groups.*

(i) *A uniformly bounded net $(x_\alpha) \in L(G)$ converges strongly to $x \in L(G)$ if and only if $\lim_\alpha \|(x_\alpha - x)\delta_e\| = 0$.*

(ii) *If $x \in L(G)$ then $x \in L(H)$ if and only if $x\delta_e \in \ell^2(H)$.*

Proof. (i) The forward direction is obvious. Suppose then that $x_\alpha \delta_e \to x\delta_e$ in norm. Then

$$(x_\alpha - x)\delta_g = (x_\alpha - x)\rho_{g^{-1}}\delta_e = \rho_{g^{-1}}(x_\alpha - x)\delta_e \to 0 \qquad (3.2.6)$$

for $g \in G$. Thus we have strong convergence on a total subset of vectors, and the uniform bound on $\|x_\alpha\|$ then implies strong convergence of (x_α) to x on all of $\ell^2(G)$.

(ii) Without loss of generality we may assume that $\|x\| = 1$. If $x \in L(H)$ then it is clear that $x\delta_e \in \ell^2(H)$, so we need only prove the converse. Thus suppose that $x\delta_e \in \ell^2(H)$. By the Kaplansky density theorem (Theorem 2.2.3) there exists a net $(x_\alpha) \in \mathbb{C}G$, $\|x_\alpha\| \leq 1$, converging strongly to x. Then the nets $(\mathbb{E}(x_\alpha))$ and $((I - \mathbb{E})(x_\alpha))$ are both uniformly bounded. Since $x\delta_e, \mathbb{E}(x_\alpha)\delta_e \in \ell^2(H)$ and $(I - \mathbb{E})(x_\alpha)\delta_e \in \ell^2(H)^\perp$, strong convergence of $x_\alpha \delta_e$ to $x\delta_e$ implies that $(I - \mathbb{E})(x_\alpha)\delta_e \to 0$. By (i), the net $((I - \mathbb{E})(x_\alpha))$ converges strongly to 0, showing that $\lim_\alpha \mathbb{E}(x_\alpha) = x$, strongly. Thus $x \in L(H)$ since $\mathbb{E}(x_\alpha) \in L(H)$ for all α. $\qquad \square$

3.3 Examples of masas from groups

In [47], Dixmier identified three types of masas A in a II_1 factor N based on the *normaliser*

$$\mathcal{N}(A) = \{u \in N : uAu^* = u^*Au = A, \ u \text{ unitary}\} \qquad (3.3.1)$$

and the von Neumann algebra $\mathcal{N}(A)''$ which it generates. They are as follows:

(i) A is *singular* if $\mathcal{N}(A)'' = A$;

(ii) A is *regular* (nowadays called *Cartan*) if $\mathcal{N}(A)'' = N$;

(iii) A is *semi-regular* if $\mathcal{N}(A)''$ is a proper subfactor of N.

In this section we will give examples of all three types based on groups and subgroups. Let us fix a proper inclusion of groups $H \subseteq G \subseteq K$, where always H is abelian and G and K are I.C.C. Appropriate choices of these groups will exhibit $L(H)$ as a singular or Cartan masa in $L(G)$, and as a semi-regular masa in $L(K)$ whose normaliser generates $L(G)$. To achieve this, certain algebraic conditions must be imposed on the groups and these we now list:

(C1) for each $g \in G \backslash H$, $\{hgh^{-1} : h \in H\}$ has infinite cardinality;

(C2) for each finite subset B of $G \backslash H$, there exists $h \in H$ such that $b_1 h b_2 \notin H$ for all $b_1, b_2 \in B$;

(C3) if $k \in K \backslash G$ is a fixed element and B is a finite subset of K, then there exists $h \in H$ such that $k^{-1} h k \notin H$ and, for each $b \in B$, either $hbh^{-1} = b$ or $hbh^{-1} \notin B$;

(C4) if $k \in K \backslash G$ is a fixed element and F is a finite subset of $K \backslash H$, then there exists $h \in H$ such that $k^{-1} h k \notin H$ and $hFh^{-1} \cap F = \emptyset$.

Lemma 3.3.1. *If $H \subseteq G$ satisfies condition (C1) then $L(H)$ is a masa in $L(G)$. If, in addition, H is a normal subgroup of G, then $L(H)$ is a Cartan masa in $L(G)$.*

Proof. Let $x \in L(H)' \cap L(G)$, and let $x\delta_e = \sum_{g \in G} \alpha_g \delta_g$ where $\sum_{g \in G} |\alpha_g|^2 < \infty$. Then, for each $h \in H$, the relation $xh\delta_e = hx\delta_e$ implies that

$$\sum_{g \in G} \alpha_g \delta_{gh} = \sum_{g \in G} \alpha_g \delta_{hg}. \tag{3.3.2}$$

Thus the change of variables $g \mapsto hgh^{-1}$ on the left gives

$$\sum_{g \in G} \alpha_{hgh^{-1}} \delta_{hg} = \sum_{g \in G} \alpha_g \delta_{hg}, \tag{3.3.3}$$

implying that $\alpha_{hgh^{-1}} = \alpha_g$ for $g \in G$. If there is a $g_0 \in G \backslash H$ with $\alpha_{g_0} \neq 0$ then $\{\alpha_{hg_0h^{-1}} : h \in H\}$ is an infinite set of numbers each equal to α_{g_0}, making it impossible for $\sum_{g \in G} |\alpha_g|^2$ to be finite. This implies that $\alpha_g = 0$ for $g \in G \backslash H$, and $x\delta_e \in \ell^2(H)$. By Lemma 3.2.2, $x \in L(H)$. Thus $L(H)$ is maximal abelian. Conversely, if $g \in G \backslash H$ is such that $\{hgh^{-1} : h \in H\}$ is finite and $\{h_i gh_i^{-1} : 1 \leq i \leq n\}$ is a list of the distinct conjugates, then $x = \sum_{i=1}^n \lambda_{h_i gh_i^{-1}} \in L(H)'$ and $L(H)$ is not a masa.

Suppose further that H is normal in G. Then each λ_g normalises $L(H)$, and thus $\mathcal{N}(L(H))'' = L(G)$. This shows that $L(H)$ is Cartan in $L(G)$. $\square$

Below, we will use the simple fact that if $x \in L(G)$ then

$$\|x\delta_e\|^2 = \langle x\delta_e, x\delta_e \rangle = \tau(x^*x) = \tau(xx^*) = \|x^*\delta_e\|^2. \qquad (3.3.4)$$

Lemma 3.3.2. *If $H \subseteq G$ satisfies (C2), then $L(H)$ is a singular masa in $L(G)$.*

Proof. We begin by showing that (C1) holds, so to obtain a contradiction we assume that (C2) is true but that (C1) is false. Then there exists $g_0 \in G \setminus H$ such that

$$B = \{hg_0h^{-1}\colon h \in H\}$$

is a finite set. If we apply (C2) for this particular B, then there exists $h_0 \in H$ such that $hg_0h^{-1}h_0kg_0k^{-1} \notin H$ for all $h, k \in H$. But this is false if we take $h = h_0$ and $k = e$. Thus (C1) follows from (C2).

Since H now satisfies (C1), $L(H)$ is a masa in $L(G)$ by Lemma 3.3.1. Let $u \in L(G)$ be a normalising unitary for $L(H)$. We will use (C2) to deduce that $u \in L(H)$. By the Kaplansky density theorem, we may choose a sequence $\{t_n\}_{n=1}^{\infty} \in \text{span}\{\lambda_g\colon g \in G\}$ with $\|t_n\| \leq 1$ such that $\|(t_n - u)\delta_e\| \to 0$. Then, for each $g \in G$, $\|(t_n - u)\delta_g\| = \|\rho_{g^{-1}}(t_n - u)\delta_e\| \to 0$, and so the uniform bound on $\|t_n\|$ implies that $t_n \to u$ strongly. Since we have that $\|(t_n^* - u^*)\delta_e\| = \|(t_n - u)\delta_e\|$ by (3.3.4), we also have that $t_n^* \to u^*$ strongly.

For each $h \in H$,

$$
\begin{aligned}
\|(u^*\lambda_h u - t_n^*\lambda_h t_n)\delta_e\| &= \|(u^* - t_n^*)\lambda_h u\delta_e + t_n^*\lambda_h(u - t_n)\delta_e\| \\
&\leq \|(u^* - t_n^*)\lambda_h u\delta_e\| + \|(u - t_n)\delta_e\| \\
&= \|u^*\lambda_h^*(u - t_n)\delta_e\| + \|(u - t_n)\delta_e\| \\
&\leq 2\|(u - t_n)\delta_e\|, \qquad (3.3.5)
\end{aligned}
$$

and this estimate is independent of h. Since $u^*\lambda_h u\delta_e \in \ell^2(H)$, the distance of the vector $t_n^*\lambda_h t_n\delta_e$ from $\ell^2(H)$ is at most $2\|(u - t_n)\delta_e\|$. Now let n be fixed but arbitrary. Write $t_n = \mathbb{E}(t_n) + (I - \mathbb{E})(t_n)$. There are only a finite number of terms in $(I - \mathbb{E})(t_n)$, all in $G \setminus H$, and so the hypothesis allows us to select $h_n \in H$ such that the product $(I - \mathbb{E})(t_n^*)\lambda_{h_n}(I - \mathbb{E})(t_n)$ lies in $\text{span}\{\lambda_g\colon g \in G \setminus H\}$. This is also true of the products $\mathbb{E}(t_n^*)\lambda_{h_n}(I - \mathbb{E})(t_n)$ and $(I - \mathbb{E})(t_n^*)\lambda_{h_n}\mathbb{E}(t_n)$. Thus $\mathbb{E}(t_n^*)\lambda_{h_n}\mathbb{E}(t_n)\delta_e$ is the nearest point in $\ell^2(H)$ to $t_n^*\lambda_{h_n}t_n\delta_e$. It follows that

$$\|\mathbb{E}(t_n^*)\lambda_{h_n}\mathbb{E}(t_n)\delta_e - t_n^*\lambda_{h_n}t_n\delta_e\| \leq 2\|(u - t_n)\delta_e\|, \qquad (3.3.6)$$

and the triangle inequality gives

$$\|\mathbb{E}(t_n^*)\lambda_{h_n}\mathbb{E}(t_n)\delta_e - u^*\lambda_{h_n}u\delta_e\| \leq 4\|(u - t_n)\delta_e\|. \qquad (3.3.7)$$

Thus $\|\mathbb{E}(t_n)\delta_e\| \to 1$, and by orthogonality

$$\|(I - \mathbb{E})(t_n)\delta_e\|^2 = \|t_n\delta_e\|^2 - \|\mathbb{E}(t_n)\delta_e\|^2 \to 0. \qquad (3.3.8)$$

It follows that $(I - \mathbb{E})(t_n) \to 0$ strongly, and so $\mathbb{E}(t_n) \to u$ strongly, showing that $u\delta_e \in \ell^2(H)$ as required. $\qquad \square$

We can now give examples of singular and Cartan masas, based on Lemmas 3.3.1 and 3.3.2.

Example 3.3.3. Let F be an infinite field (for example the rationals $\mathbb{Q}$) and let $F^\times$ denote the multiplicative group of non-zero elements. Let G denote the group

$$\left\{ \begin{pmatrix} a & x \\ 0 & 1 \end{pmatrix} : a \in F^\times, \ x \in F \right\} \tag{3.3.9}$$

under matrix multiplication. In the matrix equation

$$\begin{pmatrix} b & y \\ 0 & 1 \end{pmatrix} \begin{pmatrix} a & x \\ 0 & 1 \end{pmatrix} \begin{pmatrix} b^{-1} & -b^{-1}y \\ 0 & 1 \end{pmatrix} = \begin{pmatrix} a & bx + (1-a)y \\ 0 & 1 \end{pmatrix} \tag{3.3.10}$$

we may vary b and y to conclude that $\begin{pmatrix} a & x \\ 0 & 1 \end{pmatrix}$ has infinitely many distinct conjugates unless $x = 0$ and $a = 1$. Thus G is I.C.C. We may define two abelian subgroups H_1 and H_2 by

$$H_1 = \left\{ \begin{pmatrix} 1 & x \\ 0 & 1 \end{pmatrix} : x \in F \right\}, \quad H_2 = \left\{ \begin{pmatrix} a & 0 \\ 0 & 1 \end{pmatrix} : a \in F^\times \right\}. \tag{3.3.11}$$

It follows from equation (3.3.10) that H_1 is normal in G. The same equation with $a \neq 1$ and $b = 1$ shows that $\{hgh^{-1} : h \in H_1\}$ is infinite when $g = \begin{pmatrix} a & x \\ 0 & 1 \end{pmatrix} \in G \backslash H_1$. By Lemma 3.3.1, $L(H_1)$ is a Cartan masa in $L(G)$. The same conjugacy class condition is satisfied by H_2, taking $x \neq 0$, $y = 0$ and letting b vary. Thus $L(H_2)$ is a masa in $L(G)$, by Lemma 3.3.1. Lemma 3.3.2 shows that we need only verify (C2) to show that $L(H_2)$ is a singular masa. Given elements $\begin{pmatrix} a_i & x_i \\ 0 & 1 \end{pmatrix}$, $1 \leq i \leq n$, $x_i \neq 0$, the product

$$\begin{pmatrix} a_i & x_i \\ 1 & 0 \end{pmatrix} \begin{pmatrix} b & 0 \\ 0 & 1 \end{pmatrix} \begin{pmatrix} a_j & x_j \\ 0 & 1 \end{pmatrix} = \begin{pmatrix} ba_i a_j & a_i bx_j + x_i \\ 0 & 1 \end{pmatrix} \tag{3.3.12}$$

will always lie outside H_2 if b is chosen in the complement of the finite set of values $\{-x_i/a_i x_j : 1 \leq i, j \leq n\}$. Thus H_2 satisfies (C2), and $L(H_2)$ is an example of a singular masa.

The exact nature of the field F plays no role in this example. It might appear that different choices of field would lead to different factors $L(G)$, but this is not the case. The quotient of G by the normal abelian subgroup H_1 is isomorphic to the abelian group H_2, making G amenable [127]. Then $L(G)$ is the unique hyperfinite II$_1$ factor [36] (see Section 3.8). This is most apparent if we start with a finite field and construct F as the increasing union of a sequence of finite extensions. This exhibits $L(G)$ as the weak closure of an ascending sequence of finite dimensional subalgebras. □

We will now present an example of a semi-regular masa, based on the groups that we have already discussed. We need the following lemma to verify the required properties.

Lemma 3.3.4. *Let $H \subseteq G \subseteq K$ be discrete groups satisfying either condition (C3) or condition (C4), where G and K are I.C.C. and H is an abelian normal subgroup of G. Further suppose that condition (C1) holds for the inclusion $H \subseteq K$. Then $L(H)$ is a semi-regular masa in $L(K)$ whose normaliser generates $L(G)$.*

Proof. First suppose that (C3) is satisfied. By Lemmas 3.3.1 and 3.3.2, $L(H)$ is a masa in $L(K)$ and $L(G) \subseteq \mathcal{N}(L(H))''$. In order to show equality we consider a normalising unitary $u \in L(K)$ and we prove that $u\delta_e \in \ell^2(G)$. By Lemma 3.2.2 (ii), this will establish that $u \in L(G)$. Let $u\delta_e = \sum_{k \in K} \alpha_k \delta_k$, and suppose that there exists $k_0 \in K \backslash G$ such that $\alpha_{k_0} \neq 0$. We will show that this is impossible under the given hypotheses.

Fix an arbitrary $\varepsilon > 0$ and choose an element $t = \sum \beta_k \lambda_k \in \mathbb{C}G$ with $\|t\| \leq 1$ and $\|(u-t)\delta_e\| < \varepsilon$, possible by the Kaplansky density theorem (Theorem 2.2.3). For arbitrary elements $k_1, k_2 \in K$, $h \in H$, we have

$$|\langle (u^* \lambda_h u - t^* \lambda_h t)\delta_{k_1}, \delta_{k_2} \rangle| = |\langle ((u^* - t^*)\lambda_h u + t^* \lambda_h (u - t))\delta_{k_1}, \delta_{k_2} \rangle|$$
$$\leq |\langle \lambda_h u \delta_{k_1}, (u - t)\delta_{k_2} \rangle| + |\langle t^* \lambda_h (u - t)\delta_{k_1}, \delta_{k_2} \rangle|$$
$$< 2\varepsilon \qquad (3.3.13)$$

since the estimate $\|(u - t)\delta_k\| = \|\rho_{k^{-1}}(u - t)\delta_e\| < \varepsilon$ holds for any $k \in K$. Taking $k_1 = e$ and noting that $u^* \lambda_h u \delta_e \in \ell^2(H)$, we see that

$$|\langle t^* \lambda_h t \delta_e, \delta_k \rangle| \leq 2\varepsilon \qquad (3.3.14)$$

for all $h \in H$ and $k \in K \backslash H$. Now define a finite set

$$B = \{k_0 k^{-1} : \ \beta_k \neq 0\}. \qquad (3.3.15)$$

The condition (C3) means that we may choose $h \in H$ so that $k_0^{-1} h k_0 \notin H$, and for each $b \in B$, either $hbh^{-1} = b$ or $hbh^{-1} \notin B$. Then

$$|\langle t^* \lambda_h t \delta_e, \delta_{k_0^{-1} h k_0} \rangle| \leq 2\varepsilon. \qquad (3.3.16)$$

This quantity is $|\sum \bar{\beta}_k \beta_j|$ where the sum is taken over group elements such that $\beta_k \neq 0$, $\beta_j \neq 0$, and $k^{-1} h j = k_0^{-1} h k_0$. This is equivalent to $k_0 k^{-1} = h k_0 j^{-1} h^{-1}$, and since the term on the left is in B, this can only hold where $k_0 k^{-1} = k_0 j^{-1}$, equivalent to $k = j$. Thus we obtain that $|\langle t^* \lambda_h t \delta_e, \delta_{k_0^{-1} h k_0} \rangle|$ is a certain sum of squares containing the term $|\beta_{k_0}|^2$, and so $|\beta_{k_0}|^2 \leq 2\varepsilon$. On the other hand

$$|\alpha_{k_0} - \beta_{k_0}| = |\langle (u - t)\delta_e, \delta_{k_0} \rangle| \leq \|(u - t)\delta_e\| < \varepsilon, \qquad (3.3.17)$$

leading to $|\alpha_{k_0}| < \varepsilon + (2\varepsilon)^{1/2}$. Since $\varepsilon > 0$ was arbitrary, we now have a contradiction and we obtain $u \in L(G)$. Thus $\mathcal{N}(L(H))'' = L(G)$, and $L(H)$ is semi-regular.

Now suppose that (C4) is satisfied. The result will follow once we have shown that this condition implies (C3). Consider a fixed $k \in K \backslash G$ and a finite subset B of K, and let $F = B \cap (K \backslash H)$. By hypothesis, there exists $h \in H$ such that $k^{-1}hk \notin H$ and $hFh^{-1} \cap F = \emptyset$. If $b \in F$, then $hbh^{-1} \notin F$ and it clearly cannot lie in H, so $hbh^{-1} \notin B$. On the other hand, if $b \in B \cap H$, then $hbh^{-1} = b$. These two cases show that (C3) is implied by (C4), completing the proof. $\square$

In the following example we show that groups satisfying (C1) and (C3) exist, leading to an example of a semi-regular masa.

Example 3.3.5. As in Example 3.3.3, we let F be an infinite field and we define

$$G = \left\{ \begin{pmatrix} a & x \\ 0 & 1 \end{pmatrix} : a \in F^\times, x \in F \right\}, \quad H = \left\{ \begin{pmatrix} 1 & x \\ 0 & 1 \end{pmatrix} : x \in F \right\}. \quad (3.3.18)$$

We regard these as lying in $GL_2(F)$, the group of invertible 2×2 matrices over F, whose centre Z is the set of non-zero multiples of the identity. We then define K to be $PGL_2(F) = GL_2(F)/Z$. Since $G \cap Z$ and $H \cap Z$ are both trivial, we also have containments $H \subseteq G \subseteq K$. The situation is complicated by the fact that K is a quotient group while we prefer to work in $GL_2(F)$, so we note that two matrices $A, B \in GL_2(F)$ have equal images in K precisely when one is a scalar multiple of the other.

We first verify that K is I.C.C. Since we have already seen that this is so for G, it suffices to consider conjugates of an element $\begin{pmatrix} a & b \\ c & d \end{pmatrix} \in GL_2(F)$ with $c \neq 0$. The matrix equation

$$\begin{pmatrix} 1 & x \\ 0 & 1 \end{pmatrix} \begin{pmatrix} a & b \\ c & d \end{pmatrix} \begin{pmatrix} 1 & -x \\ 0 & 1 \end{pmatrix} = \begin{pmatrix} a + cx & -cx^2 + (d-a)x + b \\ c & d - cx \end{pmatrix} \quad (3.3.19)$$

will produce infinitely many distinct conjugates in $GL_2(F)$ as x varies, and since the (2,1) entry is constantly $c \neq 0$, these remain distinct in the quotient K. We have conjugated by elements of H, and so we have simultaneously verified (C1) and that K is I.C.C. We now verify (C3).

Let $k_0 \in K \backslash G$ have representative $\begin{pmatrix} a & b \\ c & d \end{pmatrix} \in GL_2(F)$ with $c \neq 0$. A representative for k_0^{-1} is $\begin{pmatrix} d & -b \\ -c & a \end{pmatrix}$. If $x \neq 0$, then the equation

$$\begin{pmatrix} d & -b \\ -c & a \end{pmatrix} \begin{pmatrix} 1 & x \\ 0 & 1 \end{pmatrix} \begin{pmatrix} a & b \\ c & d \end{pmatrix} = \begin{pmatrix} * & * \\ -cx^2 & * \end{pmatrix} \quad (3.3.20)$$

shows that $k_0^{-1}hk_0 \notin H$ whenever $h \in H \backslash \{e\}$, and thus the first requirement of (C3) is automatically satisfied by any element of H other than e. The second requirement is equivalent to: given distinct $k_i = \begin{pmatrix} a_i & b_i \\ c_i & d_i \end{pmatrix}$, $1 \leq i \leq n$, there exists $h = \begin{pmatrix} 1 & x \\ 0 & 1 \end{pmatrix}$ with $x \neq 0$ so that $hk_i = k_jh$ implies $i = j$. The equation $hk_i = k_jh$ is

$$\begin{pmatrix} a_i + c_i x & b_i + d_i x \\ c_i & d_i \end{pmatrix} = \begin{pmatrix} a_j & a_j x + b_j \\ c_j & c_j x + d_j \end{pmatrix} \quad (\text{mod } Z). \quad (3.3.21)$$

Consider first $c_i \neq 0$. Then $c_j \neq 0$. Proportionality of the second rows implies that $c_j d_i = c_i c_j x + c_i d_j$, and this will never be true if we avoid a finite set of values for x. Now suppose that $c_i = 0$, implying $c_j = 0$. Then the four numbers a_i, a_j, d_i, d_j are all non-zero. By scaling the matrix representatives of k_i and k_j we may assume that $d_i = d_j = 1$, giving the equality

$$\begin{pmatrix} a_i & x + b_i \\ 0 & 1 \end{pmatrix} = \begin{pmatrix} a_j & a_j x + b_j \\ 0 & 1 \end{pmatrix} \tag{3.3.22}$$

in $GL_2(F)$. When $a_j \neq 1$, a further exclusion of a finite number of x-values will make this fail, while if $a_j = 1$ then we obtain $a_i = a_j$ and $b_i = b_j$, showing that $k_i = k_j$. Thus, by choosing x from the complement of a certain finite set of values, we see that a choice of h is possible for which condition (C3) is verified. Lemma 3.3.4 then shows that $L(H)$ is a semi-regular masa in $L(K)$ whose normaliser generates the factor $L(G)$. □

Example 3.3.6. Let $\mathbb{F}_n$ be the free group on n generators $\{g_1, \ldots, g_n\}$, $n \geq 2$. Let H be the abelian subgroup generated by g_1 (or by any other generator). If w is any word containing at least one of $\{g_2, \ldots, g_n\}$ then the elements $\{g_1^k w g_1^{-k} : k \geq 1\}$ are distinct and so (C1) is satisfied. Thus $L(H)$ is a masa in $L(\mathbb{F}_n)$. The absence of relations in $\mathbb{F}_n$ means that $w_1 g_1^k w_2 \notin H$ for k sufficiently large whenever $w_1, w_2 \in \mathbb{F}_n \backslash H$. Thus (C2) is easily verified, and $L(H)$ gives another example of a singular masa in the containing group factor. □

Remark 3.3.7. The preceding examples of maximal abelian subgroups have satisfied condition (C1), leading to masas in the containing $L(G)$. However, this does not always occur. Let S_n denote the symmetric group on n letters and let $S_\infty = \bigcup_{n=1}^\infty S_n$ denote the group of finite permutations of $\{1, 2, 3, \ldots\}$. The set of transpositions $\{(2n-1, 2n) : n \geq 1\}$ generates a maximal abelian subgroup H of S_∞ but the permutation $g = (1, 2, 3, 4) \in S_\infty$ has the property that $\{hgh^{-1} : h \in H\}$ is a finite set. Consequently $L(H)$ is not a masa in $L(S_\infty)$. Up to renumbering, the maximal abelian subgroups of S_∞ can be described as follows. Fix integers $1 \leq n_1 < n_2 < n_3 \ldots$ and let σ_k be the cyclic permutation of $\{n_{k-1} + 1, \ldots, n_k\}$. Then the subgroup H generated by $\{\sigma_k : k \geq 1\}$ is maximal abelian and all such arise in this way. The cyclic permutation of $\{1, \ldots, n_2\}$ has only finitely many distinct conjugates by elements of H, and so $L(H)$ is never a masa in $L(S_\infty)$. These statements are straightforward to verify and are left to the reader. □

In the next section we will discuss further constructions to obtain masas.

3.4 Tensor products and crossed products

In this section we present some more examples of masas, based on the standard constructions of tensor products and crossed products. Let $\mathbb{M}_2$ be the 2×2 scalar matrices with diagonal masa $\mathbb{D}_2$ relative to the standard basis of $\mathbb{C}^2$. For each i we let M_i denote a copy of $\mathbb{M}_2$ with diagonal masa D_i. Then $R_k = \bigotimes_{i=1}^k M_i$ is identified with $\mathbb{M}_{2^k}$ and has diagonal masa $A_k = \bigotimes_{i=1}^k D_i$. The embedding $X \mapsto \begin{pmatrix} X & 0 \\ 0 & X \end{pmatrix}$ places $\mathbb{M}_{2^k}$ inside $\mathbb{M}_{2^{k+1}}$, and so $\bigcup_{k \geq 1} R_k$ has a natural trace. The

GNS representation leads to a copy of the hyperfinite II_1 factor R in which the weak closure A of $\bigcup_{k \geq 1} A_k$ is a masa. Since A is normalised by the unitaries in each A_k and by the permutation matrices in each R_k, there are then sufficiently many normalising unitaries of A to generate R, and thus A is a Cartan masa in R.

Another set of examples arises from *semi-direct* products of groups which are the motivating examples for crossed products of von Neumann algebras. Let H and K be groups and suppose that $\alpha \colon K \to \operatorname{Aut}(H)$ is a group homomorphism. The semidirect product $G = H \rtimes_\alpha K$ is the set of formal products hk with multiplication defined by $(h_1 k_1)(h_2 k_2) = (h_1 \alpha_{k_1}(h_2))(k_1 k_2)$. Various algebraic conditions could be placed on H and K to produce examples of masas; the following lemma contains the simplest of these. We denote the *automorphism group* of any algebraic structure by $\operatorname{Aut}(\cdot)$.

Lemma 3.4.1. *Let H and K be infinite abelian discrete groups and let $\alpha \colon K \to \operatorname{Aut}(H)$ be a group homomorphism. Further suppose that the equation $\alpha_k(h) = h$ implies either that $h = e$ or that $k = e$. Then $G = H \rtimes_\alpha K$ is I.C.C., $L(H)$ is a Cartan masa in $L(G)$, and $L(K)$ is a singular masa in $L(G)$.*

Proof. We first show that G is I.C.C. If $h \neq e$ then $khk^{-1} = \alpha_k(h)$ gives infinitely many distinct conjugates as k varies over $K \backslash \{e\}$ since $\alpha_{k_1}(h) = \alpha_{k_2}(h)$ implies that $\alpha_{k_1^{-1} k_2}(h) = h$ so that $k_1 = k_2$ by hypothesis. Now consider an element $h_0 k$ with $k \neq e$. The conjugates by $h \in H$ are distinct because this reduces to the implication that if $hh_0 kh^{-1} = h_0 k$ then $hh_0 \alpha_k(h^{-1}) = h_0$, forcing $\alpha_k(h) = h$ and $h = e$. Thus every $g \in G \backslash \{e\}$ has infinitely many conjugates and G is I.C.C. Moreover, the last calculation shows that $\{hgh^{-1} \colon h \in H\}$ is infinite for each $g \in G \backslash H$, verifying the condition for $L(H)$ to be a masa in $L(G)$. Since H is a normal subgroup of G, it follows that $L(H)$ is Cartan in $L(G)$, by Lemma 3.3.1.

If $g \in G \backslash K$ then g must have the form $g = h_0 k_0$ with $h_0 \neq e$. Then the conjugates $\{kgk^{-1} \colon k \in K\}$ are all distinct, as we now show. If $kh_0 k_0 k^{-1} = h_0 k_0$ then $kh_0 k^{-1} = h_0$ and so $\alpha_k(h_0) = h_0$. Since $h_0 \neq e$, the hypothesis forces $k = e$. From this, it is easy to deduce that $k_1 g k_1^{-1} = k_2 g k_2^{-1}$ holds precisely when $k_1 = k_2$. Thus we have verified the condition for $L(K)$ to be a masa in $L(G)$. In order to show that $L(K)$ is singular, we will check the validity of condition (C2). Let $\{g_i = h_i k_i \colon 1 \leq i \leq n\}$ be a finite set of elements in $G \backslash K$, so that $h_i \neq e$ for $1 \leq i \leq n$. We must find an element $k \in K$ such that $g_i k g_j \notin K$ for $1 \leq i, j \leq n$, and this is equivalent to $h_i \alpha_k(\alpha_{k_i}(h_j)) \neq e$ for $1 \leq i, j \leq n$. For fixed i and j, the hypothesis implies that there can be at most one $k \in K$ such that $\alpha_k(\alpha_{k_i}(h_j)) = h_i^{-1}$ since $\alpha_{k_i}(h_j) \neq e$. Thus $\{k \colon h_i \alpha_k(\alpha_{k_i}(h_j)) = e$ for some $i, j\}$ has cardinality at most n^2, and any k from the complement will satisfy $g_i k g_j \notin K$ for $1 \leq i, j \leq n$. This verifies (C2), and it follows from Lemma 3.3.2 that $L(K)$ is a singular masa in $L(G)$. $\square$

Example 3.4.2. We construct groups which satisfy the hypothesis of Lemma 3.4.1. Let K be any infinite countable discrete group containing no elements of finite order other than e ($\mathbb{Z}$ is such a group) and let H be the set of functions

$h\colon K \to \{\pm 1\}$ such that $h(x) = 1$ for all but a finite number of $x \in K$. The group operation on H is the standard multiplication of functions. For each $k \in K$, we define $\alpha_k \in \operatorname{Aut}(H)$ by $\alpha_k(h)(x) = h(k^{-1}x)$, $x \in K$. If $h \neq e$ then there exists $x_0 \in K$ with $h(x_0) = -1$. If $k \neq e$, then the equation $\alpha_k(h) = h$ implies that $h(k^n x_0) = -1$ for all $n \in \mathbb{Z}$, which is impossible since h takes the value -1 only a finite number of times and k has infinite order. We conclude from the previous lemma that $L(H)$ and $L(K)$ are respectively Cartan and singular masas in $L(H \rtimes_\alpha K)$. $\qquad\square$

This example serves as our introduction to *crossed products* since an action of K on H induces an action on $L(H)$, and $L(H \rtimes_\alpha K) = L(H) \rtimes_\alpha K$, once we have defined the latter crossed product term.

Let M be a von Neumann algebra on a Hilbert space H and let G be a discrete group with an action $\alpha\colon G \to \operatorname{Aut}(M)$. The construction we now describe is valid for all locally compact groups but the discreteness assumption avoids discussion of topologies. The crossed product $M \rtimes_\alpha G$ is generated by two types of operators, chosen so that the action α is spatially implemented. The underlying Hilbert space is $\ell^2(G, H)$, the space of functions $f\colon G \to H$ such that $\sum_{g \in G} \|f(g)\|^2 < \infty$. For each $h \in H$, the unitary λ_h is defined by

$$(\lambda_h f)(g) = f(h^{-1}g), \qquad f \in \ell^2(G, H), \quad g \in G, \tag{3.4.1}$$

while the representation $\pi\colon M \to B(\ell^2(G, H))$ is given by

$$(\pi(x)f)(g) = \alpha_{g^{-1}}(x)(f(g)), \qquad x \in M, \ g \in G, \ f \in \ell^2(G, H). \tag{3.4.2}$$

It is then simple to check that $\lambda_h \pi(x) \lambda_{h^{-1}} = \pi(\alpha_h(x))$ for $x \in M$ and $h \in G$, and the set of operators $\{\pi(x)\lambda_h\colon x \in M, \ h \in G\}$ generates the von Neumann algebra crossed product $M \rtimes_\alpha G$. If the action of G is trivial then this reduces to $M \overline{\otimes} L(G)$, so the crossed product is a generalised type of tensor product. The crossed product will not play a prominent role in these notes, and will appear infrequently. For this reason, we will not develop the theory of crossed products here but rather refer the reader to [105, Chapter 13] which has a good account of this construction. Suffice it to say that interesting type II_1 factors can arise even from abelian M and G when the action α is complicated. We explore this in the next example, where the action of $\mathbb{Z}$ on the unit circle $\mathbb{T}$ by irrational rotation gives rise to two Cartan masas in the factor $L^\infty(\mathbb{T}) \rtimes_\alpha \mathbb{Z}$.

Example 3.4.3. Let $M = L^\infty(\mathbb{T})$ acting as multiplication operators m_f on $L^2(\mathbb{T})$ (with respect to normalised Lebesgue measure $\frac{dt}{2\pi}$ on $\mathbb{T}$) and let $\theta \in \mathbb{R} \backslash \mathbb{Q}$ be fixed. We then define an action of $\mathbb{Z}$ on $L^\infty(\mathbb{T})$ by

$$\alpha_n(f)(e^{2\pi i t}) = f(e^{2\pi i(t + n\theta)}), \qquad f \in L^\infty(\mathbb{T}), \ t \in [0, 1], \ n \in \mathbb{Z}. \tag{3.4.3}$$

The crossed product $L^\infty(\mathbb{T}) \rtimes_\alpha \mathbb{Z}$ is then generated by the operators $\{\pi(f)\colon f \in L^\infty(\mathbb{T})\}$ and $\{\lambda_n\colon n \in \mathbb{Z}\}$. This is the weak closure of the C^*-algebra generated by these operators, known as the irrational rotation C^*-algebra since $\mathbb{Z}$ acts by irrational rotation. Consider the vector $\xi \in \ell^2(\mathbb{Z}, L^2(\mathbb{T}))$ whose only non-zero component is $\xi(0) = 1$. Then let τ be the vector state $\langle \cdot \xi, \xi \rangle$. For $f \in L^\infty(\mathbb{T})$ and $n \in \mathbb{Z}$,

$$\tau(\lambda_n \pi(f) - \pi(f)\lambda_n) = \tau(\pi(\alpha_n(f) - f)\lambda_n)$$

$$= \sum_{k=-\infty}^{\infty} \langle \alpha_{-k}(\alpha_n(f) - f)\xi(k-n), \xi(k) \rangle \qquad (3.4.4)$$

and this is always 0 because $\alpha_0(f) = f$, while $\xi(k-n)\overline{\xi(k)} = 0$ when $n \neq 0$. Thus τ defines a normal trace on the crossed product. Since $\|\xi\| = 1$, τ is also normalised. Now, for $f \in L^\infty(\mathbb{T})$ and $k, n \in \mathbb{Z}$,

$$(\pi(f)\lambda_n \xi)(k) = \alpha_{-k}(f)\xi(k-n) \qquad (3.4.5)$$

and so $\pi(f)\lambda_n \xi$ is the function with values $\alpha_{-n}(f)$ at n and 0 elsewhere. This shows that ξ is a cyclic vector for $L^\infty(\mathbb{T}) \rtimes_\alpha \mathbb{Z}$. In order to show that τ is faithful, we require that ξ is separating for $L^\infty(\mathbb{T}) \rtimes_\alpha \mathbb{Z}$, equivalently cyclic for the commutant. Thus we need to know some operators in the commutant.

Define a representation $\sigma \colon L^\infty(\mathbb{T}) \to B(\ell^2(\mathbb{Z}, L^2(\mathbb{T})))$ by

$$(\sigma(f)\eta)(n) = f\eta(n), \quad f \in L^\infty(\mathbb{T}), \ \eta \in \ell^2(\mathbb{Z}, L^2(\mathbb{T})), \ n \in \mathbb{Z}. \qquad (3.4.6)$$

Then, for $f, g \in L^\infty(\mathbb{T})$, $\eta \in \ell^2(\mathbb{Z}, L^2(\mathbb{T}))$ and $n \in \mathbb{Z}$,

$$(\sigma(f)\pi(g)\eta)(n) = f(\pi(g)\eta)(n) = f\alpha_{-n}(g)\eta(n)$$
$$= \alpha_{-n}(g)f\eta(n) = (\pi(g)\sigma(f)\eta)(n), \qquad (3.4.7)$$

showing that $\sigma(f)$ commutes with $\pi(g)$. Similarly, for $k \in \mathbb{Z}$,

$$(\sigma(f)\lambda_k \eta)(n) = f(\lambda_k \eta)(n) = f\eta(n-k)$$
$$= \lambda_k(\sigma(f)\eta)(n), \qquad (3.4.8)$$

and $\sigma(f)$ also commutes with each λ_k. Thus $\sigma(f) \in (L^\infty(\mathbb{T}) \rtimes_\alpha \mathbb{Z})'$ for all $f \in L^\infty(\mathbb{T})$.

Now define unitaries $u_n \in B(L^2(\mathbb{T}))$ by

$$(u_n g)(e^{2\pi i t}) = g(e^{2\pi i(t+n\theta)}), \qquad g \in L^2(\mathbb{T}), \ n \in \mathbb{Z}. \qquad (3.4.9)$$

These spatially implement the action α because

$$(u_n m_f u_n^* g)(e^{2\pi i t}) = (m_f u_n^* g)(e^{2\pi i(t+n\theta)})$$
$$= f(e^{2\pi i(t+n\theta)})(u_n^* g)(e^{2\pi i(t+n\theta)})$$
$$= (\alpha_n(f)g)(e^{2\pi i t}), \qquad (3.4.10)$$

for $f \in L^\infty(\mathbb{T})$, $g \in L^2(\mathbb{T})$. This shows that $u_n m_f u_n^* = m_{\alpha_n(f)}$ for $f \in L^\infty(\mathbb{T})$. For each $n \in \mathbb{Z}$ define a unitary $v_n \in B(\ell^2(\mathbb{Z}, L^2(\mathbb{T})))$ by

$$(v_n \eta)(k) = u_{-n} \eta(k-n), \qquad \eta \in \ell^2(\mathbb{Z}, L^2(\mathbb{T})), \ k \in \mathbb{Z}. \qquad (3.4.11)$$

Then

$$
\begin{aligned}
(v_n \pi(f) v_{-n} \eta)(k) &= u_{-n}(\pi(f) v_{-n} \eta(k - n)) \\
&= u_{-n} \alpha_{n-k}(f)(v_{-n} \eta)(k - n) \\
&= u_{-n} \alpha_{n-k}(f) u_n \eta(k) \\
&= \alpha_{-k}(f) \eta(k) = (\pi(f) \eta)(k),
\end{aligned} \tag{3.4.12}
$$

for $\eta \in \ell^2(\mathbb{Z}, L^2(\mathbb{T}))$, $n, k \in \mathbb{Z}$, $f \in L^\infty(\mathbb{T})$. Thus each v_n commutes with each $\pi(f)$. For $n, r, k \in \mathbb{Z}$ and $\eta \in \ell^2(\mathbb{Z}, L^2(\mathbb{T}))$, we have

$$
\begin{aligned}
(\lambda_{-r} v_n \lambda_r \eta)(k) &= (v_n \lambda_r \eta)(k + r) = u_{-n}(\lambda_r \eta(k + r - n)) \\
&= u_{-n}(\eta(k - n)) = v_n \eta(k).
\end{aligned} \tag{3.4.13}
$$

Thus each v_n also commutes with each λ_r, and so $v_n \in (L^\infty(\mathbb{T}) \rtimes_\alpha \mathbb{Z})'$. Now $v_n \xi(k) = 1$ when $k = n$ and is 0 otherwise, so $\sigma(f) v_n \xi(k) = f$ at $k = n$ and vanishes elsewhere. Thus ξ is cyclic for $L^\infty(\mathbb{T} \rtimes_\alpha \mathbb{Z})'$, separating for $L^\infty(\mathbb{T} \times_\alpha \mathbb{Z})$ (Lemma 2.3.1), and hence defines a faithful trace on the crossed product.

Now suppose that τ_1 is another normalised normal trace on $L^\infty(\mathbb{T}) \rtimes_\alpha \mathbb{Z}$. The functions $e^{2\pi i k t}$, $k \neq 0$, have the property that $\alpha_n(e^{2\pi i k t)}) = e^{2\pi i k n \theta} e^{2\pi i k t}$ for $n \in \mathbb{Z}$. Since $\lambda_n \pi(f) \lambda_n^* = \pi(\alpha_n(f))$, τ_1 must agree on $\pi(f)$ and $\pi(\alpha_n(f))$. Thus $\tau_1(\pi(e^{2\pi i k t})) = 0$ for $k \neq 0$ and so τ_1 and τ agree on $\pi(L^\infty(\mathbb{T}))$. If $g \in L^\infty(\mathbb{T})$ then

$$
\begin{aligned}
\tau_1(\pi(g) \lambda_n) &= \tau_1(\pi(e^{-2\pi i t}) \pi(g) \lambda_n \pi(e^{2\pi i t})) \\
&= \tau_1(e^{2\pi i n \theta} \pi(g) \lambda_n),
\end{aligned} \tag{3.4.14}
$$

showing that $\tau_1(\pi(g) \lambda_n) = 0$ for any $n \neq 0$. Thus $\tau_1 = \tau$ and $L^\infty(\mathbb{T} \rtimes_\alpha \mathbb{Z})$ is a factor, since any nontrivial central projection p would give a normalised trace $\tau(\cdot p)/\tau(p)$ different from τ.

We now turn to the question of whether $L^\infty(\mathbb{T})$ is a masa in the crossed product, and for this it is convenient to establish an orthonormal basis for $\ell^2(\mathbb{Z}, L^2(\mathbb{T}))$. For each $p \in \mathbb{Z}$, let e_p be the function $e^{2\pi i p t}$, and for each $q \in \mathbb{Z}$ let $\xi_{p,q}$ denote the function in $\ell^2(\mathbb{Z}, L^2(\mathbb{T}))$ defined by

$$
\xi_{p,q}(n) = \begin{cases} e_p, & n = q \\ 0, & n \neq q. \end{cases} \tag{3.4.15}
$$

In this notation, the cyclic and separating vector is $\xi_{0,0}$. Write β for the number $e^{2\pi i \theta}$; irrationality of θ shows that no non-zero power of β is ever equal to 1. The operators that we have already introduced act on the basis $\{\xi_{p,q}\}_{p,q \in \mathbb{Z}}$ as follows. For $n, p, q, k \in \mathbb{Z}$,

$$v_n \xi_{p,q}(k) = u_{-n}(\xi_{p,q}(k - n))$$

$$= \begin{cases} u_{-n} e_p, & k = n + q \\ 0, & k \neq n + q \end{cases}$$

$$= \begin{cases} \beta^{-np} e_p, & k = n + q \\ 0, & k \neq n + q \end{cases}$$

$$= \beta^{-np} \xi_{p,q+n}(k), \qquad (3.4.16)$$

and so $v_n \xi_{p,q} = \beta^{-np} \xi_{p,q+n}$. We also have

$$\sigma(e_n) \xi_{p,q}(k) = \begin{cases} \sigma(e_n) e_p, & k = q \\ 0, & k \neq q \end{cases}$$

$$= \begin{cases} e_{n+p}, & k = q \\ 0, & k \neq q \end{cases} \qquad (3.4.17)$$

and so $\sigma(e_n) \xi_{p,q} = \xi_{p+n,q}$. Similar calculations show that $\lambda_n \xi_{p,q} = \xi_{p,q+n}$ and $\pi(e_n) \xi_{p,q} = \beta^{-nq} \xi_{p+n,q}$.

Now consider an operator $y \in L^\infty(\mathbb{T}) \rtimes_\alpha \mathbb{Z}$ which commutes with each $\pi(e_n)$. We wish to show that $y = \pi(g)$ for some $g \in L^\infty(\mathbb{T})$, which will prove that $L^\infty(\mathbb{T})$ is maximal abelian. Write

$$y \xi_{0,0} = \sum_{p,q} \gamma_{p,q} \xi_{p,q}, \quad \sum_{p,q} |\gamma_{p,q}|^2 < \infty. \qquad (3.4.18)$$

Then

$$y\pi(e_n)\xi_{0,0} = y\xi_{n,0} = y\sigma(e_n)\xi_{0,0}$$

$$= \sigma(e_n) y \xi_{0,0} = \sum_{p,q} \gamma_{p,q} \xi_{p+n,q} \qquad (3.4.19)$$

while

$$\pi(e_n) y \xi_{0,0} = \sum_{p,q} \beta^{-nq} \gamma_{p,q} \xi_{p+n,q}. \qquad (3.4.20)$$

Comparing these two expressions leads to $(1 - \beta^{-nq}) \gamma_{p,q} = 0$ for all $p, q, n \in \mathbb{Z}$. If $q \neq 0$ then $\gamma_{p,q} = 0$, and thus $y \xi_{0,0} = \sum_p \gamma_{p,0} \xi_{p,0}$. Let $g = \sum_p \gamma_{p,0} e_p \in L^2(\mathbb{T})$. Let F be the set of functions in the unit ball of $L^2(\mathbb{T})$ which have finitely many non-zero coefficients in their Fourier expansions.

If $h_1, h_2 \in F$ with $h_1 = \sum \delta_n e_n$, $h_2 = \sum \varepsilon_n e_n$ (finite sums) then

$$\int g h_1 \bar{h}_2 = \sum_{p,n} \gamma_{p,0} \delta_n \bar{\varepsilon}_{n+p}. \qquad (3.4.21)$$

On the other hand,

$$
\left\langle y \sum_m \delta_m \xi_{m,0}, \sum_n \varepsilon_n \xi_{n,0} \right\rangle = \left\langle y \sum_m \delta_m \sigma(e_m)\xi_{0,0}, \sum_n \varepsilon_n \xi_{n,0} \right\rangle
$$

$$
= \sum_{m,n} \delta_m \bar{\varepsilon}_n \langle \sigma(e_m) y \xi_{0,0}, \xi_{n,0} \rangle
$$

$$
= \sum_{m,n,p} \delta_m \bar{\varepsilon}_n \gamma_{p,0} \langle \xi_{p+m,0}, \xi_{n,0} \rangle
$$

$$
= \sum_{m,p} \gamma_{p,0} \delta_m \bar{\varepsilon}_{m+p}. \tag{3.4.22}
$$

Thus $|\int g h_1 \bar{h}_2| \leq \|y\|$. Products of functions from F are $\|\cdot\|_1$-dense in the unit ball of $L^1(\mathbb{T})$, so we conclude that $g \in L^\infty(\mathbb{T})$. Since

$$
\pi(g)\xi_{0,0}(k) = \begin{cases} g, & k = 0 \\ 0, & k \neq 0, \end{cases} \tag{3.4.23}
$$

we have $\pi(g)\xi_{0,0} = \sum_p \gamma_{p,0}\xi_{p,0}$. The vector $\xi_{0,0}$ is separating for the crossed product, so this shows that $y = \pi(g)$, and that $L^\infty(\mathbb{T})$ is a masa. It is clearly a Cartan masa since it is normalised by the unitaries in $L^\infty(\mathbb{T})$ and by the operators $\{\lambda_n \colon n \in \mathbb{Z}\}$.

There is another natural Cartan masa B in this crossed product, generated by $\{\lambda_n \colon n \in \mathbb{Z}\}$. Since $\alpha_n(e^{2\pi imt}) = e^{2\pi imn\theta} e^{2\pi imt}$, it follows that B is normalised by each λ_n as well as each $e^{2\pi imt}$ and these generate $L^\infty(\mathbb{T}) \rtimes_\alpha \mathbb{Z}$. It remains to show that B is a masa. The subspace $K = \overline{\text{span}}\{\xi_{0,q} \colon q \in \mathbb{Z}\}$ is invariant for B and the restriction gives an isometric representation of B since the vector $\xi_{0,0}$ is separating for B. This identifies B with the masa $L(\mathbb{Z})$ on $\ell^2(\mathbb{Z})$. If $y \in L^\infty(\mathbb{T}) \rtimes_\alpha \mathbb{Z}$ commutes with each λ_n, then let $y\xi_{0,0} = \sum_{p,q} \gamma_{p,q}\xi_{p,q}$. The commutation requirement gives

$$
0 = (\lambda_n y - y\lambda_n)\xi_{0,0} = (\lambda_n y - y v_n)\xi_{0,0}
$$

$$
= (\lambda_n - v_n) \sum_{p,q} \gamma_{p,q}\xi_{p,q}
$$

$$
= \sum_{p,q} (1 - \beta^{-np})\gamma_{p,q}\xi_{p,q+n} \tag{3.4.24}
$$

and so $\gamma_{p,q} = 0$ for $p \neq 0$ since β^{np} never equals 1 when $n \neq 0$ unless $p = 0$. Thus $y\xi_{0,0} = \sum_q \gamma_{0,q}\xi_{0,q}$, and

$$
y\xi_{0,n} = y\lambda_n \xi_{0,0} = \lambda_n y\xi_{0,0} = \sum_q \gamma_{0,q}\xi_{0,q+n}. \tag{3.4.25}
$$

This shows that y leaves invariant the subspace K on which it commutes with each λ_n, so there exists $x \in L(\mathbb{Z})$ such that $x = y$ on K. Since K contains the separating vector $\xi_{0,0}$, we conclude that $x = y$, and $y \in B$. Thus B is a masa as required. $\square$

Remark 3.4.4. We have chosen to give a detailed exposition of the irrational rotation algebra to provide some insight into crossed products. If one is prepared to accept some of the underlying theory of crossed products (see [105, Chapter 13]), then a much shorter proof can be given. The algebra $L^\infty(\mathbb{T})$, acting on $L^2(\mathbb{T})$ by multiplication, is a masa. Moreover, the $\mathbb{Z}$-action α is implemented by the unitaries

$$(w_n g)(e^{2\pi i t}) = g(e^{2\pi i(t+n\theta)}), \quad n \in \mathbb{Z},\ g \in L^2(\mathbb{T}). \tag{3.4.26}$$

One checks that $L^\infty(\mathbb{T}) \cap w_n L^\infty(\mathbb{T}) = \{0\}$ for $n \neq 0$, (a *free* action), and that $\alpha_n(f) = f$ for all $n \in \mathbb{Z}$ can only occur when f is a constant function, (an *ergodic* action). Then $L^\infty(\mathbb{T})$ is a masa in the finite factor $L^\infty \rtimes_\alpha \mathbb{Z}$ by appealing to the theorems of [105, Chapter 13]. □

Remark 3.4.5. Easy examples of semi-regular masas are available from free products. A full discussion of the free product $M \star N$ of two finite factors would take us too far afield, so we refer to [15, 200] for the background. If A is a masa in M, then it remains a masa in $M \star N$, essentially because there are no relations between elements of M and N. Moreover, its normaliser in the free product is equal to its normaliser in M, for the same reason (see [62]). Thus Cartan masas in M give examples of semi-regular masas in $M \star N$, their normalisers generating the proper subfactor $M \subseteq M \star N$. □

3.5   Diffuse abelian algebras

A von Neumann algebra N is said to be *diffuse* if there are no non-zero minimal projections in N. A typical example is $L^\infty(X, \mu)$ where the measure μ is nonatomic. Type II$_1$ factors have this property, and so do their masas. In Theorem 3.5.2 we will show that abelian separable diffuse von Neumann algebras are all *-isomorphic to $L^\infty[0, 1]$ by an isomorphism that carries any specified faithful normal trace to the state $\int_0^1 f(t)\, dt$. When considered in isolation, this shows that all masas in finite von Neumann algebras are isomorphic, (Corollary 3.5.3), although the previous discussion of types of masas shows that other considerations can enter the picture when the inclusion $A \subseteq N$ is taken into account.

The first lemma deals with part of the proof of the main theorem of the section.

Lemma 3.5.1. *Let A be a diffuse abelian von Neumann subalgebra in a separable finite von Neumann algebra N with faithful normal trace τ. If $\mathcal{F}$ is a maximal totally ordered set of projections in A, then $\{\tau(f)\colon f \in \mathcal{F}\} = [0, 1]$.*

Proof. Let λ be in the closure of $\{\tau(f)\colon f \in \mathcal{F}\}$. Then there is an increasing or decreasing sequence $\tau(f_n)$, with $f_n \in \mathcal{F}$, converging to λ. Since $\mathcal{F}$ is totally ordered, the sequence (f_n) is an increasing or decreasing sequence of projections so converges strongly to a projection $e \in A$. Suppose that f_n is an increasing sequence; the decreasing case is similar. If g is in $\mathcal{F}$, then either $f_n g = f_n$ for all

n or there is an N such that $f_n g = g$ for all $n \geq N$; in the former case $eg = e$, in the latter $eg = g$. Thus the projection e is comparable with all projections in $\mathcal{F}$ so $\mathcal{F} \cup \{e\}$ is a totally ordered family of projections. By maximality, $e \in \mathcal{F}$. Hence $\{\tau(f): f \in \mathcal{F}\}$ is closed since $\tau(e) = \lambda$ by the strong continuity of the trace. If the set $\{\tau(f): f \in \mathcal{F}\}$ were disconnected by omitting the open interval $(\tau(g_1), \tau(g_2))$ with $g_1 \leq g_2$ in $\mathcal{F}$, then there is a non-zero projection h in A with $h \leq g_2 - g_1$ and

$$0 < \tau(h) < \tau(g_2 - g_1), \tag{3.5.1}$$

since A has no minimal projections. Hence

$$g_1 \leq g_1 + h \leq g_2 \tag{3.5.2}$$

and

$$\tau(g_1) < \tau(g_1 + h) < \tau(g_2), \tag{3.5.3}$$

so $\mathcal{F} \cup \{g_1 + h\}$ is a totally ordered family of projections in A properly containing $\mathcal{F}$. The resulting contradiction proves the lemma. $\qquad \square$

Theorem 3.5.2. *If A is a diffuse separable abelian von Neumann algebra with faithful normal trace τ, then there exists a $*$-isomorphism θ from A onto $L^\infty[0, 1]$ such that*

$$\tau(x) = \int_0^1 \theta(x)(t)\, dt \tag{3.5.4}$$

for all $x \in A$.

Proof. By the separability of the predual of A, there is a sequence $\mathcal{P} = \{p_n: n = 0, 1, \dots\}$ of projections in A such that the algebra B_0 generated by this set of projections is weakly dense in A. Assume that $p_0 = 1$. A totally ordered set $\mathcal{G}$ of projections in A will be constructed from $\mathcal{P}$ by induction so that the algebras generated by $\mathcal{G}$ and $\mathcal{P}$ are equal and $\mathcal{G}$ is the image of the dyadic rationals in $[0,1]$ under an order preserving mapping $t \mapsto q(t)$. Let $q(0) = 0$, $q(1) = 1$ and $q(1/2) = p_1$. Then

$$q(0) \leq q(1/2) \leq q(1) \tag{3.5.5}$$

and

$$\text{Alg}\{q(j \cdot 2^{-1}): \ j = 0, 1, 2\} = \text{Alg}\{p_0, p_1\}. \tag{3.5.6}$$

Suppose that the map q from $\{j2^{-n}: \ 0 \leq j \leq 2^n\}$ into A_0 has been defined such that $0 \leq j \leq k \leq 2^n$ implies that

$$q(j \cdot 2^{-n}) \leq q(k \cdot 2^{-n}) \tag{3.5.7}$$

and

$$\text{Alg}\{q(j \cdot 2^{-n}): \ 0 \leq j \leq 2^n\} = \text{Alg}\{p_i: \ 0 \leq i \leq n\}. \tag{3.5.8}$$

The next step in the induction requires the definition of $q((2j+1)2^{-n-1})$ which is done by splitting the difference between the projections $q((j+1)2^{-n})$ and $q(j \cdot 2^{-n})$ using p_{n+1}. Let

$$q((2j+1)2^{-n-1}) = q(j \cdot 2^{-n} + 2^{-n-1})$$
$$= q(j \cdot 2^{-n}) + p_{n+1}\{q((j+1)2^{-n}) - q(j2^{-n})\} \tag{3.5.9}$$

for $0 \le j \le n - 1$.

Then for these values of j,

$$q(j \cdot 2^{-n}) \le q((2j + 1) \cdot 2^{-n-1}) \le q((j + 1)2^{-n}) \tag{3.5.10}$$

so the map q preserves the order. Note that the inductive definition implies directly that

$$\text{Alg}\{q(j \cdot 2^{-n-1}): \ 0 \le j \le 2^{n+1}\} \subseteq \text{Alg}\{p_i: \ 0 \le i \le n + 1\}. \tag{3.5.11}$$

Observe that the equation

$$p_{n+1} = \sum_{j=0}^{2^n - 1} (q((j + 1)2^{-n}) - q(j \cdot 2^{-n}))p_{n+1}$$

$$= \sum_{j=0}^{2^n - 1} (q((2j + 1) \cdot 2^{-n-1}) - q(j \cdot 2^{-n})) \tag{3.5.12}$$

holds by (3.5.9) so that

$$p_{n+1} \in \text{Alg}\{q(j \cdot 2^{-n-1}): \ 0 \le j \le 2^{n+1}\} \tag{3.5.13}$$

and the two algebras in (3.5.11) are equal.

By Zorn's Lemma, choose a maximal totally ordered set $\mathcal{F}$ of projections in A with $\mathcal{G} \subseteq \mathcal{F}$. Then $\text{Alg}(\mathcal{F}) \supseteq \text{Alg}(\mathcal{G}) = \text{Alg}(\mathcal{P})$ so that $\text{Alg}(\mathcal{F})$ is weakly dense in A. By Lemma 3.5.1, the trace is an injective order preserving map from $\mathcal{F}$ onto $[0,1]$. This is the stage at which we move from $\mathcal{G}$ with its unusual dyadic order to the usual order of A. The $*$-isomorphism θ is defined from $\text{Alg}(\mathcal{F})$ into $L^\infty[0, 1]$ by

$$\theta(f) = \chi_{[0,\tau(f)]}, \tag{3.5.14}$$

where χ_W is the characteristic function of a subset W of $[0,1]$. Note that θ is a $*$-isomorphism because if $f_1, f_2 \in \mathcal{F}$, then

$$\tau(f_1 f_2) = \min(\tau(f_1), \tau(f_2)) \tag{3.5.15}$$

which matches the product of the characteristic functions. Clearly $\|\theta x\|_\infty = \|x\|$ for all x in $\text{Alg}(\mathcal{F})$ because the norm in A_0 is a C^*-algebra norm. The equation

$$\int_0^1 \theta(f)(t)dt = \tau(f) \tag{3.5.16}$$

for all f in $\mathcal{F}$ extends by linearity to $\text{Alg}(\mathcal{F})$ so

$$\int_0^1 \theta(x)(t)dt = \tau(x) \tag{3.5.17}$$

for all x in $\text{Alg}(\mathcal{F})$. Thus $\|\theta(x)\|_2 = \|x\|_2$ for all x in $\text{Alg}(\mathcal{F})$. Since the $\|\cdot\|$-unit ball of $\text{Alg}(\mathcal{P}) \subseteq \text{Alg}(\mathcal{F})$ is dense in the $\|\cdot\|$-unit ball B of A in the $\|\cdot\|_2$-norm by the Kaplansky density theorem and by the equivalence of the $\|\cdot\|_2$-norm and strong topologies on B, it follows that θ extends to a $*$-homomorphism θ from A into $L^\infty[0,1]$ with

$$\int_0^1 \theta(x)(t)dt = \tau(x) \tag{3.5.18}$$

for all x in A. The map θ is surjective because $\theta(\text{Alg } \mathcal{F})$ is strongly dense in $L^\infty[0,1]$. This proves the theorem. $\qquad\square$

Corollary 3.5.3. *Let N be a diffuse separable finite von Neumann algebra with a faithful normal trace τ, and let A be a masa in N. Then A is diffuse and is $*$-isomorphic to $L^\infty[0,1]$.*

Proof. Suppose that A has a minimal non-zero projection p. Then we have the decomposition $A = \mathbb{C}p \oplus A(1-p)$. Since N is diffuse, we may choose a projection $q \in N$ such that $0 < q < p$. But then $q \in A' \cap N = A$, contradicting the minimality of p. Thus A is diffuse, and the result now follows from Theorem 3.5.2. $\qquad\square$

3.6 Conditional expectations

Throughout this section N will denote a finite von Neumann algebra with a fixed faithful normal trace τ and B will denote a von Neumann subalgebra of N. The inner product $\langle \cdot, \cdot \rangle$ on $L^2(N)$ induced by the trace is defined by $\langle x, y \rangle = \tau(xy^*)$, $(x, y \in N)$. Usually an element $x \in N$ will be regarded as being in $L^2(N)$ so $N \subseteq L^2(N)$ as a dense linear subspace in $\|\cdot\|_2$-norm. When convenient to treat N as a separate entity from its image in $L^2(N)$, we shall denote 1 in $L^2(N)$ by ξ and $x \mapsto x\xi : N \to L^2(N)$ as the natural $\|\cdot\|_2$-isometric embedding. Throughout these notes we will reserve the letter ξ for this purpose.

The elements of N act naturally by left multiplication on the dense subspace $N\xi$ with unique bounded extensions to $L^2(N)$. When N is represented in this way, we say that N is in *standard form*. This is the most convenient representation of a finite von Neumann algebra, and there will only be a few occasions in these notes when we will wish to consider any different representation.

There is another useful space associated to the pair (N, τ). This is $L^1(N)$, defined as follows. The polar decomposition allows us to show that, for any $x \in N$, $\tau(|x|) = \sup\{|\tau(xy)| \colon \|y\| \le 1, \ y \in N\}$, and the latter expression is easily checked to define a norm on N, written as $\|x\|_1$. The completion of N with this norm is denoted by $L^1(N)$. A standard duality argument shows that the set $\{\theta_x(y) := \tau(xy) \colon x, y \in N\}$ of normal linear functionals is norm dense in N_*, and this identifies the predual as $L^1(N)$, since $\|\theta_x\| = \tau(|x|)$. It is also possible to define the spaces $L^p(N)$ for $p \ge 1$, but only the cases $p = 1$ and $p = 2$ will concern us subsequently, and we omit further details.

Let J be the extension to $L^2(N)$ of the conjugate linear $\|\cdot\|_2$-isometry given by $J(x) = x^*$ $(x \in N)$. Then $J^2 = I$ and J is conjugate linear self-adjoint in the sense that

$$\langle J\xi, \eta \rangle = \langle J\eta, \xi \rangle \quad \text{and} \quad \langle J\xi, J\eta \rangle = \langle \eta, \xi \rangle \quad \text{for all} \quad \xi, \eta \in L^2(N). \qquad (3.6.1)$$

The operator J is the conjugation operator for N.

Note that if $\eta \in L^2(N)$ and $x, y \in N$, then the equations

$$\langle JxJ\eta, y \rangle = \langle \eta x^*, y \rangle = \langle \eta, yx \rangle \quad \text{and} \quad \langle x\eta, y \rangle = \langle \eta, x^*y \rangle \qquad (3.6.2)$$

follow from

$$\langle JxJz, y \rangle = \langle zx^*, y \rangle = \tau(zx^*y^*) = \tau(z(yx)^*) = \langle z, yx \rangle \qquad (3.6.3)$$

and

$$\langle xz, y \rangle = \tau(xzy^*) = \langle z, x^*y \rangle \qquad (3.6.4)$$

for all $z \in N$.

Lemma 3.6.1. *Let N be a finite von Neumann algebra with a fixed faithful normal trace τ.*

(i) *The relation $JNJ = N'$ holds and the map $x \mapsto Jx^*J : N \to N'$ is an anti-isomorphism.*

(ii) *If $\eta \in L^2(N)$ and there is a constant k such that $|\langle JxJ\eta, y \rangle| \le k\|x\|_2\|y\|_2$ for all $x, y \in N$, then $\eta \in N$.*

Proof. (i)   By definition of J,

$$xJyJz = xJyz^* = xzy^* = J(y(xz)^*) = JyJxz \qquad (3.6.5)$$

for all $x, y, z \in N$. Hence $JNJ \subseteq N'$.

If $\xi = 1 \in L^2(N)$, $x' \in N'$ and $y \in N$, then by equation 3.6.1

$$
\begin{aligned}
\langle Jx'\xi, y \rangle &= \langle Jy, x'\xi \rangle = \langle y^*\xi, x'\xi \rangle \ = \langle x'^*y^*\xi, \xi \rangle \\
&= \langle y^*x'^*\xi, \xi \rangle = \langle x'^*\xi, y\xi \rangle = \langle x'^*\xi, y \rangle \qquad (3.6.6)
\end{aligned}
$$

so that $Jx'\xi = x'^*\xi$. Since ξ is a separating vector for N, it is a cyclic vector for N' (Lemma 2.3.1) so $N'\xi$ is dense in $L^2(N)$. If $x', y', z' \in N'$, then, as in equation (3.6.5),

$$x'Jy'Jz'\xi = x'Jy'z'^*\xi = x'z'y'^*\xi = J(y'(x'z')^*)\xi = Jy'Jx'z'\xi \qquad (3.6.7)$$

so that $Jy'J \in N'' = N$. Hence $JN'J \subseteq N$ so $N' \subseteq JNJ$ proving $JNJ = N'$. The anti-isomorphism of $x \mapsto Jx^*J$ follows from $J^2 = I$ and the observation that $x \mapsto x^*$ reverses products. Note that JxJ is just right multiplication on N by x^*.

(ii)   The assumed inequality implies that the linear operator $t : x \mapsto Jx^*J\eta : N \to L^2(N)$ extends to an operator in $B(L^2(N))$. If $x, y \in N$, then

$$(tJyJ)(x) = t(xy^*) = Jyx^*J\eta = JyJ\,Jx^*J\eta = (JyJt)(x) \qquad (3.6.8)$$

so that $tJyJ = JyJt$ for all $y \in N$. Hence $t \in N$ by (i), and $\eta = t\xi \in N$. $\square$

Let e_B denote the (orthogonal) projection from $L^2(N) = L^2(N, \tau)$ onto $L^2(B)$ which is the closure of B in $L^2(N)$. The trace preserving conditional expectation of $\mathbb{E}_B$ of N onto B is defined to be the restriction $e_B|_N$ of e_B to N, which is regarded as an operator from N into N by the following lemma.

As we will see in Lemma 3.6.2, conditional expectations enjoy a special property called *complete positivity*, explored first by Stinespring [183] and then by Arveson [3]. A bounded linear map $\phi\colon A \to B$ between C^*-algebras is *positive* if $\phi(a) \geq 0$ whenever $a \geq 0$. For each $n \geq 1$, there is an induced map $\phi_n = \phi \otimes id_n\colon A \otimes \mathbb{M}_n \to B \otimes \mathbb{M}_n$, and ϕ is *completely positive* if every ϕ_n is positive. The theory of such maps is extensive and may be found in the books of Effros and Ruan, [66], Paulsen, [128], Pisier, [134, 135], these authors, [172], and Takesaki, [187] as well as other sources. We mention only two results that we will use subsequently. If $\phi\colon A \to B$ is a linear map, then we may regard the range as $B(H)$ by faithfully representing B on a Hilbert space H. A theorem of Stinespring, [183], characterises completely positive maps as those that have the form $\phi(a) = V^*\pi(a)V$ for $a \in A$, where $\pi\colon A \to B(K)$ is a *-representation of A on a Hilbert space K and $V\colon H \to K$ is a bounded linear operator. There are two easy consequences of this. The first is that $\|\phi\| \leq 1$ if and only if V is a contraction. The second is the inequality

$$\phi(a)^*\phi(a) \leq \phi(a^*a), \qquad a \in A, \tag{3.6.9}$$

for all completely positive contractions ϕ, which uses the Stinespring representation and the relation $VV^* \leq I$ when $\|V\| \leq 1$.

When $\phi\colon X \to Y$ is a map between B-bimodules for an algebra B, we say that ϕ is *B-modular* if $\phi(b_1 x b_2) = b_1\phi(x)b_2$ for all $b_1, b_2 \in B$ and $x \in X$. We note the important property, expressed in Lemma 3.6.2(ii), that conditional expectations onto subalgebras B are B-modular.

Lemma 3.6.2. *Let N be a finite von Neumann algebra with a fixed faithful normal trace τ and let B be a von Neumann subalgebra of N. Then e_B and $\mathbb{E}_B$ have the following properties:*

(i) *$e_B|_N = \mathbb{E}_B$ is a norm reducing map from N onto B with $\mathbb{E}_B(1) = 1$;*

(ii) *$\mathbb{E}_B(bxc) = b\mathbb{E}_B(x)c$ for all $x \in N$, $b, c \in B$;*

(iii) *$\tau(x\mathbb{E}_B(y)) = \tau(\mathbb{E}_B(x)\mathbb{E}_B(y)) = \tau(\mathbb{E}_B(x)y)$ for all $x, y \in N$;*

(iv) *$e_B x e_B = \mathbb{E}_B(x)e_B = e_B\mathbb{E}_B(x)$ for all $x \in N$;*

(v) *$\{e_B\}' \cap N = B$ and $B' = (N' \cup \{e_B\})''$;*

(vi) *$\mathbb{E}_B$ is a normal completely positive map;*

(vii) *$e_B J = J e_B$ and $\mathbb{E}_B J = J \mathbb{E}_B$.*

If $\phi : N \to B$ with $\phi(b_1 x b_2) = b_1\phi(x)b_2$ and $\tau(\phi(x)) = \tau(x)$ for all $x \in N$ and $b_1, b_2 \in B$, then $\phi = \mathbb{E}_B$.

Proof. (i)   If $x \in N \subseteq L^2(N)$ and $b_1, b_2 \in B \subseteq L^2(B)$, then the properties $(I - e_B)L^2(N) \perp B$ and $BB = B$, and equation (3.6.1) imply that

$$\langle Jb_1^* Je_B(x), b_2 \rangle = \langle e_B(x)b_1, b_2 \rangle = \langle e_B(x), b_2 b_1^* \rangle = \langle x, e_B(b_2 b_1^*) \rangle$$
$$= \langle x, b_2 b_1^* \rangle = \langle x b_1, b_2 \rangle \qquad (3.6.10)$$

so that

$$|\langle Jb_1^* Je_B(x), b_2 \rangle| = |\langle e_B(x) b_1, b_2 \rangle|$$
$$\leq \|xb_1\|_2 \|b_2\|_2 \leq \|x\| \|b_1\|_2 \|b_2\|_2. \qquad (3.6.11)$$

By Lemma 3.6.1 $e_B(x) \in N$ and $\|e_B(x)\| \leq \|x\|$ for all $x \in N$. Hence $\mathbb{E}_B := e_B|_N : N \to N$ is a well-defined linear operator from N onto B with $\mathbb{E}_B^2 = \mathbb{E}_B$, $\|\mathbb{E}_B\| = 1$ and $\mathbb{E}_B(1) = 1$.

(ii)   Let p be a projection in B. If $x, y \in N$, then

$$\|px + (1-p)y\|^2 = \|(px + (1-p)y)^*(px + (1-p)y)\|$$
$$= \|x^* px + y^*(1-p)y\|$$
$$\leq \|x^* px\| + \|y^*(1-p)y\|$$
$$= \|px\|^2 + \|(1-p)y\|^2. \qquad (3.6.12)$$

Since $(1-p)\mathbb{E}_B = ((1-p)\mathbb{E}_B)^2$ is norm reducing on N, for each $\alpha \in \mathbb{R}$ and $x \in N$,

$$(\alpha + 1)^2 \|(1-p)\mathbb{E}_B(px)\|^2 = \|(1-p)\mathbb{E}_B(px) + \alpha(1-p)\mathbb{E}_B(px)\|^2$$
$$= \|(1-p)\mathbb{E}_B\{px + \alpha(1-p)\mathbb{E}_B(px)\}\|^2$$
$$\leq \|(px + \alpha(1-p)\mathbb{E}_B(px)\|^2$$
$$\leq \|px\|^2 + \|\alpha(1-p)\mathbb{E}_B(px)\|^2, \qquad (3.6.13)$$

so that

$$(2\alpha + 1)\|(1-p)\mathbb{E}_B(px)\| \leq \|px\|^2. \qquad (3.6.14)$$

The last inequality holds for all $\alpha \in \mathbb{R}$ so that $(1-p)\mathbb{E}_B(px) = 0$ and $\mathbb{E}_B(px) = p\mathbb{E}_B(px)$. Replacing p by $1 - p$ yields

$$\mathbb{E}_B((1-p)x) = (1-p)\mathbb{E}_B((1-p)x) = \mathbb{E}_B(x) - p\mathbb{E}_B(x) - \mathbb{E}_B(px) + p\mathbb{E}_B(px)$$
$$(3.6.15)$$

so that $\mathbb{E}_B(px) = p\mathbb{E}_B(x)$. Since B is the closed linear span of its projections,

$$\mathbb{E}_B(bx) = b\mathbb{E}_B(x) \qquad (3.6.16)$$

for all $x \in N$ and $b \in B$.

For each state ϕ on N, $\phi\mathbb{E}_B$ is a state on N, so is positive. Hence $\mathbb{E}_B$ is a positive operator and so self-adjoint. Taking the adjoint of equation (3.6.16) shows that $\mathbb{E}_B(xb) = \mathbb{E}_B(x)b$ for all $x \in N$ and all $b \in B$.

(iii)   This is the standard equality

$$\langle x, \mathbb{E}_B(y) \rangle = \langle \mathbb{E}_B(x), \mathbb{E}_B(y) \rangle = \langle \mathbb{E}_B(x), y \rangle \qquad (3.6.17)$$

for projections on a Hilbert space.

(iv)   If $x, y \in N$, then

$$
\begin{aligned}
(e_B x e_B - \mathbb{E}_B(x) e_B)(y) &= e_B x e_B y - \mathbb{E}_B(x) e_B y \\
&= e_B(x \mathbb{E}_B(y)) - \mathbb{E}_B(x) \mathbb{E}_B(y) \\
&= \mathbb{E}_B(x \mathbb{E}_B(y)) - \mathbb{E}_B(x) \mathbb{E}_B(y) = 0 \quad (3.6.18)
\end{aligned}
$$

by (ii) and $\mathbb{E}_B = e_B|_N$. Since N is dense in $L^2(N)$, $e_B x e_B = \mathbb{E}_B(x) e_B$ and conjugating gives the other equality of (iv).

(v)   By (iv), $B \subseteq \{e_B\}' \cap N$. If $x \in \{e_B\}' \cap N$, then $(x - \mathbb{E}_B(x))\xi = e_B(x - \mathbb{E}_B(x)) e_B \xi = 0$ so that $x - \mathbb{E}_B(x) = 0$ since ξ is a separating vector for N.

(vi)   The inequality

$$
|\tau(x \mathbb{E}_B(y))| = |\tau(\mathbb{E}_B(x) y)| \leq \|\mathbb{E}_B(x)\| \|y\|_1 \leq \|x\| \|y\|_1 \quad (3.6.19)
$$

for all $x, y \in N$ implies that $\mathbb{E}_B : N \to N$ defines a continuous linear map $(\mathbb{E}_B)_1 : L^1(N) \to L^1(N)$. Identifying the predual N_* of N with $L^1(N)$ via the pairing induced on $N \times L^1(N)$ by $(x, y) \mapsto \tau(xy) : N \times N \to \mathbb{C}$ induces a map $(\mathbb{E}_B)_* : N_* \to N_*$ with the property $((\mathbb{E}_B)_*)^* = \mathbb{E}_B$, which shows that $\mathbb{E}_B$ is a normal operator on N.

Let $n \in \mathbb{N}$, let $\mathbb{M}_n(N)$ denote the algebra of $n \times n$ matrices over N and let $(\mathbb{E}_B)_n$ denote the n-fold amplification of $\mathbb{E}_B$. If $(x_{ij}) \in \mathbb{M}_n(N)$ with $(x_{ij}) \geq 0$ and (b_i) is an n-column matrix from B, then

$$
\begin{aligned}
\langle (\mathbb{E}_B)_n(x_{ij})(b_j), (b_i) \rangle_n &= \langle (\mathbb{E}_B(x_{ij}))(b_j), (b_i) \rangle_n \\
&= \sum_{ij} \langle \mathbb{E}_B(x_{ij}), b_i b_j^* \rangle = \sum_{ij} \langle x_{ij}, b_i b_j^* \rangle \\
&= \sum_{ij} \langle x_{ij} b_j, b_i \rangle = \langle (x_{ij})(b_j), (b_i) \rangle_n \\
&\geq 0 \quad (3.6.20)
\end{aligned}
$$

where $\langle \cdot, \cdot \rangle_n$ is the inner product on the n-fold direct sum of $L^2(N)$. Since B is dense in $L^2(B)$ in the $\| \cdot \|_2$-norm, it follows that $(\mathbb{E}_B)_n(x_{ij}) \geq 0$ so that $\mathbb{E}_B$ is completely positive.

(vii)   The operator $\mathbb{E}_B = e_B|_N$ is self-adjoint and lifting to $L^2(N)$ implies that $J e_B = e_B J$ and $J \mathbb{E}_B = \mathbb{E}_B J$.

Let ϕ be a map from N into B with the required properties. Then

$$
\begin{aligned}
\langle (\phi - \mathbb{E}_B)(x) b_1, b_2 \rangle &= \tau(b_2^* \phi(x) b_1 - b_2^* \mathbb{E}_B(x) b_1) \\
&= \tau((\phi - \mathbb{E}_B)(b_2^* x b_1)) = 0 \quad (3.6.21)
\end{aligned}
$$

for all $x \in N$ and all $b_1, b_2 \in B$. This shows that $\phi = \mathbb{E}_B$. $\square$

Note that ϕ has not been assumed to be linear, or continuous, in the above lemma.

Definition 3.6.3. If B is a von Neumann subalgebra of a finite von Neumann algebra N with (fixed) faithful normal trace τ the (trace preserving) *conditional expectation* $\mathbb{E}_B$ from N onto B is defined by $\mathbb{E}_B = e_B|_N$ using Lemma 3.6.2. $\square$

Conditional expectations are always taken to be trace preserving in these notes.

Remark 3.6.4. (1)   In Section 3.2, we introduced the conditional expectation $\mathbb{E}$ of the group algebra $\mathbb{C}G$ onto $\mathbb{C}H$ for an inclusion $H \subseteq G$ of discrete groups. The unique bounded extension of this map to $\ell^2(G)$ is the projection onto $\ell^2(H)$, and Lemma 3.6.2 shows that there is also a normal extension of $\mathbb{E}$ to the conditional expectation $\mathbb{E}_{L(H)}$ of $L(G)$ onto $L(H)$.

(2)   A useful elementary method of constructing the conditional expectation $\mathbb{E}_B$ from N onto B is to find a natural orthonormal basis in B for $L^2(B)$ and use this to define e_B and hence $\mathbb{E}_B$. If B is separable in $\|\cdot\|_2$-norm, and (b_n) is a sequence in B that is an orthonormal basis for $L^2(B)$, then

$$\mathbb{E}_B(x) = e_B(x) = \sum \tau(xb_n^*)b_n \quad \text{for all} \quad x \in N. \qquad (3.6.22)$$

The case when B is a finite dimensional abelian self-adjoint subalgebra of N with minimal projections $f_1, \ldots, f_n$ gives

$$\mathbb{E}_B(x) = \sum_{j=1}^{n} \tau(f_j)^{-1}\tau(xf_j)f_j \text{ for all } x \in N. \qquad (3.6.23)$$

This has proved to be very useful: see, for example [136, 137, 139, 141].

A von Neumann subalgebra B is called *diffuse* if it has no minimal projections. If B is a diffuse abelian von Neumann subalgebra of a separable type II$_1$ factor N, then (B, τ) is isomorphic to $(L^\infty[0,1], \int \cdot dt)$ (see Theorem 3.5.2). This implies that there is a Haar unitary $u \in B$, that is, a unitary satisfying $\tau(u^n) = 0$ for all $n \in \mathbb{N}$ and u generates B as a von Neumann algebra. This is equivalent to $\{u^n : n \in \mathbb{Z}\}$ being an orthonormal basis for $L^2(B)$, and

$$\mathbb{E}_B(x) = \sum_{n \in \mathbb{Z}} \tau(xu^{-n})u^n \text{ for all } x \in N. \qquad (3.6.24)$$

This formula has been used recently in [125] and [174].

If B is isomorphic to the algebra $\mathbb{M}_n$ of $n \times n$ matrices over $\mathbb{C}$, then the conditional expectation onto B is easily written down in terms of matrix units, for example.

(3)   Properties of orthogonal projections on Hilbert spaces carry over to the conditional expectations onto subalgebras. For example, if M and N are finite von Neumann algebras with faithful normal traces τ_M and τ_N, and $A \subseteq M$ and $B \subseteq N$ are von Neumann subalgebras, then the equation $\mathbb{E}_{A\overline{\otimes}B} = \mathbb{E}_A \otimes \mathbb{E}_B$ follows from $e_{A\overline{\otimes}B} = e_A \otimes e_B$ when $\tau_{M\overline{\otimes}N} = \tau_M \otimes \tau_N$. This is sometimes useful with $N = \mathbb{M}_n = \mathbb{M}_n(\mathbb{C})$ and $B = \mathbb{C}1$ in the standard technique of turning an equation in n elements into an equation in one matrix element.

(4)   Let G be a discrete subgroup of unitary operators in B generating B by $B = G''$ and suppose that there is a left invariant mean μ on G. Define a map $\phi : N \to N$ by

$$\langle \phi(x)\zeta, \eta \rangle = \mu(g \mapsto \langle gxg^{-1}\zeta, \eta \rangle) \qquad (3.6.25)$$

for all $x \in N$ and all $\zeta, \eta \in L^2(N)$ by regarding $g \mapsto \langle gxg^{-1}\zeta, \eta \rangle$ as a function of g. Clearly ϕ is a $\| \cdot \|$-norm continuous linear operator from N into N (since $\phi(x)$ commutes with N') with $\phi(1) = 1$, $\|\phi\| = 1$ and $\phi(b_1' x b_2') = b_1' \phi(x) b_2'$ for all $x \in N$ and all $b_1', b_2' \in B' \cap N$. The left invariance of μ on G implies that

$$\langle h\phi(x)\zeta, \eta \rangle = \mu_g \langle hgx(hg)^{-1}h\zeta, \eta \rangle = \mu_g \langle gxg^{-1}h\zeta, \eta \rangle = \langle \phi(x)h\zeta, \eta \rangle \qquad (3.6.26)$$

for all $h \in G$, $x \in N$ and $\zeta, \eta \in L^2(N)$ so that $\phi(x) \in G' \cap N = B' \cap N$. With $\xi = 1 \in L^2(N)$ gives

$$\tau(\phi(x)) = \langle \phi(x)\xi, \xi \rangle = \mu_g \langle gxg^{-1}\xi, \xi \rangle = \mu_g \tau(gxg^{-1}) = \tau(x). \qquad (3.6.27)$$

Hence ϕ is the unique trace preserving conditional expectation from N onto $B' \cap N$ by Lemma 3.6.2.

(5)   Let B be a von Neumann subalgebra of a finite von Neumann algebra N with a faithful normal trace τ. If p is a projection in B, then the conditional expectation $\mathbb{E}_{pBp}$ from pNp onto pBp is the restriction of $\mathbb{E}_B$ to pNp. This follows from Lemma 3.6.2 because $\mathbb{E}_B$ is a pBp-module map and $\tau(\mathbb{E}_B(pxp)) = \tau(pxp)$ for all $x \in N$.

(6)   If $q \in B' \cap N$ is a projection, then the conditional expectation of qNq onto qB is given by

$$\mathbb{E}_{qB}(qxq) = q\mathbb{E}_B(\mathbb{E}_B(q)^{-1}qxq), \quad x \in N, \qquad (3.6.28)$$

where we will give a suitable interpretation of the second term in (3.6.28). When $\mathbb{E}_B(q)$ has a bounded inverse in B then it is easy to verify that this formula is correct. However, the inverse may define only an unbounded operator on the orthogonal complement of the kernel, so the right-hand side of (3.6.28) will be defined in terms of spectral projections. First observe that $\mathbb{E}_B(q)$ lies in the centre $Z(B)$, since

$$b\mathbb{E}_B(q) = \mathbb{E}_B(bq) = \mathbb{E}_B(qb) = \mathbb{E}_B(q)b \qquad (3.6.29)$$

for $b \in B$. Let $e_n \in Z(B)$ be the spectral projection of the positive operator $h = \mathbb{E}_B(q)$ corresponding to the interval $[1/n, 1]$ for each $n \geq 1$. Then $e_n h \geq e_n/n$, so $e_n h$ has an inverse b_n in the algebra $e_n Z(B)$ for each $n \geq 1$. By the functional calculus, the sequence $\{b_n\}_{n=1}^{\infty}$ is increasing. If $x \in N^+$, $\|x\| \leq 1$, then

$$\mathbb{E}_B(b_n qxq) = b_n^{1/2} \mathbb{E}_B(qxq) b_n^{1/2} \leq b_n^{1/2} \mathbb{E}_B(q) b_n^{1/2} = e_n \qquad (3.6.30)$$

for $n \geq 1$, and so $\{\mathbb{E}_B(b_n qxq)\}_{n=1}^{\infty}$ is a norm bounded sequence in B for each $x \in N$ since N is spanned by N^+. For any pair of integers $r < n$, and for $x \in N^+$,

$$\mathbb{E}_B((b_n - b_r)qxq) = (b_n - b_r)\mathbb{E}_B(qxq)$$
$$= (b_n - b_r)^{1/2}\mathbb{E}_B(qxq)(b_n - b_r)^{1/2} \geq 0, \qquad (3.6.31)$$

and we conclude that $\{\mathbb{E}_B(b_n qxq)\}_{n=1}^\infty$ converges strongly to an element of B for each $x \in N$, since the sequence is monotone increasing and norm bounded for each $x \in N^+$. We take $\mathbb{E}_B(\mathbb{E}_B(q)^{-1}qxq)$ to mean the limit of this sequence for all $x \in N$.

If we define $\mathbb{E}(qxq)$ to be the strong limit of $q\mathbb{E}_B(b_n qxq)$ for $x \in N$, then the B-modularity of $\mathbb{E}_B$ shows that $\mathbb{E}$ is qB-modular, and it is an expectation onto qB. Since

$$
\begin{aligned}
\tau(\mathbb{E}(qxq)) &= \lim_{n\to\infty} \tau(q\mathbb{E}_B(b_n qxq)) \\
&= \lim_{n\to\infty} \tau(\mathbb{E}_B(q)b_n \mathbb{E}_B(qxq)) \\
&= \lim_{n\to\infty} \tau(e_n \mathbb{E}_B(qxq)) = \tau(qxq),
\end{aligned}
\tag{3.6.32}
$$

we see that $\mathbb{E}$ is the unique trace preserving conditional expectation from qNq onto qB. The last equality in (3.6.32) required the observation that, for $x \in N^+$,

$$
\begin{aligned}
\tau((1 - e_n)\mathbb{E}_B(qxq)) &\le \tau((1 - e_n)\|x\|\mathbb{E}_B(q)) \\
&\le \|x\|\|(1 - e_n)\mathbb{E}_B(q)\| \le \|x\|/n,
\end{aligned}
\tag{3.6.33}
$$

so that $\lim_{n\to\infty} \tau((1 - e_n)\mathbb{E}_B(qxq)) = 0$ for all $x \in N$. $\qquad\square$

In part (4) of this remark, we obtained a conditional expectation from a left invariant mean on a generating group of unitaries, a special situation that we cannot expect to always occur. The following technique extends this to general von Neumann subalgebras B of N using an idea that goes back through the works of Barry Johnson and M.M. Day to John von Neumann. Erik Christensen used several versions of this technique in von Neumann algebras, where the Hilbert space is $L^2(N)$ [24]. An important observation in his proofs is that the $\|\cdot\|_2$-closure in $L^2(N)$ of a $\|\cdot\|$-norm bounded set in N is contained in N. Here is a result that is used subsequently and is implicit in [24].

Lemma 3.6.5. *Let N be a finite von Neumann algebra with a fixed faithful normal trace τ and let B be a von Neumann subalgebra of N.*

(i) *Define a map $\phi\colon N \to N$ by letting $\phi(x)$ be the unique element of minimal $\|\cdot\|_2$-norm in*

$$
K_B(x) = \overline{\mathrm{conv}}^w\{uxu^*\colon u \text{ a unitary in } B\}.
\tag{3.6.34}
$$

Then $\phi = \mathbb{E}_{B'\cap N}$.

(ii) *If $\varepsilon > 0$ and $x \in N$, then there is a unitary $u \in B$ such that*

$$
2^{1/2}\|x - \mathbb{E}_{B'\cap N}(x)\|_2 \le \|ux - xu\|_2 + \varepsilon.
\tag{3.6.35}
$$

Proof. (i) The embedding of N into $L^2(N)$ is continuous for the respective weak topologies, since if $x_\alpha \to x$ in the w^*-topology of N and $\eta \in L^2(N)$ is an arbitrary vector, then $\langle x_\alpha \xi, \eta \rangle \to \langle x\xi, \eta \rangle$. Thus the image of any norm closed ball is convex and weakly compact in the latter space. Hence $K_B(x)$ is a weakly closed convex set, and thus a $\| \cdot \|_2$-closed convex set, in $L^2(N)$ and N. Then the element $\phi(x)$ of minimal $\| \cdot \|_2$-norm in $K_B(x)$ is in N. From this definition of ϕ, the weak continuity of τ and $\tau(uxu^*) = \tau(x)$ for all unitaries u, it follows that $\|\phi(x)\| \leq \|x\|$ and $\tau(\phi(x)) = \tau(x)$ for all $x \in N$. If v is a unitary in B, then $vK_B(x)v^* = K_B(x)$ and the map $y \mapsto vyv^* : N \to N$ is a $\| \cdot \|_2$-norm isometry. The uniqueness of the element of minimal norm in a closed convex set in a Hilbert space implies that $v\phi(x)v^* = \phi(x)$, and thus $v\phi(x) = \phi(x)v$ for all $x \in N$. Hence $\phi : N \to B' \cap N$.

If w is a unitary in $B' \cap N$, then

$$K_B(wx) = wK_B(x) \quad \text{and} \quad K_B(xw) = K_B(x)w \tag{3.6.36}$$

so that

$$\phi(wx) = w\phi(x) \quad \text{and} \quad \phi(xw) = \phi(x)w \tag{3.6.37}$$

for all $x \in N$. Since any von Neumann algebra is the span of its unitaries, the equalities of (3.6.37) are also valid for all $w \in B' \cap N$. These conditions imply that $\phi = \mathbb{E}_{B \cap N'}$ by the uniqueness in Lemma 3.6.2.

(ii) Suppose that

$$\|ux - xu\|_2 = \|uxu^* - x\|_2 \leq \beta = 2^{1/2}\|x - \mathbb{E}_{B' \cap N}(x)\|_2 - \varepsilon \tag{3.6.38}$$

for all unitaries $u \in B$. Let y be in conv$\{uxu^* : u \text{ a unitary in } B\}$ with $y = \sum_{j=1}^{m} \alpha_j u_j x u_j^*$, where u_j are unitaries in B, $\alpha_j \geq 0$ for $1 \leq j \leq m$, and $\sum_{j=1}^{m} \alpha_j = 1$. By (3.6.38)

$$2\|x\|_2^2 - 2\mathrm{Re}\,\tau(yx^*) = 2\|x\|_2^2 - 2\mathrm{Re}\,\tau\left(\sum_{j=1}^{m} \alpha u_j x u_j^* x^*\right)$$

$$= \sum_{j=1}^{m} \alpha_j \left(2\|x\|_2^2 - 2\mathrm{Re}\,\tau(u_j x u_j^* x^*)\right)$$

$$= \sum_{j=1}^{m} \alpha_j \|x - u_j x u_j^*\|_2^2 \leq \beta^2. \tag{3.6.39}$$

It follows from (3.6.39) that

$$2\|x - \mathbb{E}_{B' \cap N}(x)\|_2^2 = 2\|x\|_2^2 - 2\tau(\mathbb{E}_{B' \cap N}(x)x^*)$$
$$= 2\|x\|_2^2 - 2\mathrm{Re}\,\tau(yx^*) + 2\mathrm{Re}\,\tau\big((y - \mathbb{E}_{B' \cap N}(x))x^*\big)$$
$$\leq \beta^2 + 2\|x\|_2\|y - \mathbb{E}_{B' \cap N}(x)\|_2. \tag{3.6.40}$$

By (i), $\mathbb{E}_{B' \cap N}(x)$ lies in the norm closure of conv$\{uxu^* : u \text{ a unitary in } B\}$, so we may let $y \to \mathbb{E}_{B' \cap N}(x)$ in $K_B(x)$. Equations (3.6.38) and (3.6.40) yield the contradiction

$$2\|x - \mathbb{E}_{B' \cap N}(x)\|_2^2 \leq \beta^2 < 2\|x - \mathbb{E}_{B' \cap N}(x)\|_2^2. \tag{3.6.41}$$

This completes the proof. $\qquad\square$

Remark 3.6.6. If we take $B = N$ in Lemma 3.6.5, then we obtain the conditional expectation onto the centre $Z = N' \cap N$ of N. For a fixed pair of unitaries v and w in N, averaging $uvwu^*$ over unitaries u is the same (after the change of variables $u \mapsto uv^*$) as averaging $(uv^*)(vw)(uv^*)^* = uwvu^*$, from which we see that $\mathbb{E}_Z(vw) = \mathbb{E}_Z(wv)$. We can then replace the unitaries v and w by arbitrary elements of N, and this conditional expectation now fulfills all the requirements to be the centre-valued trace discussed in Section 3.2. $\square$

There are two cases where the technique of Lemma 3.6.5 is particularly interesting. If N is a II_1 factor and B is a subfactor with $B' \cap N = \mathbb{C}1$, then $\phi(x) = \mathbb{E}_{B' \cap N}(x) = \tau(x)1$ for all $x \in N$. This was exploited by Popa in [136]. If B is a masa in N, then $B' \cap N = B$ so that $\phi = \mathbb{E}_B$.

Remark 3.6.7. Let N be a finite von Neumann algebra with a faithful normal trace τ and let B be a finite dimensional abelian von Neumann subalgebra of N with minimal projections $f_1, \cdots, f_n$. Then

$$B' \cap N = \left\{ \sum_{j=1}^{n} f_j x f_j : x \in N \right\} \quad \text{and}$$

$$\mathbb{E}_{B' \cap N}(x) = \sum_{j=1}^{n} f_j x f_j \quad \text{for all} \quad x \in N. \tag{3.6.42}$$

If $x \in B' \cap N$, then $x = \sum_{j=1}^{n} f_j x = \sum_{j=1}^{n} f_j x f_j$, and conversely this equation implies that $x \in B' \cap N$. If $\phi : N \to N$ is defined by $\phi(x) = \sum_{j=1}^{n} f_j x f_j$ for all $x \in N$, then $\tau(\phi(x)) = \tau(x)$ and $\phi(b_1' x b_2') = b_1' \phi(x) b_2'$ for all $x \in N$ and all $b_1', b_2' \in B' \cap N$. Thus $\phi = \mathbb{E}_{B' \cap N}$ by Lemma 3.6.2. This formula and the one for $\mathbb{E}_B$ in this case are used in constructive applications of Lemma 3.6.8.

Lemma 3.6.8. *Let N be a finite von Neumann algebra with a fixed faithful normal trace τ and let A_n be an increasing sequence of von Neumann subalgebras of N with A the weak closure of $\cup A_n$. Then, for all $x \in N$,*

(i) $\lim_{n \to \infty} \|\mathbb{E}_{A_n}(x) - \mathbb{E}_A(x)\|_2 = 0$;

(ii) $\lim_{n \to \infty} \|\mathbb{E}_{A_n' \cap N}(x) - \mathbb{E}_{A' \cap N}(x)\|_2 = 0$.

Proof. (i) The $\| \cdot \|_2$-norm closure of $\cup A_n$ is equal to the $\| \cdot \|_2$-norm closure of A, since $\cup A_n$ is strongly dense in A and $\|x\xi\|_2 = \|x\|_2$ for all $x \in N$, where ξ is 1 in $L^2(N)$. The following elementary Hilbert space property then proves (i) since $\mathbb{E}_{A_n}$ and $\mathbb{E}_A$ are restrictions of Hilbert space projections (e_n and e below). If $X_1 \subseteq X_2 \subseteq \cdots$ are closed linear subspaces of a Hilbert space H with norm $\| \cdot \|_2$, $X = (\cup X_n)^-$, and e_n and e are the (orthogonal) projections from H onto

X_n and X, then for all $\zeta \in H$

$$
\begin{aligned}
0 &= \inf\{\|e\zeta - \eta\|_2 : \eta \in \cup X_n\} \\
&= \inf_n \inf\{\|e\zeta - \eta\|_2 : \eta \in X_n\} \\
&= \inf_n \|e\zeta - e_n e\zeta\|_2 \\
&= \lim_n \|e\zeta - e_n\zeta\|_2,
\end{aligned}
\tag{3.6.43}
$$

since $(\|e\zeta - e_n\zeta\|_2)$ is a decreasing sequence.
(ii) Commutants give

$$
\cap(A_n' \cap N) = (\cap A_n') \cap N = (\cup A_n)' \cap N = A' \cap N. \tag{3.6.44}
$$

Let $x \in N$. Then $(\mathbb{E}_{A_n' \cap N}(x))_{n=1}^\infty$ is a bounded sequence in N in $\|\cdot\|$-norm so has a subnet converging weakly to some $y \in N$; this y is in $A' \cap N$. With this x and y, $\mathbb{E}_{A_n' \cap N}(y) = y$ for each $n \in \mathbb{N}$, so that

$$
\begin{aligned}
\|\mathbb{E}_{A_n' \cap N}(x) - y\|_2^2 &= \|\mathbb{E}_{A_n' \cap N}(x - y)\|_2^2 \\
&= \tau((x^* - y^*)\mathbb{E}_{A_n' \cap N}(x - y)).
\end{aligned}
\tag{3.6.45}
$$

Letting n run over the net for which $\mathbb{E}_{A_n' \cap N}(x) - y$ converges weakly to zero implies that (3.6.45) above tends to zero over the net. The nearest point property of projections ensures that

$$
\|\mathbb{E}_{A_n' \cap N}(x) - \mathbb{E}_{A' \cap N}(x)\|_2 \leq \|\mathbb{E}_{A_n' \cap N}(x) - y\|_2 \tag{3.6.46}
$$

tends to 0 as $n \to \infty$. $\qquad\square$

The technique above was developed and exploited by S. Popa in the form of the following corollary on the approximation of a masa by a finite dimensional subalgebra [136, 137, 139, 140, 141]. We will also use it several times subsequently.

Corollary 3.6.9. *Let N be a separable type II_1 factor, let (A_n) be an increasing sequence of finite dimensional abelian *-subalgebras of N and let A be the weak closure of $\cup A_n$.*
(i) *Then A is a masa in N if, and only if,*

$$
\|\mathbb{E}_{A_n' \cap N}(x) - \mathbb{E}_{A_n}(x)\|_2 \to 0 \quad as \quad n \to \infty \quad for \ all \quad x \in N. \tag{3.6.47}
$$

(ii) *If A is a masa in N then,*

$$
\|\mathbb{E}_{A_n' \cap N}(x) - \mathbb{E}_A(x)\|_2 \to 0 \quad as \quad n \to \infty \quad for \ all \quad x \in N. \tag{3.6.48}
$$

Proof. The algebra A is a masa if, and only if, $A = A' \cap N$. Thus Lemma 3.6.8 gives the result directly. $\qquad\square$

3.7   Group von Neumann algebras revisited

In Section 3.2, we stated Theorem 3.2.1 concerning commutants of group von Neumann algebras. With the results of Section 3.6 available, we can now give a proof.

Theorem 3.7.1 (Theorem 3.2.1). *Let G be a discrete group and let $L(G)$ and $R(G)$ be respectively the left and right group von Neumann algebras on $\ell^2(G)$.*

(i) $L(G)' = R(G)$ *and* $R(G)' = L(G)$;

(ii) $L(G)$ *and* $R(G)$ *are factors if and only if G is I.C.C.*

Proof. (i)   Let τ be the trace $\langle \cdot \delta_e, \delta_e \rangle$ on $L(G)$. Then the representations of $L(G)$ on $\ell^2(G)$ and on $L^2(L(G), \tau)$ are identical, so the operator J of Section 3.6 is given by (3.2.3).

If $g, h \in G$, then

$$J\lambda_g J\delta_h = J\lambda_g \delta_{h^{-1}} = J\delta_{gh^{-1}} = \delta_{hg^{-1}} = \rho_g \delta_h. \tag{3.7.1}$$

Keeping g fixed and varying h gives $J\lambda_g J = \rho_g$. By Lemma 3.6.1, $L(G)' = JL(G)J$, and so $L(G)' \subseteq R(G)$. The reverse containment is clear, proving equality. The second relation follows by taking commutants and applying the double commutant theorem (Theorem 2.2.1).

(ii)   If G is not I.C.C., then there is an element $g \neq e$ with only finitely many distinct conjugates which we list as $\{g_1, \dots, g_n\}$. Let $x = \sum_{i=1}^n \lambda_{g_i} \in L(G)$. For each $h \in H$,

$$\lambda_h x \lambda_h^* = \sum_{i=1}^n \lambda_{hg_i h^{-1}} = x, \tag{3.7.2}$$

since conjugation by h permutes the set $\{g_1, \dots, g_n\}$. Thus x is central in $L(G)$. Now $x \neq 0$ since

$$\tau(\lambda_{g_1^{-1}} x) = 1 + \sum_{i=2}^n \tau(\lambda_{g_1^{-1} g_i}) = 1. \tag{3.7.3}$$

However, $\tau(x) = 0$, and thus x cannot be a scalar, giving a nontrivial central element. Thus $L(G)$ is not a factor.

Now suppose that G is I.C.C., and that $x \in L(G)$ is central. We will show that $x \in \mathbb{C}1$, thus establishing that $L(G)$ is a factor. Let

$$x\delta_e = \sum_{g \in G} \alpha_g \delta_g, \quad \sum_{g \in G} |\alpha_g|^2 < \infty. \tag{3.7.4}$$

By (i), x commutes with both $L(G)$ and $R(G)$, so for any $h \in G$,

$$\sum_{g \in G} \alpha_g \delta_g = x\delta_e = x\lambda_h \rho_h \delta_e = \lambda_h \rho_h x \delta_e$$

$$= \lambda_h \rho_h \sum_{g \in G} \alpha_g \delta_g = \sum_{g \in G} \alpha_g \delta_{hgh^{-1}}$$

$$= \sum_{g \in G} \alpha_{h^{-1}gh} \delta_g. \tag{3.7.5}$$

Thus $\alpha_g = \alpha_{h^{-1}gh}$ for all $g, h \in G$. The I.C.C. hypothesis then forces $\alpha_g = 0$ for $g \neq e$, otherwise the α_g's would not be square summable. Thus $x\delta_e = \alpha_e \delta_e$. Since δ_e is separating for $L(G)$, being cyclic for the commutant $R(G)$, we conclude that $x = \alpha_e 1$ and that $L(G)$ is a factor. This proof applies equally to $R(G)$. □

3.8 Hyperfiniteness

A von Neumann algebra is *hyperfinite* if it can be expressed as the weak closure of an ascending net of finite dimensional subalgebras. In the case of a separably acting factor, this can be taken to be an ascending sequence of finite dimensional subfactors. The simplest example is $B(\ell^2(\mathbb{N}))$, where the operators are viewed as infinite matrices. The matrix algebras $\mathbb{M}_{2^n}$ embed into $B(\ell^2(\mathbb{N}))$ by repeating each matrix infinitely often down the diagonal, and the weak density of their union is easy to verify. In the II_1 case, $\bigcup_{n=1}^{\infty} \mathbb{M}_{2^n}$ has a natural trace, and the weak closure of the GNS representation with respect to this trace τ gives an example of a hyperfinite II_1 factor. This is also the case for any sequence $\mathbb{M}_{n_1} \subseteq \mathbb{M}_{n_2} \subseteq \ldots$, where n_i divides n_{i+1}, and the resulting II_1 factors were studied in the original papers of Murray and von Neumann, [116, 117, 118, 202], who showed that they were all isomorphic (although this is far from true for the C^*-algebras obtained by taking norm closures [82]). Remarkably, hyperfiniteness is equivalent to several other apparently unrelated notions, as we now describe.

In [3], Arveson showed that there is a Hahn–Banach theorem for completely positive maps into $B(H)$. If $E \subseteq F$ are self-adjoint unital subspaces of a C^*-algebra A, then each completely positive map $\phi \colon E \to B(H)$ has a completely positive extension $\psi \colon F \to B(H)$ of the same norm. Then we say that $B(H)$ is *injective*, and more generally a C^*-algebra $B \subseteq B(H)$ is *injective* if it has this extension property for completely positive maps into B. Any norm one completely positive extension $\psi \colon B(H) \to B$ of the identity map on B gives a conditional expectation of $B(H)$ onto B. On the other hand, if such a conditional expectation $\mathbb{E} \colon B(H) \to B$ exists, then B is injective because any completely positive map into B can first be extended to a map into $B(H)$ and then composed with $\mathbb{E}$. This gives a useful characterisation of injective C^*-algebras $B \subseteq B(H)$ as those for which a conditional expectation $\mathbb{E} \colon B(H) \to B$ exists.

An older concept is that of *Property P* for a von Neumann algebra $N \subseteq B(H)$, defined by the requirement that

$$\overline{\mathrm{conv}}^w \{utu^* : u \text{ unitary in } N\} \cap N' \neq \emptyset$$

for all $t \in B(H)$ [168]. Each hyperfinite von Neumann algebra is generated by an amenable group G of unitaries composed of the finite groups which generate the constituent matrix subalgebras. The invariant mean can then be used to show that Property P follows from hyperfiniteness. A *derivation* $\delta \colon N \to X$

into a dual Banach bimodule X is a linear map satisfying the relation

$$\delta(ab) = a\delta(b) + \delta(a)b, \quad a, b \in N. \tag{3.8.1}$$

If $x \in X$ is a fixed element, then $\delta(a) = ax - xa$ defines an *inner derivation* on N. We say that N is *amenable* if every derivation of N into any such bimodule X is inner.

The concept of *semidiscreteness* is defined by the requirement that the identity map on N is the limit in the point weak topology of a net of finite rank normal unital completely positive maps $\phi_\alpha \colon N \to N$. We also need to mention briefly the *minimal tensor product norm* on the algebraic tensor product of two C^*-algebras $A \subseteq B(H)$ and $B \subseteq B(K)$. This is defined by using the norm induced by allowing elements of $A \otimes B$ to act as operators on the Hibert space $H \otimes_2 K$. The completion in this norm is denoted $A \otimes_{\min} B$, and we refer to [187] for more details, including the independence from the Hilbert spaces on which A and B act. We are now ready to state the extraordinary theorem which ties together all of the quantities that we have just defined. It is primarily due to Connes [36], but with major contributions from [18, 19, 65, 205].

Theorem 3.8.1. *Let $N \subseteq B(H)$ be a von Neumann algebra. The following conditions are equivalent:*

(i) *N is hyperfinite;*

(ii) *N is injective;*

(iii) *N is semidiscrete;*

(iv) *N is amenable;*

(v) *N has Property P;*

(vi) *The map $x \otimes x' \mapsto xx' \colon N \otimes N' \to B(H)$ extends to a bounded map on $N \otimes_{\min} N'$.*

Later papers, [87, 142], gave some simplifications of parts of the proof, but it remains an extremely deep and difficult theorem, beyond the scope of these notes. There is an extended exposition of this theorem in [189, Chapter XVI], and a nice survey in [38].

There are two immediate consequences of Theorem 3.8.1 that we wish to record. The first is that there is a unique separable injective II_1 factor, which follows from the equivalence of (i) and (ii) and the known uniqueness of such hyperfinite factors. It is customary to denote this unique factor by R and we follow this convention subsequently. The second consequence is that a von Neumann algebra is hyperfinite if and only if its commutant has this property; this is most easily seen from the symmetric nature of Theorem 3.8.1 (vi).

We conclude this chapter by making the connection between amenable discrete groups and hyperfinite von Neumann algebras, as in [168].

Theorem 3.8.2. *Let G be a discrete group. Then $L(G)$ is hyperfinite if and only if G is amenable.*

Proof. Suppose that G is amenable, and let μ be a normalised invariant mean on $\ell^\infty(G)$. Fix the natural orthonormal basis $\{\delta_g : g \in G\}$ for $\ell^2(G)$, so that each $g \in G$ is represented in $B(\ell^2(G))$ by the unitary λ_g satisfying $\lambda_g(\delta_h) = \delta_{gh}$ for $h \in H$. For each $t \in B(\ell^2(G))$ and vectors $\eta_1, \eta_2 \in \ell^2(G)$, define an operator $\mathbb{E}(t)$ by

$$\langle \mathbb{E}(t)\eta_1, \eta_2 \rangle = \mu(g \mapsto \langle \lambda_g t \lambda_{g^{-1}} \eta_1, \eta_2 \rangle), \tag{3.8.2}$$

which is well defined since the function $g \mapsto \langle \lambda_g t \lambda_{g^{-1}} \eta_1, \eta_2 \rangle$ lies in $\ell^\infty(G)$. It is clear from the definition that $\mathbb{E}$ is a contraction that is the identity on $R(G) = L(G)'$, the latter equality being Theorem 3.7.1. The invariance of μ allows us to conclude from (3.8.2) that $\mathbb{E}(t) \in L(G)'$ for all $t \in B(\ell^2(G))$, and so $\mathbb{E}$ is a conditional expectation of $B(\ell^2(G))$ onto $R(G)$. Thus $R(G)$ and $L(G)$ are both hyperfinite by Theorem 3.8.1.

Conversely, suppose that $L(G)$ is hyperfinite, and let $\mathbb{E} \colon B(\ell^2(G)) \to L(G)$ be a conditional expectation. Let τ be the trace on $L(G)$ defined by the vector functional $\langle \cdot \delta_e, \delta_e \rangle$. Each function $f \in \ell^\infty(G)$ defines a bounded multiplication operator m_f by $m_f(\delta_h) = f(h)\delta_h$ for $h \in G$. We may then define $\mu \in \ell^\infty(G)^*$ by

$$\mu(f) = \tau(\mathbb{E}(m_f)), \quad f \in \ell^\infty(G). \tag{3.8.3}$$

Left translation of f is accomplished by conjugation of m_f by λ_g, since

$$\lambda_g m_f \lambda_g^*(\delta_h) = \lambda_g m_f(\delta_{g^{-1}h}) = \lambda_g(f(g^{-1}h)\delta_{g^{-1}h}) = f(g^{-1}h)\delta_h \tag{3.8.4}$$

for all $h \in G$. The modular properties of $\mathbb{E}$ and the tracial property of τ then show that μ is a left invariant normalised mean on G, and so G is amenable. $\square$

Chapter 4

The basic construction

4.1 Introduction

In Chapter 3, we discussed the theory of conditional expectations, and we now apply this to the basic construction, a von Neumann algebra $\langle N, e_B \rangle$ associated to an inclusion $B \subseteq N$ of finite von Neumann algebras where N has a faithful finite normal trace. Conditional expectations and the basic construction are the fundamental tools throughout these notes. They form the link between subalgebras, the trace and the overall structure of the algebra.

In Section 4.2, the basic construction $\langle N, e_B \rangle$ is defined and its properties are set out in Theorem 4.2.2. Most of these are straightforward to verify, but the existence of a semifinite normal trace satisfying certain conditions is a delicate matter and occupies Section 4.3 (see also Appendix C where we give a second approach to constructing this trace). Such traces are in general unbounded and only densely defined, causing standard limiting arguments to be inadmissible. This necessitates a circuitous route to the trace Tr (initially defined on a weakly dense *-subalgebra) through two other functionals $\widetilde{Tr}$ and $\mathbf{Tr}$ that we introduce. Once we have proved the equality of these three terms, the desired properties of the trace become apparent.

The basic construction has a long history in the works of Skau [179] and Christensen [24]. More recently it was employed by Jones [94] to launch the theory of finite index inclusions of subfactors, and has also been applied by researchers in subfactor theory with great success. Although easy to define, it is a remarkably powerful tool for studying subalgebras of a given finite von Neumann algebra, even in the abelian case. In Section 4.4, we present two examples to illustrate the theory, the second of which will be useful subsequently. The underlying philosophy is that knowledge can be gained about $B \subseteq N$ by working in the larger algebra $\langle N, e_B \rangle$ and projecting back into N. There is no conditional expectation down onto N in general, and the substitute for this is the pull-down map Φ onto N, which is usually unbounded and densely defined. Its properties are discussed in Section 4.5.

In these notes the emphasis is on applications of the basic construction to other problems in von Neumann algebra theory, and so we do not give an extensive exposition here of index theory. There are several good sources for this important topic, for example [83, 97, 95, 144]. However, in Chapter 10, we will briefly describe those parts of the theory that we will use subsequently.

4.2 Properties

Throughout this chapter, N will be a finite von Neumann algebra with a faithful normal trace τ, represented in standard form on the Hilbert space $L^2(N)$ (see Section 3.6). Whenever we consider a von Neumann subalgebra, we assume a common identity element. If this assumption were not in place, then many of the results below would have significantly different formulations. We begin by defining the basic construction, a von Neumann algebra associated to an inclusion $B \subseteq N$. This term is often used to refer to the circle of ideas surrounding this algebra. As we will see, the basic construction gives a very useful setting in which to study the relationship of B to N and to other subalgebras.

Definition 4.2.1. If B is a von Neumann subalgebra of a finite von Neumann algebra N with faithful normal trace τ, the *basic construction* from the inclusion $B \subseteq N$ is defined to be the von Neumann algebra $\langle N, e_B \rangle := (N \cup \{e_B\})''$. It is sometimes denoted $\langle N, B \rangle$. □

The main properties of the basic construction are summarised in the following theorem, which is really the idea behind it. It is essential for the reader to study the original papers by Christensen [24], Jones [94], Pimsner and Popa [133], Goodman, Jones and de la Harpe [83] and related works to get an overview of the basic construction and its properties.

Throughout, $Ne_B N$ denotes the linear span of the set $\{xe_B y : x, y \in N\}$. If A is a *-subalgebra of $\langle N, e_B \rangle$, then $A^+ = \{a \in A : a \geq 0\}$.

Theorem 4.2.2. *Let B be a von Neumann subalgebra of a finite von Neumann algebra N with a fixed faithful normal trace τ. Then $\langle N, e_B \rangle$ is a semifinite von Neumann algebra with a faithful semifinite normal trace Tr satisfying the following properties:*

(i) *$\langle N, e_B \rangle = JB'J$, $\langle N, e_B \rangle' = JBJ$, and the *-subalgebra $Ne_B N$ is weakly dense in $\langle N, e_B \rangle$;*

(ii) *the central support of e_B in $\langle N, e_B \rangle$ is 1;*

(iii) *$e_B \langle N, e_B \rangle e_B = Be_B$;*

(iv) *$e_B N$ and Ne_B are weakly and strongly dense in respectively $e_B \langle N, e_B \rangle$ and $\langle N, e_B \rangle e_B$;*

(v) *the map $x \mapsto xe_B : N \to Ne_B \subseteq \langle N, e_B \rangle e_B$ is injective;*

(vi) *$Tr(xe_B y) = \tau(xy)$ for all $x, y \in N$;*

(vii) Ne_BN is dense in $L^2(\langle N, e_B \rangle, Tr)$ in $\| \cdot \|_{2,Tr}$-norm.

We will prove the first five items of Theorem 4.2.2 in this section. The construction of the trace Tr requires lengthier arguments and is postponed to Section 4.3, where the seventh item is also proved (see the remarks preceding Corollary 4.3.13).

The calculations will take place in a semifinite algebra, since $\langle N, e_B \rangle$ is semifinite because $JBJ = \langle N, e_B \rangle'$ is a finite von Neumann algebra [105, Chapter 9]. The equation $e_B \langle N, e_B \rangle e_B = Be_B \cong B$ of part (iii) implies that e_B is a finite projection in $\langle N, e_B \rangle$. It is crucial that $Tr(xe_By) = \tau(xy)$ $(x, y \in N)$ defines a trace on Ne_BN, which lifts to a faithful semifinite normal trace on $\langle N, e_B \rangle$ with $Tr(e_B) = 1$. Many of the results derived from the basic construction rely on this relationship between Tr and τ.

The proof of this theorem will be contained in a number of lemmas, which also lay the foundation for the pull-down map and other results. The first of these, Lemma 4.2.3, deals with N, e_B and their related algebraic properties, and proves (i)–(v) of Theorem 4.2.2. Corollary 4.2.4 gives analogous results for N' and e_B.

Throughout this section let B be a von Neumann subalgebra of a finite von Neumann algebra N with a fixed faithful normal trace τ.

Lemma 4.2.3 (Theorem 4.2.2 (i)–(v)). *Let B be a von Neumann subalgebra of a finite von Neumann algebra N with a faithful normal trace τ. Then*

(i) *$\langle N, e_B \rangle = JB'J$, $\langle N, e_B \rangle' = JBJ$, and the *-subalgebra Ne_BN is weakly dense in $\langle N, e_B \rangle$;*

(ii) *the central support of e_B in $\langle N, e_B \rangle$ is 1;*

(iii) *$e_B \langle N, e_B \rangle e_B = Be_B$;*

(iv) *e_BN and Ne_B are weakly and strongly dense in respectively $e_B \langle N, e_B \rangle$ and $\langle N, e_B \rangle e_B$;*

(v) *the map $x \mapsto xe_B \colon N \to Ne_B \subseteq \langle N, e_B \rangle e_B$ is injective.*

Proof. (i) By Lemma 3.6.2 (v), (vii) and Lemma 3.5.1 (i),

$$JB'J = J(N' \cup \{e_B\})''J = (JN'J \cup \{Je_BJ\})'' = \langle N, e_B \rangle \qquad (4.2.1)$$

and $JBJ = \langle N, e_B \rangle'$ follows from this.

The equation

$$e_Bxe_B = \mathbb{E}_B(x)e_B = e_B\mathbb{E}_B(x) \qquad (4.2.2)$$

implies that $Ne_BN = \text{span}\{xe_By : x, y \in N\}$ is a *-subalgebra of $\langle N, e_B \rangle$, and is non-unital in general. If e is the projection onto the closed subspace $\overline{\text{span}}\{Ne_BN\, L^2(N)\}$ in $L^2(N)$, then $e = 1$ since $N\xi \subseteq Ne_BN\, L^2(N)$. If p is the identity of the weak closure of Ne_BN, then $\overline{\text{span}}\{Ne_BN\, L^2(N)\} \subseteq pL^2(N)$, and it follows that $p = 1$. Thus we may apply the double commutant theorem (Theorem 2.2.1) to conclude that

$$(Ne_BN)' = \{\{e_B\} \cup N\}' = \langle N, e_B \rangle', \tag{4.2.3}$$

and that $\langle N, e_B \rangle = (Ne_BN)''$ as required.

(ii)   The central support z of a projection p in $\langle N, e_B \rangle$ is the projection from $L^2(N)$ onto the closure of the subspace $\langle N, e_B \rangle pL^2(N)$. Thus the central support of e_B is 1, since $N\xi \subseteq \langle N, e_B \rangle e_B L^2(N)$ and $N\xi$ is dense in $L^2(N)$.

(iii)   If $x, y \in N$, then $e_B(xe_By)e_B = \mathbb{E}_B(x)\mathbb{E}_B(y)e_B \in Be_B$. This shows that $e_B Ne_B Ne_B = Be_B$. The set $\{x \in \langle N, e_B \rangle : e_B xe_B \in Be_B\}$ is weakly closed and contains Ne_BN so equals $\langle N, e_B \rangle$ by (i).

(iv)   Since $(Ne_BN)e_B = Ne_B$, with a similar relation for multiplication on the left by e_B, the weak density statement follows from (i). Strong density is a consequence of the Kaplansky density theorem (Theorem 2.2.3) and the relation $(Ne_BN)'' = \langle N, e_B \rangle$, proved in (i).

(v)   If $x \in N$ and $xe_B = 0$, then $\mathbb{E}_B(x^*x)e_B = e_Bx^*xe_B = 0$. Thus $\mathbb{E}_B(x^*x) = 0$ and $\tau(x^*x) = \tau(\mathbb{E}_B(x^*x)) = 0$, which gives $x = 0$. $\square$

Note that, by Lemma 4.2.3 (i), $\langle N, e_B \rangle$ is a factor if, and only if, B is a factor, and that $\langle N, e_B \rangle$ is finite if, and only if, B' is finite.

The following corollary of Lemma 4.2.3 has N and B replaced by N' and JBJ. The corresponding parts are deduced from Lemma 4.2.3 using the equations $JNJ = N'$, $Je_B = e_BJ$ and $J^2 = I$. It will be very useful in our construction of $\widetilde{Tr}$ on $\langle N, e_B \rangle$ in Section 4.3.

Corollary 4.2.4. *Let B be a von Neumann subalgebra of a finite von Neumann algebra N with a fixed faithful normal trace τ. Then*

(i) $\langle N', e_B \rangle = B'$, $\langle N', e_B \rangle' = B$, *and $N'e_BN'$ is a weakly dense *-subalgebra of $\langle N', e_B \rangle$;*

(ii) *the central support of e_B in $\langle N', e_B \rangle$ is 1;*

(iii) $e_B\langle N', e_B \rangle e_B = JBJe_B$;

(iv) *e_BN' and $N'e_B$ are weakly and strongly dense in respectively $e_B\langle N', e_B \rangle$ and $\langle N', e_B \rangle e_B$;*

(v)   *the map $x \mapsto xe_B \colon N' \to N'e_B \subseteq \langle N', e_B \rangle e_B$ is injective.*

This raises the following question: is the map $x \mapsto xe_B \colon N \to \langle N, e_B \rangle e_B$ surjective if, and only if, B is of finite index in N? If Ne_B is norm closed in $\langle N, e_B \rangle$, then the injective linear map $x \mapsto xe_B \colon N \to Ne_B$ has a continuous inverse by the open mapping theorem, so there is a finite k such that $\|x\| \leq k\|xe_B\|$ for all $x \in N$. Conversely if this inequality holds then Ne_B is norm closed. This inequality is equivalent to

$$\|x^*x\| \leq k^2\|e_Bx^*xe_B\| = k^2\|\mathbb{E}_B(x^*x)e_B\| = k^2\|\mathbb{E}_B(x^*x)\| \tag{4.2.4}$$

for all $x \in N$. The inequality $\|y\| \leq k\|\mathbb{E}_B(y)\|$ for all $y \in N^+$ is equivalent to B being of finite index in N *provided* that B is a subfactor of a II_1 factor N by [133, Theorem 2.22]. The answer to the question is not known in general (see [133, p. 73]).

Given $z \in \langle N, e_B \rangle$, parts (iii) and (v) of Lemma 4.2.3 show that $e_B z e_B$ has the form $b e_B$ for a unique element $b \in B$. Consequently there is a well-defined map $\phi \colon \langle N, e_B \rangle \to B$ satisfying the equation $e_B z e_B = \phi(z)e_B$.

Corollary 4.2.5. *Let B be a von Neumann subalgebra of a finite von Neumann algebra N with a fixed faithful normal trace τ. The map $\phi : \langle N, e_B \rangle \to B$ is a completely positive unital surjective B-module map whose restriction to N coincides with $\mathbb{E}_B$.*

Proof. The map ϕ is the composition of the following two completely positive maps: the compression $z \mapsto e_B z e_B$ ($z \in \langle N, e_B \rangle$), and the $*$-isomorphism $b e_B \mapsto b$ ($b \in B$). Since $\phi(b) = b$ for $b \in B$, we see that ϕ is a completely positive conditional expectation of $\langle N, e_B \rangle$ onto B. The general theory of such maps [193, 194, 195] yields the B-modularity of ϕ, and the equation $e_B x e_B = \mathbb{E}_B(x)e_B$ for $x \in N$ implies that $\phi(x) = \mathbb{E}_B(x)$ for such x.   □

Remark 4.2.6. The basic construction is related in a nice way to the algebra giving rise to the Pukánszky invariant for a masa in a separable factor. If B is a masa in a separable II_1 factor N, the Pukánszky invariant $\mathrm{Puk}(B)$ of B is derived from the exact type of the type I von Neumann algebra $\mathcal{B}' = (B \cup JBJ)'$ (see Chapter 7). For the masa B,

$$\mathcal{B}' = B' \cap JB'J = \langle N, e_B \rangle \cap \langle N', e_B \rangle$$

by Lemma 4.2.3.   □

4.3   The trace on $\langle N, e_B \rangle$

In this section we will construct the trace on $\langle N, e_B \rangle$ and verify its properties, thus completing the proof of Theorem 4.2.2. This will involve several steps, the first of which is to define a trace Tr on Ne_BN by the formula

$$Tr(xe_By) = \tau(xy), \quad x, y \in N. \tag{4.3.1}$$

The following two lemmas split off the standard parts of the argument used to deduce properties of this natural trace Tr on Ne_BN. We have anticipated the future by naming this quantity as Tr, but we will soon have the justification for this. Lemma 4.3.1 handles the question of whether it is well defined, and Lemma 4.3.2 establishes the tracial properties of Tr.

We will encounter here unbounded normal weights, and there are two reasonable definitions of normality. One is that $\rho(\sup y_\alpha) = \sup \rho(y_\alpha)$ for increasing non-negative nets (y_α). The other is to require ρ to be a sum of positive vector functionals. We adopt the first of these, which is implied by the second.

For some years it was an open question as to whether these two possible definitions were equivalent. The positive solution by Haagerup [85] is quite difficult,
so the arguments that we give below have been chosen to minimise the use of
this theorem. This appears only once, in Theorem 4.3.11. In Appendix C, the
reader will find a second and different approach to the existence of Tr.

The construction of the trace on $\langle N, e_B \rangle^+$ requires some auxiliary functionals
which we now describe. We first define Tr on Ne_BN by (4.3.1), and prove that
this is a trace in Lemma 4.3.2. We then obtain a representation of Tr as a sum
of positive vector functionals in Lemma 4.3.4. The formula given in (4.3.15)
defines a weight $\widetilde{Tr}$ on $\langle N, e_B \rangle^+$, but which is not obviously a trace, although
it will prove to be so. In Definition 4.3.6, we introduce a third map $\mathbf{Tr}$; at the
outset this is just a function. Lemmas 4.3.7 and 4.3.8 establish that this is a
normal semifinite trace on $\langle N, e_B \rangle^+$, making use of $\widetilde{Tr}$. It is also shown that
$\mathbf{Tr}$ and Tr agree on their common domain $(Ne_BN)^+$. Theorem 4.3.11 proves
that $\widetilde{Tr} = \mathbf{Tr}$, and that $\mathbf{Tr}$ is the unique normal weight on $\langle N, e_B \rangle$ which agrees
with Tr on $(Ne_BN)^+$. It is here that we use Haagerup's theorem [85].

Lemma 4.3.1 (ii) is important as it permits one to intuitively define natural
linear operators from Ne_BN using operators on N. Both the semifinite normal
trace Tr (Lemma 4.3.2) and the pull-down map Φ of Section 4.5 are examples
of this, with Φ being the universal one. The first part of this lemma is needed
in the algebraic definition of Tr; the second gives positivity and is helpful in the
proof of Theorem 4.3.15. The proof of Lemma 4.3.1 (ii) is a standard technique
going back to von Neumann (see [104, Theorem 5.5.4] or [167, p. 50]).

We note that almost all of the subsequent work can be avoided in the case
that B is a factor, due to uniqueness up to scaling of traces on semifinite factors.
The details are contained in Remark 4.3.16.

Lemma 4.3.1. *Let B be a von Neumann subalgebra of a finite von Neumann
algebra N with a faithful normal trace τ, and let $x_1, \ldots, x_n, y_1, \ldots, y_n \in N$.*

(i) *If $\sum_{j=1}^n x_j e_B y_j = 0$, then $\sum_{j=1}^n x_j y_j = 0$.*

(ii) *In $\mathbb{M}_n(B)$, consider the two matrices*

$$X = \begin{pmatrix} x_1 & \cdots & x_n \\ 0 & \cdots & 0 \\ 0 & \cdots & 0 \end{pmatrix}, \quad Y = \begin{pmatrix} y_1 & 0 & \cdots & 0 \\ \cdot & 0 & \cdots & 0 \\ y_n & 0 & \cdots & 0 \end{pmatrix}.$$

*Then $\sum_{j=1}^n x_j e_B y_j = 0$ if, and only if, there exists a projection $P \in
\mathbb{M}_n(B)$ such that $XP = X$ and $PY = 0$.*

Proof. (i) If $w \in N' e_B N'$, then $\sum_{j=1}^n x_j w y_j = 0$, since $x_j u e_B v y_j = u x_j e_B y_j v$
for all $u, v \in N'$. Hence $\sum_{j=1}^n x_j w y_j = 0$ for all $w \in \langle N', e_B \rangle = (N' e_B N')''$ by
Corollary 4.2.4 (i) so that $\sum_{j=1}^n x_j y_j = 0$. Alternatively part (i) follows directly
from (ii) since $\sum_{j=1}^n x_j y_j = 0$ corresponds to $XY = 0$.
(ii) Let $A = B \otimes \mathbb{C}I \subseteq M = N \otimes \mathbb{M}_n$ with the tensor product trace. The
equation $\sum_{j=1}^n x_j e_B y_j = 0$, corresponds to $X e_A Y = 0$. From the assumption

$Xe_AY = 0,$

$$Xe_AYY^*e_AX^* = Xe_A\mathbb{E}_A(YY^*)e_AX^*$$
$$= (Xe_A\mathbb{E}_A(YY^*)^{1/2})(Xe_A\mathbb{E}_A(YY^*)^{1/2})^* = 0, \qquad (4.3.2)$$

implying that $Xe_A\mathbb{E}_A(YY^*)^{1/2} = 0$ and $X\mathbb{E}_A(YY^*)e_A = Xe_A\mathbb{E}_A(YY^*) = 0$. Lemma 4.2.3(v) applied to A in M gives $X\mathbb{E}_A(YY^*) = 0$. If L is the weakly closed left ideal in A generated by $\mathbb{E}_A(YY^*)$, then $XL = 0$ and there is a projection $1 - P$ in A such that $L = (1 - P)A$ [105, Section 6.8]. Hence $X(1 - P) = 0$ and $(1 - P)Y = Y$ proving the $XP = X$ and $PY = 0$. The reverse implication follows from $Pe_A = e_AP$.   □

Since $e_B\langle N, e_B\rangle e_B = Be_B$ from Lemma 4.2.3 (iii), a simple algebraic calculation shows that $a_1ya_2 \in Ne_BN$ whenever $y \in \langle N, e_B\rangle$ and $a_i \in Ne_BN$, $i = 1, 2$. This observation is important in the statement of the next lemma and is also used in subsequent proofs.

To define and study the canonical trace Tr on $\langle N, e_B\rangle$ and check its properties we define and introduce two weights $\widetilde{Tr}$ and $\mathbf{Tr}$ on $\langle N, e_B\rangle$. These two weights have certain different properties that are easy to check. Here is the purely algebraic definition of Tr on Ne_BN that is the motivation and the first step.

The faithfulness and positivity of Tr could be omitted from the following lemma as they are both proved by other methods subsequently (see Lemma 4.3.4). However here is a purely algebraic account of Tr on Ne_BN, whereas the subsequent methods are analytic.

Lemma 4.3.2. *Let B be a von Neumann subalgebra of a finite von Neumann algebra N with a faithful normal trace τ. Then $Tr(xe_By) = \tau(xy)$ for all $x, y \in N$, defines a faithful trace Tr on the *-algebra Ne_BN. Moreover $Tr(aya^*) \geq 0$ for all $a \in Ne_BN$ and $y \in \langle N, e_B\rangle^+$.*

Proof. The map Tr is well defined on Ne_BN, since $\sum_{j=1}^n x_je_By_j = 0$ implies that $\sum_{j=1}^n x_jy_j = 0$ by Lemma 4.3.1 (i). The tracial property follows from the following calculation. If $x_1, x_2, y_1, y_2 \in N$, then

$$Tr((x_1e_Bx_2)(y_1e_By_2)) = Tr(x_1\mathbb{E}_B(x_2y_1)e_By_2) = \tau(x_1\mathbb{E}_B(x_2y_1)y_2)$$
$$= \tau(\mathbb{E}_B(y_2x_1)x_2y_1) = Tr(y_1\mathbb{E}_B(y_2x_1)e_Bx_2)$$
$$= Tr((y_1e_By_2)(x_1e_Bx_2)). \qquad (4.3.3)$$

Let ϕ be the completely positive map of Corollary 4.2.5 and satisfying $e_Bze_B = \phi(z)e_B$ for all $z \in \langle N, e_B\rangle$. If $x, y \in N$, $a = xe_By$ and $z \in \langle N, e_B\rangle^+$, then

$$xe_Byzy^*e_Bx^* = xe_B\phi(yzy^*)x^*. \qquad (4.3.4)$$

Since $\phi(yzy^*) \geq 0$,

$$Tr(aza^*) = Tr(xe_Byzy^*e_Bx^*) = Tr(xe_B\phi(yzy^*)x^*)$$
$$= \tau(x\phi(yzy^*)x^*) \geq 0. \qquad (4.3.5)$$

The case when $a = \sum_{j=1}^{n} x_j z y_j$ with $x_j, y_j \in N$ $(1 \leq j \leq n)$ is the same idea using $A = B \otimes \mathbb{C}1 \subseteq M = N \otimes \mathbb{M}_n$ with $\tau_M = \tau \otimes tr_n$ and $\mathbb{E}_A = \mathbb{E}_B \otimes \mathbb{E}_{\mathbb{C}1_n}$, where tr_n is the normalised trace on $\mathbb{M}_n$ and 1_n is its identity (see the proof of Lemma 4.3.1). Let

$$X = \begin{pmatrix} x_1 & \cdots & x_n \\ 0 & \cdots & 0 \\ 0 & \cdots & 0 \end{pmatrix}, \quad Y = \begin{pmatrix} y_1 & 0 & \cdots & 0 \\ \cdot & 0 & \cdots & 0 \\ y_n & 0 & \cdots & 0 \end{pmatrix},$$

and $Z = z \otimes I$ be in M. With this notation

$$Tr(aza^*) = \sum_{j,k=1}^{n} Tr(x_j e_B y_j z y_k^* e_B x_k^*) = \sum_{j,k=1}^{n} Tr(x_j e_B \phi(y_j z y_k^*) x_k^*)$$

$$= \sum_{j,k=1}^{n} \tau(x_j \phi(y_j z y_k^*) x_k^*) = \tau_M(X\phi_n(YZY^*)X^*) \geq 0, \qquad (4.3.6)$$

where $\phi_n = \phi \otimes I$ is the n-fold amplification of ϕ and $\phi_n(YZY^*) \geq 0$ by Corollary 4.2.5.

Let $a = \sum_{j=1}^{n} x_j e_B y_j \in N e_B N$ with $Tr(aa^*) = 0$. In the calculation above take $z = 1$ and $Z = 1_n$ so that $YY^* \in M$ and $\phi_n(YY^*) = (\mathbb{E}_B)_n(YY^*)$. By equation (4.3.6)

$$Tr(aa^*) = \tau_M(X\phi_n(YY^*)X^*) = \tau_M(X(\mathbb{E}_B)_n(YY^*)X^*) = 0. \qquad (4.3.7)$$

Note that $\phi_n = (\mathbb{E}_B)_n$ is distinct from $\mathbb{E}_A$. Since $(\mathbb{E}_B)_n(YY^*) \geq 0$ and τ_M is a faithful trace on M, we have $X(\mathbb{E}_B)_n(YY^*)X^* = 0$. This implies that $((\mathbb{E}_B)_n(YY^*))^{1/2}X^* = 0$, and hence that $(\mathbb{E}_B)_n(YY^*)X^* = 0$. Since X^* has only the first column non-zero, the matrix $(\mathbb{E}_B)_n(YY^*)X^*$ has first column with i^{th} element

$$\sum_{j=1}^{n} \mathbb{E}_B(y_i y_j^*)x_j^* = 0. \qquad (4.3.8)$$

Multiplying by e_B in (4.3.8) gives

$$\sum_{j=1}^{n} e_B y_i y_j^* e_B x_j^* = 0 \qquad (4.3.9)$$

for all i, so that

$$aa^* = \sum_{i,j=1}^{n} x_i e_B y_i y_j^* e_B x_j^* = 0, \qquad (4.3.10)$$

and $a = 0$. $\qquad\qquad\qquad\qquad\qquad\qquad\qquad\qquad\qquad\qquad\qquad\qquad\qquad\quad \square$

Observe that the faithful trace Tr defined above is really just defined by $Tr(x) = \tau(\Phi(x))$ on $N e_B N$, where Φ is the pull-down map of Section 4.5. No use of this can be made for extending the trace beyond $N e_B N$ because defining the pull-down map Φ on a domain larger than Ne_BN requires Tr to already be defined there. The first extension of Tr to $\langle N, e_B \rangle$ from Ne_BN in Lemma 4.3.3 yields a map that has most of the properties we require though not the tracial property on all of $\langle N, e_B \rangle$ nor uniqueness. These are obtained subsequently.

The following technical result is useful in a subsequent lemma in handling elements related to N and N'. Note the reversal of v and v^* caused by J in the next lemma.

Lemma 4.3.3. *Let B be a von Neumann subalgebra of a finite von Neumann algebra N with a faithful normal trace τ and let ξ be 1 in $L^2(N)$. If $v \in \langle N', e_B \rangle$, then*

$$\langle xv^*e_Bvy\xi, \xi \rangle = \langle xe_ByJv^*\xi, Jv^*\xi \rangle \qquad (4.3.11)$$

for all $x, y \in N$.

Proof. By Lemma 4.2.4 (ii), the central support of e_B in $\langle N', e_B \rangle$ is 1. Let $C_0 = \text{span}\{N'e_BN'\}$ and let C be the norm closure of the *-algebra C_0. Then C is a C^*-algebra that is weakly dense in $\langle N', e_B \rangle$ by Corollary 4.2.4 (i). In particular, the weak closure of C_0 contains 1. By the Kaplansky density theorem (Theorem 2.2.3) the unit ball of C, hence of C_0, is *-strongly dense in the unit ball of $\langle N', e_B \rangle$. Thus there is a net $(v_j : j \in \Lambda)$ in C_0, with $\|v_j\| \leq \|v\|$ for all $j \in \Lambda$, converging *-strongly to v. Hence $v_j e_B$ converges *-strongly to ve_B. Since $e_Bv_j \in e_BC_0 = e_BN' = e_BJNJ$, there is a $z_j \in N$ such that $e_Bv_j = e_BJz_jJ$ for each j. Then $\lim_j e_BJz_jJ = e_Bv$ *-strongly, so $Jz_j^*Je_B$ tends strongly to v^*e_B. Since J is a $\|\cdot\|_2$-isometry,

$$\lim_j \|z_j^* - Jv^*\xi\|_2 = \lim_j \|Jz_j^*\xi - v^*\xi\|_2$$
$$= \lim_j \|Jz_j^*Je_B\xi - v^*e_B\xi\|_2 = 0. \qquad (4.3.12)$$

If $x, y \in N$, then

$$\langle xv^*e_Bvy\xi, \xi \rangle = \langle e_Bvy\xi, e_Bvx^*\xi \rangle = \lim_j \langle e_Bv_jy\xi, e_Bv_jx^*\xi \rangle$$

$$= \lim_j \langle e_BJz_jJy\xi, e_BJz_jJx^*\xi \rangle = \lim_j \langle e_ByJz_jJ\xi, e_Bx^*Jz_jJ\xi \rangle$$

$$= \lim_j \langle e_Byz_j^*, x^*z_j^* \rangle = \langle e_ByJv^*\xi, x^*Jv^*\xi \rangle$$

$$= \langle xe_ByJv^*\xi, Jv^*\xi \rangle, \qquad (4.3.13)$$

where we have used $JNJ = N'$ for the fourth equality, and (4.3.12) for the penultimate one in this calculation. $\qquad\qquad\square$

By Corollary 4.2.4 (ii), the central support of e_B in $\langle N', e_B \rangle$ is 1 and so there is a net (v_j) in $\langle N', e_B \rangle$, or sequence in the case N is separable, such that

$$\sum_j v_j^*e_Bv_j = 1. \qquad (4.3.14)$$

We do not have a precise reference for this, so we sketch the argument, valid in any von Neumann algebra. Let $(v_j) \in \langle N', e_B \rangle$ be a maximal set of operators such that $(v_j^* e_B v_j)$ are orthogonal projections, and let p be their sum. If $p \neq 1$, then the product of the central supports of $1 - p$ and e_B is non-zero and so $1 - p$ and e_B have equivalent non-zero subprojections p_1 and p_2, by [105, p. 403]. Choose a partial isometry $v \in \langle N', e_B \rangle$ such that $v^* v = p_1$ and $vv^* = p_2$. Then $v^* e_B = v^*$, and so $v^* e_B v = p_1 \leq 1 - p$, contradicting the maximality of (v_j). Thus $p = 1$ as required.

We will use (4.3.14) to define the map $\widetilde{Tr}$ as an extension of Tr to $\langle N, e_B \rangle^+$. It will also be defined and agree with Tr on $N e_B N$. In the next lemma $\widetilde{Tr}$ seems to depend on the choice of the v_j's. However, it follows from Theorem 4.3.11 that any net satisfying (4.3.14) will give the same definition of $\widetilde{Tr}$. Here we make a choice once and stick with it. The symbol $\widetilde{Tr}$ for the map below will be justified later by proving that it is a trace, which agrees with Tr on the common domain of definition.

Lemma 4.3.4. *Let B be a von Neumann subalgebra of a finite von Neumann algebra N with a faithful normal trace τ. Let (v_j) be a net in $\langle N', e_B \rangle$ satisfying equation (4.3.14). Then*

$$\widetilde{Tr}(t) = \sum_j \langle t J v_j^* \xi, J v_j^* \xi \rangle \quad \text{for all} \quad t \in \langle N, e_B \rangle^+ \qquad (4.3.15)$$

defines a weight on $\langle N, e_B \rangle$. Moreover, equation (4.3.15) also defines $\widetilde{Tr}$ on $N e_B N$, and $\widetilde{Tr} = Tr$ on this space.

Proof. Assume that N is separable: if it is not separable, then the index j below runs over a general net Λ rather than over $\mathbb{N}$. Let $\xi_j = J v_j^* \xi$ for all $j \in \mathbb{N}$. Then the defining equation (4.3.15) for $\widetilde{Tr}$ becomes

$$\widetilde{Tr}(t) = \sum_j \langle t \xi_j, \xi_j \rangle \quad \text{for all} \quad t \in \langle N, e_B \rangle^+. \qquad (4.3.16)$$

Clearly $\widetilde{Tr}$ maps into $[0, \infty]$, is additive on $\langle N, e_B \rangle^+$, commutes with multiplication by positive scalars and is a sum of normal positive linear functionals on $\langle N, e_B \rangle$.

The next equation implies that $\widetilde{Tr}$ has the desired value on $N e_B N$. If x, $y \in N$, then by Lemma 4.3.3

$$\tau(xy) = \langle x1y\xi, \xi \rangle = \sum_j \langle x v_j^* e_B v_j y \xi, \xi \rangle$$

$$= \sum_j \langle x e_B y J v_j^* \xi, J v_j^* \xi \rangle = \sum_j \langle x e_B y \xi_j, \xi_j \rangle. \qquad (4.3.17)$$

This implies that $\widetilde{Tr}$ is also defined on $N e_B N$ where it agrees with Tr. $\square$

The notation $\xi_j = Jv_j^*\xi$ for all j is used from now on so that

$$\widetilde{Tr}(y) = \sum_j \langle y\xi_j, \xi_j \rangle \quad \text{for all} \quad y \in (Ne_BN) \cup \langle N, e_B \rangle^+. \qquad (4.3.18)$$

It is not clear that the weight $\widetilde{Tr}$ defined above is a trace. The trace Tr will now be extended from Ne_BN to a faithful semifinite normal trace **Tr** on $\langle N, e_B \rangle$ using the observation that $Ne_BN\langle N, e_B\rangle Ne_BN = Ne_BN$ and a supremum over a suitable set. The two weights $\widetilde{Tr}$ and **Tr** are shown to be equal in Theorem 4.3.11.

For calculations it is helpful to replace the supremum in Definition 4.3.6 by a limit over a bounded approximate identity $\{k_\lambda : \lambda \in \Lambda\}$ in Ne_BN for $\overline{Ne_BN}$ and use k_λ tends strongly to 1. These properties are provided by Lemma 4.3.5 (i); part (ii) gives a modification of the standard C^*-algebra proof for showing that there are always increasing approximate identities. This enables us to see that the net $\{k_\lambda : \lambda \in \Lambda\}$ may be chosen to be increasing, although this is not essential.

Lemma 4.3.5. *Let B be a von Neumann subalgebra of a finite von Neumann algebra N with a faithful normal trace τ.*

(i) *There is an approximate identity $\{k_\lambda : \lambda \in \Lambda\}$ in Ne_BN for the C^*-algebra $\overline{Ne_BN}$ with $0 \le k_\lambda$ and $\|k_\lambda\| \le 1$ for all $\lambda \in \Lambda$. Moreover, $k_\lambda \to 1$ strongly in $\langle N, e_B \rangle$.*

(ii) *Let $\Lambda = \{x \in Ne_BN : 0 \le x, \|x\| < 1\}$ with the order relation $\le$ induced by the C^*-algebra $\overline{Ne_BN}$. Then Λ is an increasing net and a bounded approximate identity for $\overline{Ne_BN}$.*

Proof. (i)   The norm closure $\overline{Ne_BN}$ of Ne_BN is a C^*-algebra that is weakly dense in $\langle N, e_B \rangle$, by Theorem 4.2.2 (i). Let $\{a_\gamma : \gamma \in \Gamma\}$ be an approximate identity in $\overline{Ne_BN}$ with $\|a_\gamma\| \le 1$ for all $\gamma \in \Gamma$. Define a net $\Lambda = \{(\gamma, n) : \gamma \in \Gamma, \ n \in \mathbb{N}\}$ ordered by $(\gamma, n) \le (\delta, m)$ if, and only if, $\gamma \le \delta$ and $n \le m$. For each $\lambda = (\gamma, n)$ pick a $k_\lambda \in Ne_BN$ such that

$$\|k_\lambda\| \le 1, \quad \|k_\lambda - a_\gamma\| \le \frac{1}{n}. \qquad (4.3.19)$$

Then $\{k_\lambda : \lambda \in \Lambda\}$ is an approximate identity in Ne_BN for $\overline{Ne_BN}$, with $\|k_\lambda\| \le 1$ for all $\lambda \in \Lambda$. Replacing k_λ by $((k_\lambda + k_\lambda^*)/2)^2$ means that we can also assume that $0 \le k_\lambda$ for all $\lambda \in \Lambda$.

The unit ball of the algebra Ne_BN is *-strongly dense in the unit ball of $\langle N, e_B \rangle$ by the Kaplansky Density Theorem (Theorem 2.2.3), and hence there is a net $\{z_\gamma : \gamma \in \Gamma\}$ in Ne_BN with $\|z_\gamma\| \le 1$ for all $\gamma \in \Gamma$ and $z_\gamma \to 1$ strongly over the net. Given $\varepsilon > 0$ and $\eta_1, \ldots, \eta_n \in L^2(N)$ with $\|\eta_j\| \le 1$ for all j, choose γ such that $\|z_\gamma\eta_j - \eta_j\|_2 \le \varepsilon/3$ for $1 \le j \le n$ and then choose λ_0 such that $\|k_\lambda z_\gamma - z_\gamma\| \le \varepsilon/3$ for all $\lambda \ge \lambda_0$. Then

$$\|k_\lambda\eta_j - \eta_j\|_2 \le \|k_\lambda(\eta_j - z_\gamma\eta_j) + (k_\lambda z_\gamma\eta_j - z_\gamma\eta_j) + (z_\gamma\eta_j - \eta_j)\|_2$$
$$\le 2\|z_\gamma\eta_j - \eta_j\|_2 + \|k_\lambda z_\gamma - z_\gamma\|_2 \le \varepsilon. \qquad (4.3.20)$$

Thus k_λ tends strongly to 1 over the net.

(ii)   Here we shall show that the standard proof that a C^*-algebra has an increasing bounded approximate identity bounded by 1 gives such an approximate identity in $N e_B N$ for $\overline{N e_B N}$, because

$$(N e_B N)\overline{N e_B N}(N e_B N) = (N e_B N)\langle N, e_B \rangle(N e_B N) = N e_B N, \qquad (4.3.21)$$

using Theorem 4.2.2.

If $u \in \Lambda$, then $0 \leq (1-u)^{-1}u = u + u(1-u)^{-1}u \in N e_B N$ since $u(1-u)^{-1}u \in N e_B N$. If $u, v \in \Lambda$, then $a = (1-u)^{-1}u$ and $b = (1-v)^{-1}v$ are in $(N e_B N)^+$ and

$$w = (1+a+b)^{-1}(a+b) = (a+b) - (a+b)(1+a+b)^{-1}(a+b) \in N e_B N \quad (4.3.22)$$

with $0 \leq w \leq 1$. Let

$$f_\varepsilon(t) = (\varepsilon + t)^{-1}t = (1 + t/\varepsilon)^{-1}(t/\varepsilon) \qquad (4.3.23)$$

for all $\varepsilon > 0$ and all $t \geq 0$ (see [187, Lemma 7.2] for the notation and omitted parts of the proof). Then $w = f_1(a+b) \geq f_1(a) = u$ and $w \geq v$ by [187, Lemma 7.2]. This shows that Λ is an increasing net and the remainder of the proof follows the standard C^*-algebra proof that an approximate identity exists (see [104, 131, 187]).   □

For the remainder of this section we shall assume that $\{k_\lambda : \lambda \in \Lambda\}$ is an approximate identity in $N e_B N$ for $\overline{N e_B N}$ with $0 \leq k_\lambda$ and $\|k_\lambda\| \leq 1$ for all $\lambda \in \Lambda$. If $\{k_\lambda : \lambda \in \Lambda\}$ is such an approximate identity then, for all $t \in B(L^2(N))$ and vectors $\eta, \nu \in L^2(N)$,

$$|\langle k_\lambda t k_\lambda \eta, \nu \rangle - \langle t\eta, \nu \rangle| \leq \|t\|\|\nu\|_2\|k_\lambda\eta - \eta\|_2 + \|t\|\|\eta\|_2\|k_\lambda\nu - \nu\|_2. \quad (4.3.24)$$

Since $\lim_\lambda(k_\lambda - 1) = 0$ strongly, by Lemma 4.3.5 (i), the inequality (4.3.24) shows that $k_\lambda t k_\lambda \to t$ weakly over the net Λ for all $t \in B(L^2(N))$.

The suggestive notation **Tr** is used in the following definition because subsequently **Tr** is proved to be a faithful semifinite normal trace on $\langle N, e_B \rangle$.

Definition 4.3.6. Define **Tr** $: \langle N, e_B \rangle^+ \to [0, \infty]$ by

$$\mathbf{Tr}(y) = \sup\{Tr(xyx^*): x \in N e_B N, \|x\| \leq 1\} \qquad (4.3.25)$$

for all $y \in \langle N, e_B \rangle^+$.   □

This is well defined because each term xyx^* in (4.3.25) lies in $(N e_B N)^+$ and Tr takes non-negative real values on this cone by Lemma 4.3.4. At the moment, **Tr** is just a function; additivity is established below.

Lemma 4.3.7. *Let B be a von Neumann subalgebra of a finite von Neumann algebra N with a faithful normal trace τ and let $\{k_\lambda : \lambda \in \Lambda\}$ be a bounded approximate identity for the C^*-algebra $\overline{N e_B N}$ from $N e_B N$ with $\|k_\lambda\| \leq 1$ and $0 \leq k_\lambda$ for all λ.*

(i) *For each* $y \in \langle N, e_B \rangle^+$,

$$\mathbf{Tr}(y) = \lim_\lambda Tr(k_\lambda y k_\lambda). \tag{4.3.26}$$

(ii) *If* $y \in \langle N, e_B \rangle$, *then*

$$\mathbf{Tr}(yy^*) = \sup_{\lambda, \mu \in \Lambda} Tr(k_\lambda y k_\mu^2 y^* k_\lambda). \tag{4.3.27}$$

Proof. Both parts of the lemma will be deduced from the following implication, which is tackled first: if $y \in \langle N, e_B \rangle$, $x \in Ne_B N$ with $\|x\| \leq 1$, and $\delta < Tr(xyy^*x^*)$, then $xyy^*x^* \in Ne_B N$ and there is a $\lambda_0 \in \Lambda$ such that

$$\delta < Tr(xyk_\lambda^2 y^* x^*) \leq Tr(xyy^*x^*) \quad \text{for all} \quad \lambda \geq \lambda_0. \tag{4.3.28}$$

The proof is written for separable N; for general N replace $\mathbb{N}$ by a net. By the definition (4.3.15) of $\widetilde{Tr}(xyy^*x^*) = \sum_j \langle xyy^*x^*\xi_j, \xi_j \rangle$ and the equality of $\widetilde{Tr}$ and Tr on $(Ne_B N)^+$ (Lemma 4.3.4), there is an $m \in \mathbb{N}$ such that

$$\sum_1^m \langle xyy^*x^*\xi_j, \xi_j \rangle > \delta. \tag{4.3.29}$$

Since the net $\{k_\lambda\}$ converges strongly to 1 and is bounded, $\{k_\lambda^2\}$ also converges strongly to 1 and there is a $\lambda_0 \in \Lambda$ such that $\|(k_\lambda^2 - 1)y^*x^*\xi_j\|_2$ is so small for $1 \leq j \leq m$ and $\lambda \geq \lambda_0$ that

$$\sum_1^m \langle xyk_\lambda^2 y^* x^*\xi_j, \xi_j \rangle > \delta. \tag{4.3.30}$$

for all $\lambda \geq \lambda_0$. Hence $Tr(xyk_\lambda^2 y^* x^*) = \widetilde{Tr}(xyk_\lambda^2 y^* x^*) > \delta$ for all $\lambda \geq \lambda_0$ by definition of $\widetilde{Tr}$ and the equality of Tr and $\widetilde{Tr}$ on $(Ne_B N)^+$ (Lemma 4.3.4). The map $z \mapsto Tr(xyzy^*x^*) \colon \langle N, e_B \rangle \to \mathbb{C}$ is a positive linear functional on $\langle N, e_B \rangle$ by Lemma 4.3.4, and hence $Tr(xyk_\lambda^2 y^* x^*) \leq Tr(xyy^*x^*)$, since $yk_\lambda^2 y^* \leq yy^*$. This proves the inequality (4.3.28).

(i)   Let $y \in \langle N, e_B \rangle^+$ and let $\delta < \mathbf{Tr}(y)$. Choose $x \in Ne_B N$ with $\|x\| \leq 1$ such that $\delta < Tr(xyx^*) \leq \mathbf{Tr}(y)$, possible by Definition 4.3.6. By the inequality (4.3.28) there is a λ_0 such that

$$\delta < Tr(xy^{1/2}k_\lambda^2 y^{1/2}x^*) \leq \mathbf{Tr}(y) \tag{4.3.31}$$

for all $\lambda \geq \lambda_0$. Since Tr is tracial on $Ne_B N$ and $xy^{1/2}k_\lambda \in Ne_B N$, we have

$$\delta < Tr(xy^{1/2}k_\lambda^2 y^{1/2}x^*) = Tr(k_\lambda y^{1/2}x^* xy^{1/2}k_\lambda) \tag{4.3.32}$$

for all $\lambda \geq \lambda_0$. As before, the map $z \mapsto Tr(k_\lambda y^{1/2}zy^{1/2}k_\lambda) \colon \langle N, e_B \rangle \to \mathbb{C}$ is a positive linear functional and hence

$$\delta < Tr(xy^{1/2}k_\lambda^2 y^{1/2}x^*) \leq Tr(k_\lambda y k_\lambda) \tag{4.3.33}$$

for all $\lambda \geq \lambda_0$. Because $Tr(k_\lambda y k_\lambda) \leq \mathbf{Tr}(y)$ for all λ by definition of $\mathbf{Tr}$, (i) follows.

(ii)   Let $\delta < \mathbf{Tr}(yy^*)$. By (i) choose $\lambda \in \Lambda$ such that $\delta < Tr(k_\lambda yy^* k_\lambda) \leq \mathbf{Tr}(yy^*)$ and then by the inequality (4.3.28) choose $\mu \in \Lambda$ such that

$$\delta < Tr(k_\lambda y k_\mu^2 y^* k_\lambda) \leq Tr(k_\lambda yy^* k_\lambda). \tag{4.3.34}$$

This proves (ii). $\square$

For future use, we note that the validity of Lemma 4.3.7 rests only on $\{k_\lambda\}$ being a bounded approximate identity in $Ne_B N$ with $0 \leq k_\lambda$ and $\|k_\lambda\| \leq 1$ for all λ, so we could replace $\{k_\lambda\}$ by $\{k_\lambda^2\}$ with the same conclusion. We will have occasion to do this in Lemma 4.3.10.

Lemma 4.3.8. *Let B be a von Neumann subalgebra of a finite von Neumann algebra N with a faithful normal trace τ. Then $\mathbf{Tr}$ is a faithful semifinite normal trace on $\langle N, e_B \rangle^+$ with $\mathbf{Tr} = Tr$ on $(Ne_B N)^+$.*

Proof. The properties $\mathbf{Tr}(x + y) = \mathbf{Tr}(x) + \mathbf{Tr}(y)$ and $\mathbf{Tr}(\alpha x) = \alpha \mathbf{Tr}(x)$ for all $x, y \in \langle N, e_B \rangle^+$ and all $\alpha \geq 0$ follow from the limit characterisation of $\mathbf{Tr}$ in Lemma 4.3.7 (i).

If $y \in \langle N, e_B \rangle$, then by Lemma 4.3.7 (ii)

$$\mathbf{Tr}(yy^*) = \sup_{\lambda, \mu} Tr(k_\lambda y k_\mu^2 y^* k_\lambda) = \sup_{\lambda, \mu} Tr(k_\mu y^* k_\lambda^2 y k_\mu) = \mathbf{Tr}(y^* y), \tag{4.3.35}$$

since Tr is a trace on $Ne_B N$ and $k_\lambda y k_\mu \in Ne_B N$ for all $\lambda, \mu \in \Lambda$.

Let $\{y_\gamma : \gamma \in \Gamma\}$ be an increasing net in $\langle N, e_B \rangle^+$ converging weakly to y. Let $\delta < \mathbf{Tr}(y)$. By Lemma 4.3.4, choose $\lambda \in \Lambda$ such that

$$Tr(k_\lambda y k_\lambda) = \sum_1^\infty \langle k_\lambda y k_\lambda \xi_j, \xi_j \rangle > \delta \tag{4.3.36}$$

and then choose $m \in \mathbb{N}$ such that

$$\sum_1^m \langle k_\lambda y k_\lambda \xi_j, \xi_j \rangle > \delta. \tag{4.3.37}$$

Since $y_\gamma \to y$ weakly, there is a γ_0 such that

$$\sum_1^m \langle k_\lambda y_\gamma k_\lambda \xi_j, \xi_j \rangle > \delta. \tag{4.3.38}$$

for all $\gamma \geq \gamma_0$. Hence

$$\mathbf{Tr}(y_\gamma) \geq Tr(k_\lambda y_\gamma k_\lambda) = \widetilde{Tr}(k_\lambda y_\gamma k_\lambda)$$
$$= \sum_1^\infty \langle k_\lambda y_\gamma k_\lambda \xi_j, \xi_j \rangle \geq \sum_1^m \langle k_\lambda y_\gamma k_\lambda \xi_j, \xi_j \rangle > \delta \tag{4.3.39}$$

for all $\gamma \geq \gamma_0$. This proves that $\mathbf{Tr}$ is normal.

In order to show that $\mathbf{Tr} = Tr$ on $(Ne_B N)^+$, their common domain of definition, we introduce an auxiliary linear functional ψ on a subspace Y containing $Ne_B N$. Let

$$Y = \{y \in \langle N, e_B \rangle : \lim_\lambda Tr(k_\lambda y k_\lambda) \text{ exists and is finite}\}.$$

This is a linear subspace on which we define the linear functional $\psi \colon Y \to \mathbb{C}$ by

$$\psi(y) = \lim_\lambda Tr(k_\lambda y k_\lambda) \quad \text{for all} \quad y \in Y. \tag{4.3.40}$$

If $x, z \in N$ then, for all λ,

$$Tr(k_\lambda x e_B z k_\lambda) = Tr(e_B z k_\lambda^2 x e_B), \tag{4.3.41}$$

and this converges to $Tr(e_B z x e_B) = Tr(x e_B z)$, since $Tr(e_B \cdot e_B)$ is a bounded linear functional on $\langle N, e_B \rangle$ and

$$\lim_\lambda \|e_B z(1 - k_\lambda)\| = \lim_\lambda \|(1 - k_\lambda) x e_B\| = 0. \tag{4.3.42}$$

This shows that $Ne_B N \subseteq Y$, and that ψ and Tr agree on $Ne_B N$. By Lemma 4.3.7 (i), $\mathbf{Tr}$ and ψ agree on $(Ne_B N)^+$, and the same is then true for $\mathbf{Tr}$ and Tr.

Let $y \in \langle N, e_B \rangle^+$ with $\mathbf{Tr}(y) = \mathbf{Tr}(y^{1/2} y^{1/2}) = 0$. Then $Tr(k_\lambda y^{1/2} k_\mu^2 y^{1/2} k_\lambda) = 0$ for all $\lambda, \mu \in \Lambda$ by Lemma 4.3.7 (ii). Since Tr is faithful on $Ne_B N$ by Lemma 4.3.2 and $k_\lambda y^{1/2} k_\mu \in Ne_B N$, it follows that $k_\lambda y^{1/2} k_\mu = 0$ for all λ, μ. Since k_λ converges strongly to 1, $y^{1/2} = 0$ and hence $\mathbf{Tr}$ is faithful.

Finally we show that $\mathbf{Tr}$ is semifinite. This entails showing that any non-zero $y \in \langle N, e_B \rangle^+$ dominates a positive element of non-zero finite trace. Since

$$\mathbf{Tr}(y^{1/2} k_\lambda^2 y^{1/2}) = \mathbf{Tr}((k_\lambda y^{1/2})^*(k_\lambda y^{1/2})) = \mathbf{Tr}(k_\lambda y k_\lambda) = Tr(k_\lambda y k_\lambda) \tag{4.3.43}$$

and $\lim_\lambda Tr(k_\lambda y k_\lambda) = \mathbf{Tr}(y)$, we may choose λ so large that $Tr(k_\lambda y k_\lambda) > 0$. For this choice of λ ,

$$0 < \mathbf{Tr}(y^{1/2} k_\lambda^2 y^{1/2}) < \infty, \tag{4.3.44}$$

while the inequality $y^{1/2} k_\lambda^2 y^{1/2} \leq y$ follows from $k_\lambda^2 \leq 1$. $\qquad\square$

Remark 4.3.9. Note that once we have proved the faithfulness of $\mathbf{Tr}$, and that $\mathbf{Tr}$ and Tr agree on $(Ne_B N)^+$ the faithfulness of Tr follows, so providing another proof of this (see Lemma 4.3.2). We now show the faithfulness of $\mathbf{Tr}$ without using Lemma 4.3.2, which will also establish the same property for Tr. Consider $y \in \langle N, e_B \rangle^+$ with $y \neq 0$. Since $Ne_B N$ is weakly dense in $\langle N, e_B \rangle$, there must exist $x \in N$ such that $y^{1/2} x e_B \neq 0$, otherwise it would follow that $y^{1/2}\langle N, e_B \rangle = \{0\}$, showing that $y = 0$. Then $e_B x^* y x e_B \neq 0$, and this positive element lies in $e_B \langle N, e_B \rangle e_B = Be_B$, so that $e_B x^* y x e_B = be_B$ for some $b \in B$ with $b \neq 0$. Since $b \mapsto be_B : B \to Be_B$ is a *-isomorphism we have $b \geq 0$, so that $Tr(be_B) = \tau(b) > 0$. Then

$$\mathbf{Tr}(y^{1/2} x e_B x^* y^{1/2}) = \mathbf{Tr}(e_B x^* y x e_B) = Tr(be_B) = \tau(b) > 0, \tag{4.3.45}$$

where we have used the agreement of **Tr** and Tr on $(Ne_BN)^+$ proved above. Since

$$y^{1/2} x e_B x^* y^{1/2} \leq \|x\|^2 y, \qquad (4.3.46)$$

the inequality $\mathbf{Tr}(y) > 0$ follows. $\square$

Lemma 4.3.10. *Let B be a von Neumann subalgebra of a finite von Neumann algebra N with a faithful normal trace τ. For the faithful normal semifinite trace* **Tr**, *the *-algebra Ne_BN is dense in $L^2(\langle N, e_B \rangle, \mathbf{Tr})$.*

Proof. Let $y \in \langle N, e_B \rangle$ with $\mathbf{Tr}(y^*y)$ finite and let $\{k_\lambda : \lambda \in \Lambda\}$ be an approximate identity in Ne_BN with $0 \leq k_\lambda$ and $\|k_\lambda\| \leq 1$ for all λ. Because $\mathbf{Tr} = Tr$ on $(Ne_BN)^+$, **Tr** is tracial and $k_\lambda \langle N, e_B \rangle k_\lambda \subseteq Ne_BN$ for all λ, we have

$$\|y - yk_\lambda^2\|_{2,\mathbf{Tr}}^2 = \mathbf{Tr}(y^*y + k_\lambda^2 y^* y k_\lambda^2 - k_\lambda y^* y k_\lambda - k_\lambda y y^* k_\lambda)$$
$$= \mathbf{Tr}(y^*y) + Tr(k_\lambda^2 y^* y k_\lambda^2 - k_\lambda y^* y k_\lambda - k_\lambda y y^* k_\lambda) \qquad (4.3.47)$$

The final three terms in equation (4.3.47) converge to $\mathbf{Tr}(y^*y) = \mathbf{Tr}(yy^*)$ over the net Λ by Lemma 4.3.7, since $\{k_\lambda\}$ and $\{k_\lambda^2\}$ are approximate identities in Ne_BN with $0 \leq k_\lambda$ and $\|k_\lambda\| \leq 1$. Thus $\lim_\lambda \|y - yk_\lambda^2\|_{2,\mathbf{Tr}} = 0$. The adjoint J is a conjugate linear isometry in $\|\cdot\|_{2,\mathbf{Tr}}$-norm, so $\|y - k_\lambda^2 y\|_{2,\mathbf{Tr}} = \|y^* - y^* k_\lambda^2\|_{2,\mathbf{Tr}}$ also tends to zero over Λ. From the triangle inequality, we obtain

$$\|y - k_\lambda^2 y k_\lambda^2\|_{2,\mathbf{Tr}} \leq \|y - k_\lambda^2 y\|_{2,\mathbf{Tr}} + \|k_\lambda^2 (y - y k_\lambda^2)\|_{2,\mathbf{Tr}}$$
$$\leq \|y - k_\lambda^2 y\|_{2,\mathbf{Tr}} + \|y - y k_\lambda^2\|_{2,\mathbf{Tr}}, \qquad (4.3.48)$$

and the norm density of Ne_BN in $L^2(\langle N, e_B \rangle, \mathbf{Tr})$ follows by taking the limit over λ, since $k_\lambda^2 y k_\lambda^2 \in Ne_BN$. $\square$

The following theorem shows that **Tr** and $\widetilde{Tr}$ are equal on $\langle N, e_B \rangle^+$, and so **Tr** is an infinite sum of normal positive linear functionals on $\langle N, e_B \rangle$ without appealing to Haagerup's Theorem on the equivalent properties of normal weights [85, 188]. If Haagerup's Theorem is assumed, then Theorem 4.3.11 shows that a normal weight on $\langle N, e_B \rangle$ equal to Tr on $(Ne_BN)^+$ is also equal to **Tr**.

The idea of the following proof is to show that $F(x) = \rho(x^*x) = \|x\|_{2,\rho}^2$ is a continuous function on the normed space

$$Y = \{x \in \langle N, e_B \rangle : \mathbf{Tr}(x^*x) < \infty\} = \langle N, e_B \rangle \cap L^2(\langle N, e_B \rangle, \mathbf{Tr})$$

with norm $\|\cdot\|_{2,\mathbf{Tr}}$, which agrees with the continuous function $G(x) = \mathbf{Tr}(x^*x) = \|x\|_{2,\mathbf{Tr}}$ on the dense set $Ne_BN \subseteq Y$. The equality of continuous functions on dense sets finishes the proof once we have handled $\mathbf{Tr}(y) = \infty$ is equivalent to $\rho(y) = \infty$ for $y \in \langle N, e_B \rangle^+$ at the end of the proof.

Theorem 4.3.11. *Let B be a von Neumann subalgebra of a finite von Neumann algebra N with a faithful normal trace τ.*

(i) *Let ρ be a weight on $\langle N, e_B \rangle$ with $\rho = Tr$ on $(Ne_BN)^+$. If ρ is a sum of normal positive linear functionals on $\langle N, e_B \rangle$, or ρ is normal, then $\rho = \mathbf{Tr}$.*

(ii) *The faithful normal semifinite trace* **Tr** *and the weight* $\widetilde{Tr}$ *are equal on* $\langle N, e_B \rangle^+$.

Proof. (i) Let $y \in \langle N, e_B \rangle^+$ and let $\alpha < \rho(y)$. By the assumption there are normal positive linear functionals ψ_j on $\langle N, e_B \rangle$ such that $\rho(y) = \sum_j \psi_j(y)$ for all $y \in \langle N, e_B \rangle^+$, where the sum is over the index set of the ψ_j. Hence there is a finite set F such that

$$\rho(y) \geq \sum_{j \in F} \psi_j(y) > \alpha. \tag{4.3.49}$$

Since $\|k_\lambda\| \leq 1$ for all λ and $k_\lambda \to 1$ strongly over Λ, the net $k_\lambda y k_\lambda \to y$ weakly over Λ. Since the ψ_j are normal linear functionals on $\langle N, e_B \rangle$ and F is finite, there is a $\lambda \in \Lambda$ such that $|\psi_j(k_\lambda y k_\lambda) - \psi_j(y)|$ is so small for $j \in F$ that

$$\sum_{j \in F} \psi_j(k_\lambda y k_\lambda) > \alpha, \tag{4.3.50}$$

from (4.3.49). By the equality of ρ and Tr on $(Ne_B N)^+$,

$$\begin{aligned} \mathbf{Tr}(y) &\geq \mathbf{Tr}(y^{1/2} k_\lambda^2 y^{1/2}) = \mathbf{Tr}(k_\lambda y k_\lambda) = Tr(k_\lambda y k_\lambda) \\ &= \rho(k_\lambda y k_\lambda) = \sum_j \psi_j(k_\lambda y k_\lambda) \geq \sum_{j \in F} \psi_j(k_\lambda y k_\lambda) > \alpha, \end{aligned} \tag{4.3.51}$$

where the last inequality is (4.3.50). Hence $\mathbf{Tr}(y) \geq \rho(y)$ and

$$\mathbf{Tr} \geq \rho \quad \text{on} \quad \langle N, e_B \rangle^+. \tag{4.3.52}$$

This shows that if $y \in \langle N, e_B \rangle^+$ with $\rho(y) = \infty$, then $\mathbf{Tr}(y) = \infty$. The reverse of this infinite case implication is handled after the finite case, which is required in its proof.

The inequality (4.3.52) shows that

$$\|y\|_{2,\rho}^2 = \rho(y^* y) \leq \mathbf{Tr}(y^* y) = \|y\|_{2,\mathbf{Tr}}^2 \tag{4.3.53}$$

for all $y \in Y = \{y \in \langle N, e_B \rangle : \mathbf{Tr}(y^* y) < \infty\}$. This implies that $\|\cdot\|_{2,\rho} \leq \|\cdot\|_{2,\mathbf{Tr}}$ on $\langle N, e_B \rangle \cap L^2(\langle N, e_B \rangle, \mathbf{Tr})$ so that the norm $\|\cdot\|_{2,\rho}$ is continuous with respect to $\|\cdot\|_{2,\mathbf{Tr}}$. By Lemma 4.3.8 and the hypothesis,

$$\|y\|_{2,\rho}^2 = \rho(y^* y) = Tr(y^* y) = \mathbf{Tr}(y^* y) = \|y\|_{2,\mathbf{Tr}}^2 \tag{4.3.54}$$

for all $y \in Ne_B N$, so the norms $\|\cdot\|_{2,\rho}$ and $\|\cdot\|_{2,\mathbf{Tr}}$ are equal on the linear space $Ne_B N$, and hence on its closure in Y by continuity. Since $Ne_B N$ is dense in Y in the $\|\cdot\|_{2,\mathbf{Tr}}$-norm by Lemma 4.3.10,

$$\rho(y^* y) = \|y\|_{2,\rho}^2 = \|y\|_{2,\mathbf{Tr}}^2 = \mathbf{Tr}(y^* y) \tag{4.3.55}$$

for all $y \in \langle N, e_B \rangle$ with $\mathbf{Tr}(y^* y) < \infty$.

Hence

$$\rho = \mathbf{Tr} \quad \text{on} \quad \{y \in \langle N, e_B \rangle^+ : \mathbf{Tr}(y) < \infty\}. \tag{4.3.56}$$

The proof will be complete once we have shown that $\mathbf{Tr}(y) = \infty$ implies that $\rho(y) = \infty$. Let $y \in \langle N, e_B \rangle^+$ with $\mathbf{Tr}(y) = \infty$. Let $m \in \mathbb{N}$. By Lemma 4.3.7 (i), there is a $\lambda \in \Lambda$ such that $Tr(k_\lambda y k_\lambda) \geq m$. Since $k_\lambda y k_\lambda \in Ne_B N$ on which $\mathbf{Tr}$ and Tr are equal,

$$\mathbf{Tr}(y^{1/2} k_\lambda^2 y^{1/2}) = Tr(k_\lambda y k_\lambda) \geq m \qquad (4.3.57)$$

and all of these terms are finite. By equation (4.3.56), $\mathbf{Tr}(y^{1/2} k_\lambda^2 y^{1/2}) = \rho(y^{1/2} k_\lambda^2 y^{1/2})$ so that

$$\rho(y) \geq \rho(y^{1/2} k_\lambda^2 y^{1/2}) \geq m, \qquad (4.3.58)$$

from (4.3.57). This implies that $\rho(y) = \infty$ and finishes the proof of (i) for ρ of this type.

If ρ is normal, then ρ is a sum of normal positive linear functionals on $\langle N, e_B N \rangle$ by Haagerup's Theorem [85, 188]. The proof above now applies.

(ii)   The weight $\widetilde{Tr}$ satisfies the hypotheses of (i) by Lemma 4.3.4. $\qquad \square$

Theorem 4.3.11 thus completes the proof of Theorem 4.2.2 by showing that the two constructions yield the same faithful semifinite normal trace $\widetilde{Tr} = \mathbf{Tr}$, which is an infinite sum of normal positive linear functionals on $\langle N, e_B \rangle^+$.

Definition 4.3.12. Let B be a von Neumann subalgebra of a finite von Neumann algebra N with a faithful normal trace τ. The faithful normal semifinite trace $\mathbf{Tr} = \widetilde{Tr}$ on $\langle N, e_B \rangle$ defined in Definition 4.3.6 and Lemma 4.3.4 is denoted by Tr from now onwards. $\qquad \square$

At this point we have achieved the main goal of this section, the construction of a faithful normal semifinite trace on $\langle N, e_B \rangle$, satisfying the requirements of Theorem 4.2.2 (vi)–(vii).

Corollary 4.3.13. *Let B be a von Neumann subalgebra of a finite von Neumann algebra N with a faithful normal trace τ.*

(i) *The map $b \mapsto be_B : (B, \tau) \to (e_B \langle N, e_B \rangle e_B, Tr)$ is a trace preserving isomorphism.*

(ii) *If $Z(B)$ is the centre of B and $Z(\langle N, e_B \rangle)$ is the centre of $\langle N, e_B \rangle$, then the map $z \mapsto Jz^* J : Z(B) \to Z(\langle N, e_B \rangle)$ is an isomorphism with $Tr(Jz^* J e_B) = \tau(z)$ for all $z \in Z(B)$, and $Jz^* J$ is the unique element z_0 in $Z(\langle N, e_B \rangle)$ such that $z_0 e_B = z e_B$.*

(iii) *$\{e_B\}' \cap \langle N, e_B \rangle = J \langle B, e_B \rangle' J$.*

Proof. (i)   This is a direct consequence of Theorem 4.2.2 (iii) and Lemma 4.3.2.
(ii)   By Theorem 4.2.2 (i)

$$Z(\langle N, e_B \rangle) = \langle N, e_B \rangle \cap \langle N, e_B \rangle' = JB'J \cap JBJ = JZ(B)J, \qquad (4.3.59)$$

which shows that the map $z \mapsto Jz^* J : Z(B) \to Z(\langle N, e_B \rangle)$ is a *-isomorphism. If $z_0 \in Z(\langle N, e_B \rangle)$ with $z_0 e_B = z e_B$, then $(Jz^* J - z_0) e_B = 0$ so that $Jz^* J - z_0 = 0$, since the central support of e_B is 1.

Let $z \in Z(B)$. If $x \in N$, then

$$Jz^*Je_Bx\xi = Jz^*Je_Bxe_B\xi = Jz^*J\mathbb{E}_B(x)\xi$$
$$= \mathbb{E}_B(x)z\xi = z\mathbb{E}_B(x)\xi = ze_Bx\xi \qquad (4.3.60)$$

so that $Jz^*Je_B = ze_B$, since $N\xi$ is dense in $L^2(N)$. Hence

$$Tr(Jz^*Je_B) = Tr(ze_B) = \tau(z) \qquad (4.3.61)$$

for all $z \in Z(B)$.

(iii)   By Theorem 4.2.2,

$$\{e_B\}' \cap \langle N, e_B \rangle = \{Je_BJ\}' \cap (JBJ)' = \langle Je_BJ, JBJ \rangle' = J\langle e_B, B \rangle'J. \quad (4.3.62)$$

$\qquad\qquad\qquad\qquad\qquad\qquad\qquad\qquad\qquad\qquad\qquad\qquad\qquad\qquad\square$

Remark 4.3.14. If B is a masa in a separable II_1 factor N, then there is a *-isomorphism θ from $L^\infty[0,1]$ onto the centre $Z(\langle N, e_B \rangle)$ of $\langle N, e_B \rangle$ such that

$$Tr(\theta(f)e_B) = \int_0^1 f(t)dt \quad \text{for all} \quad f \in L^\infty[0,1]. \qquad (4.3.63)$$

This follows from the corresponding *-isomorphism from $L^\infty[0,1]$ onto B with the integral corresponding to the trace (Theorem 3.5.2) and Corollary 4.3.13 (ii).

Note that in the case of a masa B in a separable II_1 factor N the basic construction algebra $\langle N, e_B \rangle$ is always isomorphic to $L^\infty[0,1]\overline{\otimes}B(H)$ with the faithful semifinite normal trace that arises from the tensor of the usual traces of the two algebras $L^\infty[0,1]$ and $B(H)$. Thus the algebra $\langle N, e_B \rangle$ is independent of the masa or separable factor in this case. It is how $B \subseteq N$ lies in $\langle N, e_B \rangle$ that encodes the information. This is similar to the situation of the hyperfinite factor R contained in a non-hyperfinite separable factor, when $\langle N, e_R \rangle = (JRJ)' \cong B(H)\overline{\otimes}R$.

$\qquad\qquad\qquad\qquad\qquad\qquad\qquad\qquad\qquad\qquad\qquad\qquad\qquad\qquad\square$

The unitary normaliser $\mathcal{N}(B)$ of B in N is linked with e_B in a nice way. Let $u \in \mathcal{N}(B)$. Since $uBu^* = B$ and $uB^\perp u^* = B^\perp$, where $B^\perp = L^2(N) \ominus L^2(B)$, we have $e_BuJuJe_B = uJuJe_B$ and

$$(I - e_B)uJuJ(I - e_B) = e_{B^\perp}uJuJe_{B^\perp}$$
$$= uJuJe_{B^\perp} = uJuJ(I - e_B) \qquad (4.3.64)$$

so that $e_BuJuJ = uJuJe_B$. Hence

$$uJuJ \in \{e_B\}' \cap \langle N, e_B \rangle = J\langle B, e_B \rangle'J. \qquad (4.3.65)$$

The converse also holds so that

$$\mathcal{N}(B) = \{u \in \mathcal{U}(N) : uJuJ \in J\langle B, e_B \rangle'J\}$$
$$= \{u \in \mathcal{U}(N) : uJuJ \in \langle B, e_B \rangle'\}$$
$$= \{u \in \mathcal{U}(N) : uJuJ \in \langle B, e_B \rangle' \cap J\langle B, e_B \rangle'J\}, \qquad (4.3.66)$$

where the second equality results from the commutation of u and JuJ. Note that with the notation $\mathcal{B} = (B \cup JBJ)''$ used in Chapter 7 on the Pukánszky invariant, we have

$$\langle B, e_B \rangle' \cap J \langle B, e_B \rangle' J = ((B \cup JBJ)'' \cup \{e_B\})'$$
$$= \langle \mathcal{B}, e_B \rangle' = \mathcal{B}' \cap \{e_B\}'. \qquad (4.3.67)$$

Here we record a useful technical observation (see the proof of Theorem 4.5.3). If $\{y_\lambda : \lambda \in \Lambda\}$ is a bounded net in $\langle N, e_B \rangle$ converging strongly to 0, then

$$\|y_\lambda e_B\|_{2,Tr} \to 0 \quad \text{over} \quad \Lambda. \qquad (4.3.68)$$

The equality of the strong and σ-strong topologies on closed balls in a von Neumann algebra [187, p. 69] implies that y_λ converges σ-strongly to 0. If ξ_j is as in the definition of $\widetilde{Tr}$, (4.3.15), then

$$\sum_j \|e_B \xi_j\|_2^2 = \sum_j \langle e_B \xi_j, \xi_j \rangle = Tr(e_B) = 1. \qquad (4.3.69)$$

Thus

$$\|y_\lambda e_B\|_{2,Tr}^2 = Tr(e_B y_\lambda^* y_\lambda e_B) = \sum_j \|y_\lambda e_B \xi_j\|_2^2 \to 0 \qquad (4.3.70)$$

over Λ.

Parts of Theorem 4.2.2 characterise the basic construction as in the following theorem (see [144] or [206, Theorem 1.1]).

Theorem 4.3.15. *Let M be a von Neumann algebra with a semifinite faithful normal trace Tr_M and von Neumann subalgebras $B \subseteq N \subseteq M$. If there is a projection $e \in M$ such that*

(i) *M is the weak closure of the *-subspace NeN;*

(ii) *$Tr_M(e) = 1$ and $\tau(x) := Tr_M(xe)$ defines a faithful normal trace τ on N;*

(iii) *$eMe = Be = eB$;*

then there is a trace preserving isomorphism θ from $\langle N, e_B \rangle$ onto M with $\theta(x) = x$ for all $x \in N$ and $\theta(e_B) = e$.

Proof. Firstly note that (i) and (iii) imply that NeN is a *-subalgebra of M. The map $x \mapsto xe : N \to Ne$ is injective, since $xe = 0$ implies that $\tau(x^*x) = Tr_M(ex^*x) = Tr_M(ex^*xe) = 0$ and τ is faithful. This is used in the proof of Lemma 4.3.1 (ii) as can be seen by examining that proof. Hence if $x_1, \ldots, x_n, y_1, \ldots, y_n \in N$, and

$$X = \begin{pmatrix} x_1 & \cdots & x_n \\ 0 & \cdots & 0 \\ 0 & \cdots & 0 \end{pmatrix}, \quad Y = \begin{pmatrix} y_1 & 0 & \cdots & 0 \\ \cdot & 0 & \cdots & 0 \\ y_n & 0 & \cdots & 0 \end{pmatrix},$$

then $\sum_{j=1}^{n} x_j e_B y_j = 0$ if and only if there is a projection $P \in \mathbb{M}_n(B)$ such that $XP = X$ and $PY = 0$. This shows that the map $\theta : x e_B y \mapsto x e y : N e_B N \to NeN$ is well defined, injective and surjective with

$$Tr_M(\theta(x e_B y)) = Tr_M(x e y) = \tau(xy) = Tr(x e_B y) \qquad (4.3.71)$$

for all $x, y \in N$. Thus if $x \in N$, there is a unique $\phi(x) \in B$ such $exe = \phi(x)e$. Then $\tau(\phi(x)) = Tr_M(\phi(x)e) = Tr_M(exe) = \tau(x)$ for all $x \in N$.

If $b \in B$ with $b = b^*$ and $be = ec$ for some $c \in B$, then $ebe = ec$ is self-adjoint, so be is self-adjoint and $be = eb$. It follows that

$$\phi(b_1 x b_2)e = e b_1 x b_2 e = b_1 e x e b_2 = b_1 \phi(x) b_2 e, \qquad (4.3.72)$$

and so $\phi(b_1 x b_2) = b_1 \phi(x) b_2$ for all $b_1, b_2 \in B$. The uniqueness conditions of Lemma 3.6.2 are satisfied and $\phi = \mathbb{E}_B$. Consequently

$$\begin{aligned}
\theta((x_1 e_B y_1)(x_2 e_B y_2)) &= \theta(x_1 \mathbb{E}_B(y_1 x_2) e_B y_2) \\
&= x_1 \mathbb{E}_B(y_1 x_2) e y_2 = (x_1 e y_1)(x_2 e y_2). \qquad (4.3.73)
\end{aligned}$$

Thus θ is a trace preserving *-isomorphism from the weakly dense *-subalgebra $N e_B N$ of $(\langle N, e_B \rangle, Tr)$ onto the weakly dense *-subalgebra NeN of (M, Tr_M). Since θ is a trace preserving *-isomorphism,

$$\|\theta(x)\|_{2,Tr_M}^2 = Tr_M(\theta(x^*x)) = Tr(x^*x) = \|x\|_{2,Tr}^2 \qquad (4.3.74)$$

for all $x \in N e_B N$. Hence θ lifts to a unitary operator $U : L^2(\langle N, e_B \rangle, Tr) \to L^2(M, Tr_M)$ with $U(x) = \theta(x)$ for all $x \in N e_B N$. If $\langle N, e_B \rangle$ is represented using its left action on $L^2(\langle N, e_B \rangle, Tr)$ and M using its left action on $L^2(M, Tr_M)$, then $\theta(x)U(y) = U(xy)$ for all $x, y \in N e_B N$ so that $\theta(x)U = Ux$ for all $x \in N e_B N$. The map $\theta(x) := U^*xU$ $(x \in \langle N, e_B \rangle)$ induces the required *-isomorphism from $\langle N, e_B \rangle$ onto M. $\qquad \square$

In this theorem, if $Tr_M(e)$ is finite but not 1, then just rescale Tr_M to be 1 on e. The same result then holds.

Remark 4.3.16. The technical complications of constructing the trace for $\langle N, e_B \rangle$ disappear if B is a factor. In this case the basic construction is a semifinite factor since its commutant is the finite factor JBJ, so standard von Neumann algebra theory gives a faithful normal semifinite trace which is unique up to scaling. The projection e_B is finite, so there is a unique trace with $Tr(e_B) = 1$, and whose restriction to the algebra $Be_B \cong B$ must then give the trace τ on B. Thus $Tr(be_B) = \tau(b)$ for $b \in B$. For $x, y \in N$, we have

$$\begin{aligned}
Tr(x e_B y) &= Tr(e_B y x e_B) = Tr(\mathbb{E}_B(yx)e_B) \\
&= \tau(\mathbb{E}_B(yx)) = \tau(yx) = \tau(xy). \qquad (4.3.75)
\end{aligned}$$

Thus Tr automatically has this crucial property $Tr(x e_B y) = \tau(xy)$, which required considerable work in the general case. This was unavoidable as many of the applications of the basic construction occur when B is not a factor. $\qquad \square$

4.4 Examples

In this section we will present two examples. The first is to illustrate that in simple cases of the basic construction, the operator e_B and the trace Tr can be computed explicitly. The second example, (Theorem 4.4.4), serves two purposes. It is a result which will be needed when we discuss general perturbations of subalgebras in Chapter 10, but it is also intended to show the power of the basic construction. A particular case would be the following. Let $\delta > 0$ be given (assumed close to 0) and let B be a weakly closed subalgebra of $L^\infty[0,1]$ with the property that, for each $f \in L^\infty[0,1]$ with $\|f\|_\infty \leq 1$, there exists $g \in B$ with $\|f - g\|_2 \leq \delta$. Then there is a measurable subset F of $[0,1]$ of large measure such that B and $L^\infty[0,1]$ agree on F. At first sight it should be possible to resolve this by elementary measure theory, but it seems to require the basic construction.

Example 4.4.1. Let $N = \ell^\infty(3)$, viewed in its standard representation as 3×3 diagonal matrices with minimal projections p_1, p_2, p_3. Let ξ_1, ξ_2, ξ_3 be the standard orthonormal basis for $\mathbb{C}^3$. Let $B \subseteq \ell^\infty(3)$ be the algebra span$\{p_1 + p_2, p_3\}$, and define a normalised trace τ on $\ell^\infty(3)$ by

$$\tau\left(\sum_{i=1}^{3} \alpha_i p_i\right) = \sum_{i=1}^{3} \alpha_i \lambda_i, \quad \alpha_i \in \mathbb{C}, \tag{4.4.1}$$

where $\lambda_i > 0$, $1 \leq i \leq 3$, are fixed scalars summing to 1. In the GNS representation for this trace, p_i corresponds to $\sqrt{\lambda_i}\xi_i$ for $1 \leq i \leq 3$, and so e_B is the projection onto span$\{\sqrt{\lambda_1}\xi_1 + \sqrt{\lambda_2}\xi_2, \xi_3\}$. Thus

$$e_B = \begin{pmatrix} \frac{\lambda_1}{\lambda_1+\lambda_2} & \frac{\sqrt{\lambda_1\lambda_2}}{\lambda_1+\lambda_2} & 0 \\ \frac{\sqrt{\lambda_1\lambda_2}}{\lambda_1+\lambda_2} & \frac{\lambda_2}{\lambda_1+\lambda_2} & 0 \\ 0 & 0 & 1 \end{pmatrix}.$$

Then $\langle N, e_B \rangle = \mathbb{M}_2 \oplus \mathbb{C}$, and the set of all faithful traces on this algebra are combinations of the traces on $\mathbb{M}_2$ and $\mathbb{C}$, the former being multiples of the unique trace τ_2 for which $\tau_2(I_2) = 2$. The requirements $Tr(e_B) = 1$ and $Tr(p_1 e_B p_1) = \tau(p_1)$ easily lead to

$$Tr(X + \mu p_3) = (\lambda_1 + \lambda_2)\tau_2(X) + \lambda_3\mu \tag{4.4.2}$$

for $X \in \mathbb{M}_2$ and $\mu \in \mathbb{C}$. In this case $Tr(I_3) = 2 - \lambda_3$. Thus abelian inclusions can lead to nonabelian algebras, and the operator e_B can vary with the choice of trace in the non-factor case. □

We now require two preliminary lemmas. The first of these is an inequality which might be useful in other contexts, and is valid in any von Neumann algebra.

Lemma 4.4.2. *Let N be a von Neumann algebra containing two projections e and f. If $\lambda \geq 0$, then $efe \leq \lambda e$ if and only if $fef \leq \lambda f$.*

Proof. The stated inequalities are equivalent respectively to $\|ef\| \le \sqrt{\lambda}$ and $\|fe\| \le \sqrt{\lambda}$. The result follows because $\|ef\| = \|(ef)^*\| = \|fe\|$.   □

Lemma 4.4.3. *Let $A \subseteq B$ be an inclusion of abelian von Neumann algebras, and let $\mathbb{E} \colon B \to A$ be a faithful conditional expectation with the property that if $p_1, p_2 \in B$ are orthogonal projections, then the supports of $\mathbb{E}(p_1)$ and $\mathbb{E}(p_2)$ are disjoint. Then $A = B$ and $\mathbb{E}$ is the identity.*

Proof. Let $p \in B$ be a projection. If we apply the hypothesis to p and $1 - p$, then $\mathbb{E}(p)\mathbb{E}(1 - p) = 0$, implying that $\mathbb{E}(p) = \mathbb{E}(p)^2$. Thus $\mathbb{E}(p)$ is a projection since this element is self-adjoint. Using the A-modularity of $\mathbb{E}$, we see that

$$\mathbb{E}(\mathbb{E}(p)(1 - p)) = \mathbb{E}(p)(1 - \mathbb{E}(p)) = 0, \qquad (4.4.3)$$

so faithfulness of $\mathbb{E}$ gives $\mathbb{E}(p)(1 - p) = 0$. Thus $\mathbb{E}(p) \le p$. This applies equally to $1 - p$, so $1 - \mathbb{E}(p) \le 1 - p$. These two inequalities imply that $\mathbb{E}(p) = p$, and the result follows since the projections span a norm dense subspace of B.   □

The following result could be proved more efficiently by using the techniques developed in Chapter 9, but we include it now to illustrate the basic construction. With a somewhat different statement, it is [152, Lemma 3.2]. The proof given here fills a gap in the proof from [152].

Theorem 4.4.4. *Let $A \subseteq B$ be an inclusion of abelian von Neumann algebras where B has a faithful normal normalised trace τ. Let $\delta > 0$ be given, and suppose that $\|b - \mathbb{E}_A(b)\|_{2,\tau} \le \delta$ for all $b \in B$ with $\|b\| \le 1$. Then there exists a projection $p \in B$ such that $Ap = Bp$ and $\tau(p) \ge 1 - 4\delta^2$.*

Proof. Consider the basic construction $A \subseteq B \subseteq \langle B, e_A \rangle$, and let Tr be the semifinite normal faithful trace on $\langle B, e_A \rangle$ from Theorem 4.2.2. The unitary group $\mathcal{U}(B)$ is abelian and so amenable when viewed with the discrete topology. Let μ be an invariant mean on $\ell^\infty(\mathcal{U}(B))$ with $\mu(1) = 1$. For each $\phi \in \langle B, e_A \rangle_*$, the equation

$$\phi(h) = \mu(u \mapsto \phi(ue_A u^*)), \quad u \in \mathcal{U}(B), \qquad (4.4.4)$$

defines an element $h \in \langle B, e_A \rangle^+$ satisfying $0 \le h \le 1$, and the invariance of μ shows that $h \in B' \cap \langle B, e_A \rangle$. Since B is maximal abelian in $B(L^2(B, \tau))$, we also have $h \in B$, and all of its spectral projections have this property as well. For any unitary $u \in \mathcal{U}(B)$,

$$
\begin{aligned}
1 - Tr(e_A u e_A u^*) &= 1 - Tr(e_A \mathbb{E}_A(u) u^*) = 1 - \tau(\mathbb{E}_A(u) u^*) \\
&= 1 - \tau(\mathbb{E}_A(u)\mathbb{E}_A(u^*)) = 1 - \|\mathbb{E}_A(u)\|_{2,\tau}^2 \\
&= \|u - \mathbb{E}_A(u)\|_{2,\tau}^2 \le \delta^2, \qquad (4.4.5)
\end{aligned}
$$

and so $|Tr(e_A - e_A u e_A u^*)| \le \delta^2$. Since $Tr(e_A \cdot e_A)$ is a normal bounded linear functional on $\langle B, e_A \rangle$, we may apply μ to the function $u \mapsto Tr(e_A - e_A u e_A u^*)$ to obtain

$$1 - Tr(e_A h) = 1 - Tr(e_A h e_A) \le \delta^2. \qquad (4.4.6)$$

For each $x \in Be_A B$, $\|x\| \leq 1$, $Tr(x \cdot x^*)$ defines a bounded normal linear functional on $\langle B, e_A \rangle$, and so

$$Tr(xue_A u^* x^*) = Tr(e_A u^* x^* xue_A) \leq Tr(e_A) = 1, \quad u \in \mathcal{U}(B). \tag{4.4.7}$$

If we apply μ, regarding the terms in (4.4.7) as functions of u, then we obtain $Tr(xhx^*) \leq 1$. The definition of Tr on $\langle B, e_A \rangle^+$ then implies that $Tr(h) \leq 1$. Thus $Tr(h^{1/2} \cdot h^{1/2})$ defines a bounded normal linear functional on $\langle B, e_A \rangle$, and applying μ to the function

$$u \mapsto Tr(h^{1/2} u e_A u^* h^{1/2}) = Tr(e_A u^* h u e_A) = Tr(e_A h)$$

gives $Tr(h^2) = Tr(e_A h)$, and also shows that $h \in L^2(\langle B, e_A \rangle, Tr)$. Then

$$\|h - e_A\|_{2,Tr}^2 = Tr(h^2 - 2he_A + e_A) = Tr(e_A - he_A)$$
$$= 1 - Tr(he_A) \leq \delta^2. \tag{4.4.8}$$

For each $\lambda \in (2^{-1}, 1)$, let $f_\lambda \in B$ be the spectral projection of h for the interval $[\lambda, 1]$. Fix an arbitrary λ in the specified range.

We first show that, for every projection $q \leq f_\lambda$, the inequality

$$\mathbb{E}_A(q) \geq \lambda \operatorname{supp}(\mathbb{E}_A(q)) \tag{4.4.9}$$

holds. If not, then there exists a projection $q \leq f_\lambda$ and $\varepsilon > 0$ so that the spectral projection $q_0 \in A$ of $\mathbb{E}_A(q)$ for the interval $[0, \lambda - \varepsilon]$ is non-zero. Then

$$0 \neq \mathbb{E}_A(qq_0) \leq \lambda - \varepsilon. \tag{4.4.10}$$

From this it follows that $e_A(qq_0)e_A \leq (\lambda - \varepsilon)e_A$, and so $qq_0 e_A qq_0 \leq (\lambda - \varepsilon)qq_0$, applying Lemma 4.4.2 to the projections e_A and qq_0. If we average this inequality over unitaries $uqq_0 \in Bqq_0$ where $u \in \mathcal{U}(B)$, then we obtain

$$hqq_0 \leq (\lambda - \varepsilon)qq_0. \tag{4.4.11}$$

The inequality $hf_\lambda \geq \lambda f_\lambda$ implies that

$$hqq_0 = hf_\lambda qq_0 \geq \lambda f_\lambda qq_0 = \lambda qq_0, \tag{4.4.12}$$

and this contradicts (4.4.11), establishing (4.4.9).

Let $g_\lambda \in A$ be the support projection of $\mathbb{E}_A(f_\lambda)$. Then, by (4.4.9), $\mathbb{E}_A(f_\lambda) \geq \lambda g_\lambda$, and so there exists $a_\lambda \in A^+$ such that $\mathbb{E}_A(f_\lambda)a_\lambda = g_\lambda$. Moreover, the A-modularity of $\mathbb{E}_A$ gives $\mathbb{E}_A(f_\lambda(1 - g_\lambda)) = \mathbb{E}_A(f_\lambda)(1 - g_\lambda) = 0$, and faithfulness of this conditional expectation then gives $f_\lambda = f_\lambda g_\lambda$. Define $\mathbb{E}: Bf_\lambda \to Af_\lambda$ by $\mathbb{E}(x) = \mathbb{E}_A(x)a_\lambda f_\lambda$ for $x \in Bf_\lambda$. We will show below that $\mathbb{E}$ is a faithful conditional expectation. Note that it is clearly completely positive since $\mathbb{E}_A$ has this property.

Consider two orthogonal projections q_1 and q_2 in Bf_λ, which both lie below f_λ. From (4.4.9) we obtain

$$1 \geq \mathbb{E}_A(q_1 + q_2) \geq \lambda(\operatorname{supp}\mathbb{E}_A(q_1) + \operatorname{supp}\mathbb{E}_A(q_2))$$
$$\geq 2\lambda(\operatorname{supp}\mathbb{E}_A(q_1) \cdot \operatorname{supp}\mathbb{E}_A(q_2)). \tag{4.4.13}$$

Since $2\lambda > 1$, this forces $\mathbb{E}_A(q_1)$ and $\mathbb{E}_A(q_2)$ to have disjoint support projections, and the definition of $\mathbb{E}$ shows that this is also the case for $\mathbb{E}(q_1)$ and $\mathbb{E}(q_2)$.

For $a \in A$, the A-modularity of $\mathbb{E}_A$ implies that

$$\mathbb{E}(af_\lambda) = \mathbb{E}_A(af_\lambda)a_\lambda f_\lambda = a\mathbb{E}_A(f_\lambda)a_\lambda f_\lambda = ag_\lambda f_\lambda = af_\lambda, \qquad (4.4.14)$$

showing that $\mathbb{E}$ is the identity on Af_λ and thus is a conditional expectation. If $b \in B^+$ and $\mathbb{E}(bf_\lambda) = 0$, then $\mathbb{E}_A(bf_\lambda)a_\lambda f_\lambda = 0$. Multiply by b and apply $\mathbb{E}_A$. This yields $a_\lambda \mathbb{E}_A(bf_\lambda)^2 = 0$. Multiply by $\mathbb{E}_A(f_\lambda)$ and use $\mathbb{E}_A(f_\lambda)a_\lambda = g_\lambda$ to obtain $g_\lambda \mathbb{E}_A(bf_\lambda)^2 = 0$. Move g_λ inside $\mathbb{E}_A(bf_\lambda)$, and the result is $\mathbb{E}_A(bf_\lambda)^2 = 0$. The faithfulness of $\mathbb{E}_A$ shows that $bf_\lambda = 0$, proving faithfulness of $\mathbb{E}$. We may now apply Lemma 4.4.3 to the containment $Af_\lambda \subseteq Bf_\lambda$ to obtain equality of these algebras.

Let p be the spectral projection of h for the interval $(2^{-1}, 1]$. By taking the limit $\lambda \to 2^{-1}+$, we obtain $Ap = Bp$, and the estimate $\tau(p) \geq 1 - 4\delta^2$ follows by taking limits in the inequality

$$\begin{aligned}
(1-\lambda)^2(1 - \tau(f_\lambda)) = \tau(((1-\lambda)(1-f_\lambda))^2) &\leq \tau((1-h)^2) \\
&= Tr(e_A(1-h)^2) = \|e_A(1-h)\|_{2,Tr}^2 \\
&= \|e_A(e_A - h)\|_{2,Tr}^2 \leq \|e_A - h\|_{2,Tr}^2 \leq \delta^2, \qquad (4.4.15)
\end{aligned}$$

the last inequality being (4.4.8). This completes the proof. $\square$

4.5   The pull-down map

This section is devoted to the pull-down map Φ defined and exploited by Popa (see [144]). This map Φ plays an important role in certain calculations, since it is an N-module map which is norm reducing in the $\|\cdot\|_1$-norms and which is an isometry from Ne_B to N for $\|\cdot\|_2$-norms. Here, as with the trace Tr, we start with a purely algebraic definition before giving the full definition and it properties.

Lemma 4.5.1. *Let B be a von Neumann subalgebra of a finite von Neumann algebra N with a faithful normal trace τ. There is a well defined N-bimodule $*$-map Φ from $Ne_BN \to N$ defined by*

$$\Phi(xe_By) = xy \quad for\ all \quad x, y \in N \qquad (4.5.1)$$

satisfying:

(i) $\tau(\Phi(x)) = Tr(x)$ *for all* $x \in Ne_BN$;

(ii) $e_B\Phi(e_Bx) = e_Bx$ *for all* $x \in Ne_BN$;

(iii) $\|\Phi(x)\|_1 \leq \|x\|_{1,Tr}$ *for all* $x \in Ne_BN$;

(iv) Φ *maps* e_BN *onto* N *with* $\|\Phi(x)\|_2 = \|x\|_{2,Tr}$ *for all* $x \in e_BN$.

Proof. Note that Φ is well defined by Lemma 4.3.1 and is an N-module map by definition. Equation (i) follows from Theorem 4.2.2. Equation (ii) is a consequence of the following. If $x = ye_Bz$ for $y, z \in N$, then

$$e_B\Phi(e_Bx) = e_B\Phi(e_Bye_Bz) = e_B\Phi(e_B\mathbb{E}_B(y)z)$$
$$= e_B\mathbb{E}_B(y)z = e_Bye_Bz = e_Bx. \quad (4.5.2)$$

(iii) Let $y \in Ne_BN$. If $x \in N$ with $\|x\| \leq 1$, then

$$|\tau(x\Phi(y))| = |\tau(\Phi(xy)| = |Tr(xy)| \leq \|xy\|_{1,Tr} \leq \|y\|_{1,Tr} \quad (4.5.3)$$

since Φ is an N-module map. Taking the supremum over all such $x \in N$ shows that $\|\Phi(y)\|_1 \leq \|y\|_{1,Tr}$.
(iv) If $x = ye_B$ for $y \in N$, then

$$\|x\|_{2,Tr}^2 = Tr(e_By^*ye_B) = Tr(e_B\mathbb{E}_B(y^*y))$$
$$= \tau(\mathbb{E}_B(y^*y)) = \tau(y^*y) = \|\Phi(x)\|_2^2. \quad (4.5.4)$$

$\square$

There are two ideals associated to $\langle N, e_B \rangle$ and the trace, and these can be defined in any semifinite von Neumann algebra:

$$\mathfrak{N}_{Tr} = \{x \in \langle N, e_B \rangle \colon Tr(x^*x) < \infty\}$$

and

$$\mathfrak{M}_{Tr} = \left\{ \sum_{i=1}^n x_iy_i \colon x_i, y_i \in \mathfrak{N}_{Tr}, \ n \in \mathbb{N} \right\},$$

and both contain Ne_BN.

Before lifting the algebraic definition of Φ from Ne_BN to $\mathfrak{M}_{Tr}$ here is a lemma on the density of Ne_BN in $L^1(\langle N, e_B \rangle, Tr)$ in the $\| \cdot \|_{1,Tr}$-norm, which uses the standard duality between $\langle N, e_B \rangle$ and $L^1(\langle N, e_B \rangle, Tr)$ (see [187]).

Lemma 4.5.2. *Let B be a von Neumann subalgebra of a finite von Neumann algebra N with a faithful normal trace τ. The *-subalgebra Ne_BN is dense in $\mathfrak{M}_{Tr}$ in $\| \cdot \|_{1,Tr}$.*

Proof. It is sufficient to consider $y \in \langle N, e_B \rangle^+$ with $Tr(y) \leq 1$, since $\mathfrak{M}_{Tr}$ is the linear span of $\mathfrak{M}_{Tr} \cap \langle N, e_B \rangle^+ = \{y \in \langle N, e_B \rangle^+ : Tr(y) < \infty\}$ by [187, Lemma V.2.16].

Let $\varepsilon > 0$. By the density of Ne_BN in $L^2(\langle N, e_B \rangle, Tr)$ there is an $x \in Ne_BN$ with $\|y^{1/2} - x\|_{2,Tr} < \varepsilon$. For each $w \in \langle N, e_B \rangle$ with $\|w\| \leq 1$,

$$|Tr((y - xx^*)w)| \leq |Tr((y^{1/2} - x)y^{1/2}w)| + |Tr(wx(y^{1/2} - x^*))|$$
$$\leq \|y^{1/2} - x\|_{2,Tr}\|y^{1/2}w\|_{2,Tr} + \|y^{1/2} - x\|_{2,Tr}\|wx\|_{2,Tr}$$
$$\leq \|y^{1/2} - x\|_{2,Tr}(Tr(y)^{1/2} + \|x\|_{2,Tr}) < 2\varepsilon + \varepsilon^2, \quad (4.5.5)$$

using the inequalities

$$\|y^{1/2}w\|_{2,Tr}^2 = Tr(y^{1/2}ww^*y^{1/2}) \leq \|w\|^2 Tr(y) \leq Tr(y) \qquad (4.5.6)$$

and

$$\|wx\|_{2,Tr}^2 = Tr(x^*w^*wx) \leq \|w\|^2 Tr(x^*x) \leq \|x\|_{2,Tr}^2. \qquad (4.5.7)$$

Taking the supremum over $w \in \langle N, e_B \rangle$ with $\|w\| \leq 1$ gives $\|y - xx^*\|_{1,Tr} \leq 2\varepsilon + \varepsilon^2$ by the duality between $\langle N, e_B \rangle$ and $L^1(\langle N, e_B \rangle, Tr)$ (see [187, Theorem V.2.18]). This proves the lemma since $xx^* \in Ne_BN$. □

The pull-down map Φ is lifted from a map of Ne_BN into N to a map of $L^1(\langle N, e_B \rangle, Tr)$ into $L^1(N)$ by continuity, completeness and density. Note that the map $x \mapsto xe_B : N \to Ne_BN$ lifts to an isometry from $L^1(N)$ into $L^1(\langle N, e_B \rangle, Tr)$.

Theorem 4.5.3. *Let B be a von Neumann subalgebra of a finite von Neumann algebra N with a faithful normal trace τ. There is an N-bimodule *-map Φ from $L^1(\langle N, e_B \rangle, Tr)$ to $L^1(N)$ such that*

$$\Phi(xe_By) = xy \quad \text{for all} \quad x, y \in N, \qquad (4.5.8)$$

satisfying:

(i) *$\tau(\Phi(x)) = Tr(x)$ for all $x \in L^1(\langle N, e_B \rangle, Tr)$;*

(ii) *$e_B\Phi(e_Bx) = e_Bx$ for all $x \in L^1(\langle N, e_B \rangle, Tr)$;*

(iii) *$\|\Phi(x)\|_1 \leq \|x\|_{1,Tr}$ for all $x \in L^1(\langle N, e_B \rangle, Tr)$;*

(iv) *Φ maps e_BN onto N with $\|\Phi(x)\|_2 = \|x\|_{2,Tr}$ for all $x \in e_BN$;*

(v) *the equations $\Phi(xe_By) = xJy^*\xi$ and $\Phi(ye_Bx) = Jx^*Jy\xi$ hold and both are in $L^2(N)$ for all $x \in N$ and $y \in \langle N, e_B \rangle$.*

Proof. The map Φ lifts from Ne_BN to $L^1(\langle N, e_B \rangle, Tr)$ by continuity, density and the completeness of $L^1(N)$ (Lemmas 4.5.1 and 4.5.2). The properties (i) to (iv) follow directly from Lemma 4.5.1.
(v)   The algebraic case when $y \in Ne_BN$ is proved first. Let $x \in N$ and let $y = ve_Bw$ with $v, w \in N$. Then

$$\Phi(xe_By) = \Phi(xe_Bve_Bw) = \Phi(x\mathbb{E}_B(v)e_Bw) = x\mathbb{E}_B(v)w$$
$$= xJw^*\mathbb{E}_B(v^*)e_B\xi = xJw^*e_Bv^*e_B\xi = xJy^*\xi, \qquad (4.5.9)$$

since $z = Jz^* = Jz^*\xi = Jz^*e_B\xi$ for all $z \in N$. Similarly

$$\Phi(ye_Bx) = \Phi(ve_Bwe_Bx) = \Phi(v\mathbb{E}_B(w)e_Bx)$$
$$= v\mathbb{E}_B(w) = Jx^*\mathbb{E}_B(w)^*v^* = Jx^*Jv\mathbb{E}_B(w)$$
$$= Jx^*Jve_Bw\xi = Jx^*Jy\xi. \qquad (4.5.10)$$

Let $x \in N$ and $y \in \langle N, e_B \rangle$. By the Kaplansky Density Theorem (Theorem 2.2.3), there is a bounded net $\{y_\lambda : \lambda \in \Lambda\}$ in $N e_B N$ with y_λ converging in the strong-* topology to y. By Remark 4.3.14 based on the equivalence of the strong and σ-strong topologies on closed balls and on the $\widetilde{Tr}$ definition of the trace, $\|(y_\lambda - y)e_B\|_{2,Tr}$ and $\|e_B(y_\lambda - y)\|_{2,Tr} = \|(y_\lambda^* - y^*)e_B\|_{2,Tr}$ tend to 0 over the net. If $z \in \langle N, e_B \rangle$ with $\|z\| \leq 1$, then by the Cauchy–Schwarz inequality,

$$|Tr(zxe_B(y_\lambda - y))| \leq \|zxe_B\|_{2,Tr}\|e_B(y_\lambda - y)\|_{2,Tr}$$
$$\leq \|xe_B\|_{2,Tr}\|e_B(y_\lambda - y)\|_{2,Tr}. \qquad (4.5.11)$$

Taking the supremum over all such z gives

$$\|xe_B(y_\lambda - y)\|_{1,Tr} \leq \|xe_B\|_{2,Tr}\|e_B(y_\lambda - y)\|_{2,Tr} \to 0 \qquad (4.5.12)$$

over Λ. Similarly $\|(y_\lambda - y)e_B x\|_{1,Tr} \to 0$ over Λ. Finally

$$\|\Phi(xe_B y_\lambda) - xJy^*\xi\|_2 = \|xJ(y_\lambda^* - y^*)\xi\|_2 \leq \|x\|\|(y_\lambda^* - y^*)\xi\|_2 \qquad (4.5.13)$$

tends to zero, since y_λ converges strong-* to y. Since $\|\cdot\|_1 \leq \|\cdot\|_2$ on N, we have $\Phi(xe_B y) = xJy^*\xi \in L^2(N)$. In a similar way $\Phi(ye_B x) = Jx^*Jy\xi \in L^2(N)$. $\square$

Note that the ideal $\mathfrak{M}_{Tr}$ is contained in $L^1(\langle N, e_B \rangle, Tr)$. We have not found a single formula for Φ from $\langle N, e_B \rangle e_B \langle N, e_B \rangle$ to N, nor a general formula that defines Φ on $\langle N, e_B \rangle^+$. In general the space $\Phi(\langle N, e_B \rangle e_B \langle N, e_B \rangle)$ contains unbounded operators affiliated with N. Such operators are discussed in Appendix B, particularly in Sections B.4 and B.5.

Chapter 5

Projections and partial isometries

5.1 Introduction

In this chapter we consider various useful technical results concerning projections, partial isometries and $\|\cdot\|_2$-norm estimates arising from the polar decomposition. Many of these will be crucial when we come to discuss perturbations in Chapters 9 and 10. Everything in this chapter is well known and can be found in [36, 188], although the proofs are sometimes different from the originals.

We begin by considering two projections, and we show that the algebra they generate has only irreducible representations of dimensions 1 and 2. This reduces the study of such a pair to the 2×2 matrices, and we use this in obtaining Theorem 5.2.5. We also draw attention to Lemma 5.2.7 which contains estimates that we use repeatedly in the sequel. The third section is devoted to estimates concerning pairs of projections.

The concluding Sections 5.5 and 5.6 are devoted to some work of Kadison, [99, 100], which constructs abelian projections with desirable properties in type I von Neumann algebras. These will be relevant in Chapter 9 since masas in II_1 factors can lead to problems in type I von Neumann algebras.

5.2 Comparison of two projections

We begin by proving that the von Neumann algebra generated by two projections is, in general, a direct sum of an abelian algebra and one of type I_2. The first few lemmas lead in this direction.

Lemma 5.2.1. *Let H be a Hilbert space and let p and q be projections that generate $B(H)$. Then the dimension of H is at most 2.*

Proof. Suppose that the dimension of H is at least 3. The projections $p \wedge q$, $(1-p) \wedge q$, $p \wedge (1-q)$ and $(1-p) \wedge (1-q)$ all commute with both p and q, and thus must all be 0. This means that the intersections of $\ker p$ and $\ker(1 - p)$ with $\ker q$ and $\ker(1 - q)$ are all trivial. Now consider the operator $(1 - p)qp$. If ξ is in its kernel, then $qp\xi$ is in $\ker(1 - p)$ and $\ker(1 - q)$, so $qp\xi = 0$. Thus $p\xi \in \ker(1-p) \cap \ker q$, so $p\xi = 0$. This shows that $(1-p)qp$ maps pH injectively into $(1-p)H$. Any vector η orthogonal to the range of $(1-p)qp$ is in the kernel of the adjoint $pq(1 - p)$, and we may repeat the previous argument to conclude that $(1 - p)\eta = 0$. Thus no non-zero vector in $(1 - p)H$ is orthogonal to the range of $(1 - p)qp$, showing that this range is dense in $(1 - p)H$. Thus the source and range projections of $(1 - p)qp$ are respectively p and $1 - p$, so these are equivalent projections in $B(H)$. If we let K be the range of p then, up to unitary isomorphism, $H = K \oplus K$, $p = \left(\begin{smallmatrix} 1 & 0 \\ 0 & 0 \end{smallmatrix}\right)$ and $q = \left(\begin{smallmatrix} a & b \\ b^* & c \end{smallmatrix}\right)$ with $a, b, c \in B(K)$. Note that the dimension of K is at least 2. The injectivity of $(1 - p)qp$ on pH implies that $b^* \in B(K)$ is injective, and injectivity of b is implied by a similar argument for $pq(1 - p)$. Thus b is injective with dense range so the partial isometry in its polar decomposition is a unitary, allowing us to write $b = du$ with $d \geq 0$ and u unitary. Thus q has the form $\left(\begin{smallmatrix} a & du \\ u^*d & c \end{smallmatrix}\right)$. Conjugation by the unitary $\left(\begin{smallmatrix} 1 & 0 \\ 0 & u \end{smallmatrix}\right)$ shows that $B(H)$ is generated by the projections $p = \left(\begin{smallmatrix} 1 & 0 \\ 0 & 0 \end{smallmatrix}\right)$ and $\tilde{q} = \left(\begin{smallmatrix} a & d \\ d & e \end{smallmatrix}\right)$ where $e = ucu^*$ and $a, d, e \geq 0$. If we express $\tilde{q}^2 = \tilde{q}$ in terms of the matrix entries, then we find that $d = (a - a^2)^{1/2}$ and $e = 1 - a$. Since K has dimension at least 2, there is a nontrivial projection $f \in B(K)$ which commutes with a. If a is a multiple of the identity, then f can be any rank one projection; otherwise take f to be a nontrivial spectral projection for a. The element $\left(\begin{smallmatrix} f & 0 \\ 0 & f \end{smallmatrix}\right) \in B(H)$ then commutes with p and $\tilde{q}$, and hence with $B(H)$. This contradiction proves that $\dim H$ is at most 2. $\qquad\square$

Lemma 5.2.2. *Let p and q be projections in a von Neumann algebra M, and let N be the von Neumann algebra generated by p, q and 1. Then there exist abelian von Neumann algebras A_1 and A_2 such that N is isomorphic to $A_1 \oplus (A_2 \otimes \mathbb{M}_2)$.*

Proof. Let B be the C^*-algebra generated by p, q and 1. Then $B'' = N$. By Lemma 5.2.1, any irreducible representation of B must be on a Hilbert space of dimension at most 2. If we take a countable dense set of pure states of B and let π be the direct sum of their GNS representations, then π embeds B into a countable direct sum of copies of $\mathbb{M}_2$, equivalently $\ell^\infty \overline{\otimes} \mathbb{M}_2$. Thus π^{**} embeds B^{**} into $(\ell^\infty)^{**} \overline{\otimes} \mathbb{M}_2$. It follows that B^{**} has only irreducible representations of dimensions at most 2, and the same is true of N which is isomorphic to $B^{**}z$ for a central projection $z \in B^{**}$. Thus N is a direct sum of von Neumann algebras of types I_1 and I_2. $\qquad\square$

The next lemma constitutes part of the proof of the following theorem.

Lemma 5.2.3. *In $\mathbb{M}_2$, let $P = \left(\begin{smallmatrix} 1 & 0 \\ 0 & 0 \end{smallmatrix}\right)$ and $Q = \left(\begin{smallmatrix} c^2 & cse^{-i\phi} \\ cse^{i\phi} & s^2 \end{smallmatrix}\right)$, where $c = \cos\theta$, $s = \sin\theta$, $0 < \theta < \pi/2$. Then there exist a partial isometry $W \in \mathbb{M}_2$ and a unitary $U \in \mathbb{M}_2$ satisfying*

(i) $W^*W = P$, $WW^* = Q$;

(ii) $W|P - Q| = |P - Q|W$ and $U|P - Q| = |P - Q|U$;

(iii) $|W - P|, |W - Q| \leq \sqrt{2}|P - Q|$;

(iv) $UPU^* = Q$;

(v) $|I_2 - U| \leq \sqrt{2}|P - Q|$.

Proof. Let $W = \left(\begin{smallmatrix} c & 0 \\ se^{i\phi} & 0 \end{smallmatrix} \right)$. Then (i) is a simple matrix computation. Since $P - Q = \left(\begin{smallmatrix} s^2 & -cse^{-i\phi} \\ -cse^{i\phi} & -s^2 \end{smallmatrix} \right)$, we see that $P - Q = sV$ for a unitary V, and so $|P - Q| = sI_2$. Thus (ii) holds.

Now $W - P = \left(\begin{smallmatrix} c-1 & 0 \\ se^{i\phi} & 0 \end{smallmatrix} \right)$, and

$$\|W - P\|^2 = (c - 1)^2 + s^2 = 2 - 2c \leq 2 - 2c^2 = 2s^2, \tag{5.2.1}$$

and so

$$|W - P| \leq \|W - P\|I_2 \leq \sqrt{2}\, sI_2 = \sqrt{2}|P - Q|. \tag{5.2.2}$$

This proves the first inequality in (iii). For the second, observe that

$$W - Q = WP - WW^* = W(P - W^*), \tag{5.2.3}$$

and so

$$\|W - Q\| \leq \|P - W^*\| = \|P - W\| \leq \sqrt{2}\, s \tag{5.2.4}$$

from (5.2.1). This gives $|W - Q| \leq \sqrt{2}\, sI_2$, as required.

Define $U = \left(\begin{smallmatrix} c & -se^{-i\phi} \\ se^{i\phi} & c \end{smallmatrix} \right)$. Then $UP = W$, so $UPU^* = WW^* = Q$, showing (iv). The eigenvalues of U are $c \pm is$ and so

$$\|I_2 - U\|^2 = (1 - c)^2 + s^2 = 2 - 2c \leq 2s^2, \tag{5.2.5}$$

and (v) follows since $|P - Q| = sI_2$. □

Remark 5.2.4. It is useful to note that, in Lemma 5.2.3, the matrix W is $QP/\|QP\|$, while U can be obtained from W by $U = W + U_1WU_2$, where

$$U_1 = \begin{pmatrix} 0 & -e^{-2i\phi} \\ 1 & 0 \end{pmatrix}, \qquad U_2 = \begin{pmatrix} 0 & 1 \\ 1 & 0 \end{pmatrix}. \tag{5.2.6}$$

Consequently, if s, c and ϕ are regarded as measurable functions on some measure space, then the matrices U and W are also measurable. □

Theorem 5.2.5. *Let M be a von Neumann algebra, and let $e, f \in M$ be finite equivalent projections. Then there exist a partial isometry $w \in M$ and a unitary $u \in M$ satisfying*

(i) $w^*w = e$, $ww^* = f$;

(ii) $w|e - f| = |e - f|w$ and $u|e - f| = |e - f|u$;

(iii) $|w - e|, |w - f| \leq \sqrt{2}|e - f|$;

(iv) $ueu^* = f$;

(v) $|1 - u| \leq \sqrt{2}|e - f|$.

Proof. By cutting M by the projection $e \vee f$, we may assume that M is a finite von Neumann algebra. Let N be the von Neumann algebra generated by e and f. By Lemma 5.2.2, there is a central projection $z \in N$ such that zN is abelian and $(1 - z)N$ is type I_2. We will first consider the latter summand, and so we will assume that $z = 0$, but drop the assumption that e and f are equivalent, since in general this is not obviously so for $(1 - z)e$ and $(1 - z)f$ (z need not be central in M).

Since N is type I_2, there is a measure space (Ω, μ) such that N is isomorphic to $L^\infty(\Omega, M_2)$, the space of M_2-valued bounded measurable functions. The projections e and f are then projection valued functions $e(\omega)$ and $f(\omega)$. After discarding a null set, we have that $e(\omega)$ and $f(\omega)$ generate M_2 for all $\omega \in \Omega$, and thus $e(\omega)$ and $f(\omega)$ must be noncommuting rank 1 projections. The general form of a rank 1 projection in M_2 is $\begin{pmatrix} c^2 & cse^{-i\phi} \\ cse^{i\phi} & s^2 \end{pmatrix}$, as in Lemma 5.2.3. Thus $e(\omega)$ has this form where c, s and ϕ are measurable functions of ω. If we define a unitary by

$$u = \begin{pmatrix} c & -se^{-i\phi} \\ se^{i\phi} & c \end{pmatrix}$$

then $u(\omega)^*e(\omega)u(\omega) = \begin{pmatrix} 1 & 0 \\ 0 & 0 \end{pmatrix}$, and we may thus assume that $e(\omega)$ is the constant function $\omega \mapsto \begin{pmatrix} 1 & 0 \\ 0 & 0 \end{pmatrix}$. We are now in the situation of Lemma 5.2.3 with $P = e(\omega)$ and $Q = f(\omega)$. In view of Remark 5.2.4, we see that Lemma 5.2.3 constructs the operators u and w satisfying (i)–(v) pointwise on Ω, and hence globally.

In the general case, the above discussion shows that $(1 - z)e$ and $(1 - z)f$ are equivalent projections in N. By applying the centre-valued trace for M, we see that ze and zf are equivalent in M. Since z commutes with e and f we also have equivalence in zMz. Restricting to zMz, we may now assume that e and f are commuting equivalent projections. On that part of the space where $e = f$ we may take $u = w = 1$, and this reduces to the case when $ef = 0$. If we decompose 1 as $e + f + (1 - e - f)$, then relative to these projections we may write

$$e = \begin{pmatrix} 1 & 0 & 0 \\ 0 & 0 & 0 \\ 0 & 0 & 0 \end{pmatrix}, \qquad f = \begin{pmatrix} 0 & 0 & 0 \\ 0 & 1 & 0 \\ 0 & 0 & 0 \end{pmatrix}. \tag{5.2.7}$$

We may then define

$$w = \begin{pmatrix} 0 & 0 & 0 \\ 1 & 0 & 0 \\ 0 & 0 & 0 \end{pmatrix}, \qquad u = \begin{pmatrix} 0 & -1 & 0 \\ 1 & 0 & 0 \\ 0 & 0 & 1 \end{pmatrix}. \tag{5.2.8}$$

Since

$$|e - f| = \begin{pmatrix} 1 & 0 & 0 \\ 0 & 1 & 0 \\ 0 & 0 & 0 \end{pmatrix}, \tag{5.2.9}$$

the verifications of (i)–(v) are routine matrix calculations based on (5.2.7)–(5.2.9). □

In [36], Connes comments that this result is particularly important as it permits good estimates of $norm(u - 1)$ and $norm(v - e)$ where $norm(\cdot)$ is any norm on M satisfying $norm(x) = norm(|x|)$ for $x \in M$, and $norm(x) \leq norm(y)$ for commuting elements $x, y \in M$ with $0 \leq x \leq y$. For a II_1 factor M, this applies to $\| \cdot \|_p$, $p \geq 1$, and particularly to the special case $p = 2$. The constant $\sqrt{2}$ in Theorem 5.2.5 (iii), (v) is best possible, as may be seen by considering the projections

$$e = \begin{pmatrix} 1 & 0 \\ 0 & 0 \end{pmatrix}, \qquad f = \begin{pmatrix} 1/2 & 1/2 \\ 1/2 & 1/2 \end{pmatrix}$$

in $\mathbb{M}_2$.

We will return to Theorem 5.2.5 shortly, but we first establish some estimates for the polar decomposition of an operator. These will be used several times subsequently when we discuss perturbations in Chapter 9. We begin with an estimate in the operator norm.

Lemma 5.2.6. *Let N be a von Neumann algebra and let $k \in [0, 1)$. If $x \in N$ has polar decomposition $x = uh$ and satisfies $\|1 - x\| \leq k$, then*

$$\|1 - u\| \leq 2^{1/2}(1 - (1 - k^2)^{1/2})^{1/2} \leq 2^{1/2}k. \tag{5.2.10}$$

Proof. The hypothesis implies that x is invertible and so $u \in N$ is a unitary. Then

$$\|u - h\| = \|u^* - h\| = \|1 - uh\| \leq k. \tag{5.2.11}$$

Since u generates an abelian C^*-algebra, the norm of $1 - u$ is determined by states on N that are extensions of characters on $C^*(u, 1)$. If ϕ is such a state and $\phi(u) = \alpha + i\beta$ then $\alpha^2 + \beta^2 = 1$ and

$$|\beta| \leq |\phi(u) - \phi(h)| \leq \|u - h\| \leq k \tag{5.2.12}$$

from (5.2.11) and the fact that $\phi(h) \in \mathbb{R}$ since $h \geq 0$. Using (5.2.12), we see that

$$|1 - \phi(u)|^2 = |1 - \alpha - i\beta|^2 = (1 - \alpha)^2 + \beta^2 = 1 - 2\alpha + \alpha^2 + \beta^2$$
$$= 2 - 2\alpha = 2(1 - (1 - \beta^2)^{1/2})$$
$$\leq 2(1 - (1 - k^2)^{1/2}) \leq 2k^2, \tag{5.2.13}$$

and (5.2.10) follows from (5.2.13). □

The following gives estimates in the $\| \cdot \|_2$-norm for a finite von Neumann algebra. The subsequent remark explains the modifications necessary to obtain the same results in a semifinite von Neumann algebra.

Lemma 5.2.7. *Let N be a finite von Neumann algebra with a faithful normal trace τ. Let $w \in N$ have polar decomposition $w = vk$ where $k = (w^*w)^{1/2}$, and let $p = v^*v$ and $q = vv^*$ be the initial and final projections of v. If $e \in N$ is a projection satisfying $ew = w$, then*

(i) $\|p - k\|_2 \leq \|e - w\|_2$;

(ii) $\|e - q\|_2 \leq \|e - w\|_2$;

(iii) $\|e - v\|_2 \leq 2\|e - w\|_2$.

Proof. The first inequality is equivalent to

$$\tau(p + k^2 - 2pk) \leq \tau(e + w^*w - w - w^*), \tag{5.2.14}$$

since $ew = w$. Since $p = v^*v$, we have

$$k^2 = w^*w = kpk \leq k^2, \tag{5.2.15}$$

showing that $(1 - p)k = 0$ and $pk = k$. Thus (5.2.14) is equivalent to

$$\tau(w + w^*) \leq \tau(e - p + 2k). \tag{5.2.16}$$

The map $x \mapsto \tau(k^{1/2}xk^{1/2})$ is a positive linear functional on N whose norm is $\tau(k)$. Thus

$$|\tau(w)| = |\tau(w^*)| = |\tau(vk)| = |\tau(k^{1/2}vk^{1/2})| \leq \tau(k). \tag{5.2.17}$$

The range of e contains the range of w, so $e \geq q$. Thus

$$\tau(e) \geq \tau(q) = \tau(p), \tag{5.2.18}$$

and so (5.2.16) follows from (5.2.17) and (5.2.18), proving (i).

Since $eq = q$, the second inequality is equivalent to

$$-\tau(q) \leq \tau(w^*w - ew - w^*e) = \tau(k^2 - w - w^*), \tag{5.2.19}$$

or to

$$\tau(w + w^*) \leq \tau(k^2 + p), \tag{5.2.20}$$

since $\tau(q) = \tau(p)$. By (5.2.17),

$$\tau(w + w^*) \leq 2\tau(k) = \tau(p + k^2 - (k - p)^2) \leq \tau(p + k^2), \tag{5.2.21}$$

proving (5.2.20) and establishing (ii).

The last inequality is

$$\|e - v\|_2 \leq \|e - vk\|_2 + \|v(k - p)\|_2$$
$$\leq \|e - w\|_2 + \|k - p\|_2 \leq 2\|e - w\|_2, \tag{5.2.22}$$

by (i). $\qquad \square$

Remark 5.2.8. (i) We point out two special cases of Lemma 5.2.7 in which e is taken to be either q or 1. We also note that 2 is the best possible constant in (iii), as may be seen by taking $e = 1$ and $w = -\varepsilon 1$ for $\varepsilon > 0$.

(ii) Lemma 5.2.7 is also valid for semifinite von Neumann algebras with a faithful normal semifinite trace under the additional hypothesis that p, q and e are all finite projections. Then fNf is a finite von Neumann algebra, where $f = e \vee p \vee q$, and the proof of Lemma 5.2.7 applies to fNf which contains w, v, k, p, q, and e.   □

The next lemma gives an estimate on the operators appearing in the polar decomposition of a product of projections, and it forms part of the argument of Theorem 10.4.1.

Lemma 5.2.9. *Let N be a von Neumann algebra with a faithful normal semifinite trace Tr. Let $e, f \in N$ be finite projections and let $ef = (efe)^{1/2}v$ be the polar decomposition of ef. Then*

(i) $|e - v|^2, |f - v|^2 \le 2|e - f|^2$;

(ii) $\|e - v\|_{2,Tr}, \|f - v\|_{2,Tr} \le \sqrt{2}\|e - f\|_{2,Tr}$.

Proof. (i)   If we cut N by the finite projection $e \vee f$, then we may assume that Tr is a finite trace. By Lemma 5.2.2, the von Neumann algebra generated by e and f is a direct sum of an abelian algebra and one of type I_2. The inequalities of (i) are obvious in the abelian summand, and so we may reduce to the case when the algebra is $A \otimes M_2$ for an abelian von Neumann algebra A. As in Theorem 5.2.5, we may assume that $A \otimes M_2$ is $L^\infty(\Omega, M_2)$ for some measure space (Ω, μ), and that

$$e(\omega) = \begin{pmatrix} 1 & 0 \\ 0 & 0 \end{pmatrix}, \quad f(\omega) = \begin{pmatrix} c^2 & cse^{-i\phi} \\ cse^{i\phi} & s^2 \end{pmatrix},$$

where c, s and ϕ are measurable functions of ω. By Lemma 5.2.3, there is a partial isometry $w \in L^\infty(\Omega, M_2)$ such that

$$|w(\omega) - e(\omega)|, |w(\omega) - f(\omega)| \le \sqrt{2}|e(\omega) - f(\omega)| = \sqrt{2}sI_2. \qquad (5.2.23)$$

The proof of that lemma shows that

$$w(\omega) = \begin{pmatrix} c & 0 \\ se^{i\phi} & 0 \end{pmatrix},$$

and it is a direct matrix calculation to show that the polar decomposition of ef is $ef = (efe)^{1/2}w^*$, since $(efe)(\omega) = c(\omega)^2 e(\omega)$. Thus $v(\omega) = w(\omega)^*$. From (5.2.2) and (5.2.4), we obtain

$$|e(\omega) - v(\omega)|^2, |f(\omega) - v(\omega)|^2 \le 2s(\omega)^2 I_2 = 2|e(\omega) - f(\omega)|^2. \qquad (5.2.24)$$

The inequalities of (i) follow, since (5.2.24) shows that they hold pointwise over Ω.

(ii)   Apply the trace Tr to the inequalities of (i).   □

The following is a special case of Theorem 5.2.5, proved using an elementary algebraic method from Banach algebra K-theory [163, p. 22]. Note that the estimate is not as good, but the method is simpler.

Lemma 5.2.10. (i) *Let $k \in [0,1)$ and let p and q be projections in a von Neumann algebra N satisfying $\|p - q\| \leq k$. Then there is a unitary $v \in N$ with $p = vqv^*$ and*

$$\|1 - v\| \leq 2^{1/2}(1 - (1 - k^2)^{1/2})^{1/2} \leq 2^{1/2}k. \tag{5.2.25}$$

(ii) *Let p and q be projections in a semifinite von Neumann algebra N with a faithful normal semifinite trace Tr. If $\text{Tr}(p)$ and $\text{Tr}(q)$ are finite and if $\|p - q\|_{2,\text{Tr}} \leq k < 1$, then there is a partial isometry $v \in N$ satisfying*

$$pv = vq, \qquad vv^*p = pvv^*, \qquad v^*vq = qv^*v, \tag{5.2.26}$$
$$\|1 - vv^*\|_{2,\text{Tr}} \leq k, \qquad \|1 - v\|_{2,\text{Tr}} \leq 2k. \tag{5.2.27}$$

Proof. The first part of the proof is common to both cases. Let $x = pq + (1 - p)(1 - q)$. Then $px = pq = xq$ and

$$1 - x = p + q - 2pq = (2p - 1)(p - q) = (p - q)(2q - 1). \tag{5.2.28}$$

The operator $2q - 1$ is a unitary, so from (5.2.28) we obtain the two estimates

$$\|p + q - 2pq\| = \|1 - x\| \leq \|p - q\|, \tag{5.2.29}$$
$$\|p + q - 2pq\|_{2,\text{Tr}} = \|1 - x\|_{2,\text{Tr}} \leq \|p - q\|_{2,\text{Tr}}. \tag{5.2.30}$$

Let $x = hv$ be the polar decomposition of x, where $h^2 = xx^*$ and v is a partial isometry. Then

$$ph^2 = pxx^* = xqx^* = xx^*p = h^2p, \tag{5.2.31}$$

and so p commutes with h^2 and thus with h. Since vv^* is the range projection of x and of xx^*, it follows that $pvv^* = vv^*p$. A similar calculation shows that $v^*vq = qv^*v$. We now consider the two cases separately.

(i) In this case x is invertible and so v is a unitary. Then

$$q = x^{-1}px = v^*h^{-1}phv = v^*pv, \tag{5.2.32}$$

since p commutes with h. Thus $p = vqv^*$. Moreover, by Lemma 5.2.6 applied to $x^* = v^*h$, we obtain (5.2.25) from (5.2.29).

(ii) Let us first suppose that N is finite. From properties of the polar decomposition, v is the strong limit of the sequence $\{(h + \frac{1}{n})^{-1}x\}_{n=1}^{\infty}$. Since p commutes with h and $px = xq$, we may multiply by $(h + \frac{1}{n})^{-1}$ and let $n \to \infty$ to obtain $pv = vq$. Lemma 5.2.7 (iii) gives $\|1 - v\|_{2,\text{Tr}} \leq 2\|1 - x\|_{2,\text{Tr}}$, so $\|1 - v\|_{2,\text{Tr}} \leq 2k$ follows from (5.2.30). The other estimate of (5.2.27) is $\|1 - vv^*\|_{2,\text{Tr}} \leq k$, which is a consequence of (5.2.30) and Lemma 5.2.7 (ii).

For the general case, let f be the finite projection $p \vee q$. Then the preceding discussion gives a partial isometry $v \in fNf$ satisfying (5.2.26) and (5.2.27) when 1 is replaced by f. The result follows by replacing v by $v + (1 - f) \in N$. $\square$

We close this section with a result concerning the spatial implementation of *-homomorphisms which are close in norm to the identity. This will be needed in Theorem 9.3.1.

Theorem 5.2.11. *Let N be a II_1 factor with a unital von Neumann subalgebra A. If $\alpha \in [0, 1)$ and $\phi\colon A \to N$ is a unital *-homomorphism satisfying $\|\phi - I\| \leq \alpha$, then there exists a unitary $v \in N$ such that $\phi(a) = vav^*$ for $a \in A$, and $\|1 - v\| \leq \sqrt{2}(1 - (1 - \alpha^2)^{1/2})^{1/2}$.*

Proof. Denote the unitary group of A by $\mathcal{U}(A)$ and consider the set

$$K = \overline{\mathrm{conv}}^w\{\phi(u^*)u\colon u \in \mathcal{U}(A)\} \subset N.$$

For each $u \in \mathcal{U}(A)$, $\|\phi(u^*) - u^*\| \leq \alpha$, and so $\|\phi(u^*)u - 1\| \leq \alpha$. Thus K is contained in the norm closed ball of radius α centred at 1, and so every operator in K is invertible in N. This set is weakly compact when viewed as a subset of $L^2(N)$, and so there is a unique element $t \in K$ of minimal $\| \cdot \|_2$-norm. Since each $u \in \mathcal{U}(A)$ induces a $\| \cdot \|_2$-isometric map $k \mapsto \phi(u^*)ku$ of K to itself, the uniqueness shows that each such map fixes t. Thus $\phi(u)t = tu$ for $u \in \mathcal{U}(A)$ and, by linearity, we also have

$$\phi(a)t = ta \quad \text{and} \quad t^*\phi(a)t = t^*ta, \quad a \in A. \tag{5.2.33}$$

Taking $a \in A^+$ in the second of these equations shows that t^*t lies in $A' \cap N$, and the same conclusion holds for $(t^*t)^{1/2}$. Let $t = v(t^*t)^{1/2}$ be the polar decomposition of t in N, where v is a unitary since t is invertible. Substitution for t in the first equality of (5.2.33) yields

$$\phi(a)v(t^*t)^{1/2} = v(t^*t)^{1/2}a = va(t^*t)^{1/2}, \quad a \in A, \tag{5.2.34}$$

and so $\phi(a)v = va$. Hence $\phi(a) = vav^*$ for $a \in A$. The norm estimate for $1 - v$ follows immediately from (5.2.10) with k replaced by α. $\qquad \square$

5.3 Approximations of projections

The following lemma on approximating a positive element by a projection is well known: see [21, 36, 48, 139]. The proof below is that of Christensen [21].

The following two functions appear frequently in estimations: the first is

$$G(\delta) = \delta + 2^{-1}(1 - (1 - 4\delta)^{1/2}), \quad 0 \leq \delta \leq 1/4, \tag{5.3.1}$$

while the second is

$$H(k) = 2^{1/2}(1 - (1 - k^2)^{1/2})^{1/2}, \quad 0 \leq k \leq 1. \tag{5.3.2}$$

The first is used immediately and both will appear in Chapter 9 where their properties will be discussed.

Lemma 5.3.1. *Let N be a semifinite von Neumann algebra with faithful normal semifinite trace Tr. Let h be in N with $0 \le h \le 1$ and let f be the spectral projection of h corresponding to the interval $[1/2, 1]$.*

(i) *If $Tr(h^*h) < \infty$, then*

$$\|e - f\|_{2,Tr} \le 2\|e - h\|_{2,Tr} \tag{5.3.3}$$

for all projections $e \in N$ of finite trace.

(ii) *The increasing function*

$$G(\delta) = \delta + 2^{-1}(1 - (1 - 4\delta)^{1/2}) \tag{5.3.4}$$

on the interval $[0, 1/4]$ satisfies

$$\|f - p\| \le G(\|h - p\|) \tag{5.3.5}$$

for all projections $p \in N$ with $\|h - p\| \le 1/4$.

Proof. (i) The functional calculus for the projection f associated with the interval $[1/2, 1]$ for h implies that

$$\frac{1}{2}f \le h \quad \text{and} \quad \frac{1}{2}(1 - f) \le 1 - h \tag{5.3.6}$$

so that

$$f \le 4h^2 \quad \text{and} \quad (1 - f) \le 4(1 - h)^2. \tag{5.3.7}$$

This inequality, combined with orthogonality and the positivity of Tr, implies that

$$
\begin{aligned}
\|e - f\|_{2,Tr}^2 &= \|(1 - e)f - e(1 - f)\|_{2,Tr}^2 \\
&= \|(1 - e)f\|_{2,Tr}^2 + \|e(1 - f)\|_{2,Tr}^2 \\
&= Tr((1 - e)f(1 - e)) + Tr(e(1 - f)e) \\
&\le 4Tr((1 - e)h^2(1 - e)) + 4Tr(e(1 - h)^2e) \\
&= 4\|(1 - e)h\|_{2,Tr}^2 + 4\|e(1 - h)\|_{2,Tr}^2 \\
&= 4\|(1 - e)h - e(1 - h)\|_{2,Tr}^2 \\
&= 4\|e - h\|_{2,Tr}^2.
\end{aligned}
\tag{5.3.8}
$$

(ii)   Let $h = p + x$ so $\|x\| = \|h - p\| = \delta$. Since $0 \le h \le 1$, properties of self-adjoint operators yield

$$
\begin{aligned}
0 \le h - h^2 &= x - xp - px - x^2 \\
&= x(1/2 - p) + (1/2 - p)x - x^2 \\
&\le x(1/2 - p) + (1/2 - p)x,
\end{aligned}
\tag{5.3.9}
$$

implying that $\|h - h^2\| \leq \delta$, since $\|1/2 - p\| \leq 1/2$. The spectrum $\sigma(h)$ of h lies in the set $[0, \alpha] \cup [1 - \alpha, 1]$, where α is the smaller root $2^{-1}(1 - (1 - 4\delta)^{1/2})$ of the equation $\lambda^2 - \lambda + \delta = 0$. Then the inequality $\|h - f\| \leq \alpha$ is a simple consequence of the functional calculus. Hence $\|f - p\| \leq \delta + 2^{-1}(1 - (1 - 4\delta)^{1/2}) = G(\delta)$. $\square$

The constant 2 in Lemma 5.3.1 is best possible as the following elementary example shows. Let e and f be orthogonal projections in a finite factor N with $e + f = 1$, let $0 < \lambda < 1/2$ and let $h = \lambda e + (1 - \lambda)f$. The spectral projection of h corresponding to $[1/2, 1]$ is f. Here $\|e - f\|^2_{2,Tr} = 1$ and $\|e - h\|^2_{2,Tr} = (1 - \lambda)\|e - f\|^2_{2,Tr} = 1 - \lambda$ so the constant in Lemma 5.3.1 is no smaller than $(1 - \lambda)^{-1}$ for $0 < \lambda \leq 1/2$ and 2 is best possible.

Replacing $1/2$ in the hypothesis of Lemma 5.3.1 by k for $0 < k < 1$ gives a similar result. In this case f corresponds to the projection for h on $[k, 1]$, and the inequality

$$\|e - f\|_{2,Tr} \leq \max\{k^{-1}, (1 - k)^{-1}\}\|e - h\|_{2,Tr} \tag{5.3.10}$$

holds by the same method.

It is sometimes helpful to approximate a projection by a projection in a masa with the same trace.

Lemma 5.3.2. *Let B be a masa in a type II_1 factor N with trace τ. If e is a projection in N, then there is a projection f in B with $\tau(f) = \tau(e)$ and*

$$\|e - f\|_2 \leq 2^{3/2}\|e - \mathbb{E}_B(e)\|_2. \tag{5.3.11}$$

Proof. Let f_0 be the projection corresponding to the interval $[1/2, 1]$ for $\mathbb{E}_B(e)$ given by Lemma 5.3.1. If $\tau(f_0) \geq \tau(e)$, then there is a projection p in $f_0 B$ with $\tau(p) = \tau(f_0 - e)$, since $f_0 B$ is a masa in $f_0 N f_0$. Let $f = f_0 - p$ so $\tau(f) = \tau(e)$. Here

$$\|e - f_0\|^2_2 = \tau(f_0 - e) + 2\tau(e(1 - f_0)e) \geq \tau(f_0 - e) = \tau(p). \tag{5.3.12}$$

This equation and the relation $pf_0 = p$ imply that

$$\begin{aligned}\|e - f\|^2_2 &= \|e - f_0\|^2_2 + 2\tau((e - f_0)p) + \tau(p) \\ &= \|e - f_0\|^2_2 + \tau(p(2e - 1)p) \\ &\leq \|e - f_0\|^2_2 + \tau(p) \\ &\leq 2\|e - f_0\|^2_2 \leq 8\|e - \mathbb{E}_B(e)\|^2_2.\end{aligned} \tag{5.3.13}$$

The case $\tau(e) \geq \tau(f)$ is similar but is handled by working in $(1 - f_0)B$. $\square$

We do not know if the constant $2^{3/2}$ in Lemma 5.3.2 is best possible.

It is useful to be able to approximate a pair of projections by a pair of orthogonal projections as in the following lemma.

Lemma 5.3.3. *Let B be a von Neumann subalgebra of a finite von Neumann algebra with trace τ. If e_1 and e_2 are projections in N, then there are orthogonal projections f_1 and f_2 in B such that*

$$\|e_j - f_j\|_2 \leq 2(2^{1/2} + 1)\|e_j - \mathbb{E}_B(e_j)\|_2 + 2^{1/2}\|e_1 e_2\|_2 \tag{5.3.14}$$

for $j = 1, 2$.

Proof. Suppose that $\|e_1 - \mathbb{E}_B(e_1)\|_2 \leq \|e_2 - \mathbb{E}_B(e_2)\|_2$. Let f_1 be the spectral projection for the self-adjoint element $\mathbb{E}_B(e_1)$ corresponding to $[1/2, 1]$. By (5.3.3),

$$\|f_1 - e_1\|_2 \leq 2\|e_1 - \mathbb{E}_B(e_1)\|_2. \tag{5.3.15}$$

By orthogonality,

$$\begin{aligned}
\|e_2 - (1 - f_1)e_2(1 - f_1)\|_2^2 &= \|f_1 e_2(1 - f_1) + e_2 f_1\|_2^2 \\
&= \|f_1 e_2(1 - f_1)\|_2^2 + \|e_2 f_1\|_2^2 \\
&\leq \|f_1 e_2\|_2^2 + \|e_2 f_1\|_2^2 \\
&= 2\|f_1 e_2\|_2^2
\end{aligned} \tag{5.3.16}$$

since $\|x^*\|_2 = \|x\|_2$ for all $x \in N$. By (5.3.16) and (5.3.3)

$$\begin{aligned}
\|e_2 - (1 - f_1)e_2(1 - f_1)\|_2 &= 2^{1/2}\|(f_1 - e_1)e_2 + e_1 e_2\|_2 \\
&\leq 2^{1/2}\|(f_1 - e_1)e_2\|_2 + 2^{1/2}\|e_1 e_2\|_2 \\
&\leq 2^{1/2}\|(f_1 - e_1)\|_2 + 2^{1/2}\|e_1 e_2\|_2 \\
&\leq 2^{3/2}\|e_1 - \mathbb{E}_B(e_1)\|_2 + 2^{1/2}\|e_1 e_2\|_2.
\end{aligned} \tag{5.3.17}$$

Let f_2 be the spectral projection for

$$h = (1 - f_1)\mathbb{E}_B(e_2)(1 - f_1) = \mathbb{E}_B((1 - f_1)(e_2)(1 - f_1)). \tag{5.3.18}$$

Then

$$\begin{aligned}
\|e_2 - f_2\|_2 &\leq \|e_2 - (1 - f_1)e_2(1 - f_1)\|_2 + \|(1 - f_1)e_2(1 - f_1) - f_2\|_2 \\
&\leq 2^{3/2}\|e_1 - \mathbb{E}_B(e_1)\| + 2^{1/2}\|e_1 e_2\|_2 \\
&\quad + 2\|(1 - f_1)e_2(1 - f_1) - \mathbb{E}_B((1 - f_1)e_2(1 - f_1))\|_2 \\
&\leq (2^{3/2} + 2)\|e_2 - \mathbb{E}_B(e_2)\|_2 + 2^{1/2}\|e_1 e_2\|_2,
\end{aligned} \tag{5.3.19}$$

by the supposition on $\|e_2 - \mathbb{E}_B(e_2)\|_2$ and (5.3.3). These are the required inequalities. $\qquad\square$

5.4 Commutants of compressions

In this section we describe some basic results concerning the interplay between compressions of von Neumann algebras and of their commutants.

Theorem 5.4.1.   (i) *Let $M \subseteq B(H)$ be a von Neumann algebra and let $p \in M$ be a projection. Then*

$$(pMp)' \cap pB(H)p = M'p.$$

(ii) *Let $B \subseteq N \subseteq B(H)$ be von Neumann algebras and let $p \in B$ be a projection. Then*

$$(pBp)' \cap pNp = (B' \cap N)p.$$

(iii) *With the hypothesis of (ii), let $q \in (pBp)' \cap pNp$ be a projection. Then there exists a projection $p' \in B' \cap N$ such that $q = pp'$.*

Proof. (i)   The containment $M'p \subseteq (pMp)' \cap pB(H)p$ is clear. The reverse containment is equivalent, by the double commutant theorem on pH (Theorem 2.2.1), to the statement $(M'p)' \cap pB(H)p \subseteq pMp$. To establish this, consider $t \in (M'p)' \cap pB(H)p$, and extend this operator to $\tilde{t} \in B(H)$ by setting it equal to 0 on $(1-p)H$. Since $M' = M'p + M'(1-p)$, we have $t \in (M')' = M$, and $t = p\tilde{t}p \in pMp$ as required.

(ii)   Since $p \in B \subseteq (B' \cap N)'$, we may apply (i) with $M = (B' \cap N)'$ to obtain

$$(p(B' \cap N)'p)' \cap pB(H)p = (B' \cap N)p. \tag{5.4.1}$$

We may also apply (i) with $M = N$, yielding

$$(pNp)' \cap pB(H)p = N'p. \tag{5.4.2}$$

If we denote the von Neumann algebra generated by two algebras as $W^*(\cdot, \cdot)$, then

$$((pBp)' \cap pNp)' \cap pB(H)p = W^*(pBp, (pNp)' \cap pB(H)p)$$
$$= W^*(pBp, N'p), \tag{5.4.3}$$

using (5.4.2). Since B and N' commute, it is easy to check that $W^*(pBp, N'p) = pW^*(B, N')p$. Taking commutants in (5.4.3) gives

$$(pBp)' \cap pNp = (pW^*(B, N')p)' \cap pB(H)p$$
$$= (W^*(B, N')'p) = (B' \cap N)p, \tag{5.4.4}$$

using (i) with $M = W^*(B, N')$, which contains p.

(iii)   By (ii), $x \mapsto xp\colon B' \cap N \to (B' \cap N)p$ is a normal *-homomorphism of $B' \cap N$ onto $(pBp)' \cap pNp$. It follows that there is a projection $p' \in B' \cap N$ such that $q = pp'$.   □

Remark 5.4.2. If we take B to be N in Theorem 5.4.1, and $p \in N$ is a projection, then we obtain $Z(pNp) = Z(N)p$, where $Z(\cdot)$ denotes the centre. This follows because $Z(M) = M' \cap M$ for any von Neumann algebra M.   □

   The next result is the counterpart of Theorem 5.4.1 when the projection lies in the commutant.

Theorem 5.4.3. *Let $M \subseteq B(H)$ be a von Neumann algebra with centre $Z(M)$, and let $p \in M'$ be a projection.*

(i) *The commutant of Mp in $pB(H)p = B(pH)$ is $pM'p$.*

(ii) *The centre of Mp is $Z(M)p$.*

Proof. (i) It is clear that $pM'p \subseteq (Mp)' \cap pB(H)p$. Let $t \in B(H)$ be such that $ptp \in (Mp)'$. Since each $m \in M$ can be written $m = pmp + (1-p)m(1-p)$, we see that $ptp \in M'$, showing that $ptp \in pM'p$.

(ii) The containment $Z(M)p \subseteq Z(Mp)$ is clear. If $z \in Z(Mp)$, then $z = yp$ for some $y \in M$. For each unitary $u \in M$,

$$z = (up)yp((up)^* = (uyu^*)p. \tag{5.4.5}$$

The Dixmier approximation theorem (Theorem 2.2.2) then shows that there exists $\tilde{z} \in Z(M)$ such that $z = \tilde{z}p$, proving the reverse containment. $\qquad\square$

5.5 Basic lemmas for Kadison's results

This section contains the basic lemmas on which the main results of the next two sections are built. We begin by recalling two standard definitions from Section 2.4.

Definition 5.5.1. (i) A projection p in a von Neumann algebra M is said to be *abelian* if the algebra pMp is abelian. The prototypical example is a rank one projection in $B(H)$.

(ii) The *central support* (also known as the *central carrier*) z of an element $x \in M$ is the smallest central projection satisfying $zx = x$. It is denoted c_x and clearly depends on the containing algebra M.

The rest of this chapter, and a number of calculations in Chapter 9, depend on studying the central support of the projections under discussion. One simple property is the following: if q is a central projection in M, then $c_{qx} = qc_x$ for all $x \in M$. The standard cut down results on commutants are also required. The ones stated here are special cases of more general results in Theorem 5.4.1.

If M is a von Neumann algebra acting on a Hilbert space H with centre Z, q is a projection in M' and p is a projection in M, then

(1) *Mq is a von Neumann algebra acting on qH with centre Zq and commutant $qM'q$ and*

(2) *pMp is a von Neumann algebra acting on pH with centre Zp and commutant pM'.*

The proofs of the next two lemmas require this simple observation: the algebra generated by the centre and any given self-adjoint element in M is abelian, so the centre is a masa precisely when it contains every self-adjoint element. This forces the algebra M to be abelian.

Lemma 5.5.2. *Let B be a masa in a von Neumann algebra M and let p be a projection in B with central support c_p in M. If p is minimal in B among projections with central support c_p, then p is abelian.*

Proof. If Z is the centre of M, then $Z \subseteq B$ and the centre of pMp is Zp as noted above (see Theorem 5.4.1). Additionally, Bp is a masa in pMp. Let q be a non-zero projection in pB. Then $q \in B$ and $q \leq p$, and so $q = c_q q \leq c_q p$. If $q < c_q p$, then $(1 - c_q)p + q = p + q - c_q p < p$ and $(1 - c_q)p + q$ is a projection in B with central support $c_p = (1 - c_q)c_p + c_q$. This contradicts the minimality of p. Hence $q = c_q p$. Thus each projection in Bp is in Zp so that $Bp = Zp$; the centre Zp of pMp is the masa Bp. Thus $pMp = Zp$ is abelian, since if the centre of a von Neumann algebra is a masa, then the algebra is abelian and equal to its centre. $\square$

Lemma 5.5.3. *If M is a non-abelian von Neumann algebra and B is a masa in M, then B contains two non-zero projections p and q with $pq = 0$, $c_p = c_q$ and $p \preccurlyeq q$.*

Proof. If f is a projection in B such that $c_f c_{1-f} = 0$, then $f = c_f$, since $c_f \leq 1 - c_{1-f}$ so that

$$f \leq c_f \leq 1 - c_{1-f} \leq 1 - (1 - f) = f. \tag{5.5.1}$$

Because M is non-abelian there is a non-central projection f in B so that $z = c_f c_{1-f} \neq 0$. Thus the central supports of zf and $z(1 - f)$ are equal since

$$c_{zf} = zc_f = c_f c_{1-f} = z = zc_{1-f} = c_{z(1-f)}. \tag{5.5.2}$$

By the comparison theorem for projections [105, Theorem 6.2.7], there is a non-zero central projection z_0 such that $z_0 \leq z$ and either

$$z_0 f \preccurlyeq z_0(1 - f) \quad \text{or} \quad z_0(1 - f) \preccurlyeq z_0 f. \tag{5.5.3}$$

Choosing $p = z_0 f$ and $q = z_0(1 - f)$ in the first case (or $p = z_0(1 - f)$ and $q = z_0 f$ in the second) gives the required projections with

$$c_p = z_0 c_f = z_0 = z_0 c_{1-f} = c_{z_0(1-f)} = c_q. \tag{5.5.4}$$

$\square$

Lemma 5.5.4. *If M is a von Neumann algebra that has no abelian central direct summand and if B is a masa in M, then B contains a non-zero projection p such that $c_p = c_{1-p} = 1$ and $p \preccurlyeq 1 - p$.*

Proof. Let $\{p_\lambda : \lambda \in \Lambda\}$ be a family of non-zero projections in B that is maximal with respect to the properties that $\{c_{p_\lambda} : \lambda \in \Lambda\}$ is an orthogonal family and $p_\lambda \preccurlyeq 1 - p_\lambda$ for each $\lambda \in \Lambda$. By Lemma 5.5.3, B contains non-zero projections p_0 and q_0 such that $p_0 \preccurlyeq q_0 \leq 1 - p_0$. Thus there is a family $\{p_0\}$ with a single projection in it of the required type and hence a maximal family exists by Zorn's lemma.

Let $p = \sum p_\lambda$. Then $c_p = \sum c_{p_\lambda}$, by a standard result on central supports, [104, Proposition 5.5.3], applied to finite partial sums. If $z = c_p \neq 1$, then $(1-z)M$ is a non-abelian von Neumann algebra by hypothesis and $(1-z)B$ is a masa in $(1-z)M$. By Lemma 5.5.3 again, there is a non-zero projection p_1 in $(1-z)B$ such that $p_1 \preccurlyeq (1-z) - p_1$. This contradicts the maximality of the family $\{p_\lambda : \lambda \in \Lambda\}$, since $\{p_1\} \cup \{p_\lambda : \lambda \in \Lambda\}$ is a larger family with the required properties. Hence $z = 1$.

For each λ,

$$p_\lambda = c_{p_\lambda} p_\lambda \preccurlyeq c_{p_\lambda}(1 - p_\lambda) = c_{p_\lambda} - p_\lambda \tag{5.5.5}$$

so that

$$p = \sum p_\lambda \preccurlyeq \sum (c_{p_\lambda} - p_\lambda) = 1 - p \tag{5.5.6}$$

and $c_p = z = 1 = c_{1-p}$ as required. $\qquad\square$

5.6 Range of the centre-valued trace

Theorem 5.6.2 is the result that the centre-valued trace fills up its range in the strongest way in a II_1 von Neumann algebra.

Recall the following standard method from conditional expectations, which is required later in the next lemma 5.6.1. If M is a type II_1 von Neumann algebra with a faithful normal trace τ and centre Z, then the centre-valued trace $\mathbb{T}_M$ is $\mathbb{E}_Z$. This follows easily from Remark 3.6.6 applied to the whole algebra M as a subalgebra of itself, as the following little calculation shows. If $x \in M$, let

$$K_M(x) = \overline{\mathrm{conv}}^w \{uxu^* : u \text{ a unitary in } M\} \tag{5.6.1}$$

be the weakly closed convex hull of $\{uxu^* : u \text{ a unitary in } M\}$. If v is a unitary in M, then

$$\{uxu^* : u \text{ a unitary in } M\} = \{u(vxv^*)u^* : u \text{ a unitary in } M\} \tag{5.6.2}$$

so that $K_M(x) = K_M(vxv^*)$. Thus $K_M(x)$ and $K_M(vxv^*)$ have a common element h of minimal $\| \cdot \|_2$-norm. By Lemma 3.6.5 and Remark 3.6.6, $h = \mathbb{E}_Z(x) = \mathbb{E}_Z(vxv^*)$, since $Z = M \cap M'$. Replacing x by xv, and using the observation that the unitaries span M, gives $\mathbb{E}_M(xy) = \mathbb{E}_M(yx)$ for all $x, y \in M$. The Z-module property of $\mathbb{E}_Z$ and $\tau(\mathbb{E}_M(x)) = \tau(x)$ for all $x \in M$ imply that $\mathbb{E}_Z$ is the unique centre-valued trace $\mathbb{T}_M$, [187, p. 312]. If p is a projection in M, then $\mathbb{T}_M(p) \leq c_p$, since $p \leq c_p$ and $c_p \in Z$.

Lemma 5.6.1. *Let B be a masa in a II_1 von Neumann algebra M and let p be a non-zero projection in B. If $\varepsilon > 0$ and z is a non-zero central projection in M with $z \leq c_p$, then there is a non-zero projection q in B with*

$$q \leq p, \quad c_q = z \quad \text{and} \quad \mathbb{T}_M(q) \leq \varepsilon z \mathbb{T}_M(p). \tag{5.6.3}$$

Proof. Assume that $c_p = 1$ by replacing M by $c_p M$, B by $c_p B$ and $\mathbb{T}_M$ by $\mathbb{T}_{c_p M}$. By induction on n construct a sequence of non-zero projections $p_n \in B$ such that $p_1 = p$,

$$p_n \leq p, \quad p_{n+1} \precsim p_n - p_{n+1} \quad \text{and} \quad c_{p_n} = 1 \tag{5.6.4}$$

for all $n \in \mathbb{N}$. The same calculation that works form n to $n+1$ gives p_2. Suppose that $p_1, \ldots, p_n$ have been constructed. Apply Lemma 5.5.4 to the masa $p_n B$ in the II_1 von Neumann algebra $p_n M p_n$ to obtain a projection p_{n+1} in B with

$$p_{n+1} \leq p_n, \quad p_{n+1} \precsim p_n - p_{n+1} \quad \text{and} \quad c_{p_{n+1}}(p_n M p_n) = p_n, \tag{5.6.5}$$

where $c_{p_{n+1}}(p_n M p_n)$ is the central support of p_{n+1} in the von Neumann algebra $p_n M p_n$. Note that the order relation $\precsim$ is calculated initially in $p_n M p_n$ and then holds in M. If z_0 is a central projection in M with $p_{n+1} \leq z_0$, then $z_0 p_n$ is a central projection in $p_n M p_n$ with $p_{n+1} \leq z_0 p_n$ so that $p_n = c_{p_{n+1}}(p_n M p_n) \leq z_0 p_n$. Hence $p_n \leq z_0$ and $1 = c_{p_n} \leq z_0$. Thus $c_{p_{n+1}} = 1$ as required in the induction. From equation (5.6.4) it follows that $\mathbb{T}_M(p_n) \leq \mathbb{T}_M(p_{n-1}) - \mathbb{T}_M(p_n)$ so that $\mathbb{T}_M(p_n) \leq 2^{-1}\mathbb{T}_M(p_{n-1})$ for all $n \geq 2$. This implies that $\mathbb{T}_M(p_n) \leq 2^{-n+1}\mathbb{T}_M(p)$ for all $n \in \mathbb{N}$, because $p_1 = p$.

Choose $n \in \mathbb{N}$ with $2^{-n+1} < \varepsilon$ and let $q = z p_n$. Then $c_q = c_{z p_n} = z c_{p_n} = z$, by equation (5.6.4) and $\mathbb{T}_M(q) = z\mathbb{T}(p_n) \leq \varepsilon z\mathbb{T}_M(p)$ as required. $\square$

Theorem 5.6.2. *Let B be a masa in a II_1 von Neumann algebra M and let p be a non-zero projection in B. If h is in the centre of M with $0 \leq h \leq \mathbb{T}_M(p)$, then there exists a projection $q \in B$ with $q \leq p$ and $\mathbb{T}_M(q) = h$.*

Proof. Assume $h \neq 0$. By the spectral theorem there is a non-zero central projection z and a positive ε such that $0 \leq \varepsilon z \leq h$. By Lemma 5.6.1, there is a non-zero projection $q_0 \in B$ with $q_0 \leq p$ and $\mathbb{T}_M(q_0) \leq \varepsilon z$. Thus there is at least one non-zero projection $q_0 \in B$ with $q_0 \leq p$ and $\mathbb{T}_M(q_0) \leq h$. Let $\mathcal{F}$ be the set of all non-zero projections q_1 in B with $q_1 \leq p$ and $\mathbb{T}_M(q_1) \leq h$ with order $\leq$ on $\mathcal{F}$. If $\{q_\lambda : \lambda \in \Lambda\}$ is an increasing totally ordered net in $\mathcal{F}$, then $q_1 = \sup\{q_\lambda : \lambda \in \Lambda\}$ is a projection in B with $q_1 \leq p$ and $\mathbb{T}_M(q_1) = \mathbb{T}_M(\sup q_\lambda) = \sup \mathbb{T}_M(q_\lambda) \leq h$ by the normality of the centre-valued trace $\mathbb{T}_M$. Hence there is a maximal element $q \in \mathcal{F}$. If $\mathbb{T}_M(q) \neq h$, then by the spectral theorem there is a $\varepsilon > 0$ and a non-zero central projection z such that

$$\varepsilon z \leq h - \mathbb{T}_M(q). \tag{5.6.6}$$

Considering centre-valued traces and central supports,

$$\varepsilon z \leq z(h - \mathbb{T}_M(q)) \leq z(\mathbb{T}_M(p) - \mathbb{T}_M(q))$$
$$= \mathbb{T}_M(z(p-q)) \leq c_{z(p-q)} = z c_{p-q} \tag{5.6.7}$$

so that $z \leq c_{p-q}$. By Lemma 5.6.1 applied to B in M, there is a non-zero projection $q_1 \in B$ with

$$q_1 \leq p - q, \ c_{q_1} = c_{p-q} \text{ and } \mathbb{T}_M(q_1) \leq \varepsilon z\mathbb{T}_M(p-q) \leq h. \tag{5.6.8}$$

Then $q+q_1$ is a projection in B with $q+q_1 \leq p$ and $\mathbb{T}_M(q+q_1) \leq \mathbb{T}_M(q)+\varepsilon z \leq h$, so $q + q_1 \in \mathcal{F}$. This contradicts the maximality of q in $\mathcal{F}$ and implies that $\mathbb{T}_M(q) = h$ as required. □

Chapter 6

Normalisers, orthogonality and distances

6.1 Introduction

This chapter is devoted to technical results that apply to von Neumann subalgebras of a finite von Neumann algebra with a fixed faithful trace τ. Section 6.2 contains results on the unitary, and partial isometry, normalisers of a masa that are due to Dye [55], although they are presented here as in Popa [141] and Jones and Popa [96]. We also discuss Cartan masas, and give a theorem on the unitary conjugacy of two such subalgebras that will prove useful subsequently. Section 6.3 on orthogonality of von Neumann subalgebras of a finite von Neumann algebra is due to Popa [138]; this idea has played an important role in the theory. Section 6.4 contains some properties of the set of all von Neumann subalgebras of a II_1 factor when this is given a natural metric arising from the $\|\cdot\|_2$-norm, by $d_2(A, B) = \|\mathbb{E}_A - \mathbb{E}_B\|_{\infty,2}$ for all von Neumann subalgebras A and B of N. These results are due to Christensen [24], although his formulations are worded in terms of an equivalent metric d_1, which is defined by

$$d_1(A, B) = \inf\{\delta > 0 \colon A \subset_\delta B \quad \text{and} \quad B \subset_\delta A\}$$

for all von Neumann subalgebras A, B of N. Here $A \subset_\delta B$ is defined by the requirement that

$$\|a - \mathbb{E}_B(a)\|_2 \leq \delta \text{ for all } a \in A \text{ with } \|a\| \leq 1.$$

The advantage of these equivalent metrics is that they are much smaller than that induced by the uniform norm.

6.2 Normalisers of masas

Let $\mathcal{U}(N)$ denote the unitary group of a von Neumann algebra N. We begin by recalling from (3.3.1) the definition of the unitary normaliser of a subalgebra.

Definition 6.2.1. If M is a von Neumann subalgebra of a von Neumann algebra N, then $\mathcal{N}(M)$, or $\mathcal{N}(M, N)$, denotes the unitary normaliser, defined by

$$\{u \in \mathcal{U}(N) : uMu^* = M\}.$$

□

Thus $\mathcal{N}(M)$ is a subgroup of $\mathcal{U}(N)$ that is stable under automorphisms of N that fix M as a set, and the von Neumann algebra $\mathcal{N}(M)''$ is the $\| \cdot \|_2$-closure in N of the linear span of $\mathcal{N}(M)$. There is a related notion of a groupoid normaliser defined as follows.

Definition 6.2.2. If A is an abelian von Neumann subalgebra of N, let $\mathcal{GN}(A)$, or $\mathcal{GN}(A, N)$, be the normalising groupoid, consisting of those partial isometries $v \in N$ which satisfy

$$vv^*, \; v^*v \in A, \quad vAv^* = Avv^* = vv^*A.$$

□

The results in this section are taken from [55], [141, Lemma 2.1] and [96, Lemma 2.2]. The following two lemmas summarise the basic properties of the normaliser and groupoid normaliser of abelian von Neumann subalgebras and masas in N. Observe that the abelian hypothesis is used in the initial two parts of the next lemma.

Lemma 6.2.3. *Let N be a von Neumann algebra with a faithful normal trace τ and let A be an abelian von Neumann subalgebra of N.*

(i) *If v_1, $v_2 \in \mathcal{GN}(A)$, then $v_1 v_2$, $v_1^* \in \mathcal{GN}(A)$.*

(ii) *If v_1, $v_2 \in \mathcal{GN}(A)$ and if $v_1 v_2^* = v_2^* v_1 = 0$, then $v_1 + v_2 \in \mathcal{GN}(A)$.*

(iii) *A partial isometry v is in $\mathcal{GN}(A)$ if, and only if, there is a projection $e \in A$ and a unitary $u \in \mathcal{N}(A)$ such that $v = ue = (ueu^*)u$.*

(iv) *The set $\mathcal{GN}(A)$ is in the linear span of $\mathcal{N}(A)$ so that $\mathcal{GN}(A)'' = \mathcal{N}(A)''$.*

Proof. (i) Clearly $v_1 v_2 A (v_1 v_2)^*$ and $v_1^* A v_1$ are contained in A and $v_1 v_2 (v_1 v_2)^*$ is a projection in A, since

$$\begin{aligned} v_1 v_2 (v_1 v_2)^* v_1 v_2 (v_1 v_2)^* &= v_1 (v_2 v_2^*)(v_1^* v_1)(v_2 v_2^*) v_1^* \\ &= v_1 (v_1^* v_1)(v_2 v_2^*) v_1^* = v_1 v_2 v_2^* v_1^* \end{aligned} \tag{6.2.1}$$

from the commutativity of A.

(ii) Let v_1, v_2 be in $\mathcal{GN}(A)$ with $v_1 v_2^* = v_2^* v_1 = 0$ and let $v = v_1 + v_2$. Then v is a partial isometry and $vAv^* = v_1 A v_1^* + v_2 A v_2*$, since $v_1 v_1^* v_2 v_2^* = 0$ and $v_1^* v_1 v_2^* v_2 = 0$.

(iii)   By (i), only one implication need be proved. Let v be in $\mathcal{GN}(A)$ with $v^*v = e \in A$ and $vv^* = f \in A$. We begin by constructing a partial isometry w in N such that

$$w^*w = e - ef, \quad ww^* = f - ef \quad \text{and} \tag{6.2.2}$$
$$wA(e - ef)w^* = A(f - ef)$$

from which the unitary u is subsequently constructed.

By (i), the partial isometries $w_n = (1 - e)v^n(1 - f) = (f - ef)v^n(e - ef)$ are in $\mathcal{GN}(A)$ for all positive integers n, since $1 - e$, $1 - f$ and v are in $\mathcal{GN}(A)$. Let $e_n = w_n^*w_n$ and $f_n = w_nw_n^*$ be the initial and final projections of w_n. From this, if $n > m$,

$$v^n(1 - f)v^{*n}v^m(1 - f)v^{*m}$$
$$= v^n(1 - f)v^{*(n-m)}[v^{*m}v^m](1 - f)v^{*m}$$
$$= v^n(1 - f)v^{*(n-m)}(1 - f)[v^mv^{*m}]v^{*m} = 0, \tag{6.2.3}$$

since $v^mv^{*m} \in A$ and $v^*(1 - f) = 0$, and

$$w_nw_m^* = (f - ef)v^n(e - ef)v^{*m}(f - ef)$$
$$= (f - ef)v^{n-m}[v^m(e - ef)v^{*m}](f - ef)$$
$$= (f - ef)v^{n-m}(1 - e)[v^m(e - ef)v^{*m}]f = 0, \tag{6.2.4}$$

since $v^m(1 - f)v^{*m} \in A$, and $v(1 - e) = 0$. The equation $w_n^*w_m = 0$ is derived in the same way. If $n < m$, a similar calculation yields $w_nw_m^* = w_n^*w_m = 0$. If $m \neq n$, then

$$e_ne_m = w_n^*w_nw_m^*w_m = 0 \quad \text{and} \quad f_nf_m = w_nw_n^*w_mw_m^* = 0, \tag{6.2.5}$$

so that $\{e_n : n \in \mathbb{N}\}$ and $\{f_n : n \in \mathbb{N}\}$ are sequences of orthogonal projections in A.

Let $e_0 = e - ef - \sum_{n \geq 1} e_n$. The projections $v^ne_0v^{*n}$ are pairwise orthogonal in A for $n \geq 1$, since $v^ne_0v^{*n} \leq v^n(1 - f)v^{*n}$ and the latter projections are orthogonal by equation (6.2.3).

By definition, $e_0 \leq 1 - f$ and, for all $n \in \mathbb{N}$,

$$0 = e_0e_n = e_0(1 - f)v^{*n}(1 - e)v^n(1 - f)e_0$$
$$= e_0v^{*n}(1 - e)v^ne_0 \tag{6.2.6}$$

so that

$$(1 - e)v^ne_0 = 0 = (1 - e)v^ne_0v^{*n} \quad \text{and} \quad v^ne_0v^{*n} \leq e. \tag{6.2.7}$$

From (6.2.7), we obtain

$$\tau(v^{(n+1)}e_0v^{*(n+1)}) = \tau(v^ne_0v^{*n}v^*v) = \tau(v^ne_0v^{*n}e) = \tau(v^ne_0v^{*n}), \tag{6.2.8}$$

so that the projections $v^n e_0 v^{*n}$ are all equivalent for $n \geq 0$. The orthogonality of these projections implies that they are all zero since N is finite. Thus $e - ef = \sum_{n \geq 1} e_n$, and similarly $f - ef = \sum_{n \geq 1} f_n$. If $w = \sum_{n \geq 1} w_n$, then

$$ww^* = \sum_{m,\, n} w_m^* w_n = \sum_{n \geq 1} e_n = e - ef \qquad (6.2.9)$$

since $w_m^* w_n = 0$ for $m \neq n$ by equations (6.2.5). Similarly $ww^* = f - ef$. Also

$$wA(e - ef)w^* = \sum_{m,\, n} w_m A(e - ef)w_n^* = \sum_{n \geq 1} w_n A e_n w_n^*$$
$$= \sum_{n \geq 1} A f_n = A(f - ef). \qquad (6.2.10)$$

Let

$$u = v + w^* + 1 - e - f + ef$$
$$= v + w^* + (1 - e)(1 - f). \qquad (6.2.11)$$

Then direct calculations show that u is a unitary in N, that $uAu^* = A$, and that $v = ue$.

(iv)   If v is in $\mathcal{GN}(A)$, then $v = ue$ by (iii) so that $-2v = u(1 - 2e) - u$ with u and $(1 - 2e)$ both in $\mathcal{N}(A)$. Hence $\mathcal{GN}(A)$ is in the linear span of $\mathcal{N}(A)$ so $\mathcal{GN}(A)'' = \mathcal{N}(A)''$. $\qquad\square$

See [96, Remark 2.2] for a discussion of Lemma 6.2.3. The masa hypothesis $A' \cap N = A$ is used at a crucial stage in each part of the proof of the following lemma.

Lemma 6.2.4. *Let N be a von Neumann algebra with a faithful normal trace τ and let A be a masa in N.*

(i) *Let M be a von Neumann subalgebra of N with $A \subseteq M \subseteq N$. If $v \in \mathcal{GN}(A, N)$, then there is a projection e in A satisfying $\mathbb{E}_M(v) = ve$ and $e \leq v^*v$. Moreover, e is the maximal projection in A with ve in M and $e \leq v^*v$. Hence*

$$\mathbb{E}_M(\mathcal{GN}(A, N)) = \mathcal{GN}(A, M). \qquad (6.2.12)$$

(ii) *Suppose that θ is an automorphism of A, that v is a partial isometry in N satisfying*

$$va = \theta(a)v \quad \text{for all} \quad a \in A, \qquad (6.2.13)$$

and that $e = v^*v$. Then $v \in \mathcal{GN}(A)$, $\theta(e) = vv^*$, and there is a unitary u in N such that $v = ue$.

Proof. (i)   Let $v = uf$ be the decomposition of $v \in \mathcal{GN}(A, N)$, where $u \in \mathcal{N}(A, N)$ and f is a projection in A. If a is in A, then $uau^* \in A$, and

$$\mathbb{E}_M(u)a = \mathbb{E}_M(ua) = \mathbb{E}_M(uau^*u) = uau^*\mathbb{E}_M(u), \qquad (6.2.14)$$

so that $u^*\mathbb{E}_M(u)a = au^*\mathbb{E}_M(u)$. Hence $e = u^*\mathbb{E}_M(u)$ is in $A' \cap N = A$. Moreover,

$$\mathbb{E}_M(e) = e = \mathbb{E}_M(u^*\mathbb{E}_M(u)) = \mathbb{E}_M(u)^*\mathbb{E}_M(u) \geq 0, \qquad (6.2.15)$$

and

$$e^2 = e^*e = \mathbb{E}_M(u)^*uu^*\mathbb{E}_M(u) = \mathbb{E}_M(u)^*\mathbb{E}_M(u) = e. \qquad (6.2.16)$$

Hence $\mathbb{E}_M(u) = ue$ so that $\mathbb{E}_M(v) = \mathbb{E}_M(u)f = uef = ve$ is in $M \cap \mathcal{GN}(A, N)$ and thus lies in $\mathcal{GN}(A, M)$.

If e_0 is a projection in A with ve_0 in M, $e_0 \leq v^*v$ and $e_0e = 0$, then

$$v(e + e_0) = \mathbb{E}_M(v(e + e_0)) = \mathbb{E}_M(v)(e + e_0) = ve(e + e_0) = ve, \qquad (6.2.17)$$

so that $ve_0 = 0$ and $e_0 = 0$, since $e_0 \leq v^*v$. This proves the maximality of e.

(ii)   Let $f = vv^*$. Repeated use will be made of equation (6.2.13). Since

$$v^*va = v^*\theta(a)v = (\theta(a^*)v)^*v = (va^*)^*v = av^*v \quad \text{and} \qquad (6.2.18)$$

$$vv^*\theta(a) = v(\theta(a^*)v)^* = v(va^*)^* = vav^* = \theta(a)vv*, \qquad (6.2.19)$$

e and f are in $A' \cap N = A$. Hence

$$f = vev^* = \theta(e)vv^* = \theta(e)f \qquad (6.2.20)$$

and

$$e\theta^{-1}(f) = v^*v\theta^{-1}(f) = v^*\theta(\theta^{-1}(f))v = v^*fv = e. \qquad (6.2.21)$$

We conclude that the relations $f \leq \theta(e)$, $e \leq \theta^{-1}(f)$ and $\theta(e) = f$ all hold. It then follows that

$$\theta(ae) = \theta(a)f = \theta(a)vv^* = vav^*, \qquad (6.2.22)$$

for all $a \in A$, so $v \in \mathcal{GN}(A)$ and u exists by Lemma 6.2.3 (iii). $\qquad \square$

Corollary 6.2.5. *Let N be a von Neumann algebra with a faithful normal trace τ, let A be a masa in N with unitary normaliser $\mathcal{N}(A)$ and let M be a von Neumann subalgebra of N.*

(i) *If $M \subseteq \mathcal{N}(A)''$, then $M = \mathcal{N}(A, M)''$.*

(ii) *If A is a Cartan masa in N, then A is a Cartan masa in M.*

Proof. Let $\overline{\mathrm{span}}^w(X)$ denote the weak closure of the linear span of X. Since $M \subseteq \mathcal{N}(A, N)''$ and $\mathbb{E}_M$ is a normal map, Lemmas 6.2.3 and 6.2.4 imply that

$$M = \mathbb{E}_M(\overline{\mathrm{span}}^w(\mathcal{N}(A, N))) = \overline{\mathrm{span}}^w(\mathbb{E}_M(\mathcal{GN}(A, N)))$$
$$= \overline{\mathrm{span}}^w(\mathcal{GN}(A, M)) = \overline{\mathrm{span}}^w(\mathcal{N}(A, M)) = (\mathcal{N}(A, M))''. \qquad (6.2.23)$$

This proves (i). If A is Cartan, then $\mathcal{N}(A, N)'' = N$ and so (i) applies to M. We conclude from (i) that A is Cartan in M. $\qquad \square$

Corollary 6.2.5 shows that being Cartan is a hereditary property of masas in finite von Neumann algebras. Clearly being singular is also a hereditary property. It is of interest that semi-regularity is NOT a hereditary property. For an example of a masa A with $A \subseteq M \subseteq N$, A semi-regular in N and singular in M, see [96, Example 2.6].

We now turn to the topic of unitary equivalence of Cartan masas and of semi-regular masas. For non-conjugate masas A and B in a II_1 factor N, it may be possible to find a partial isometry v such that $v^* v \in A$, $vv^* \in B$ and $vAv^* = Bvv^*$. An example of this is given by choosing masas A_1 and A_2 in the hyperfinite factor R which are respectively Cartan and singular, and then letting $A = A_1 \oplus A_1$ and $B = A_1 \oplus A_2$ in $R \otimes \mathbb{M}_2$. However, we will show that if both masas are taken to be Cartan or semi-regular, then the existence of such a partial isometry v implies unitary conjugacy. We require the following lemma. The two projections in the statement are equivalent in N; the point is that the partial isometry implementing the equivalence can be chosen from the groupoid normaliser $\mathcal{GN}(A)$. Note that the hypothesis applies to both Cartan and semi-regular masas.

Lemma 6.2.6. *Let A be a masa in a II_1 factor N and let $p, q \in A$ be projections with equal traces. If $\mathcal{N}(A)''$ is a factor, then there exists a partial isometry $v_0 \in \mathcal{GN}(A)$ such that $p = v_0 v_0^*$ and $q = v_0^* v_0$.*

Proof. The masa A is Cartan in the factor $\mathcal{N}(A)''$. Then replacing N by $\mathcal{N}(A)''$, if necessary, allows us to assume without loss of generality that A is Cartan in N.

The weak density of the span of $\mathcal{N}(A)$ in N shows that there must exist $u \in \mathcal{N}(A)$ such that $puq \neq 0$, otherwise $pNq = \{0\}$. Then $w = puq$ is a non-zero partial isometry in $\mathcal{GN}(A)$ satisfying $ww^* \leq p$ and $w^* w \leq q$. Thus the set
$$S = \{v \in \mathcal{GN}(A) \colon vv^* \leq p, \ v^* v \leq q, \ v \neq 0\}$$
is a non-empty set of partial isometries. By Zorn's lemma, we may select from S a subset S_0 which is maximal with respect to having pairwise orthogonal initial projections and pairwise orthogonal final projections. The trace dictates that this subset is at most countable, so we may list the elements as $S_0 = \{v_i \colon i \geq 1\}$ without loss of generality. Let $v_0 = \sum_{i=1}^{\infty} v_i$. This sum converges strongly and defines a partial isometry in $\mathcal{GN}(A)$ such that $v_0 v_0^* \leq p$ and $v_0^* v_0 \leq q$, using the orthogonality conditions for the v_i's. Let $p_0 = v_0 v_0^*$ and $q_0 = v_0^* v_0$. If $p_0 < p$ then the trace shows that $q_0 < q$, and so we may apply the previous discussion to the non-zero projections $p - p_0$ and $q - q_0$ to obtain a non-zero partial isometry $\tilde{v} \in S$ with $\tilde{v}\tilde{v}^* \leq p - p_0$ and $\tilde{v}^* \tilde{v} \leq q - q_0$. This contradicts the maximality of S_0, and proves that $v_0 v_0^* = p$, $v_0^* v_0 = q$. $\square$

The following theorem applies when we have that A and B are Cartan, are semi-regular, or are one of each type.

Theorem 6.2.7. *Let A and B be masas in a II_1 factor N, and suppose that there exists a non-zero partial isometry $v \in N$ such that $vv^* \in B$, $v^* v \in A$, and*

$vAv^* = Bvv^*$. If $\mathcal{N}(A)''$ and $\mathcal{N}(B)''$ are factors, then there exists a unitary $u \in N$ such that $uAu^* = B$.

Proof. Let $p = v^*v \in A$. Let n be an integer such that $n^{-1} \leq \tau(p)$, and select a projection $p_1 \in A$ lying below p and having $\tau(p_1) = n^{-1}$. Then let $q_1 = vp_1v^* \in B$. Now choose orthogonal sets of projections $\{p_i\}_{i=1}^n \in A$ and $\{q_i\}_{i=1}^n \in B$ which all have trace n^{-1} with both sets summing to 1. By Lemma 6.2.6, we may choose partial isometries $u_i \in \mathcal{GN}(A)$, $w_i \in \mathcal{GN}(B)$, $1 \leq i \leq n$, with $u_1 = p_1$, $w_1 = q_1$, and such that $u_iu_i^* = p_1$, $u_i^*u_i = p_i$, $w_iw_i^* = q_1$, and $w_i^*w_i = q_i$. Then the partial isometries $v_i = w_i^*vu_i$, $1 \leq i \leq n$, satisfy $v_iv_i^* = q_i$, $v_i^*v_i = p_i$, and $v_iAv_i^* = Bq_i$. Now define $u = \sum_{i=1}^n v_i$, which is a unitary satisfying $u(Ap_i)u^* = Bq_i$ for $1 \leq i \leq n$. From this it follows that $uAu^* = B$, completing the proof. $\square$

6.3 Orthogonality of von Neumann subalgebras

The results in this section are from [138] (but see also [73, Lemma 2.4]). Throughout this section N will be a finite von Neumann algebra with a faithful normal trace τ with $\tau(1) = 1$ and 2-norm given by $\|x\|_2 = (\tau(xx^*))^{1/2}$ for all $x \in N$. If S is a non-empty subset of N, let

$$S^\perp = \{x \in N : \tau(xs^*) = 0 \quad \text{for all} \quad s \in S\}. \tag{6.3.1}$$

This orthogonal subspace notation is to hold in this section only in that $S^\perp$ is a subset of N, whereas the orthogonal subspace would usually be taken in $L^2(N, \tau)$. Note that if M is a von Neumann subalgebra of N, then $(M^\perp)^\perp = M$, since if x is in $(M^\perp)^\perp$ then

$$\|x - \mathbb{E}_M(x)\|_2^2 = \tau(x(x - \mathbb{E}_M(x))) - \tau(\mathbb{E}_M(x)(x - \mathbb{E}_M(x)))$$
$$= 0, \tag{6.3.2}$$

so $x \in M$. From this it follows that if M_1 and M_2 are von Neumann algebras of N, then $M_1 \subseteq M_2$ if, and only if, $M_1^\perp \supseteq M_2^\perp$. For von Neumann subalgebras to be orthogonal, it is clear that the identity must be removed from consideration; here are some of the equivalences.

Lemma 6.3.1. Let N be a von Neumann algebra with a faithful normal trace τ and let M_1, M_2 be von Neumann subalgebras of N. The following conditions on M_1 and M_2 are equivalent, and are equivalent to the same conditions with the order of M_1 and M_2 interchanged.

(i) If $x \in M_1$ and $y \in M_2$ with $\tau(x) = \tau(y) = 0$, then $\tau(xy) = 0$.

(ii) If $x \in M_1$ and $y \in M_2$, then $\tau(xy) = \tau(x)\tau(y)$.

(iii) If $x \in M_1$ and $y \in M_2$, then $\|xy\|_2 = \|x\|_2\|y\|_2$.

(iv) If $z \in N$, then $\mathbb{E}_{M_1}\mathbb{E}_{M_2}(z) = \tau(z)1$.

(v) $\mathbb{E}_{M_1}(M_2) = \mathbb{C}1$.

Proof. Let $x \in M_1$, $y \in M_2$ and $z \in N$ and let these elements vary when necessary. The equivalence of the first three conditions follows from the equations

$$\tau((x - \tau(x)1)(y - \tau(y)1)) = \tau(xy) - \tau(x)\tau(y) \quad \text{and}$$
$$\|xy\|_2^2 = \tau(x^*xyy^*), \tag{6.3.3}$$

using linearity in (ii) to pass from positive x and y to all elements in the relevant von Neumann algebras.

Condition (ii) implies (iv) by letting $x \in M_1$ be arbitrary in the equations

$$\tau(\mathbb{E}_{M_1}(\mathbb{E}_{M_2}(z))x) = \tau(\mathbb{E}_{M_1}(\mathbb{E}_{M_2}(z)x)) = \tau(\mathbb{E}_{M_2}(z)x)$$
$$= \tau(\mathbb{E}_{M_2}(z))\tau(x) = \tau(z)\tau(x)$$
$$= \tau(\tau(z)x). \tag{6.3.4}$$

If (iv) holds, then, for $x \in M_1$ and $y \in M_2$, the equation

$$\tau(x)\tau(y) = \tau(x\tau(y)) = \tau(x\mathbb{E}_{M_1}(\mathbb{E}_{M_2}(y)))$$
$$= \tau(\mathbb{E}_{M_1}(xy)) = \tau(xy) \tag{6.3.5}$$

proves (ii).

If (v) holds, then the equation

$$\tau(\mathbb{E}_{M_1}\mathbb{E}_{M_2}(z) - \tau(z)1) = \tau(z) - \tau(z) = 0 \tag{6.3.6}$$

implies that $\mathbb{E}_{M_1}(\mathbb{E}_{M_2}(z)) = \tau(z)1$, since τ is one-to-one on $\mathbb{C}1$. Clearly (iv) implies (v). The symmetry of (ii) in M_1 and M_2 gives the symmetry of the conditions. □

If M_1 and M_2 are von Neumann subalgebras of N with faithful normal trace τ, then M_1 is said to be *orthogonal* to M_2 if, and only if, the equivalent conditions of Lemma 6.3.1 hold. Observe that orthogonality depends on the trace chosen but is unique in type II_1 factors due to the uniqueness of the normalised trace. Note that orthogonality is preserved by *-automorphisms of N that preserve the trace.

Discrete group von Neumann algebras provide a nice example of the orthogonality of von Neumann subalgebras that illustrates this definition.

Let H and K be subgroups of a countable discrete group G : then $L(H)$ is orthogonal to $L(K)$ in $L(G)$ if, and only if, $H \cap K = \{e\}$. This follows from Lemma 6.3.1 (v) since $\mathbb{E}_{L(H)}(L(K)) = \mathbb{C}1$ if, and only if, $H \cap K = \{e\}$.

Lemma 6.3.2. *Let M be a von Neumann subalgebra of the finite von Neumann algebra N with faithful normal trace τ and let u be a unitary element in N. If there is a diffuse von Neumann subalgebra B of M with uBu^* orthogonal to M, then u is orthogonal to $\mathcal{N}(M)''$.*

Proof. Let v be in $\mathcal{N}(M)$ and let $\varepsilon > 0$. If A_0 is a masa in B, then A_0 is diffuse by Theorem 3.5.2 and by repeated subdivision there is a positive integer n with $1/n < \varepsilon$ and projections $e_1, \ldots, e_n$ in A_0 with $\sum_j e_j = 1$ and $\tau(e_j) < \varepsilon$ for $1 \leq j \leq n$. Intuitively this depends on A_0 with τ being isomorphic to $L^\infty[0, 1]$ with its integral. Let A be the finite dimensional von Neumann subalgebra generated by $e_1, \ldots, e_n$. Since $vuAu^*v^* \subseteq vuBu^*v^*$ is orthogonal to $vMv^* = M$,

$$\tau(vue_ju^*v^*e_j) = \tau(vue_ju^*v^*)\tau(e_j) = \tau(e_j)^2 \tag{6.3.7}$$

by Lemma 6.3.1 (ii).

Since $|\tau(x)| \leq \|\mathbb{E}_{A' \cap N}x\|_2$ for all $x \in N$, we may sum equation (6.3.7) over j and apply the definition of the conditional expectation $\mathbb{E}_{A' \cap N}$ (see Section 3.6) to obtain

$$|\tau(vu)|^2 \leq \|\mathbb{E}_{A' \cap N}(vu)\|_2^2 = \|\sum_j e_jvue_j\|_2^2$$

$$= \sum_j \|e_jvue_j\|_2^2 = \sum_j \tau(vue_ju^*v^*e_j) = \sum_j \tau(e_j)^2$$

$$\leq \max\{\tau(e_j) : 1 \leq j \leq n\} \sum_j \tau(e_j) < \varepsilon. \tag{6.3.8}$$

It follows that $\tau(vu) = 0$ for all $v \in \mathcal{N}(M)$ and so u is orthogonal to $\mathcal{N}(M)$, and hence is orthogonal to $\mathcal{N}(M)''$ as required. □

Remark 6.3.3. We have already noted that orthogonality of subgroup von Neumann subalgebras corresponds to the subgroups having trivial intersection. If G_1 and G_2 are subgroups of a discrete group G, with $G_1 \cap G_2 = \{e\}$, then $L(G_1)$ and $L(G_2)$ are orthogonal von Neumann subalgebras of $L(G)$. For group elements $g_1 \in G_1$ and $g_2 \in G_2$ such that $g_1g_2 \neq e$, we have $\tau(g_1g_2) = 0$. This formula extends by linearity and continuity to orthogonality of the relevant von Neumann subalgebras by Lemma 6.3.1 (i). The converse is clear.

If H is a subgroup of the group G and $g \in G$ with $gHg^{-1} \cap H = \{e\}$, then $gL(H)g^{-1}$ is orthogonal to $L(H)$. This explains the singularity of $L(H)$ in $L(G)$ if $gHg^{-1} \cap H = \{e\}$ for all $g \in G \setminus H$ (see Lemma 3.3.2).

Orthogonality is a weaker notion than freeness (see [200]) and consequently has wider applications. □

6.4 Distances between subalgebras

A natural way to handle the distance between two subalgebras A and B of a type II_1 factor N is to consider a norm, or metric on $\mathbb{E}_A - \mathbb{E}_B$. The operator norm is rather large though it leads to a nice perturbation result (Theorem 6.4.2) due to Christensen [24]. On the other hand, strong operator, or weak operator convergence is too weak for these results. The infinity-two norm $\| \cdot \|_{\infty,2}$, or the equivalent metric derived from $\subset_\delta$, is the natural metric in estimating distances between subalgebras.

The infinity-two norm $\| \cdot \|_{\infty,2}$ of an operator ϕ on a II_1 factor N is defined by

$$\|\phi\|_{\infty,2} = \sup\{\|\phi(x)\|_2 : \|x\| \leq 1, \quad x \in N\}. \tag{6.4.1}$$

This norm induces a metric, denoted by d_2, on the von Neumann subalgebras of a type II_1 factor N. Christensen [24] defined a similar metric d to this; ours is different from but equivalent to d, as we will see below. Letting $\mathbb{E}_M$ denote, as usual, the unique trace preserving conditional expectation of N onto a subalgebra M, we define

$$d_2(L, M) = \|\mathbb{E}_L - \mathbb{E}_M\|_{\infty,2} \tag{6.4.2}$$

for any pair of von Neumann subalgebras L and M. It is clearly a metric, and we begin by showing completeness of the resulting metric space.

Lemma 6.4.1. *Let N be a type II_1 factor. The set of all von Neumann subalgebras of N is complete in the metric d_2.*

Proof. Let $\{L_n\}_{n=1}^\infty$ be a Cauchy sequence for d_2. Then, for each $x \in N$, $\{\mathbb{E}_{L_n}(x)\}_{n=1}^\infty$ is a Cauchy sequence in the $\| \cdot \|_2$-norm. Since the unit ball of N in the operator norm is also complete in the $\| \cdot \|_2$-norm, this sequence converges to an element of N. Thus we may define

$$\phi(x) = \lim_{n \to \infty} \mathbb{E}_{L_n}(x), \qquad x \in N, \tag{6.4.3}$$

where the limit is taken in the $\| \cdot \|_2$-norm, and we note that convergence in the w^*-topology also holds. From this it is easy to verify that ϕ is a unital completely positive trace preserving map, since each $\mathbb{E}_{L_n}$ has these properties. The identity

$$\phi^2 - \phi = \phi(\phi - \mathbb{E}_{L_n}) + (\phi - \mathbb{E}_{L_n})\mathbb{E}_{L_n} + (\mathbb{E}_{L_n} - \phi) \tag{6.4.4}$$

shows that ϕ is a projection, by letting $n \to \infty$ in (6.4.4).

Let L be the range of ϕ. It is clear that L is self-adjoint, and we now show that it is a von Neumann algebra. Each L_n is an algebra so, for all $x, y \in N$,

$$\mathbb{E}_{L_n}(\mathbb{E}_{L_n}(x)\mathbb{E}_{L_n}(y)) = \mathbb{E}_{L_n}(x)\mathbb{E}_{L_n}(y). \tag{6.4.5}$$

The right-hand side of the identity

$$\begin{aligned}
\phi(\phi(x)\phi(y)) - \phi(x)\phi(y) = &\ (\phi - \mathbb{E}_{L_n})(\phi(x)\phi(y)) \\
&+ \mathbb{E}_{L_n}((\phi(x) - \mathbb{E}_{L_n}(x))\phi(y)) \\
&+ \mathbb{E}_{L_n}(\mathbb{E}_{L_n}(x)(\phi(y) - \mathbb{E}_{L_n}(y))) \\
&+ \mathbb{E}_{L_n}(x)(\mathbb{E}_{L_n}(y) - \phi(y)) \\
&+ (\mathbb{E}_{L_n}(x) - \phi(x))\phi(y) \tag{6.4.6}
\end{aligned}$$

tends to 0 as $n \to \infty$, showing that $\phi(x)\phi(y) \in L$ for all $x, y \in N$. Thus L is an algebra.

Now consider a net $\{x_\alpha\}$ from the unit ball of L converging to $x \in L''$ in the w*-topology. Then $\phi(x_\alpha) = x_\alpha$, so

$$x_\alpha - \phi(x) = (\phi - \mathbb{E}_{L_n})(x_\alpha - x) + \mathbb{E}_{L_n}(x_\alpha - x), \qquad (6.4.7)$$

for each $n \geq 1$. The terms in (6.4.7) are uniformly bounded in norm, so to show w*-convergence of (x_α) to $\phi(x)$, it suffices to consider normal linear functionals of the form $\omega(\cdot) = \tau(\cdot y)$ for a fixed $y \in M$. Applying ω to (6.4.7) gives the estimate

$$|\omega(x_\alpha - \phi(x))| \leq \|\phi - \mathbb{E}_{L_n}\|_{\infty,2}\|x_\alpha - x\| \ \|y\|_2 + |\omega(\mathbb{E}_{L_n}(x_\alpha - x))|. \qquad (6.4.8)$$

Each $\omega\mathbb{E}_{L_n}$ is normal so, from (6.4.8),

$$\varlimsup_\alpha |\omega(x_\alpha - \phi(x))| \leq 2\|\phi - \mathbb{E}_{L_n}\|_{\infty,2}\|y\|_2. \qquad (6.4.9)$$

Then let $n \to \infty$ in (6.4.9) to obtain $\lim_\alpha \omega(x_\alpha - \phi(x)) = 0$. We conclude that $x = \phi(x) \in L$. By the Kaplansky density theorem (Theorem 2.2.3), $L = L''$, and L is a von Neumann algebra. This proves completeness. $\qquad \square$

We now present some useful inequalities concerning conditional expectations and the $\| \cdot \|_{\infty,2}$-norm, which are motivated by ideas in [24].

Theorem 6.4.2. *Let A, B and M be von Neumann subalgebras of a type II_1 factor N. Then the following inequalities hold:*

(i) $\|\mathbb{E}_A(I - \mathbb{E}_B)\|_{\infty,2}^2 \leq 2\|(I - \mathbb{E}_B)\mathbb{E}_A\|_{\infty,2}$;

(ii) $\|(I - \mathbb{E}_B)\mathbb{E}_A\|_{\infty,2}^2 \leq \|\mathbb{E}_A(I - \mathbb{E}_B)\|_{\infty,2}$;

(iii) $\|(I - \mathbb{E}_{B' \cap N})\mathbb{E}_{A' \cap N}\|_{\infty,2} \leq 2\|(I - \mathbb{E}_A)\mathbb{E}_B\|_{\infty,2}$;

(iv) $\|\mathbb{E}_{A' \cap M}(I - \mathbb{E}_{B' \cap N})\|_{\infty,2} \leq 2\|(I - \mathbb{E}_A)\mathbb{E}_B\|_{\infty,2}$;

(v) $\|\mathbb{E}_{A' \cap N} - \mathbb{E}_{B' \cap N}\|_{\infty,2} \leq 4\|\mathbb{E}_A - \mathbb{E}_B\|_{\infty,2}$;

(vi) $\|\mathbb{E}_A - \mathbb{E}_B\|_{\infty,2}^2 \leq \|(I - \mathbb{E}_B)\mathbb{E}_A\|_{\infty,2} + \|(I - \mathbb{E}_A)\mathbb{E}_B\|_{\infty,2}$;

(vii) $\|\mathbb{E}_{A' \cap A} - \mathbb{E}_{B' \cap B}\|_{\infty,2} \leq 5\|\mathbb{E}_A - \mathbb{E}_B\|_{\infty,2}$.

Proof. (i) Let $x \in N$, $\|x\| = 1$, be arbitrary. Then, using the Cauchy–Schwarz inequality, we have

$$\begin{aligned}
\|\mathbb{E}_A(I - \mathbb{E}_B)x\|_2^2 &= \langle (I - \mathbb{E}_B)\mathbb{E}_A(I - \mathbb{E}_B)x, x \rangle \\
&\leq \|(I - \mathbb{E}_B)\mathbb{E}_A\|_{\infty,2}\|(I - \mathbb{E}_B)x\| \ \|x\|_2 \\
&\leq 2\|(I - \mathbb{E}_B)\mathbb{E}_A\|_{\infty,2}. \qquad (6.4.10)
\end{aligned}$$

The result follows by taking the supremum over all such x in (6.4.10).

(ii)   As in (i),

$$\|(I - \mathbb{E}_B)\mathbb{E}_A x\|_2^2 = \langle \mathbb{E}_A(I - \mathbb{E}_B)\mathbb{E}_A x, x \rangle$$
$$\leq \|\mathbb{E}_A(I - \mathbb{E}_B)\|_{\infty,2}\|\mathbb{E}_A x\|\|x\|_2. \tag{6.4.11}$$

The inequality follows immediately.

(iii)   Consider $x \in N$, $\|x\| \leq 1$, and write $w = \mathbb{E}_{A' \cap N}(x)$. Let u be in the unitary group $\mathcal{U}(B)$ of B. Then

$$\|w - uwu^*\|_2 = \|wu - uw\|_2$$
$$\leq \|w(u - \mathbb{E}_A(u))\|_2 + \|(u - \mathbb{E}_A(u))w\|_2 + \|w\mathbb{E}_A(u) - \mathbb{E}_A(u)w\|_2, \tag{6.4.12}$$

and so

$$\|w - uwu^*\|_2 \leq 2\|(I - \mathbb{E}_A)(u)\|_2, \tag{6.4.13}$$

since w commutes with $\mathbb{E}_A(u)$ in (6.4.12). It follows that

$$\|w - uwu^*\|_2 \leq 2\|(I - \mathbb{E}_A)\mathbb{E}_B\|_{\infty,2}, \tag{6.4.14}$$

and the same inequality holds if uwu^* is replaced by any element of

$$\overline{\text{conv}}^{\|\cdot\|_2}\{uwu^*: \ u \in \mathcal{U}(B)\}.$$

By Lemma 3.6.5,

$$\|w - \mathbb{E}_{B' \cap N}(w)\|_2 \leq 2\|(I - \mathbb{E}_A)\mathbb{E}_B\|_{\infty,2}. \tag{6.4.15}$$

Replacing w by $\mathbb{E}_{A' \cap N}(x)$ in (6.4.15) gives the required inequality.

(iv)   Let $x, y \in N$ with $\|x\|, \|y\|_2 \leq 1$. Write $w = \mathbb{E}_{A' \cap M}(y)$ and let $u \in \mathcal{U}(B)$. Note that $w \in A'$ and $\|w\|_2 \leq 1$. Then

$$|\langle \mathbb{E}_{A' \cap M}(x - uxu^*), y \rangle| = |\langle x - uxu^*, w \rangle|$$
$$= |\tau((x - uxu^*)w^*)|$$
$$= |\tau(xw^* - xu^*w^*u)|$$
$$\leq |\tau(xw^* - xu^*\mathbb{E}_A(u)w^*| + |\tau(xu^*w^*(u - \mathbb{E}_A(u)))|. \tag{6.4.16}$$

Here we have used the module properties of conditional expectations and that w^* and $\mathbb{E}_A(u)$ commute. The last expression in (6.4.16) is no greater than

$$|\tau(xu^*(u - \mathbb{E}_A(u))w^*)| + |\tau(xu^*w^*(u - \mathbb{E}_A(u)))|$$
$$\leq 2\|(I - \mathbb{E}_A)(u)\|_2$$
$$\leq 2\|(I - \mathbb{E}_A)\mathbb{E}_B\|_{\infty,2} \tag{6.4.17}$$

since $u = \mathbb{E}_B(u)$. The estimates (6.4.16) and (6.4.17) combine to yield

$$\|E_{A' \cap M}(x - uxu^*)\|_2 \leq 2\|(I - \mathbb{E}_A)\mathbb{E}_B\|_{\infty,2}, \tag{6.4.18}$$

letting y vary over the unit ball of $L^2(N, \tau)$. The argument which led from (6.4.14) to (6.4.15) is also applicable here, giving

$$\|E_{A' \cap M}(I - \mathbb{E}_{B' \cap N})(x)\|_2 \le 2\|(I - \mathbb{E}_A)\mathbb{E}_B\|_{\infty,2}. \quad (6.4.19)$$

The result follows by letting x vary over the unit ball of N in (6.4.19).

(v) Since we have the simple algebraic identity

$$(I - \mathbb{E}_A)\mathbb{E}_B = (\mathbb{E}_B - \mathbb{E}_A)\mathbb{E}_B, \quad (6.4.20)$$

the inequality

$$\|(I - \mathbb{E}_A)\mathbb{E}_B\|_{\infty,2} \le \|\mathbb{E}_A - \mathbb{E}_B\|_{\infty,2} \quad (6.4.21)$$

is immediate. The inequality of (iv) then implies that

$$\|E_{A' \cap N}(I - \mathbb{E}_{B' \cap M})\|_{\infty,2} \le 2\|\mathbb{E}_A - \mathbb{E}_B\|_{\infty,2}, \quad (6.4.22)$$

and this is also valid with A and B interchanged. For any two idempotents P and Q, we have

$$P - Q = (P - Q)(Q + Q^\perp) = PQ + PQ^\perp - Q = PQ^\perp - P^\perp Q. \quad (6.4.23)$$

Letting P and Q be respectively $\mathbb{E}_{A' \cap N}$ and $\mathbb{E}_{B' \cap N}$ in (6.4.23) gives

$$\|\mathbb{E}_{A' \cap N} - \mathbb{E}_{B' \cap M}\|_{\infty,2} \le \|\mathbb{E}_{A' \cap N}(I - \mathbb{E}_{B' \cap N})\|_{\infty,2} + \|(I - \mathbb{E}_{A' \cap N})\mathbb{E}_{B' \cap N}\|_{\infty,2}$$
$$\le 4\|\mathbb{E}_A - \mathbb{E}_B\|_{\infty,2}, \quad (6.4.24)$$

using (6.4.22), (iii) and (iv).

(vi) For any $x \in N, \|x\| \le 1$,

$$\|(\mathbb{E}_A - \mathbb{E}_B)(x)\|_2^2 = \langle \mathbb{E}_A(x), (\mathbb{E}_A - \mathbb{E}_B)(x)\rangle - \langle \mathbb{E}_B(x), (\mathbb{E}_A - \mathbb{E}_B)(x)\rangle$$
$$= \langle(I - \mathbb{E}_B)\mathbb{E}_A(x), x\rangle + \langle(I - \mathbb{E}_A)\mathbb{E}_B(x), x\rangle$$
$$\le \|(I - \mathbb{E}_B)\mathbb{E}_A\|_{\infty,2} + \|(I - \mathbb{E}_A)\mathbb{E}_B\|_{\infty,2}, \quad (6.4.25)$$

from which the result follows.

(vii) Suppose that $x \in A' \cap N$ and $a \in A$. The module properties of $\mathbb{E}_A$ imply that

$$a\mathbb{E}_A(x) = \mathbb{E}_A(ax) = \mathbb{E}_A(xa) = \mathbb{E}_A(x)a. \quad (6.4.26)$$

Thus $\mathbb{E}_A(x) \in A \cap A'$. Replacement of x by $\mathbb{E}_{A' \cap N}(y)$, $y \in N$, gives $\mathbb{E}_A(\mathbb{E}_{A' \cap N}(y)) \in A \cap A'$. From this it follows that $\mathbb{E}_{A \cap A'} = \mathbb{E}_A \mathbb{E}_{A' \cap N}$, since this product of conditional expectations is the identity on $A \cap A'$. Then

$$\|\mathbb{E}_{A \cap A'} - \mathbb{E}_{B \cap B'}\|_{\infty,2} = \|\mathbb{E}_A \mathbb{E}_{A' \cap N} - \mathbb{E}_B \mathbb{E}_{B' \cap N}\|_{\infty,2}$$
$$\le \|(\mathbb{E}_A - \mathbb{E}_B)\mathbb{E}_{A' \cap N}\|_{\infty,2} + \|\mathbb{E}_B(\mathbb{E}_{A' \cap N} - \mathbb{E}_{B' \cap N})\|_{\infty,2}$$
$$\le \|\mathbb{E}_A - \mathbb{E}_B\|_{\infty,2} + \|\mathbb{E}_{A' \cap N} - \mathbb{E}_{B' \cap N}\|_{\infty,2}$$
$$\le 5\|\mathbb{E}_A - \mathbb{E}_B\|_{\infty,2}, \quad (6.4.27)$$

using the inequality of (v) to estimate the penultimate term in (6.4.27). $\square$

Corollary 6.4.3. *If A and B are masas in a type II_1 factor N, then*

$$\max\{\|(I - \mathbb{E}_A)\mathbb{E}_B\|_{\infty,2}, \ \|(I - \mathbb{E}_B)\mathbb{E}_A\|_{\infty,2}\}$$
$$\leq \|\mathbb{E}_A - \mathbb{E}_B\|_{\infty,2}$$
$$\leq 3\min\{\|(I - \mathbb{E}_A)\mathbb{E}_B\|_{\infty,2}, \ \|(I - \mathbb{E}_B)\mathbb{E}_A\|_{\infty,2}\} \tag{6.4.28}$$

Proof. The first inequality is immediate. The second one follows from Theorem 6.4.2 (iv) since symmetry, $A = A' \cap N$ and $B = B' \cap N$ give

$$\|\mathbb{E}_A - \mathbb{E}_B\|_{\infty,2} \leq \|\mathbb{E}_A(I - \mathbb{E}_B)\|_{\infty,2} + \|(I - \mathbb{E}_A)\mathbb{E}_B\|_{\infty,2}$$
$$\leq 3\|(I - \mathbb{E}_A)\mathbb{E}_B\|_{\infty,2}. \tag{6.4.29}$$

$\square$

Remark 6.4.4. For each pair of von Neumann subalgebras A and B of a II_1 factor N and for each $\delta > 0$, recall from Section 6.1 that $A \subset_\delta B$ means that $\|a - \mathbb{E}_B(a)\|_2 \leq \delta$ for all $a \in A$ with $\|a\| \leq 1$. In terms of the $\|\cdot\|_{\infty,2}$ norm, $A \subset_\delta B$ if, and only if, $\|(I - \mathbb{E}_B)\mathbb{E}_A\|_{\infty,2} \leq \delta$.

The metric d_1 of Christensen [24] is defined by

$$d_1(A, B) = \inf\{\delta : \delta > 0, \ A \subset_\delta B \quad \text{and} \quad B \subset_\delta A\}.$$

This corresponds, in our notation, to

$$d_1(A, B) = \max\{\|(I - \mathbb{E}_A)\mathbb{E}_B\|_{\infty,2}, \quad \|(I - \mathbb{E}_B)\mathbb{E}_A\|_{\infty,2}\} \tag{6.4.30}$$

for all von Neumann subalgebras A and B. It follows from Theorem 6.4.2 (vi) that

$$d_1(A, B) \leq d_2(A, B) \leq (2d_1(A, B))^{1/2} \tag{6.4.31}$$

and the two metrics are equivalent.

Note that the estimate between the metrics is not linear for general von Neumann subalgebras but is linear for masas by Corollary 6.4.3. $\square$

All but part (iii) in the following corollary are due to Christensen in [24]. The idea of the proofs is the same but the technical results of Theorem 6.4.2 reduce the calculations to an elementary level. Several other subsets are shown to be closed in [24]. We anticipate the result of Chapter 11 which says that a masa A in a II_1 factor is singular if and only if it satisfies the inequality

$$\|\mathbb{E}_A - \mathbb{E}_{uAu^*}\|_{\infty,2} \geq \|u - \mathbb{E}_A(u)\|_{2,\tau} \tag{6.4.32}$$

for all unitaries $u \in N$. This will be used in part (iii) below.

Corollary 6.4.5. *Let N be a type II_1 factor. The following sets of von Neumann subalgebras are closed in the d_2-metric:*

(i) *abelian subalgebras;*

(ii) *masas;*

(iii) *singular masas;*

(iv) *subfactors;*

(v) *subfactors with trivial relative commutant.*

Proof. (i) A subalgebra A is abelian if and only if $A \subseteq A' \cap N$, which is equivalent to $(I - \mathbb{E}_{A' \cap N})\mathbb{E}_A = 0$. Suppose that $\{A_n\}_{n=1}^{\infty}$ is a sequence of abelian algebras converging to A in the d_2-metric. The map $B \to B' \cap N$ is continuous, by Theorem 6.4.2 (v), so the result follows by taking limits in the identity

$$(I - \mathbb{E}_{A' \cap N})\mathbb{E}_A = (I - \mathbb{E}_{A' \cap N})(\mathbb{E}_A - \mathbb{E}_{A_n}) + (I - \mathbb{E}_{A' \cap N})\mathbb{E}_{A_n}$$
$$= (I - \mathbb{E}_{A' \cap N})(\mathbb{E}_A - \mathbb{E}_{A_n}) + (I - \mathbb{E}_{A'_n \cap N})\mathbb{E}_{A_n}$$
$$+ (\mathbb{E}_{A'_n \cap N} - \mathbb{E}_{A' \cap N})\mathbb{E}_{A_n}. \tag{6.4.33}$$

(ii) A subalgebra A is a masa if and only if $A = A' \cap N$. If a sequence of masas $\{A_n\}_{n=1}^{\infty}$ converges to a subalgebra A then this sequence also converges to $A' \cap N$, by Theorem 6.4.2 (v). This proves the result.

(iii) If a sequence of singular masas $\{A_n\}_{n=1}^{\infty}$ converges to a subalgebra A, then A is a masa by (ii). From (6.4.32), for each unitary $u \in N$ and $n \geq 1$, we have

$$\|\mathbb{E}_{uA_n u^*} - \mathbb{E}_{A_n}\|_{\infty,2} \geq \|u - \mathbb{E}_{A_n}(u)\|_2, \tag{6.4.34}$$

and this inequality persists after letting $n \to \infty$. This shows that A is also singular, by (6.4.32).

(iv) The condition for M to be a subfactor is $M \cap M' = \mathbb{C}1$. The result then follows from the continuity of the map $A \to A \cap A'$, which is Theorem 6.4.2 (vii).

(v) The condition $M' \cap N = \mathbb{C}1$ already entails that M is a factor, so the result follows from continuity of the map $M \to M' \cap N$, which is proved in Theorem 6.4.2 (v). $\qquad \square$

Chapter 7

The Pukánszky invariant

7.1 Introduction

Pukánszky [154] defined an invariant for a masa A in a separable type II_1 factor N based on the type I decomposition of $\mathcal{A}' = (A \cup JAJ)'$. The latter algebra is the commutant of an abelian algebra and so is type I. The standard theory then gives a decomposition as a direct sum of n-homogeneous algebras (where $n = \infty$ is allowed). The Pukánszky invariant is essentially the union of these numbers; the exact definition is given in Definition 7.1.2 whose wording takes account of a subtle point that we explain subsequently. The only previous invariants for masas were Cartan (= regularity), semi-regularity and singularity due to Dixmier [47]. Attention is restricted to separable type II_1 factors to avoid pathologies and so there is only one infinite cardinal, which is denoted ∞. See the paper by Popa [139] for a discussion of some of the unusual problems that can arise in non-separable type II_1 factors. In this chapter we define the Pukánszky invariant, prove the basic theorems concerning it and give some examples.

We briefly review our standard notation from earlier chapters. Let A be a masa in a separable type II_1 factor N with a faithful normal trace τ, let J be the conjugate linear isometry $Jx = x^*$ on $L^2(N)$ and let ξ be the vector in $L^2(N)$ corresponding to 1. We let e_A be the projection of $L^2(N)$ onto $L^2(A)$. We denote by $\mathcal{A}$ the abelian von Neumann algebra generated by A and JAJ. The following lemma is required to justify the definition of the Pukánszky invariant which we will give shortly.

Lemma 7.1.1. *Let N be a II_1 factor represented on $L^2(N)$ and let A be a masa in N. Then $e_A \in \mathcal{A}$ and is a central projection in $\mathcal{A}'$.*

Proof. Let U be the unitary group of A, viewed as a discrete abelian group, and let μ be an invariant mean on $\ell^\infty(U)$. We will exhibit e_A as a suitable average of the operators $uJuJ \in \mathcal{A}$. Define a bounded sesquilinear form on $L^2(N)$ by

$$\Psi(\eta_1, \eta_2) = \mu(u \mapsto \langle uJuJ\eta_1, \eta_2 \rangle), \quad \eta_1, \eta_2 \in L^2(N), \ u \in U. \tag{7.1.1}$$

Then there is an operator $t \in B(L^2(N))$ such that $\langle t\eta_1, \eta_2 \rangle$ is equal to this last expression. If $x \in \mathcal{A}'$, then

$$\langle uJuJx\eta_1, \eta_2 \rangle = \langle uJuJ\eta_1, x^*\eta_2 \rangle \tag{7.1.2}$$

for all $u \in U$ and $\eta_1, \eta_2 \in L^2(N)$. If we apply μ to (7.1.2), the result is

$$\langle tx\eta_1, \eta_2 \rangle = \langle t\eta_1, x^*\eta_2 \rangle, \tag{7.1.3}$$

showing that $t \in (\mathcal{A}')' = \mathcal{A}$ since $x \in \mathcal{A}'$ was arbitrary. In particular, t commutes with both A and JAJ.

We now show that $t = e_A$. If $\eta_1 = a\xi$ for a fixed but arbitrary $a \in A$, then $uJuJa\xi = uau^*\xi = a\xi$, and so $\langle ta\xi, \eta_2 \rangle = \langle a\xi, \eta_2 \rangle$ for all $\eta_2 \in L^2(N)$, by applying μ. Thus $ta\xi = a\xi$, and in particular, $t\xi = \xi$. For each $v \in U$, the invariance of μ gives

$$\begin{aligned}
\langle t\eta_1, \eta_2 \rangle &= \mu(u \mapsto \langle vuJvuJ\eta_1, \eta_2 \rangle) \\
&= \mu(u \mapsto \langle vJvJuJuJ\eta_1, \eta_2 \rangle) \\
&= \langle vJvJt\eta_1, \eta_2 \rangle, \quad \eta_1, \eta_2 \in L^2(N). \tag{7.1.4}
\end{aligned}$$

If we let η_1 be $x\xi$ for a fixed but arbitrary $x \in N$, then (7.1.4) becomes

$$\langle tx\xi, \eta_2 \rangle = \langle tvxv^*\xi, \eta_2 \rangle, \quad v \in U. \tag{7.1.5}$$

Taking convex combinations and a weak limit in (7.1.5) gives

$$\begin{aligned}
\langle tx\xi, \eta_2 \rangle &= \langle t\mathbb{E}_A(x)\xi, \eta_2 \rangle = \langle \mathbb{E}_A(x)t\xi, \eta_2 \rangle \\
&= \langle \mathbb{E}_A(x)\xi, \eta_2 \rangle = \langle e_A x\xi, \eta_2 \rangle, \tag{7.1.6}
\end{aligned}$$

using $t\xi = \xi$. Since $x \in N$ and $\eta_2 \in L^2(N)$ were arbitrary, we conclude that $t = e_A$, and that $e_A \in \mathcal{A}$. $\qquad \square$

Since $\mathcal{A}$ is an abelian von Neumann algebra on $L^2(N)$, its commutant $\mathcal{A}'$ is a type I von Neumann algebra, because type is preserved under commutation [105, Chapter 9]. The projection e_A is not I and is in $\mathcal{A}$ by Lemma 7.1.1, so $\mathcal{A}'(1 - e_A)$ is a type I von Neumann algebra with centre $\mathcal{A}(1 - e_A)$.

Definition 7.1.2. The *Pukánszky invariant* Puk(A) of a masa A in a separable type II$_1$ factor N is the set of $n \in \mathbb{N} \cup \{\infty\}$ such that $(1 - e_A)\mathcal{A}'$ has a non-zero type I$_n$ part. Equivalently it is the set

$$\{n : \text{there is a projection } q \in \mathcal{A}, \ 0 \neq q \leq 1 - e_A, \ q\mathcal{A}' \text{ is homogeneous of type I}_n\}.$$

In cases of potential ambiguity, the ambient factor will be indicated by the notation Puk(A, N). $\qquad \square$

Removing the projection e_A from $\mathcal{A}'$ cuts off the part $\mathcal{A}'e_A = \mathcal{A}e_A$, which is abelian for all masas A and leaves a better invariant for A. Its inclusion would add the number 1 to all Pukánszky invariants, so masas with invariants $\{2\}$

and $\{1,2\}$, for example, could not be distinguished. Note that this projection e_A plays an important role in the Pukánszky invariant in tensor products (see Theorem 7.3.1).

The Pukánszky invariant is always a non-empty subset of $\mathbb{N} \cup \{\infty\}$, since $e_A \neq I$. Pukánszky [154] defined this invariant and gave examples of singular masas A in the hyperfinite type II_1 factor R that have Pukánszky invariant $\{n\}$ for all n in $\mathbb{N} \cup \{\infty\}$. The notation $\mathrm{Puk}(A)$ comes from [176].

This invariant is an isomorphism invariant for masas in separable type II_1 factors, because isomorphisms are spatially induced. If ϕ is an isomorphism from a type II_1 factor M onto N with $\phi(A) = B$, where A is a masa in M, then this isomorphism is spatially induced by the unitary w from $L^2(M)$ to $L^2(N)$, where $w(m) = \phi(m)$ for all $m \in M$. The spatial isomorphism $x \mapsto wxw^*$ from $B(L^2(M))$ to $B(L^2(N))$ sends A to B, e_A to e_B, and J_M to J_N. Thus it maps $e_A\mathcal{A}'$ to $e_B\mathcal{B}'$, and consequently $\mathrm{Puk}(A) = \mathrm{Puk}(B)$. It should be noted, however, that it is quite possible to have non-conjugate masas with the same $\mathrm{Puk}(\cdot)$. Examples will be given below.

The Pukánszky invariant is well-behaved with respect to tensor products of factors and masas inside them (Theorem 7.3.1) but is badly behaved with respect to passing from the full factor to a subfactor containing the masa (Example 7.5.5). The Pukánszky invariant is upper semi-continuous for convergence in the $\|\cdot\|_{\infty,2}$ norm, as we discuss in Section 7.6. We will give many values below of $\mathrm{Puk}(\cdot)$ in the hyperfinite and free group factors in Section 7.5. Recently White [207] discovered that all possible values can occur in the hyperfinite factor and also in certain McDuff factors. A discussion of his methods, which are quite different from the ones presented here, would take us too far afield, so we refer the reader to [207].

Section 7.2 contains an important technical theorem of Popa [141] relating A, $\mathcal{A}$ and $\mathcal{A}'$ on $\overline{A v}$ for all v in the normaliser $\mathcal{N}(A)$. Section 7.3 contains some general results on the Pukánszky invariant and a discussion of known results of the values it attains in various algebras. Methods for computing the invariant in group factors are given in Section 7.4, and examples are presented in Section 7.5.

While this invariant was developed by Pukánszky, it should be observed that Ambrose and Singer also discovered a similar version but did not publish their results; their notes seem to be unavailable. However, there is a discussion of their ideas in the survey paper by Kadison [101]. They introduced *simple* masas as those for which $\mathcal{A}$ is a masa in $B(L^2(N))$; equivalently as those for which $\mathrm{Puk}(A) = \{1\}$.

The following values of the Pukánszky invariant $\mathrm{Puk}(A)$ of a masa A in a separable type II_1 factor are known. Some of these mentioned here will be shown in the examples of Section 7.5.

(a) For each $n \in \mathbb{N} \cup \{\infty\}$, Pukánszky constructed a singular masa A in the hyperfinite type II_1 factor R such that $\mathrm{Puk}(A) = \{n\}$ [154].

(b) Popa [141] proved that $\mathrm{Puk}(A) = \{1\}$ for all Cartan subalgebras A and that the generator masa in the free group factor $L(\mathbb{F}_n)$, $n \geq 2$, has invariant $\{\infty\}$.

(c) Rădulescu [156] proved that $\mathrm{Puk}(A) = \{\infty\}$ for the radial (= Laplacian masa) masa A in $L(\mathbb{F}_n)$ for $2 \leq n < \infty$. He and Boca [8] proved a corresponding result for certain masas in free products.

(d) Roberston and Steger [162] showed that ∞ is in $\mathrm{Puk}(A)$, or $\mathrm{Puk}(A) = \{\infty\}$, for certain masas arising in factors from $\tilde{A}_2$-groups. Their results also give the result of (b) that $\mathrm{Puk}(A) = \{\infty\}$ for A a generator masa of $L(\mathbb{F}_n)$.

(e) Niţică and Török [123] showed that $\mathrm{Puk}(A) = \{\infty\}$ for certain Laplacian style masas in $L(\mathbb{F}_n)$.

(f) Neshveyev and Størmer [119] showed that if $1 \in W \subseteq \mathbb{N} \cup \{\infty\}$, then there is a masa A in the hyperfinite type II_1 factor R with $\mathrm{Puk}(A) = W$. Their result is based on an ergodic theory example due to Kwiatkowski and Lemańczyk [110].

(g) Ge [75] showed that each masa A in $L(\mathbb{F}_n)$ $(2 \leq n \leq \infty)$ has $\mathrm{Puk}(A) \neq \{1\}$ by showing that $A\xi A$ is not dense in $L^2(L(\mathbb{F}_n))$ using important methods of Voiculescu [199], who has shown that $L(\mathbb{F}_n)$ has no Cartan masa.

(h) Dykema [59] showed that each masa A in $L(\mathbb{F}_n)$ has $\mathrm{Puk}(A)$ not contained in any finite subset of $\mathbb{N}$ by checking the hypotheses of Lemma 7.3.3, and proved the same result for the interpolated free group factors. Stefan [181] extended this to subfactors of finite index in these free group factors.

(i) The authors [176] gave various examples of Pukánszky invariants based on groups. Some of these will be discussed in Section 7.5.

(j) White [208] has shown that $\mathrm{Puk}(A) = \{1\}$ for each Tauer masa A in the hyperfinite type II_1 factor R, which shows that there are many non-conjugate masas with the same Pukánszky invariant.

(k) Dykema and the authors [62] showed that if $\{\infty\} \subseteq W \subseteq \mathbb{N} \cup \{\infty\}$, then W is the Pukánszky invariant for masas in the hyperfinite factor and in any free group factor $L(\mathbb{F}_n)$, $2 \leq n \leq \infty$.

(l) White [207] showed that all possible values occur for the hyperfinite type II_1 factor R, and for certain McDuff factors.

In the last section of this chapter, we list a number of open problems concerning the Pukánszky invariant.

7.2 The algebras A, $\mathcal{A}$, $\mathcal{A}'$ and $\mathcal{N}(A)''$ interact

This section contains an important technical result, due to Popa [141], that links together the algebras A, $\mathcal{A} = (A \cup JAJ)''$, $\mathcal{A}'$ and $\mathcal{N}(A)''$ when they act on elements of the groupoid normaliser $\mathcal{GN}(A)$ in $L^2(N)$. Basically it says that the three algebras A, $\mathcal{A}$ and $\mathcal{A}'$ agree when cut down to $vL^2(A)$ for all v in $\mathcal{GN}(A)$, and $\mathcal{A}$ and $\mathcal{A}'$ agree on $L^2(\mathcal{N}(A)'')$.

The following notation is convenient in the statement and proof of Theorem 7.2.1 and applies in this section only. If X is a linear subspace of N, or of $L^2(N)$, let e_X denote the orthogonal projection from $L^2(N)$ onto the $\|\cdot\|_2$-closure of X.

Theorem 7.2.1. *Let A be a masa in a separable type II_1 factor N and let*

$$\mathcal{A} = (A \cup JAJ)''.$$

(i) *If $v \in \mathcal{GN}(A)$, then*

$$e_{Av} = e_{\mathcal{A}v} = e_{\mathcal{A}'v} \in \mathcal{A} \quad and \tag{7.2.1}$$
$$Ae_{Av} = \mathcal{A}e_{\mathcal{A}v} = A'e_{Av} = \mathcal{A}'e_{\mathcal{A}'v}. \tag{7.2.2}$$

(ii) *If $A \subseteq M \subseteq \mathcal{N}(A)''$, then $e_M \in \mathcal{A}$ and $\mathcal{A}e_M = \mathcal{A}'e_M$.*

Proof. (i)   If v is in $\mathcal{GN}(A)$, then

$$Av\xi = A(vv^*)v\xi = v(v^*Av)\xi = vA\xi \tag{7.2.3}$$

so that

$$A\,JAJv\xi = AvA\xi = AAv\xi = Av\xi \quad and \quad e_{Av\xi} = e_{\mathcal{A}v\xi} = e_{\mathcal{A}vA\xi}. \tag{7.2.4}$$

We first consider the case $v = 1$, from which the general case will be deduced. By Lemma 7.1.1, $e_A \in \mathcal{A}$. The projection $e_{\mathcal{A}'\xi}$ is the smallest projection p in $\mathcal{A}$ with $p\xi = \xi$ so $e_{\mathcal{A}'\xi} \leq e_A$, because $e_A \in \mathcal{A}$ and $e_A\xi = \xi$. The inclusions $\mathcal{A}'\xi \supseteq \mathcal{A}\xi \supseteq A\xi$ imply that

$$e_{\mathcal{A}'\xi} = e_{\mathcal{A}\xi} = e_A. \tag{7.2.5}$$

The action of A on $L^2(A)$ has a cyclic vector ξ, so A is a masa in $B(L^2(A))$, by Theorem 2.3.4. Hence $\mathcal{A}e_A = Ae_A$, since $\mathcal{A}e_A$ is abelian and contains Ae_A. Regarding $\mathcal{A}'e_A$ as the commutant of $\mathcal{A}e_A$ in $B(L^2(A))$ gives

$$A'e_A = \mathcal{A}'e_A = \mathcal{A}e_A = Ae_A, \tag{7.2.6}$$

which proves (7.2.1) and (7.2.2) with $v = 1$.

Let u be in $\mathcal{N}(A)$. If K is a closed linear subspace of $L^2(A)$, then $ue_K u^* = e_{uK}$. The (spatial) automorphism $x \mapsto uxu^*$ on $B(L^2(N))$ leaves A invariant and is the identity automorphism on JAJ so leaves $\mathcal{A}$ and $\mathcal{A}'$ invariant. Also $ue_A u^* = e_{uA} = e_{Au}$ and similarly for $\mathcal{A}$ and $\mathcal{A}'$. This proves (7.2.1) and (7.2.2) for a unitary $u \in \mathcal{N}(A)$.

If v is in $\mathcal{GN}(A)$, then $v = pu$ with p a projection in A and u a unitary in $\mathcal{N}(A)$ by Lemma 6.2.3 (iii). Observe that

$$e_{\mathcal{A}'v} = e_{p\mathcal{A}'u} = pe_{\mathcal{A}'u} \tag{7.2.7}$$

since both $e_{p\mathcal{A}'u}$ and $pe_{\mathcal{A}'u}$ are the orthogonal projections onto the $\|\cdot\|_2$-norm closure of $p\mathcal{A}'u$. A similar equation holds with A and $\mathcal{A}$ in place of $\mathcal{A}'$ so proving (7.2.1) and (7.2.2).

(ii) Let $A \subseteq M \subseteq \mathcal{N}(A)''$. By Corollary 6.2.5 (i), $M = \mathcal{N}(A, M)''$ so that e_M is the supremum of the projections e_{Au} over u in $\mathcal{N}(A, M)$, since $L^2(M)$ equals the $\|\cdot\|_2$-norm closed linear span of

$$\cup\{Au : u \in \mathcal{N}(A, M)\}.$$

Hence e_M is in the abelian algebra $\mathcal{A}$ and $\mathcal{A}e_M = \mathcal{A}'e_M$, since $\mathcal{A}e_{Au} = \mathcal{A}'e_{Au}$ for all u in $\mathcal{N}(A)$. $\qquad\square$

Is the converse of Theorem 7.2.1 (ii) above true in some sense? If $A \subseteq M \subseteq N$ with A a masa in N and $e_M \in \mathcal{A}$, is $M \subseteq \mathcal{N}(A, N)''$? Is this true with the additional assumption that $\mathcal{A}e_M = \mathcal{A}'e_M$?

The following corollary has proved useful in showing that certain masas are singular: see, for example [123, 141, 156, 162].

Corollary 7.2.2. *Let A be a masa in a separable type* II_1 *factor N.*

(i) *The following statements are equivalent:*

(a) *the algebra A has a cyclic vector in $L^2(N)$;*

(b) *the algebra $\mathcal{A}$ is a masa in $B(L^2(N))$;*

(c) $\mathcal{A}'$ *is abelian;*

(d) $\mathrm{Puk}(A) = \{1\}$.

(ii) *If A is a Cartan subalgebra of N, then* $\mathrm{Puk}(A) = \{1\}$.

(iii) *If* $\mathrm{Puk}(A) \subseteq \{2, 3, \cdots, \infty\}$, *then A is singular.*

(iv) *If* $\mathcal{N}(A)'' \neq A$, *then* $1 \in \mathrm{Puk}(A)$.

Proof. (i)   This is the definition of $\mathrm{Puk}(A)$, the result that, on a separable Hilbert space, an abelian subalgebra of $B(H)$ is a masa if, and only if, it has a cyclic vector (Theorem 2.3.4), and the standard fact that an abelian von Neumann algebra is a masa if and only if its commutant is abelian.

(ii)   If A is a Cartan subalgebra of N, then $\mathcal{N}(A)'' = N$ and $e_N = I$. Thus $\mathcal{A}' = \mathcal{A}$ by Theorem 7.2.1.

(iii)   If $\mathrm{Puk}(A) \subseteq \{2, 3, \cdots, \infty\}$, then e_A is the maximal projection in $\mathcal{A}$ such that $\mathcal{A}'e_A$ is abelian, since $(1 - e_A)\mathcal{A}'$ has no abelian (= type I_1) central direct summand. By Theorem 7.2.1 (ii), $e_M\mathcal{A}' = e_M\mathcal{A}$ is abelian so $e_M = e_A$, where $M = \mathcal{N}(A)''$. Hence $\mathcal{N}(A)'' = A$ and A is singular.

(iv)   If $\mathcal{N}(A)'' = M \neq A$, then $e_M \neq e_A$ so

$$(1 - e_A)e_M\mathcal{A}' = (e_M - e_A)\mathcal{A}' \tag{7.2.8}$$

is a non-zero central type I_1 direct summand of $(1 - e_A)\mathcal{A}'$. $\qquad\square$

Remark 7.2.3. (1) Pukánszky [154] gave an example of a singular masa A in the hyperfinite type II_1 factor R with $\mathrm{Puk}(A) = \{1\}$ (see Example 7.5.3). This prohibits certain obvious potential strengthenings of (ii), (iii) and (iv) in Corollary 7.2.2.

(2) Masas A in arbitrary separable II_1 factors N exist with $1 \in \mathrm{Puk}(A)$. Take a projection $p \in N$ of trace $1/2$, and let $M = pNp$, so that N is isomorphic to $M \otimes \mathbb{M}_2$. Then choose a masa B in M, and let $A = B \otimes \mathbb{D}_2$ where $\mathbb{D}_2$ is the diagonal masa in $\mathbb{M}_2$. Then A is not singular because it is normalised by $1 \otimes u$ where the unitary u permutes the canonical basis vectors of $\mathbb{C}^2$. Thus $1 \in \mathrm{Puk}(A)$ by Corollary 7.2.2 (iii).

(3) Theorem 7.2.1 and Corollary 7.2.2 are due to Popa [141]. □

To this point, the Cartan masas are the only ones for which we can determine the Pukánszky invariants, and they are always $\{1\}$. Since we must develop a considerable amount of theory before we can present classes of examples with varying $\mathrm{Puk}(\cdot)$, we insert here an example of a masa A with $\mathrm{Puk}(A) \neq \{1\}$. It will reappear in Section 7.5, where we will be able to show that its invariant is $\{2\}$ (Example 7.5.3).

Corollary 7.2.2 (i) shows that a masa A has $\mathrm{Puk}(A) = \{1\}$ if and only if $\mathcal{A}'$ is abelian, so we will exhibit a masa A for which $\mathcal{A}'$ is non-abelian, guaranteeing that $\mathrm{Puk}(A) \neq \{1\}$.

Example 7.2.4. Let $\mathbb{Q}$ denote the rationals and let $\mathbb{Z}_{\mathrm{odd}}$ denote the set of odd integers. Define a countable discrete group as the set of matrices

$$G = \left\{ \begin{pmatrix} 4^n p/q & x \\ 0 & 1 \end{pmatrix} : x \in \mathbb{Q}, \ n \in \mathbb{Z}, \ p, q \in \mathbb{Z}_{\mathrm{odd}} \right\},$$

under matrix multiplication. Let H be the abelian subgroup of diagonal matrices in G, and let k be the diagonal matrix $\begin{pmatrix} 2 & 0 \\ 0 & 1 \end{pmatrix}$, which is not in G, but which defines an outer automorphism ϕ of G by $\phi(g) = kgk^{-1}$, $g \in G$. Then ϕ has the effect of doubling the $(1,2)$ entry of each matrix in G. Note that k commutes with each matrix $h \in H$, so that

$$\phi(h_1 g h_2) = h_1 \phi(g) h_2, \quad h_1, h_2 \in H, \ g \in G. \tag{7.2.9}$$

Routine matrix calculations show that G is I.C.C., and that H satisfies the algebraic conditions of Lemma 3.3.1 for $A = L(H)$ to be a masa in $L(G)$.

Let $g_0 = \begin{pmatrix} 1 & 1 \\ 0 & 1 \end{pmatrix} \in G$, and define a subset of G by

$$F = \{h_1 g_0 h_2 : h_1, h_2 \in H\},$$

the double coset of g_0 with respect to H. The standard basis for $\ell^2(G)$ is $\{\delta_g : g \in G\}$, and we may define a partial isometry v by specifying the values of v on the basis vectors by

$$v\delta_g = \begin{cases} \delta_{\phi(g)}, & g \in F, \\ 0, & g \notin F. \end{cases}$$

Equation (7.2.9) shows that v commutes with left and right multiplications by elements of H, so $v \in \mathcal{A}'$. Since the initial and final spaces for v are respectively

$$\overline{\text{span}}\{\delta_g : g \in F\} \quad \text{and} \quad \overline{\text{span}}\{\delta_{\phi(g)} : g \in F\},$$

and these are unequal (δ_{g_0} is in the first but not the second), it follows that $vv^* \neq v^*v$, so $\mathcal{A}'$ is non-abelian. As we pointed out before this example, this ensures that $\text{Puk}(A) \neq \{1\}$. In fact, the invariant for this masa is $\{2\}$, a special case of Example 7.5.3. $\qquad\square$

7.3 Properties of the Pukánszky invariant

This section contains some basic general properties of the Pukánszky invariant, firstly with respect to tensor products. Note that although $e_A \mathcal{A}'$ is removed from the definition of the Pukánszky invariant of A, the projection e_A plays an important role when A interacts with other objects. In the next theorem it contributes the term $\text{Puk}(A) \cup \text{Puk}(B)$ to the invariant of the tensor product. In order to include the possibility that one of the factors is finite dimensional, we phrase the next result in terms of finite factors.

The statement of the next theorem contains a product of two sets. If S and T are two subsets of $\mathbb{N} \cup \{\infty\}$, then their product ST is defined to be

$$ST = \{mn : m \in S, \, n \in T\}.$$

Theorem 7.3.1. *If A and B are masas in separable finite factors M and N respectively, then $A \overline{\otimes} B$ is a masa in $M \overline{\otimes} N$ and*

$$\text{Puk}(A \overline{\otimes} B) = \text{Puk}(A) \cup \text{Puk}(B) \cup \text{Puk}(A)\text{Puk}(B).$$

In particular, if N is a matrix factor, then $\text{Puk}(A \overline{\otimes} B) = \text{Puk}(A) \cup \{1\}$.

Proof. Tomita's commutant theorem, (see Theorem 2.2.4), implies that $(A \overline{\otimes} B)' = A' \overline{\otimes} B'$, and

$$(A' \overline{\otimes} B') \cap (M \overline{\otimes} N) = (A' \cap M) \overline{\otimes} (B' \cap N) = A \overline{\otimes} B, \qquad (7.3.1)$$

so that $A \overline{\otimes} B$ is a masa in $M \overline{\otimes} N$. Let $C = A \overline{\otimes} B$, $\mathcal{A} = (A \cup JAJ)''$, $\mathcal{B} = (B \cup JBJ)''$, and $\mathcal{C} = (C \cup JCJ)''$. The properties of J in relation to the tensor product show that $J(A \overline{\otimes} B)J = JAJ \overline{\otimes} JBJ$ since $J_{M \overline{\otimes} N} = J_M \otimes J_N$. This implies that

$$\mathcal{C} = ((A \overline{\otimes} B) \cup (JAJ \overline{\otimes} JBJ))'' = (A \cup JAJ)'' \overline{\otimes} (B \cup JBJ)'' = \mathcal{A} \overline{\otimes} \mathcal{B}. \quad (7.3.2)$$

If e_X is the projection onto the closure of X in the relevant L^2-space, then

$$1 \otimes 1 - e_C = e_C{}^{\perp} = e_A{}^{\perp} \otimes 1 + e_A \otimes e_B{}^{\perp} \qquad (7.3.3)$$

so that

$$e_C{}^\perp \mathcal{C}' = (e_A{}^\perp \otimes 1)\mathcal{C}' \oplus (e_A \otimes e_B{}^\perp)\mathcal{C}'$$
$$= (e_A{}^\perp \mathcal{A}'\overline{\otimes}\mathcal{B}') \oplus (e_A \mathcal{A}'\overline{\otimes}e_B{}^\perp \mathcal{B}')$$
$$= (e_A{}^\perp \mathcal{A}'\overline{\otimes}e_B \mathcal{B}') \oplus (e_A \mathcal{A}'\overline{\otimes}e_B{}^\perp \mathcal{B}') \oplus (e_A{}^\perp \mathcal{A}'\overline{\otimes}e_B{}^\perp \mathcal{B}'). \qquad (7.3.4)$$

Here $e_A \mathcal{A}' = e_A A$ and $e_B \mathcal{B}' = e_B B$, by Theorem 7.2.1, so both are abelian. Since the tensor product of a type I_m von Neumann algebra with a type I_n von Neumann algebra is type I_{mn}, the three terms in the direct sum contribute the three terms expressing $\mathrm{Puk}(A\overline{\otimes}B)$ in the statement of the theorem. □

Remark 7.3.2. Theorem 7.3.1 and induction give the formula for the Pukánszky invariant for a finite tensor product. If A_j is a masa in a separable type II_1 factor N_j for $1 \le j \le k$, then

$$\mathrm{Puk}(A_1\overline{\otimes}\ldots\overline{\otimes}A_k) = \{n_{j_1}\cdots n_{j_m} : 1 \le j_1 < j_2 < \cdots < j_m \le k,$$
$$\text{and } n_{j_i} \in \mathrm{Puk}(A_{j_i}), \quad 1 \le m \le k\} \qquad (7.3.5)$$

is the set of all products of the Pukánszky invariants of A_j up to k factors with at most one from each factor. Once we have given examples of masas A_n with $\mathrm{Puk}(A_n) = \{n\}$, $1 \le n \le \infty$, equation (7.3.5) will generate many new values of $\mathrm{Puk}(\cdot)$. □

Lemma 7.3.3. *Let A be a masa in a separable type II_1 factor N. There are vectors $\xi_1, \ldots, \xi_m$ in $L^2(N)$ such that $A\xi_1 A + \ldots + A\xi_m A$ is dense in $L^2(N)$ if, and only if $\mathrm{Puk}(A) \subseteq \{1, 2, \ldots, m\}$.*

Proof. The density of $A\xi_1 A + \ldots + A\xi_m A$ in $L^2(N)$ is equivalent to the density of $\mathcal{A}\xi_1 + \ldots + \mathcal{A}\xi_m$ in $L^2(N)$.

If there were a non-zero projection $p \in \mathcal{A}$ such that $p\mathcal{A}'$ is of type I_n for a value of n in the range $m + 1 \le n \le \infty$, then $p\mathcal{A}'$ is isomorphic to $\mathbb{M}_n(p\mathcal{A})$, since $p\mathcal{A}$ is the centre of the homogeneous algebra $p\mathcal{A}'$. Thus at least n vectors $\eta_1, \ldots, \eta_n$ are required in $L^2(pN)$ for $p\mathcal{A}\eta_1 + \ldots + p\mathcal{A}\eta_n$ to be dense in $L^2(pN)$. This contradicts

$$p(\mathcal{A}\xi_1 + \ldots + \mathcal{A}\xi_m) = p\mathcal{A}p\xi_1 + \ldots + p\mathcal{A}p\xi_m$$

being dense in $L^2(N)$.

Conversely, suppose that $\mathrm{Puk}(A) \subseteq \{1, 2, \ldots, m\}$. Then the von Neumann algebra $\mathcal{A}'$ is subhomogeneous of degree m so that there are m vectors $\xi_1, \ldots, \xi_m$ such that $\mathcal{A}\xi_1 + \ldots + \mathcal{A}\xi_m$ is dense in $L^2(N)$. This proves the result. □

Remark 7.3.4. If p is a projection in a masa A in a separable type II_1 factor N, then the Pukánszky invariant $\mathrm{Puk}(pA, pNp)$ of pA in pNp is contained in $\mathrm{Puk}(A)$. The union of $\mathrm{Puk}(pA, pNp)$ over all projections p in A is $\mathrm{Puk}(A)$. □

The following theorem gives upper semi-continuity of the Pukánszky invariant. An example in the hyperfinite II_1 factor R after the theorem shows that this result is the best one can obtain in general. In order to obtain this result we will need to make a forward reference to Theorem 9.6.5, which is proved without reference to the present chapter.

Theorem 7.3.5. *Let A_n be a sequence of masas in a separable II_1 factor N and let B be a masa in N. If*

$$\lim_{n \to \infty} \|\mathbb{E}_{A_n} - \mathbb{E}_B\|_{\infty,2} = 0,$$

then

$$\text{Puk}(B) \subseteq \bigcup_{r \geq 1} \bigcap_{n \geq r} \text{Puk}(A_n).$$

Proof. By omitting the first terms of the sequence $\{A_n\}_{n=1}^{\infty}$ we can assume that $\|\mathbb{E}_{A_n} - \mathbb{E}_B\|_{\infty,2} < \delta$ for all n so that for each n there is a unitary u_n in N and projections $p_n \in A_n$ and $q_n \in B$ such that $u_n p_n A u_n^* = B q_n$ and $\|1 - q_n\|_2 \leq \sqrt{3}\|\mathbb{E}_{A_n} - \mathbb{E}_B\|_{\infty,2}$ by Theorem 9.6.5. Let $B_n = u_n A u_n^*$ for all $n \in \mathbb{N}$. Then $\text{Puk}(B_n) = \text{Puk}(A_n)$, $q_n \in B_n \cap B$ and $q_n B_n = q_n B$ for all n.

Let $k \in \text{Puk}(B)$. Then there is a non-zero central projection $q \leq 1 - e_B$ in B', so in B, such that qB' is of type I_k. The projection $q_n J q_n J \in B_n \cap B$ tends strongly to 1 as $n \to \infty$, since

$$\|q_n J q_n J x - x\|_2 \leq \|q_n x - x\|_2 + \|q_n (x q_n - x)\|_2 \leq 2\|q_n - 1\|_2 \|x\| \quad (7.3.6)$$

for all $n \in \mathbb{N}$ and $x \in N$. Thus there is an integer $r \in \mathbb{N}$ such that $q_n J q_n J q \neq 0$ for all $n \geq r$, because $q_j J q_j J q \to q$ strongly as $j \to \infty$.

Since $q_n \in B_n \cap B$ with $q_n B_n = q_n B$ we have $q_n J q_n J \in B_n \cap B$ and so $q_n J q_n J B_n' = q_n J q_n J B'$. The von Neumann algebra $q_n J q_n J B'$ is of type I and the projection $q_n J q_n J q$ is in its centre $q_n J q_n J B$. The algebra $q_n J q_n J q B'$ is of type I_k, being a non-zero central cutdown of qB', which is of type I_k. Hence $k \in \text{Puk}(B_n)$ for all $n \geq r$, and so

$$\text{Puk}(B) \subseteq \bigcup_{r \geq 1} \bigcap_{n \geq r} \text{Puk}(B_n)$$

$$= \bigcup_{r \geq 1} \bigcap_{n \geq r} \text{Puk}(A_n). \quad (7.3.7)$$

$\square$

The following example shows that equality need not hold in Theorem 7.3.5.

Example 7.3.6. We give an example to show that one can have

$$\text{Puk}(B) \neq \bigcup_{r \geq 1} \bigcap_{n \geq r} \text{Puk}(A_n) \quad (7.3.8)$$

in the above situation. We need to assume for the moment that the hyperfinite II_1 factor contains a masa with $\mathrm{Puk}(\cdot) = \{2\}$. This will be constructed subsequently in Example 7.5.3.

Let B be a Cartan masa in the hyperfinite II_1 factor R, or a singular masa with $\mathrm{Puk}(B) = \{1\}$, let $p_1 \leq p_2 \leq \cdots$ be a sequence of projections in B with $0 < \|p_n - 1\|_2 \to 0$ as $n \to \infty$. Let A_n be a masa in the hyperfinite II_1 factor $(1 - p_n)R(1 - p_n)$ with $\mathrm{Puk}(A_n) = \{2\}$. Here one could choose any fixed nonempty subset of $\mathbb{N}$ not equal to $\{1\}$ in place of $\{2\}$ that was the Pukánszky invariant of a masa in the hyperfinite II_1 factor.

Let $B_n = p_n B + A_n$ for all n. This is a masa in R that is just a direct sum. Then

$$\|\mathbb{E}_{B_n} - \mathbb{E}_B\|_{\infty,2} \leq 2\|1 - p_n\|_2 \to 0 \quad \text{as} \quad n \to \infty, \tag{7.3.9}$$

$\mathrm{Puk}(B) = \{1\}$ and

$$\mathrm{Puk}(B_n) \supseteq \mathrm{Puk}(p_n B) \cup \mathrm{Puk}(A_n) = \{1, 2\} \tag{7.3.10}$$

for all $n \in \mathbb{N}$. $\square$

7.4   The Pukánszky invariant in group factors

As we have seen in Sections 3.2 and 3.3, the von Neumann algebra generated by an abelian subgroup of an I.C.C. group is a masa, or a singular masa, depending on the cardinality of certain conjugacy classes in the group associated with the subgroup. There is a strong connection between the algebraic structure of the group and the properties of the associated von Neumann algebra, and this is particularly apparent for the Pukánszky invariant. Under quite general hypotheses, if an abelian subgroup H of an I.C.C. group G generates a masa $L(H) \subseteq L(G)$, then $\mathrm{Puk}(L(H))$ can be expressed in terms of the cardinality of certain equivalence classes in the double coset space for the abelian subgroup.

If G is a countable I.C.C. group and H is an abelian subgroup such that $\{hgh^{-1} : h \in H\}$ is infinite for all $g \in G \setminus H$, then $A = L(H)$ is a masa in the separable type II_1 factor $N = L(G)$. The further assumption, that $gHg^{-1} \cap H = \{e\}$ for all $g \in G \setminus H$, implies that A is a singular masa in $L(G)$. To see this, note that for given elements $b_1, b_2 \in G \setminus H$ at most one element $h \in H$ can satisfy $b_1 h b_2 \in H$. If this relation were true for $h_1 \neq h_2 \in H$, then we would have $b_1(h_1 h_2^{-1})b_1^{-1} \in H$, contrary to assumption. Thus there can only be a finite number of elements of H for which at least one of the relations $b_i h b_j \in H$ holds, where $B = \{b_1, \ldots, b_n\}$ is a given finite subset of $G \setminus H$. This shows that condition (C2) of Section 3.3 holds and A is then singular, by Lemma 3.3.2. This raises the question of what group structure determines the Pukánszky invariant $\mathrm{Puk}(A)$ of A in this situation. The invariant is determined by the structure of the double coset space $_H\backslash G /_H$ and the action of $H \times H$ on individual double cosets HgH for $g \in G \setminus H$. In general this relationship is complicated but under an additional hypothesis on the action of $H \times H$ on the double cosets, the exact type of the algebra $\mathcal{A}'$ is determined by the cardinality of certain equivalence classes in $_H\backslash G/_H$, Theorem 7.4.5. If this additional hypothesis is not satisfied, as may happen, then the Pukánszky invariant can still be described, but in a less satisfactory manner (see [176]).

Throughout this section H will be assumed to be an abelian subgroup of a countable I.C.C. group G such that $\{hgh^{-1} : h \in H\}$ is infinite for all $g \in G \backslash H$. Let the *stabiliser subgroup* of an element $g \in G$ be defined by

$$K_g = \{(h_1, h_2) \in H \times H : h_1 g h_2 = g\}. \tag{7.4.1}$$

If K is a subgroup of the group G, let

$$\mathcal{C}(K) = \{g \in G : gk = kg \quad \text{for all} \quad k \in K\} \tag{7.4.2}$$

be the commutant of K.

Definition 7.4.1. The (*stabiliser*) *equivalence relation* $\sim$ is defined on $_H\backslash G/_H$ by $HgH \sim HkH$ if and only if $K_g = K_k$. □

A second equivalence relation, '*commensurable equivalence relation*', on the double coset space plays an important role in the Pukánszky invariant. The Pukánszky invariant is easy to calculate when the two equivalence relations coincide, so we shall restrict to that situation. However, to motivate the hypothesis that ensures the relations are equal, we introduce the second equivalence relation, which is a stronger version of commensurability of two subgroups of a group. In group theory, two subgroups F_1 and F_2 of a group G are said to be *commensurable* if they have isomorphic finite index subgroups $L_i \subseteq F_i$, $i = 1, 2$. Just for this chapter, let us say two subgroups F_1 and F_2 of $H \times H$ are *nearly equal* if $F_1 \cap F_2$ is a subgroup of finite index in F_1 and F_2; this is equivalent to $F_1 \cap F_2$ being of finite index in $F_1 F_2$. The second equivalence relation $\sim_{ne}$ is defined on $_H\backslash G/_H$ by $HcH \sim_{ne} HdH$ if, and only if, $K_c \cap K_d$ is of finite index in K_c and K_d. The two relations $\sim$ and $\sim_{ne}$ are well-defined because if $HcH = HdH$, then $K_c = K_d$. The following elementary group theory lemma implies that $\sim_{ne}$ is an equivalence relation on $_H\backslash G/_H$.

Lemma 7.4.2. Let F_1, F_2, F_3 be subgroups of an abelian group L and suppose that $F_1 F_2/F_1 \cap F_2$ and $F_2 F_3/F_2 \cap F_3$ are finite groups. Then $F_1 F_3/F_1 \cap F_3$ is a finite group.

Proof. The hypotheses imply that the orders of $F_1/F_1 \cap F_2$, $F_2/F_1 \cap F_2$, $F_2/F_2 \cap F_3$, and $F_3/F_2 \cap F_3$ are all finite. Let $\pi: L \to L/F_2 \cap F_3$ be the quotient homomorphism and let ρ be its restriction to $F_1 \cap F_2$. Then ρ maps $F_1 \cap F_2$ into $F_2/F_2 \cap F_3$ with kernel $F_1 \cap F_2 \cap F_3$. Thus $F_1 \cap F_2/F_1 \cap F_2 \cap F_3$ is a finite group. Then each of the inclusions

$$F_1 \cap F_2 \cap F_3 \subseteq F_1 \cap F_2 \subseteq F_1 \tag{7.4.3}$$

is of finite index, so $F_1/F_1 \cap F_2 \cap F_3$ is a finite group, as is $F_1 \cap F_3/F_1 \cap F_2 \cap F_3$. Finiteness of $F_1/F_1 \cap F_3$ now follows from the inclusions

$$F_1 \cap F_2 \cap F_3 \subseteq F_1 \cap F_3 \subseteq F_1, \tag{7.4.4}$$

and similarly $F_3/F_1 \cap F_3$ is a finite group. Since F_1F_3/F_1 is isomorphic to $F_3/F_1 \cap F_3$, finiteness of $F_1F_3/F_1 \cap F_3$ is a consequence of the finite index inclusions

$$F_1 \cap F_3 \subseteq F_1 \subseteq F_1F_3. \qquad (7.4.5)$$

$\square$

Note that the stabiliser equivalence relation $\sim$ implies the nearly equal equivalence relation $\sim_{ne}$. When these two equivalence relations are equal there is a nice characterisation of the type of $\mathcal{A}'$ in terms of the cardinality of their common equivalence classes in $_H\backslash G/_H$ as Theorem 7.4.5 shows.

Lemma 7.4.3. *Let G, H and K_g be as defined above and let c, $d \in G$. The following conditions are equivalent:*

(i) *the map*

$$\psi : HcH \to HdH : h_1ch_2 \mapsto h_1dh_2$$

is a well-defined bijection;

(ii) $K_c = K_d$;

(iii) $H \cap cHc^{-1} = H \cap dHd^{-1}$ *and* $d = kc$ *for some* $k \in \mathcal{C}(H \cap dHd^{-1})$.

Proof. Note that

$$\begin{aligned}
K_c : &= \{(h_1, h_2) : h_1, h_2 \in H,\ h_1ch_2 = c\} \\
&= \{(h, c^{-1}h^{-1}c) : h \in H \text{ and } c^{-1}h^{-1}c \in H\} \\
&= \{(h, c^{-1}h^{-1}c) : h \in H \cap cHc^{-1}\} \qquad (7.4.6)
\end{aligned}$$

and that $h \mapsto (h, c^{-1}h^{-1}c)$ is an isomorphism from $H \cap cHc^{-1}$ onto the group K_c.

Observe that (i) is equivalent to (ii) since equality in (ii) is equivalent to ψ being well-defined and one-to-one.

If (ii) holds, then $c^{-1}hc = d^{-1}hd$ and $dc^{-1}h = hdc^{-1}$ for all $h \in H \cap cHc^{-1} = H \cap dHd^{-1}$ so that $dc^{-1} \in \mathcal{C}(H \cap cHc^{-1})$ as required. The implication from (iii) to (ii) is clear by the reverse argument. $\square$

If W is a subset of G, let p_W be the projection from $\ell^2(G) = L^2(N)$ onto $\ell^2(W)$, which is the closed linear span of W in $L^2(N)$. To simplify notation, we will use $g \in G$ to denote the basis vector $\delta_g \in \ell^2(G)$.

Theorem 7.4.4. *Let G, H and K_g be as defined above and let c, $d \in G$. Then the following statements hold:*

(i) p_{HcH} *is an abelian projection in $\mathcal{A}'$ with $p_{HcH}\mathcal{A} = p_{HcH}\mathcal{A}'p_{HcH}$ a masa in $B(\ell^2(HcH))$;*

(ii) $HcH \neq HdH$ *if and only if* $p_{HcH}p_{HdH} = 0$;

(iii) *the projections p_{HcH} and p_{HdH} are equivalent in $\mathcal{A}'$ if and only if $K_c = K_d$, that is, if and only if $HcH \sim HdH$;*

(iv) *there exists an operator $t \in \mathcal{A}'$ such that $p_{HdH}\, t\, p_{HcH} \neq 0$ if and only if $K_c \cap K_d$ is of finite index in K_c and K_d, that is, if and only if $HcH \sim_{ne} HdH$;*

(v) *if q is the projection onto the closed subspace spanned by all the group elements in a* nearly equal $\sim_{ne}$ *equivalence class of nontrivial double cosets, then $q \in \mathcal{A}$.*

Proof. (i)   If $h_1, h_2 \in H$, then $h_1 HcH h_2^{-1} = HcH$ so that $h_1 J h_2 J$ leaves the closed linear span of HcH in $L^2(N)$ invariant and thus

$$h_1 J h_2 J p_{HcH} = p_{HcH} h_1 J h_2 J. \tag{7.4.7}$$

Hence p_{HcH} is in $\mathcal{A}' \cap (JAJ)' = (A \cup JAJ)' = \mathcal{A}'$. Note that c is a cyclic vector for A on $\ell^2(HcH)$, since $\mathcal{A} = (A \cup JAJ)''$ and $AJAJc = AcA$ is dense in $\ell^2(HcH)$. Thus Ap_{HcH} is a masa in $B(\ell^2(HcH))$ so that $p_{HcH}\mathcal{A} = p_{HcH}\mathcal{A}'p_{HcH}$ and p_{HcH} is an abelian projection in $\mathcal{A}'$ (see Definition 2.4.5).

(ii)   If $HcH \neq HdH$, then $HcH \cap HdH = \emptyset$ and $p_{HcH}p_{HdH} = 0$, and conversely.

(iii)   If v is in $\mathcal{A}'$ with $v^*v = p_{HcH}$ and $vv^* = p_{HdH}$, then there is a y in the closed linear span $\ell^2(HcH)$ of HcH such that $vy = d$. If $h_1, h_2 \in H$ with $h_1 c h_2 = c$, then $h_1 y h_2 = y$ and

$$h_1 d h_2 = h_1 J h_2^{-1} J v y = v h_1 J h_2^{-1} J y = v h_1 y h_2 = vy = d \tag{7.4.8}$$

so that $H_c \subseteq H_d$. The reverse containment follows from considering v^*, so $K_c = K_d$ and $HcH \sim HdH$.

If $HcH \sim HdH$, then there is a bijection ψ from HcH onto HdH defined by $\psi(h_1 c h_2) = h_1 d h_2$ $(h_1, h_2 \in H)$ by Lemma 7.4.3. Define a partial isometry v on $\ell^2(G)$ by

$$v(x) = \begin{cases} \psi(x) & \text{if } x \in HcH \\ 0 & \text{if } x \in G \setminus HcH. \end{cases} \tag{7.4.9}$$

Then $v^*v = p_{HcH}$ and $vv^* = p_{HdH}$, since

$$v^*(x) = \begin{cases} \psi^{-1}(x) & \text{if } x \in HdH \\ 0 & \text{if } x \in G \setminus HdH. \end{cases} \tag{7.4.10}$$

If $h \in H$ and $x \in HcH$, then $v(hx) = \psi(hx) = h\psi(x) = hv(x)$ and similarly $v(JhJx) = JhJv(x)$. If $h \in H$ and $x \in G \setminus HcH$, then $v(hx) = 0 = hv(x)$ and $v(JhJx) = 0 = JhJv(x)$. Thus $v \in \mathcal{A}' \cap (JAJ)' = \mathcal{A}$ so that p_{HcH} is equivalent to p_{HdH} in $\mathcal{A}'$.

(iv)   Let $t \in \mathcal{A}'$ be such that $p_{HdH}\, t\, p_{HcH} \neq 0$. To obtain a contradiction, suppose that $K_d/K_c \cap K_d$ is infinite, and let $\{(h_n, k_n)K_c \cap K_d\}_{n=1}^\infty$ be a listing of the $K_c \cap K_d$-cosets in K_d. Then the group elements $\{h_n c k_n\}_{n=1}^\infty$ are distinct, since equality of $h_n c k_n$ and $h_m c k_m$ would imply that $(h_n h_m^{-1}, k_n k_m^{-1}) \in K_c$ and (h_n, k_n) and (h_m, k_m) would define the same $K_c \cap K_d$-coset. Thus $\{h_n c k_n\}_{n=1}^{\infty}$ are distinct as orthonormal vectors in $\ell^2(G)$. Each vector in HdH is left invariant by all elements of K_d, and in particular $p_{HdH} t(h_n c k_n) = p_{HdH} t(c)$ for $n \geq 1$ since $t \in \mathcal{A}'$. For each $m \geq 1$, the vector $\sum_{n=1}^{m} n^{-1} h_n c k_n$, whose norm is bounded by $\left(\sum_{n=1}^{\infty} n^{-2} \right)^{1/2} = \pi/\sqrt{6}$, is thus mapped by $p_{HdH} t$ to $\sum_{n=1}^{m} n^{-1} h_n p_{HdH} t(c) k_n = \left(\sum_{n=1}^{m} n^{-1} \right) p_{HdH} t(c)$, which forces $p_{HdH} t(c) = 0$, otherwise $p_{HdH} t$ is an unbounded operator. But then $p_{HdH} t(hck) = h p_{HdH} t(c) k = 0$ for $h, k \in H$ and so $p_{HdH} t p_{HcH} = 0$, a contradiction. Thus $K_d / K_c \cap K_d$ is finite, and the same conclusion holds for $K_c / K_c \cap K_d$ by considering t^*. This proves that K_c and K_d are nearly equal and $HcH \sim_{ne} HdH$.

Conversely, suppose that $K_c \cap K_d$ is of finite index in K_c and K_d. Then there is an action of the finite group $K_c K_d / K_c \cap K_d$ on both HcH and HdH. Let $\{(h_i, k_i) K_c \cap K_d\}_{i=1}^{n}$ be a listing of the cosets of $K_c \cap K_d$ in K_c and define $t \colon HcH \to HdH$ on the vectors arising from group elements by

$$t(xcy) = \sum_{i=1}^{n} h_i x d y k_i, \qquad x, y \in H. \tag{7.4.11}$$

This is well-defined because if $xcy = wcz$ for $w, x, y, z \in H$ then $(w^{-1}x, yz^{-1}) \in K_c$ and $(w^{-1}x, yz^{-1}) = (h_j r, k_j s)$ for some $j \in \{1, \ldots, n\}$ and $(r, s) \in K_c \cap K_d$, leading to

$$t(xcy) = \sum_{i=1}^{n} h_i x d y k_i = \sum_{i=1}^{n} h_i w h_j r d k_j s z k_i = t(wcz) \tag{7.4.12}$$

since $rds = d$ and $\{(h_i h_j, k_i k_j)\}_{i=1}^{n}$ gives another listing of the $K_c \cap K_d$-cosets. Note that the final equality in (7.4.12) uses the commutativity of H to reorder the products on either side of the group element d.

If $\{(x_j, y_j)\}_{i=1}^{\infty}$ are representatives of the cosets of K_c in $H \times H$, then $\{x_j c y_j\}_{j=1}^{m}$ are distinct in G and $\left\| \sum_{j=1}^{m} \alpha_j x_j c y_j \right\|_2^2 = \sum_{j=1}^{m} |\alpha_j|^2$. Then

$$t \left(\sum_{j=1}^{m} \alpha_j x_j c y_j \right) = \sum_{i=1}^{n} \sum_{j=1}^{m} \alpha_j h_i x_j d y_j k_i \tag{7.4.13}$$

and the right-hand side of (7.4.13) has norm at most $n \left\| \sum_{j=1}^{m} \alpha_j x_j d y_j \right\|_2$, by the triangle inequality. The vectors of the form $x_j d y_j$ are either equal or orthogonal and so

$$n \left\| \sum_{j=1}^{m} \alpha_j x_j d y_j \right\|_2 \leq n \left\| \sum_{j=1}^{m} |\alpha_j| x_j d y_j \right\|_2. \tag{7.4.14}$$

To estimate the latter sum, define an equivalence relation on $\{1, \ldots, m\}$ by $r \sim s$ if $x_r dy_r = x_s dy_s$ and note that each equivalence class has at most $|K_d/K_c \cap K_d|$ elements. To see this, let $\{(a_i, b_i)K_c \cap K_d\}_{i=1}^{\ell}$ be a listing of the cosets of $K_c \cap K_d$ in K_d and fix s. Then $r \sim s$ if and only if $(x_r x_s^{-1}, y_r y_s^{-1}) \in K_d$ and so (x_r, y_r) has the form $(x_s, y_s)(a_i, b_i)(h, k)$ for some $i \in \{1, \ldots, \ell\}$ and $(h, k) \in K_c \cap K_d$ so can lie in at most ℓ cosets of K_c in $H \times H$. If we replace each $|\alpha_j|$ in (7.4.14) by $\max\{|\alpha_p| : \ p \sim j\}$, then we obtain the estimate

$$\left\| t\left(\sum_{j=1}^{m} \alpha_j x_j c y_j\right)\right\|_2 \leq n \left\|\sum_{j=1}^{m} |\alpha_j| x_j dy_j\right\|_2$$

$$\leq n\ell \left\|\sum_{j=1}^{m} |\alpha_j|^2\right\|_2^{1/2} \tag{7.4.15}$$

and $\|t\| \leq |K_c/K_c \cap K_d| \cdot |K_d/K_c \cap K_d|$ by letting m and the α_j's vary. Thus t is a bounded operator and it extends with the same norm to the whole space by setting it equal to 0 on the orthogonal complement of HcH. It is clear from the definition that this extension, also denoted by t, commutes with left and right multiplications by elements of H and so $t \in \mathcal{A}'$. It is also clear from (7.4.11) that $t(c) \neq 0$ so $p_{HdH}\, t\, p_{HcH} \neq 0$.

(v)   The projection q commutes with left and right multiplications by group elements from H, and so $q \in \mathcal{A}'$. If HcH is in the equivalence class but HdH is not, then (iv) shows that any $t \in \mathcal{A}'$ mapping HcH to HdH must be 0. Thus the range of q is invariant for $\mathcal{A}'$ and $q \in \mathcal{A}'' = \mathcal{A}$.   □

An alternative proof of part of Theorem 7.4.4 may be given by abelian harmonic analysis, using the fact that $\ell^2(HgH)$ is isomorphic to $\ell^2(H \times H/K_g)$ as a Hilbert space on which $H \times H/K_g$ acts. One uses the observation that a closed subgroup of a compact abelian group with positive measure has finite index in the whole group.

If $g \in G$, let $[HgH]$ denote the $\sim$ equivalence class containing HgH and let $|[HgH]|$ denote the number of elements in this equivalence class in $H \backslash G / H$. A convenient notation in Theorem 7.4.5 is to denote the algebra of all bounded linear operators $B(\ell^2)$ on ℓ^2 by $\mathbb{M}_\infty$.

The hypothesis on G and H in the following theorem is exactly that to ensure the stabiliser equivalence relation $\sim$ and the *nearly equal* equivalence relation $\sim_{ne}$ are equal so that all parts of Theorem 7.4.4 apply.

Theorem 7.4.5. *Let G be a countable I.C.C. group, let H be an abelian subgroup such that $\{hgh^{-1} : h \in H\}$ is infinite for all $g \in G \setminus H$, let $A = L(H)$ in $N = L(G)$ and let $\mathcal{A} = (A \cup JAJ)''$. Suppose that, for each $c, d \in G$, either $K_c = K_d$ or $K_c \cap K_d$ is of infinite index in $K_c K_d$. Then*

(i)

$$\mathcal{A}' \cong \bigoplus \mathbb{M}_{|[HgH]|}\overline{\otimes}Ap_{HgH},$$

where the direct sum is over the equivalence classes $[HgH]$ in $H \backslash G / H$;

(ii) *the Pukánszky invariant* $\text{Puk}(A)$ *of* A *in* N *is*

$$\{|[HgH]| : [HgH] \quad \text{is an equivalence class in} \quad _H\backslash G/_H \setminus \{H\}\}.$$

Proof. The projections p_{HdH} for $HdH \sim HgH$ and the partial isometries defined in Theorem 7.4.4 (iii) form a set of matrix units in $\mathcal{A}'$ for a subalgebra that is isomorphic to $\mathbb{M}_{|[HgH]|}$ for each g in G.

If there is a non-zero partial isometry v in $\mathcal{A}'$ with $v^*v \le p_{HvH}$ and $vv^* \le p_{HgH}$, then $K_c \cap K_g$ is of finite index in K_cK_g by Theorem 7.4.4 (iv) and hence $K_c = K_g$ by hypothesis. This shows that if $q = p_{\cup[HgH]}$ is the projection onto the subspace $\ell^2(\cup HgH)$ associated with the whole equivalence class $\cup[HgH]$, then

$$q = p_{\cup[HgH]} = \sum\{p_{HcH} : HcH \sim HgH\} \qquad (7.4.16)$$

is a central projection in $\mathcal{A}'$, hence is in $\mathcal{A}$, and is the central support of p_{HgH}.

Since G is the disjoint union of the equivalence classes in $_H\backslash G/_H$, $1 = \sum_{g \in G} p_{\cup HgH}$ where the summation is over all equivalence classes. Hence

$$\mathcal{A}' = \bigoplus p_{\cup HgH} \mathcal{A}'. \qquad (7.4.17)$$

Part (i) follows from this since

$$\mathcal{A}' p_{\cup[HgH]} \cong \mathbb{M}_{|[HgH]|} \overline{\otimes} \mathcal{A} p_{HgH} \qquad (7.4.18)$$

by standard type I results (see [105, Theorem 6.6.5]).

Note that $\{H\}$ is an equivalence class in $_H\backslash G/_H$ and $e_A = p_H$ so that

$$\mathcal{A}'(1 - e_A) \cong \bigoplus \mathbb{M}_{|[HgH]|} \overline{\otimes} \mathcal{A} p_{HgH}, \qquad (7.4.19)$$

where the direct sum is over all equivalence classes in $(_H\backslash G/_H) \setminus \{H\}$. The set of values giving non-zero type I_n parts of $\mathcal{A}'(1 - e_A)$ is thus

$$\text{Puk}(A) = \{|[HgH]| : [HgH] \quad \text{is an equivalence class in} (_H\backslash G/_H) \setminus \{H\}\}. \tag{7.4.20}$$

This proves the result. $\qquad\qquad\qquad\qquad\qquad\qquad\qquad\qquad\qquad\qquad\square$

In the following corollary, adding ∞ or an infinite sum of integers is interpreted as ∞. The corollary follows directly from Theorem 7.4.5 (ii) as $_H\backslash G/_H \setminus \{H\}$ is the union of all the equivalence classes counted in $\text{Puk}(A)$.

Corollary 7.4.6. *Let* G *be a countable I.C.C. group, let* H *be an abelian subgroup such that* $\{hgh^{-1} : h \in H\}$ *is infinite for all* $g \in G \setminus H$ *and let* $A = L(H)$ *in* $N = L(G)$. *If the stabiliser and 'nearly equal' equivalence relations* $\sim$ *and* $\sim_{ne}$ *are equal on* $_H\backslash G/_H$, *then*

$$\sum\{n : n \in \text{Puk}(A)\} = |_H\backslash G/_H| - 1.$$

Examples of masas and their Pukánszky invariants may be constructed by finding I.C.C. groups G and abelian subgroups H satisfying Theorem 7.4.5. This is particularly interesting when the masa A is singular. Corollary 7.4.6 is a restriction on the example when $\mathrm{Puk}(A)$ is required to be a finite subset of $\mathbb{N}$. The strategy for finding examples like this goes back to the work of Pukánszky [154] and Dixmier [47]. The answer to the following question is surely known.

If an abelian group H in an I.C.C. group G has $\{hgh^{-1} : h \in H\}$ infinite for all $g \in G \setminus H$ and if there are only a finite number of H double cosets in G (i.e., $_H\backslash G/_H$ is finite) is G a soluble group?

7.5 Examples of the Pukánszky invariant

This section contains examples of I.C.C. groups G and abelian subgroups H giving rise to various values of the Pukánszky invariant using Theorem 7.4.5. The first example is that of the generator masa in the free group factor, and the second is that of Pukánszky [154] based on matrix groups over fields.

Example 7.5.1. The Pukánszky invariant of a generator masa in a free group factor is $\{\infty\}$.

Let $2 \leq n \leq \infty$, let $G = \mathbb{F}_n$ be the free group on n generators, let h be one of the generators in G and let $H = \langle h \rangle$ be the subgroup generated by h. Then $A = L(H) = \{h, h^{-1}\}''$ is a singular masa in $L(\mathbb{F}_n)$, by Example 3.3.6. If $g \in \mathbb{F}_n \setminus H$, then

$$K_g := \{(h_1, h_2) : h_1 g h_2 = g, \ h_1, h_2 \in H\} = \{(e, e)\}. \tag{7.5.1}$$

Then $_H\backslash G/_H \setminus \{H\}$ is a single infinite equivalence class because on $_H\backslash G/_H$ the $\sim$ and $\sim_{ne}$ equivalence relations are equal. Hence $\mathrm{Puk}(A) = \{\infty\}$. This also applies if h is a prime element in $\mathbb{F}_n$. $\square$

Related to this free group generator masa but not proved in these notes is the following theorem of Rădulescu [156]. If $2 \leq n < \infty$, then the abelian von Neumann subalgebra B of $L(\mathbb{F}_n)$ generated by the hermitian element

$$h = \sum_{j=1}^{n} (g_j + g_j^{-1}),$$

where g_j are the generators of $\mathbb{F}_n$, is a singular masa with $\mathrm{Puk}(B) = \{\infty\}$. $\square$

The remaining examples in both this section and the next are based on Pukánszky's approach of considering certain matrix groups over fields. The next lemma gives a general framework for the computations. Many variations are possible, some of which are presented subsequently.

Lemma 7.5.2. *Let F be a countably infinite field with multiplicative group $F^\times$, and let P be an infinite subgroup of $F^\times$. Let G be the multiplicative matrix group*

$$\left\{ \begin{pmatrix} a & b \\ 0 & 1 \end{pmatrix} : a \in P, \ b \in F \right\},$$

and let

$$H = \left\{ \begin{pmatrix} a & 0 \\ 0 & 1 \end{pmatrix} : a \in P \right\}, \quad K = \left\{ \begin{pmatrix} 1 & b \\ 0 & 1 \end{pmatrix} : b \in F \right\}.$$

Then $L(G)$ is the hyperfinite II_1 factor, $L(H)$ is a singular masa in $L(G)$, and $\mathrm{Puk}(L(H)) = \{n\}$, where n is the index $[F^\times : P]$ of P in $F^\times$.

Proof. Let

$$h = \begin{pmatrix} x & 0 \\ 0 & 1 \end{pmatrix}, \quad k = \begin{pmatrix} 1 & y \\ 0 & 1 \end{pmatrix}, \quad g = \begin{pmatrix} a & b \\ 0 & 1 \end{pmatrix}, \qquad (7.5.2)$$

be general elements of H, K and G respectively, where $a, x \in P$ and $b, y \in F$. Varying first x and then y, the equations

$$hgh^{-1} = \begin{pmatrix} a & xb \\ 0 & 1 \end{pmatrix}, \quad khk^{-1} = \begin{pmatrix} x & y(1-x) \\ 0 & 1 \end{pmatrix}, \qquad (7.5.3)$$

show respectively that $\{hgh^{-1} : h \in H\}$ is infinite for $g \in G \setminus H$ and that $\{khk^{-1} : k \in K\}$ is infinite for $h \in H \setminus \{e\}$. Thus G is I.C.C. The group K is an abelian normal subgroup and G/K is isomorphic to the abelian group H, and thus G is amenable [127, p. 31]. Hence $L(G)$ is the hyperfinite II_1 factor, by Theorem 3.8.2. With g and h as in (7.5.2),

$$ghg^{-1} = \begin{pmatrix} x & b(1-x) \\ 0 & 1 \end{pmatrix}, \qquad (7.5.4)$$

and so lies outside H when $h \neq e$ and $g \notin H$. Thus, by Lemma 3.3.2, $L(H)$ is a singular masa in $L(G)$. Let $y_1, y_2, \ldots \in F$ be such that $Py_1, Py_2, \ldots$ is a listing of the distinct cosets of P in $F^\times$, and let $g_i = \begin{pmatrix} 1 & y_i \\ 0 & 1 \end{pmatrix}$. Then $H, Hg_1H, Hg_2H, \ldots$ is a listing of the distinct double cosets in G. A simple matrix calculation shows that each stabiliser subgroup K_{g_i} is trivial, and so there is one equivalence class of double cosets. By Theorem 7.4.5, $\mathrm{Puk}(L(H)) = \{n\}$, where $n = [F^\times : P]$. $\square$

Example 7.5.3. With the notation of Lemma 7.5.2, let $F = \mathbb{Q}$, and define subgroups of $F^\times$ by

$$P_\infty = \{p/q : p, q \in \mathbb{Z}_{\mathrm{odd}}\},$$
$$P_n = \{r2^{nk} : k \in \mathbb{Z}, \ r \in P_\infty\}, \ 1 \leq n < \infty.$$

Then $[F^\times : P_n] = n$ for $1 \leq n \leq \infty$, and Lemma 7.5.2, with P taken to be P_n, gives singular masas A_n in the hyperfinite II_1 factor R with $\mathrm{Puk}(A_n) = \{n\}$, $1 \leq n \leq \infty$. For each n in $\mathbb{N} \cup \{\infty\}$ there is singular masa in the hyperfinite type II_1 factor R with Pukánszky invariant $\{n\}$.

Since A_1 is singular, it cannot be conjugate to a Cartan masa, even though they share the same invariant $\{1\}$. $\square$

With the A_n's as in Example 7.5.3, the masa

$$A_n \overline{\otimes} A_\infty \subseteq R \overline{\otimes} R \cong R$$

has $\text{Puk}(A_n \overline{\otimes} A_\infty) = \{n, \infty\}$, by Theorem 7.3.1. We obtain the same result in the next example, a variant of Example 7.5.3, the objective being to obtain the same Pukánszky invariant even after taking a free product. The calculations are very similar, so we will just point out the steps.

Example 7.5.4. Let G_n, $n \geq 1$, be the matrix group

$$G_n = \left\{ \begin{pmatrix} 1 & x & y \\ 0 & f & 0 \\ 0 & 0 & g \end{pmatrix} : x, y \in \mathbb{Q}, \ f \in P_n, \ g \in P_\infty \right\}$$

with abelian subgroup H_n consisting of the diagonal matrices in G_n. Then one checks that G_n is an amenable I.C.C. group, and that H_n satisfies the conditions for $L(H_n)$ to be a singular masa in $L(G_n)$. There are three types of generators for the non-trivial double cosets of H_n:

$$\begin{pmatrix} 1 & 2^r & 0 \\ 0 & 1 & 0 \\ 0 & 0 & 1 \end{pmatrix}, \quad \begin{pmatrix} 1 & 0 & 2^s \\ 0 & 1 & 0 \\ 0 & 0 & 1 \end{pmatrix}, \quad \begin{pmatrix} 1 & 2^r & 2^s \\ 0 & 1 & 0 \\ 0 & 0 & 1 \end{pmatrix},$$

where $1 \leq r \leq n$ and $s \geq 1$. The corresponding stabiliser subgroups are respectively

$$\left\{ \left(\begin{pmatrix} 1 & 0 & 0 \\ 0 & 1 & 0 \\ 0 & 0 & g \end{pmatrix}, \begin{pmatrix} 1 & 0 & 0 \\ 0 & 1 & 0 \\ 0 & 0 & g^{-1} \end{pmatrix} \right) : g \in P_\infty \right\},$$

$$\left\{ \left(\begin{pmatrix} 1 & 0 & 0 \\ 0 & f & 0 \\ 0 & 0 & 1 \end{pmatrix}, \begin{pmatrix} 1 & 0 & 0 \\ 0 & f^{-1} & 0 \\ 0 & 0 & 1 \end{pmatrix} \right) : f \in P_n \right\},$$

$$\left\{ \left(\begin{pmatrix} 1 & 0 & 0 \\ 0 & 1 & 0 \\ 0 & 0 & 1 \end{pmatrix}, \begin{pmatrix} 1 & 0 & 0 \\ 0 & 1 & 0 \\ 0 & 0 & 1 \end{pmatrix} \right) \right\}. \tag{7.5.5}$$

Thus there are three equivalence classes of double cosets with respectively n, ∞, and ∞ elements, and so $\text{Puk}(L(H_n)) = \{n, \infty\}$.

If Γ is a countable discrete group, then H_n also satisfies the conditions to generate a singular masa in $L(G_n \star \Gamma) \cong L(G_n) \star L(\Gamma)$, and the stabiliser subgroups for group elements from $(G_n \star \Gamma) \setminus G_n$ are all trivial. Thus the extra cosets from these elements fall into the third equivalence class above, showing that the invariant is unchanged by passing to the larger algebra, in this case. The following example shows that this may not always happen.   □

Example 7.5.5. We will continue to use the notation of Lemma 7.5.2 and Example 7.5.3. For each $n \in \mathbb{N} \cup \{\infty\}$, there is a masa A in the hyperfinite type II_1 factor R and a subfactor R_0 of R with $A \subset R_0$ such that $\text{Puk}(A, R) = \{n\}$ and $\text{Puk}(A, R_0) = \{1\}$.

Let $F_0 \subseteq F_1 \subseteq F$ be two infinite subfields of a countable field F and let $P = F_0^\times$ be the multiplicative group of F_0. Let G be the multiplicative group of 2×2 matrices of the form

$$g = \begin{pmatrix} a & b \\ 0 & 1 \end{pmatrix}$$

with $b \in F$ and $a \in P$, and let H be the diagonal matrices in G as in Example 7.5.3. Let G_1 be the subgroup of G with $b \in F_1$. Let $A = L(H)$, $M = L(G_1)$ and $N = L(G)$. Using Lemma 7.5.2 and the observation that the index $[F^\times : F_0^\times]$ is the vector space dimension of F over F_0, it follows that

$$\mathrm{Puk}(A, N) = \left\{ [F^\times : F_0^\times] \right\}$$

and

$$\mathrm{Puk}(A, M) = \left\{ [F_1^\times : F_0^\times] \right\},$$

these quantities being respectively the dimensions of F over F_0 and of F_1 over F_0. Choosing $F_0 = F_1$ and F an extension of F_0 of dimension n over F_0 gives an example with $\mathrm{Puk}(A, N) = \{n\}$ and $\mathrm{Puk}(A, M) = \{1\}$. One could take $F_0 = \mathbb{Q}$ and $F = \mathbb{Q}(\sqrt[n]{2})$. $\square$

The following example shows that there are uncountably many possible values of $\mathrm{Puk}(A)$ in the hyperfinite factor R. This was first observed by Neshveyev and Størmer [119]. Using ergodic theory techniques, they were able to show that any subset $W \subseteq \mathbb{N} \cup \{\infty\}$ containing 1 was the invariant of some masa in R. We apply the group theory techniques already developed to get a similar result with 1 replaced by ∞. In particular, the next example gives an uncountable set of possible values.

Example 7.5.6. Let S be an arbitrary subset of $\mathbb{N}$. We construct a singular masa A in the hyperfinite factor R with $\mathrm{Puk}(A) = S \cup \{\infty\}$ as follows. We will need below the group P_∞, defined by

$$P_\infty = \left\{ pq^{-1} : p, q \in \mathbb{Z}_{\mathrm{odd}} \right\},$$

as in Example 7.5.3.

We will assume that S is non-empty since $\{\infty\}$ is already a known value (see Example 7.5.3 with $n = \infty$). Let $\{n_1, n_2, \ldots\}$ be a listing of the numbers in S, where each is repeated infinitely often to ensure that the list is infinite. Then define a matrix group G by specifying the general group elements to be

$$\begin{pmatrix} 1 & x_1 & x_2 & x_3 & \cdots \\ 0 & f_1 2^{n_1 k} & 0 & 0 & \cdots \\ 0 & 0 & f_2 2^{n_2 k} & 0 & \cdots \\ 0 & 0 & 0 & f_3 2^{n_3 k} & \\ \vdots & \vdots & \vdots & & \ddots \end{pmatrix}, \tag{7.5.6}$$

where $k \in \mathbb{Z}$, $x_j \in \mathbb{Q}$, $f_j \in P_\infty$, and the relations $x_i \neq 0$ and $f_j \neq 1$ occur only finitely often. This makes G a countable group which is easily checked to be I.C.C. Moreover G is amenable since there is an abelian normal subgroup N (those matrices with only 1's on the diagonal) so that the quotient G/N is isomorphic to the abelian subgroup H consisting of the diagonal matrices in G. Thus $L(G)$ is the hyperfinite factor R, by Theorem 3.8.2, and $L(H)$ is a singular masa A in R, just as in Example 7.5.3. For any finite nonempty subset T of $\mathbb{N}$, let H_T be the subgroup of H obtained by the requirements that $k = 0$ and that $f_i = 1$ for $i \in T$.

Each nontrivial double coset is generated by a nontrivial element

$$x = \begin{pmatrix} 1 & x_1 & x_2 & x_3 & \cdots \\ 0 & 1 & 0 & 0 & \cdots \\ 0 & 0 & 1 & 0 & \cdots \\ 0 & 0 & 0 & 1 & \\ \vdots & \vdots & \vdots & & \ddots \end{pmatrix},$$

of N, and we split into two cases according to whether the number of non-zero x_i is exactly 1 or is greater than 1. In the first case, suppose that x_1 is the sole non-zero value. We obtain n_1 distinct cosets generated by the elements

$$\begin{pmatrix} 1 & 2^k & 0 & 0 & \cdots \\ 0 & 1 & 0 & 0 & \cdots \\ 0 & 0 & 1 & 0 & \\ 0 & 0 & 0 & 1 & \\ \vdots & \vdots & & & \ddots \end{pmatrix}, \quad 1 \leq k \leq n_1, \tag{7.5.7}$$

respectively, whose stabiliser subgroups are all $\{(h, h^{-1}) : h \in H_{\{1\}}\}$. A similar result holds when the non-zero entry occurs in the ith position: n_i distinct cosets with stabiliser subgroups $\{(h, h^{-1}) : h \in H_{\{i\}}\}$.

If T is a finite subset of $\mathbb{N}$ with $|T| > 1$, then two elements of N, having non-zero entries respectively x_i and y_i for $i \in T$, generate the same double coset precisely when there exist $k \in \mathbb{Z}$ and $f_i \in P_\infty$ such that $x_i = y_i f_i 2^{n_i k}$, $i \in T$. The stabiliser subgroup in this case is $\{(h, h^{-1}) : h \in H_T\}$. Thus the distinct stabiliser subgroups are pairwise noncommensurable and Theorem 7.4.5 allows us to determine the Pukánszky invariant by counting the equivalence classes of double cosets. In the first case we obtain the integers $n_i \in S$. In the second case each equivalence class has infinitely many elements and so the contribution is $\{\infty\}$, showing that $\text{Puk}(A) = S \cup \{\infty\}$, as required. $\square$

The results of the last few sections do allow us to generate other Pukánszky invariants not already discussed. For example, tensoring two masas, whose invariants are $\{m\}$ and $\{n\}$ respectively, gives a new masa whose invariant is $\{m, n, mn\}$, by Theorem 7.3.1.

The last example that we present allows us to obtain many new values of the Pukánszky invariant in the free group factors. These are taken from [62]. To do full justice would require us to develop free products of von Neumann algebras, so we will just describe the approach and refer to [62].

Example 7.5.7. If S is any subset of $\mathbb{N}$ and $2 \leq k \leq \infty$, then $L(\mathbb{F}_k)$ contains a masa whose Pukánszky invariant is $S \cup \{\infty\}$. This is already covered by Example 7.5.1 if S is empty, so we assume $S \neq \emptyset$.

For a set $S \subseteq \mathbb{N}$, repeat each element an infinite number of times to ensure that S is infinite. By Example 7.5.4, there are masas $A_n \subseteq R_n$ (where R_n is a copy of R) with $\mathrm{Puk}(A_n) = \{n, \infty\}$, for $n \geq 1$. Form the direct sum

$$\bigoplus_{n \in S} A_n \subseteq \bigoplus_{n \in S} R_n,$$

to which we may assign a trace as a suitable convex combination of the traces on the R_n's. Denote these direct sums as $A \subseteq M$. As a sum of hyperfinite factors, M is hyperfinite, so $M \star L(\mathbb{F}_{k-1}) \cong L(\mathbb{F}_k)$ [56]. Then one checks that A is a masa in $L(\mathbb{F}_k)$, and that the Pukánszky invariant is $S \cup \{\infty\}$. The special properties of the groups in Example 7.5.4 are needed to ensure that the finite integers are not lost in passing from M to the free product $M \star L(\mathbb{F}_{k-1})$. $\square$

The groups that we have used in these examples have really been disguised forms of semidirect products. We end this section with a general result, although it should be noted that the direct calculations above are easier to carry out.

Proposition 7.5.8. *Let $G = H \rtimes_\alpha K$ be the semidirect product of a countable infinite group H by a countable infinite abelian group K. Let $\{h^{-1}\alpha_k(h) : h \in H\}$ be infinite for all $k \in K \setminus \{e\}$ and $\{\alpha_k(h) : k \in K\}$ be infinite for all $h \in H \setminus \{e\}$. Assume that $\alpha_{k^n}(h) = h$ for $h \in H \setminus \{e\}$, $k \in K$ and $n \in \mathbb{N}$ implies that $\alpha_k(h) = h$; this says each stabiliser group is its own radical in K. The equivalence relation $\sim$ is defined on H by $h_1 \sim h_2$ if, and only if, the stabiliser groups $\{k \in K : \alpha_k(h_1) = h_1\}$ and $\{k \in K : \alpha_k(h_2) = h_2\}$ are equal. Then $L(K)$ is a singular masa in $L(G)$ with Pukánszky invariant given by the set of $n \in \mathbb{N} \cup \{\infty\}$ for which an $h \in H \setminus \{e\}$ exists with n orbits in its $\sim$ equivalence class.*

Proof. By Lemma 3.3.2 $L(K)$ is a singular masa in the type II_1 factor $L(G)$. The double cosets KgK in $_K \backslash G /_K$ may be indexed by elements of H by the definition of $H \rtimes_\alpha K$, and this is done below. Two double cosets Kh_1K and Kh_2K are equal for $h_1, h_2 \in H$ if, and only if, $h_2 = \alpha_k(h_1)$ for some $k \in K$ so h_2 is in the orbit of h_1. A double coset in $_K \backslash G /_K \setminus \{K\}$ is represented by the element h of $H \setminus \{e\}$ it contains. Then

$$K_h = \{(k_1, k_2) : k_1 h k_2 = h; \ k_1, k_2 \in K\} = \{(k_1, k_1^{-1}) : \alpha_{k_1}(h) = h, k_1 \in K\}$$

and K_h is determined by the stabiliser subgroup $\{k \in K : \alpha_k(h) = h\}$ of h. The stabiliser equivalence relation $\sim$ is given by $K h_{h_1} K \sim K h_{h_2} K$ if, and only if,

$$\{k \in K : \alpha_k(h_1) = h_1\} = \{k \in K : \alpha_k(h_2) = h_2\},$$

and is the same as the equivalence relation defined here. The number of double cosets KhK in a $\sim$ equivalence class is thus the number of orbits in this equivalence class. $\square$

7.6   Open problems

In this concluding section, we present a list of open problems concerning the Pukánszky invariant.

(1) If N is a separable type II_1 factor, what is the range of the Pukánszky invariant on N; what is $\{\mathrm{Puk}(A) : A \text{ is a masa in } N\}$?

(2) The same question with A restricted to *singular* masas in N.

(3) Can $\mathrm{Puk}(A)$ be equal to an infinite set of integers for some masa A in $L(\mathbb{F}_n)$, $(2 \le n \le \infty)$? We only know that any set containing $\{\infty\}$ can occur, and that a finite set of integers is impossible.

(4) Is the range of $\mathrm{Puk}(\cdot)$ countable over all masas in a property T separable type II_1 factor?

(5) What happens to the Pukánszky invariant under suitable crossed products?

Chapter 8

Operators in $L^\infty[0,1]\overline{\otimes}B(H)$

8.1 Introduction

In this section we establish a result, Theorem 8.3.5, which is the key to obtaining good numerical estimates in the subsequent work on perturbations of masas, Chapters 9 and 10. When B is a masa in a separable type II$_1$ von Neumann algebra N with a faithful normal trace τ, the algebra $\langle N, e_B \rangle \subseteq B(L^2(N))$, introduced in Chapter 4, has commutant JBJ (Theorem 4.2.2) which is isomorphic to $L^\infty[0,1]$ (Corollary 3.5.3). Thus $\langle N, e_B \rangle$ is identified with $L^\infty[0,1]\overline{\otimes}B(H)$, and operators in this algebra are regarded as measurable $B(H)$-valued functions on $[0,1]$. Construction of particular operators then becomes a matter of choosing appropriate measurable selections from $B(H)$, and measure-theoretic considerations require careful handling (see, for example, [105, Ch. 14], where direct integrals are discussed). Thus we begin in Section 8.2 with some technical results designed to deal with measurability questions that arise in the main results of Section 8.3.

When an algebra has a trace τ then $\|\cdot\|_2$ is defined by $\|x\|_2 = (\tau(x^*x))^{1/2}$. In this chapter several traces will occur, so in cases of possible ambiguity we will adopt the notation $\|\cdot\|_{2,\tau}$ to indicate which trace is being used.

8.2 Matrix computations

In this section we will establish a number of technical lemmas concerning estimates in matrix algebras. These will be needed for the main theorems in Section 8.3.

Lemma 8.2.1. *Let $\alpha_1 \geq \alpha_2 \geq \alpha_3 \geq 0$, let $\lambda_1, \lambda_2, \lambda_3 \geq 0$ satisfy $\lambda_1^2+\lambda_2^2+\lambda_3^2 = 1$, and let $\mu = \max\{\lambda_1, \lambda_2\}$. Then*

$$1 + \sum_{i=1}^{3}\alpha_i^2 - 2\left(\sum_{i=1}^{3}\alpha_i^2\lambda_i^2\right)^{1/2} \geq (2/3)\min\{2 - 2\mu, 1\}. \qquad (8.2.1)$$

Proof. We consider two ranges for the value of μ.

(i) $0 \le \mu \le 3^{-1/2}$.

In this case $\lambda_1^2, \lambda_2^2 \le 1/3$ and so $\lambda_3^2 \ge 1/3$. Reducing λ_3^2 to $1/3$ and distributing $\lambda_3^2 - 1/3$ to λ_1^2 and λ_2^2 to bring them to $1/3$ decreases the left-hand side of (8.2.1), and so it suffices to show that

$$1 + \sum_{i=1}^{3} \alpha_i^2 - (2/\sqrt{3})\left(\sum_{i=1}^{3}\alpha_i^2\right)^{1/2} \ge 2/3. \qquad (8.2.2)$$

This inequality is verified by observing that the function $1 + t - (2/\sqrt{3})t^{1/2}$ has a minimum value of $2/3$ on $[0,\infty)$, occurring at $t = 1/3$.

(ii) $3^{-1/2} < \mu \le 1$.

In this case $\mu^2 \ge 1/3$, and so $1 - 2\mu^2 \le \mu^2$. Then

$$\begin{aligned}
\sum_{i=1}^{3} \alpha_i^2 \lambda_i^2 &= \lambda_1^2(\alpha_1^2 - \alpha_3^2) + \lambda_2^2(\alpha_2^2 - \alpha_3^2) + \alpha_3^2 \\
&\le \mu^2(\alpha_1^2 - \alpha_3^2) + \mu^2(\alpha_2^2 - \alpha_3^2) + \alpha_3^2 \\
&= \mu^2(\alpha_1^2 + \alpha_2^2) + (1 - 2\mu^2)\alpha_3^2 \\
&\le \mu^2(\alpha_1^2 + \alpha_2^2 + \alpha_3^2). \qquad (8.2.3)
\end{aligned}$$

Thus the left-hand side of (8.2.1) is no smaller than $1 + \sum_{i=1}^{3}\alpha_i^2 - 2\mu\left(\sum_{i=1}^{3}\alpha_i^2\right)^{1/2}$.
The function $1 + t - 2\mu t^{1/2}$ has a minimum value of $1 - \mu^2$ on $[0,\infty)$, occurring at $t = \mu^2$. The right-hand side of (8.2.1) is $(4/3)(1 - \mu)$ and so it suffices to show that

$$1 - \mu^2 \ge (4/3)(1 - \mu). \qquad (8.2.4)$$

The case $\mu = 1$ is trivial while, for $\mu < 1$, (8.2.4) is equivalent to $\mu \ge 1/3$, which is true on the range that we are considering. $\square$

Remark 8.2.2. The choices of $\alpha_1 = \alpha_2 = \alpha_3 = 1/3$ and $\lambda_1 = \lambda_2 = \lambda_3 = 1/\sqrt{3}$ give equality in (8.2.1), showing that this inequality is sharp. $\square$

We will now use the inequality of Lemma 8.2.1 to obtain an estimate that will be needed subsequently. We let τ denote the trace on $\mathbb{M}_3$ which assigns the value 1 to each rank one projection, while noting that (8.2.7) below is valid for any scaling of τ.

Lemma 8.2.3. *Let $\alpha_1 \ge \alpha_2 \ge \alpha_3 \ge 0$ and consider the three matrices*

$$p_1 = \begin{pmatrix} 1 & 0 & 0 \\ 0 & 0 & 0 \\ 0 & 0 & 0 \end{pmatrix}, \quad p_2 = \begin{pmatrix} 0 & 0 & 0 \\ 0 & 1 & 0 \\ 0 & 0 & 0 \end{pmatrix}, \quad x = \begin{pmatrix} \alpha_1 & 0 & 0 \\ 0 & \alpha_2 & 0 \\ 0 & 0 & \alpha_3 \end{pmatrix}.$$

Let $v_0 \in \mathbb{M}_3$ be a rank one partial isometry. Then there exists a partial isometry $v \in \mathbb{M}_3$ with the following properties:

(i) $vv^* \leq v_0 v_0^*$; (8.2.5)

(ii) $v^* v \in \{p_1, p_2, 0\}$; (8.2.6)

(iii) $\|v - v_0\|_2^2 \leq (3/2)\|x - v_0\|_2^2$. (8.2.7)

Proof. Since v_0 has rank 1 and $\|v_0\|_2 = 1$, there exist a unit vector $\xi \in \mathbb{C}^3$ and three constants λ_i, $1 \leq i \leq 3$, satisfying $\sum_{i=1}^{3} |\lambda_i|^2 = 1$ such that the columns of v_0 are respectively $\lambda_i \xi$, $1 \leq i \leq 3$. Interpreting the quantity $\lambda/|\lambda|$ to be 1 when $\lambda = 0$, we define three partial isometries in $\mathbb{M}_3$ by specifying their columns as

$$v_1 = (\lambda_1 \xi / |\lambda_1|, 0, 0), \quad v_2 = (0, \lambda_2 \xi / |\lambda_2|, 0), \quad v_3 = (0, 0, 0).$$

We will show that

$$\min\{\|v_i - v_0\|_2^2 \colon 1 \leq i \leq 3\} \leq (3/2)\|x - v_0\|_2^2. \quad (8.2.8)$$

Then (8.2.7) will follow by choosing v to be that v_i which minimises the left-hand side of (8.2.8).

Let $\{\xi_i\}_{i=1}^3$ be the standard basis for $\mathbb{C}^3$. Then

$$\|x - v_0\|_2^2 = \sum_{i=1}^{3} \langle \lambda_i \xi - \alpha_i \xi_i, \lambda_i \xi - \alpha_i \xi_i \rangle$$

$$= \sum_{i=1}^{3} (|\lambda_i|^2 + \alpha_i^2) - 2 \operatorname{Re} \sum_{i=1}^{3} \alpha_i \lambda_i \langle \xi, \xi_i \rangle, \quad (8.2.9)$$

and this is minimised over all unit vectors ξ by taking ξ parallel to the vector $(\alpha_1 \bar{\lambda}_1, \alpha_2 \bar{\lambda}_2, \alpha_3 \bar{\lambda}_3)^t$ and suitably normalised. From (8.2.9) we obtain

$$\|x - v_0\|_2^2 \geq 1 + \sum_{i=1}^{3} |\alpha_i|^2 - 2 \left(\sum_{i=1}^{3} \alpha_i^2 |\lambda_i|^2 \right)^{1/2}$$

$$\geq (2/3) \min\{2 - 2 \max\{|\lambda_1|, |\lambda_2|\}, 1\}, \quad (8.2.10)$$

where the latter inequality is Lemma 8.2.1. Now

$$\|v_1 - v_0\|_2^2 = \|(\lambda_1 (1 - |\lambda_1|)/|\lambda_1|\xi, \lambda_2 \xi, \lambda_3 \xi)\|_2^2$$

$$= (1 - |\lambda_1|)^2 + |\lambda_2|^2 + |\lambda_3|^2$$

$$= 2 - 2|\lambda_1|, \quad (8.2.11)$$

and the equation

$$\|v_2 - v_0\|_2^2 = 2 - 2|\lambda_2| \quad (8.2.12)$$

is derived in the same way. Since $v_3 = 0$, the equality

$$\|v_3 - v_0\|_2^2 = 1 \quad (8.2.13)$$

is immediate. From (8.2.10) to (8.2.13), it follows that

$$\min\{\|v_i - v_0\|_2^2 \colon 1 \le i \le 3\} = \min\{2 - 2|\lambda_1|, 2 - 2|\lambda_2|, 1\}$$
$$= \min\{2 - 2\max\{|\lambda_1|, |\lambda_2|\}, 1\}$$
$$\le (3/2)\|x - v_0\|_2^2, \tag{8.2.14}$$

proving (8.2.8) and thus establishing (iii). Parts (i) and (ii) are straightforward from the definitions of v_1, v_2 and v_3. $\qquad\square$

Remark 8.2.4. (i) We could introduce a parameter ω from a measure space (Ω, Σ) and require the entries of the matrices x and v_0 in Lemma 8.2.3 to be measurable functions of ω. Then v_1, v_2, v_3 are measurable functions, as are $\|v_i - v_0\|_2^2$, $1 \le i \le 3$. We can then define three measurable sets F_i, $1 \le i \le 3$, as the sets of values of ω where $\|v_i(\omega) - v_0(\omega)\|_2^2$ gives the minimum on the left-hand side of (8.2.14), and we can then choose $v(\omega)$ measurably by letting it be $v_1(\omega)$ on F_1, $v_2(\omega)$ on $F_2 \cap F_1^c$, and $v_3(\omega)$ on $(F_1 \cup F_2)^c$.
(ii) When we apply this lemma, v_0 will be a rank one projection, a special case of a partial isometry. $\qquad\square$

We will also require a similar inequality in the algebra $\mathbb{M}_2$ of 2×2 matrices. Below, c and s will denote $\cos\theta$ and $\sin\theta$. The general form for a rank 1 projection $p \in \mathbb{M}_2$ is

$$p = \begin{pmatrix} c^2 & cse^{-i\phi} \\ cse^{i\phi} & s^2 \end{pmatrix},$$

where $\phi \in [0, 2\pi]$ and $\theta \in [0, \pi/2]$ so that $c, s \in [0, 1]$. The trace on $\mathbb{M}_2$ which assigns the value 1 to each rank 1 projection is denoted by τ, and is used to define $\|\cdot\|_2$.

Lemma 8.2.5. *Let $\alpha \ge 0$, let $x = \left(\begin{smallmatrix} \alpha & 0 \\ 0 & 0 \end{smallmatrix}\right) \in \mathbb{M}_2$, and let $p = \left(\begin{smallmatrix} c^2 & cse^{-i\phi} \\ cse^{i\phi} & s^2 \end{smallmatrix}\right) \in \mathbb{M}_2$ be a rank 1 projection. Let $v = \left(\begin{smallmatrix} c & 0 \\ se^{i\phi} & 0 \end{smallmatrix}\right)$ be a partial isometry. Then*

$$\|x - p\|_2^2 \ge (15/16)\min\{\|v - p\|_2^2, 1\} \ge (2/3)\min\{\|v - p\|_2^2, 1\}. \tag{8.2.15}$$

Proof. In terms of c, we have

$$\|x - p\|_2^2 = (\alpha - c^2)^2 + s^4 + 2c^2 s^2$$
$$= \alpha^2 - 2\alpha c^2 + (c^2 + s^2)^2$$
$$= 1 + \alpha^2 - 2\alpha c^2 \ge 1 - c^4 \tag{8.2.16}$$

by taking the minimum over α, occurring at $\alpha = c^2$. A similar computation shows that

$$\|v - p\|_2^2 = (c^2 - c)^2 + (cs - s)^2 + c^2 s^2 + s^4$$
$$= 2 - 2c. \tag{8.2.17}$$

We now consider two cases.

(i) $0 \le c < 2^{-1}$.

From (8.2.16) and (8.2.17), we see that $\|x - p\|_2^2 \ge 15/16$ while $\min\{\|v - p\|_2^2, 1\} = 1$. Thus (8.2.15) is true in this case.

(ii) $2^{-1} \le c \le 1$.

On this interval, $1 + c + c^2 + c^3 \ge 15/8$, so multiplication by $1 - c$ gives $1 - c^4 \ge (15/8)(1 - c)$. Then, from (8.2.16) and (8.2.17), it follows that

$$\|x - p\|_2^2 \ge (15/8)(1 - c) = (15/16)\|v - p\|_2^2$$
$$= (15/16)\min\{\|v - p\|_2^2, 1\}. \tag{8.2.18}$$

Thus (8.2.15) is also established in this case. □

Corollary 8.2.6. *Let (Ω, Σ) be a measure space, let $x(\omega) = \left(\begin{smallmatrix} \alpha(\omega) & 0 \\ 0 & 0 \end{smallmatrix}\right)$, $\omega \in \Omega$, be a measurable function with $\alpha(\omega) \ge 0$, and let $p(\omega)$ be a measurable function with $p(\omega)$ a rank 1 projection in $\mathbb{M}_2$ for each $\omega \in \Omega$. Let $p_1 = \left(\begin{smallmatrix} 1 & 0 \\ 0 & 0 \end{smallmatrix}\right)$. Then there exists a measurable function $v(\omega)$, where each $v(\omega)$ is a partial isometry in $\mathbb{M}_2$, satisfying*

(i) $v(\omega)v(\omega)^* \le p(\omega), \quad v(\omega)^*v(\omega) \le p_1,$ \hfill (8.2.19)

(ii) $\|v(\omega) - p(\omega)\|_2^2 \le (3/2)\|x(\omega) - p(\omega)\|_2^2, \quad \omega \in \Omega.$ \hfill (8.2.20)

Proof. Let $y \colon \Omega \to \mathbb{M}_2$ be a measurable function, and fix a dense set $\{\xi_i\}_{i=1}^{\infty}$ of vectors in the unit ball of $\mathbb{C}^2$. Then

$$\|y(\omega)\| = \sup\{|\langle y(\omega)\xi_i, \xi_j \rangle| \colon i, j \in \mathbb{N}\} \tag{8.2.21}$$

and is thus a measurable function. Applying this to $y(\omega) = p_1 p(\omega) p_1$, we see that the sets

$$F_1 = \{\omega \colon \|p_1 p(\omega) p_1\| < 2^{-1}\}, \quad F_2 = \{\omega \colon \|p_1 p(\omega) p_1\| \ge 2^{-1}\}$$

are measurable, and they correspond to the two cases in the proof of Lemma 8.2.5. We may then define partial isometries by

$$v(\omega) = \begin{cases} 0, & \omega \in F_1, \\ p(\omega) p_1 / \|p(\omega) p_1\|, & \omega \in F_2, \end{cases}$$

and this is measurable in ω. Then (8.2.19) is immediate from the definition of $v(\omega)$. The inequality (8.2.20) follows from Lemma 8.2.5, noticing that if $p(\omega)$ has the form $\left(\begin{smallmatrix} c^2 & cse^{-i\phi} \\ cse^{i\phi} & s^2 \end{smallmatrix}\right)$, then $p(\omega)p_1/\|p(\omega)p_1\| = \left(\begin{smallmatrix} c & 0 \\ se^{i\phi} & 0 \end{smallmatrix}\right)$. The sharper estimate with $16/15$ replacing $3/2$ is also valid but is not needed subsequently. □

8.3 Main results

Let (Ω, Σ, μ) be a probability space and let H be a fixed separable infinite dimensional Hilbert space. We now investigate the von Neumann algebra $L^\infty(\Omega) \overline{\otimes} B(H)$, which can be identified with the algebra of weakly measurable uniformly bounded functions $x\colon \Omega \to B(H)$. By this we mean that $\omega \mapsto \langle x(\omega)\xi, \eta\rangle$ is a $\mathbb{C}$-valued measurable function for each pair of vectors $\xi, \eta \in H$. It is useful to fix a norm dense sequence $\{\xi_i\}_{i=1}^{\infty}$ in the unit ball of H and an orthonormal basis $\{\eta_i\}_{i=1}^{\infty}$ for H. We denote by τ the standard normal semifinite trace on $B(H)$ defined by $\tau(\cdot) = \sum_{i=1}^{\infty}\langle \cdot\,\eta_i, \eta_i\rangle$, and we let $T(H)$ be the space of trace class operators on H. We begin by establishing the measurability of certain quantities associated to measurable $B(H)$- or $T(H)$-valued functions. When $x(\omega) \in T(H)^+$, $\lambda_1(\omega) > \lambda_2(\omega) \ldots$ denote the eigenvalues with associated finite rank projections $q_1(\omega), q_2(\omega), \ldots$ onto the eigenspaces.

Proposition 8.3.1. *Let (Ω, Σ) be a measure space and let $x\colon \Omega \to B(H)$ be measurable.*

 (i) $\|x(\omega)\|$ *is a measurable function;*

 (ii) *if $x(\omega) \in T(H)^+$ for each $\omega \in \Omega$, then the functions $\lambda_j(\omega)$ and $q_j(\omega)$ are measurable, for $j \geq 1$;*

 (iii) *if $x(\omega) \in T(H)^+$ for each $\omega \in \Omega$ and if $d_j(\omega) \in \mathbb{N} \cup \{\infty\}$ denotes the dimension of the eigenspace for $\lambda_j(\omega)$, then each $d_j(\omega)$, $j \geq 1$, is measurable.*

Proof. (i) As in Corollary 8.2.6, measurability of $\|x(\omega)\|$ follows from the formula

$$\|x(\omega)\| = \sup\{|\langle x(\omega)\xi_i, \xi_j\rangle|\colon\ i,j \in \mathbb{N}\}. \tag{8.3.1}$$

(ii) If $x(\omega) \in T(H)^+$ then $\lambda_1(\omega)$, the largest eigenvalue, is equal to the norm and measurability follows from (i). Then $F = \{\omega\colon\ \lambda_1(\omega) = 0\}$ is measurable, and on this set $q_1(\omega) = 1$. By considering F and its complement separately, we may assume that $\lambda_1(\omega) > 0$ for all $\omega \in \Omega$. By dividing by this function, we may then assume that $\lambda_1(\omega) = 1$ everywhere, which will not change $q_1(\omega)$. For each ω, the sequence $\{x(\omega)^n\}_{n=1}^{\infty}$ converges strongly to $q_1(\omega)$ and so, for each pair $\xi, \eta \in H$,

$$\langle q_1(\omega)\xi, \eta\rangle = \lim_{n\to\infty} \langle x(\omega)^n\xi, \eta\rangle, \tag{8.3.2}$$

showing measurability of $q_1(\omega)$.

Now consider $x(\omega) - \lambda_1(\omega)q_1(\omega)$, which is trace class and has largest eigenvalue $\lambda_2(\omega)$. Arguing as above, $\lambda_2(\omega)$ and $q_2(\omega)$ will be measurable. In general, measurability of $\lambda_{n+1}(\omega)$ and $q_{n+1}(\omega)$ will follow from that of $x(\omega) - \sum_{j=1}^{n} \lambda_j(\omega)q_j(\omega)$, since this operator has largest eigenvalue $\lambda_{n+1}(\omega)$ and associated projection $q_{n+1}(\omega)$.

(iii) Since

$$d_j(\omega) = \tau(q_j(\omega)) = \sum_{n=1}^{\infty}\langle q_j(\omega)\eta_n, \eta_n\rangle, \qquad j \geq 1, \tag{8.3.3}$$

measurability of these functions follows from (ii). $\square$

We point out that the following argument will be used several times below. If we want to establish that $\|x\|_2 \geq \|y\|_2$ for certain operators x and y, and if we have a projection p such that $pyp = y$, then it suffices to prove that $\|pxp\|_2 \geq \|y\|_2$ since $\|x\|_2 \geq \|pxp\|_2$.

Lemma 8.3.2. *Let (Ω, Σ) be a measure space and let $x \colon \Omega \to T(H)^+$ be a measurable function such that the projection $q_1(\omega)$ onto the eigenspace of the largest eigenvalue $\lambda_1(\omega)$ of $x(\omega)$ has rank at least 3 for each $\omega \in \Omega$. Let $p \colon \Omega \to T(H)^+$ be a measurable function such that $p(\omega)$ is a rank 1 projection for all $\omega \in \Omega$. Then, for $\omega \in \Omega$,*

$$\|p(\omega)\|_2^2 \leq (3/2)\|p(\omega) - x(\omega)\|_2^2. \tag{8.3.4}$$

Proof. Fix an element $\omega \in \Omega$, let $n \geq 3$ be the rank of $q_1(\omega)$, and choose an $(n+1)$-dimensional subspace K of H containing the ranges of $p(\omega)$ and $q_1(\omega)$. It suffices to prove (8.3.4) for the compressions of these operators to K and so we need only consider the following situation: $P \in \mathbb{M}_{n+1}$ is a rank 1 projection, $Q \in \mathbb{M}_{n+1}$ is a rank n projection, and $X = \lambda Q + \alpha(I - Q)$ with $\lambda \geq \alpha \geq 0$. Then $\|P\|_2^2 = 1$, while

$$
\begin{aligned}
\|X - P\|_2^2 &= \|\lambda Q + \alpha(I - Q) - P\|_2^2 \\
&= \|\lambda I + (\alpha - \lambda)(I - Q) - P\|_2^2 \\
&= \lambda^2(n+1) + (\alpha - \lambda)^2 + 1 + 2\lambda(\alpha - \lambda) \\
&\quad - 2\lambda - 2(\alpha - \lambda)\tau((I - Q)P) \\
&\geq \lambda^2(n+1) + \alpha^2 - 2\alpha\lambda + \lambda^2 + 1 + 2\alpha\lambda - 2\lambda^2 - 2\lambda \\
&= n\lambda^2 - 2\lambda + \alpha^2 + 1 \\
&\geq n\lambda^2 - 2\lambda + 1 \\
&\geq 1 - n^{-1}, \tag{8.3.5}
\end{aligned}
$$

by minimising over λ. Since $n \geq 3$, we obtain $\|X - P\|_2^2 \geq 2/3$, and the result follows. In the above inequalities, the condition $\lambda \geq \alpha$ was used to neglect the positive term $-2(\alpha - \lambda)\tau((I - Q)P)$. $\qquad\square$

The following result, concerning diagonalisation of matrices over an abelian von Neumann algebra, is a special case of a more general theorem in [99, 100] where no restrictions are placed on the von Neumann algebra. We give here a measure-theoretic proof to complement the one of Theorem 2.4.3 which was formulated in terms of stonean spaces. Theorem 8.3.3 is needed in the course of proving Theorem 8.3.4.

Theorem 8.3.3. *Let (Ω, Σ, μ) be a probability space such that $L^1(\Omega)$ is norm separable, and let $n \in \mathbb{N}$. If A is a masa in $L^\infty(\Omega) \otimes \mathbb{M}_n$, then A is unitarily equivalent to the masa $L^\infty(\Omega) \otimes \mathbb{D}_n$, where $\mathbb{D}_n$ is the algebra of diagonal matrices in $\mathbb{M}_n$. In particular, any set of commuting normal operators in $L^\infty(\Omega) \otimes \mathbb{M}_n$ can be simultaneously diagonalised.*

Proof. The algebra $L^\infty(\Omega) \otimes \mathbb{M}_n$ may be viewed as acting on the Hilbert space $L^2(\Omega)\otimes\mathbb{C}^n$, where its commutant is $L^\infty(\Omega)\otimes I$, the centre of $L^\infty(\Omega)\otimes\mathbb{M}_n$. Since A must contain $L^\infty(\Omega) \otimes I$ (otherwise A would not be a masa) we see that any operator in $B(L^2(\Omega))$ commuting with A must lie in $L^\infty(\Omega)\otimes\mathbb{M}_n$, by the double commutant theorem, Theorem 2.2.1, and so is isomorphic to $\ell^\infty(k) \oplus L^\infty[0,1]$ for some $k \in \mathbb{N} \cup \{\infty\}$, by Theorem 2.3.7 (either summand might be absent). In all cases, A is generated by a single unitary. Taking the logarithm shows that A is generated by a single self-adjoint operator x, and we may assume that x is positive and invertible by adding a suitable multiple of the identity. Thus the result will follow if we can diagonalise x. This amounts to selecting vectors $\eta_1,\ldots,\eta_n \in L^2(\Omega,\mathbb{C}^n)$ such that, for each $\omega \in \Omega$, $\{\eta_1(\omega),\ldots,\eta_n(\omega)\}$ is an orthonormal basis for $\mathbb{C}^n$ consisting of eigenvectors for $x(\omega)$.

Let $\{\xi_1,\ldots,\xi_n\}$ be the standard basis for $\mathbb{C}^n$. With the notation of Proposition 8.3.1, $\lambda_1(\omega)$ is the largest eigenvalue of $x(\omega)$ and $q_1(\omega)$ is the projection onto the corresponding eigenspace, and both are measurable functions. Since $q_1(\omega) \neq 0$, we cannot have $q_1(\omega)\xi_i = 0$ for all i in the range $1 \leq i \leq n$. Thus we may define measurable sets $F_i \subseteq \Omega$, $1 \leq i \leq n$, by

$$F_i = \{\omega : q_1(\omega)\xi_i \neq 0\},$$

and Ω will be the union of these sets. There could be overlap, so we make them disjoint by defining $G_1 = F_1$, $G_2 = F_2 \setminus F_1$, and so on. We then define a measurable function $\eta_1 : \Omega \to \mathbb{C}^n$ by

$$\eta_1(\omega) = q_1(\omega)\xi_i/\|q_1(\omega)\xi_i\| \text{ for } \omega \in G_i, \ 1 \leq i \leq n,$$

and η_1 is then a unit vector in $L^2(\Omega,\mathbb{C}^n)$ such that $\eta_1(\omega)$ is a unit eigenvector for $x(\omega)$ for each $\omega \in \Omega$. Define $v_1 \in L^\infty(\Omega)\otimes\mathbb{M}_n$ by letting $v_1(\omega)$ be the matrix whose first column is $\eta_1(\omega)$, all other columns being 0. This is a partial isometry, and we define a projection $p_1 \in L^\infty(\Omega) \otimes \mathbb{M}_n$ by $p_1 = v_1v_1^*$. Then $p_1(\omega)$ is a rank one projection onto the span of an eigenvector for $x(\omega)$, so that x and p_1 commute. The vector η_2 is obtained by repeating this argument for the positive operator $(1-p_1)x$, with a corresponding projection p_2. In general, η_i is obtained by considering $(1 - p_1 - \cdots - p_{i-1})x$. Now form a unitary $u \in L^\infty(\Omega) \otimes \mathbb{M}_n$ by specifying the columns of $u(\omega)$ to be the vectors $\eta_1(\omega),\ldots,\eta_n(\omega)$. Since these are eigenvectors for $x(\omega)$, it is apparent that u^*xu is diagonal. Thus $u^*Au \subseteq L^\infty(\Omega) \otimes \mathbb{D}_n$, since x generates A, and equality must hold since both algebras are masas.

The last statement of the theorem is immediate by observing that any set of normal commuting elements generates an abelian von Neumann subalgebra which is then contained in a masa. $\qquad\square$

We now come to the main results of the section, for which we require some notation. We let (Ω, Σ, μ) be a probability space for which $L^\infty(\Omega)\overline{\otimes}B(H)$ has a separable predual. We let A denote an abelian von Neumann subalgebra of $L^\infty(\Omega)\overline{\otimes}B(H)$, and we fix a positive operator $h \in A' \cap (L^\infty(\Omega)\overline{\otimes}B(H))$ such that $h(\omega) \in T(H)$ for all $\omega \in \Omega$. We let $\{a_n\}_{n=1}^\infty$ be a sequence in A which is weakly dense, and we let $A(\omega) \subseteq B(H)$ be the abelian von Neumann algebra generated by $\{a_n(\omega)\}_{n=1}^{\infty}$ for $\omega \in \Omega$. Note that, to ensure that $A(\omega)$ is abelian, we may have to neglect a countable number of null sets which has no effect.

Theorem 8.3.4. *Let* $p\colon \Omega \to T(H)$ *be a measurable function such that each* $p(\omega)$ *is a rank 1 projection. Then there exists a partial isometry* $v = v(\omega) \in L^{\infty}(\Omega)\overline{\otimes}B(H)$ *such that*

(i) $vv^* \leq p, \quad v^*v \in A' \cap (L^{\infty}(\Omega)\overline{\otimes}B(H));$

(ii) $\|v(\omega) - p(\omega)\|_2^2 \leq (3/2)\|p(\omega) - h(\omega)\|_2^2, \qquad \omega \in \Omega.$ \hfill (8.3.6)

Proof. Since $h \in A'$, we may assume without loss of generality that A is a masa in $L^{\infty}(\Omega)\overline{\otimes}B(H)$ containing h, and also containing the centre $L^{\infty}(\Omega) \otimes 1$. Then $h(\omega) \in A(\omega)'$ for $\omega \in \Omega$, and the same holds for the spectral projections of $h(\omega)$, denoted $q_j(\omega)$ for the jth eigenvalue $\lambda_j(\omega)$. We will construct v pointwise over Ω, the main difficulty being to do this measurably. Consequently we will decompose the measure space into measurable subsets and define $v(\omega)$ separately on each. We carry this out in several steps.

(1) $F_1 = \{\omega\colon h(\omega) = 0\}$.

This set is $\{\omega\colon \|h(\omega)\| = 0\}$, and is measurable. Here we take $v(\omega) = 0$, and (8.3.6) is clearly satisfied.

(2) $F_2 = \{\omega\colon \text{rank } q_1(\omega) = 1, \quad \|h(\omega) - \lambda_1(\omega)q_1(\omega)\| = 0\}$.

By Proposition 8.3.1 (i), (iii), this is a measurable set. Replacing Ω by F_2, we may assume that the conditions defining F_2 hold for all $\omega \in \Omega$. For a fixed $\omega \in \Omega$, choose a two-dimensional subspace K of H containing the ranges of $p(\omega)$ and $q_1(\omega)$. Choose a basis for K where the first vector is an eigenvector for $q_1(\omega)$. Then the compression of h to K has the form $\left(\begin{smallmatrix} \lambda_1(\omega) & 0 \\ 0 & 0 \end{smallmatrix}\right)$ and Corollary 8.2.6 applies. On the measurable set $G = \{\omega\colon \|p(\omega)q_1(\omega)\| < 2^{-1}\}$, we let $v(\omega) = 0$ (corresponding to the first case in the proof of Lemma 8.2.5). On the complement of G we take $v(\omega) = p(\omega)q_1(\omega)/\|p(\omega)q_1(\omega)\|$. Then (8.3.6) follows from (8.2.15), while $v(\omega)v^*(\omega)$ is either 0 or $p(\omega)$ and $v(\omega)^*v(\omega)$ is either 0 or $q_1(\omega)$, both in $A(\omega)'$.

(3) $F_{3,n} = \{\omega\colon \text{rank } q_1(\omega) = 1, \quad \lambda_2(\omega) > 0, \text{ rank } q_2(\omega) = n\}, \quad n \geq 1$.

These are measurable sets by Proposition 8.3.1 (ii), (iii). Replacing Ω by $F_{3,n}$ for a fixed n, we may assume that the conditions defining $F_{3,n}$ hold on all of Ω. If we compress $L^{\infty}(\Omega)\overline{\otimes}B(H)$ by q_2, then Aq_2 is a masa in $L^{\infty}(\Omega)\overline{\otimes}\mathbb{M}_n$, and thus Aq_2 is unitarily equivalent to the algebra of diagonal matrices with entries from $L^{\infty}(\Omega)$, by Theorem 8.3.3. Consequently there is a projection $p_2 \in A$ such that $p_2 \leq q_2$ and $p_2(\omega)$ has rank 1 for all $\omega \in \Omega$. For a fixed ω, pick a three-dimensional subspace K of H containing the ranges of $q_1(\omega)$, $p_2(\omega)$ and $p(\omega)$. Then there is a basis with respect to which $q_1(\omega) = \left(\begin{smallmatrix} 1 & 0 & 0 \\ 0 & 0 & 0 \\ 0 & 0 & 0 \end{smallmatrix}\right)$, $p_2(\omega) = \left(\begin{smallmatrix} 0 & 0 & 0 \\ 0 & 1 & 0 \\ 0 & 0 & 0 \end{smallmatrix}\right)$ and the compression of $h(\omega)$ to K is $\left(\begin{smallmatrix} \lambda_1(\omega) & 0 & 0 \\ 0 & \lambda_2(\omega) & 0 \\ 0 & 0 & \alpha \end{smallmatrix}\right)$ with $\alpha \leq \lambda_2(\omega)$. Lemma 8.2.3 now applies, and the choice of $v(\omega)$ is made measurably by Remark 8.2.4. The inequality (8.3.6) follows from (8.2.7) while the proof of Lemma 8.2.3 shows that $v(\omega)v(\omega)^*$ is either $p(\omega)$ or 0, while $v(\omega)^*v(\omega)$ is $q_1(\omega)$, $p_2(\omega)$ or 0, all three lying in $A(\omega) = A(\omega)'$.

(4) $F_4 = \{\omega\colon \text{ rank } q_1(\omega) = 2\}$.

As before, we may assume that this condition holds on all of Ω. Compression by q_1 gives us a masa Aq_1 in $L^\infty(\Omega) \otimes \mathbb{M}_2$, and thus orthogonal projections $p_1, p_2 \in Aq_1$ with $p_1(\omega)$, $p_2(\omega)$ of rank 1 for all $\omega \in \Omega$. The argument now is that of the previous part, with p_1 replacing q_1.

(5) $F_5 = \{\omega\colon \text{ rank } q_1(\omega) \geq 3\}$.

This is a measurable set, and again we assume it is Ω. By Lemma 8.3.2, we may let $v(\omega) = 0$ on this set.

These five parts exhaust the possibilities and so we have a partial isometry satisfying the conclusions of the theorem. $\square$

Now consider a type II_1 von Neumann algebra N with separable predual and faithful normal trace τ, and let B be a masa in N. Let Tr denote the canonical normal semifinite trace on $\langle N, e_B \rangle$ such that $Tr(e_B) = 1$ (see Theorem 4.2.2). In the statement of the next theorem, A will denote an abelian subalgebra of N. In the proof, the standard trace on $B(H)$ will be denoted by tr.

Theorem 8.3.5. *If $h \in A' \cap \langle N, e_B \rangle$, $h \geq 0$ and $Tr(h) < \infty$, then there exists a partial isometry $V \in \langle N, e_B \rangle$ such that $VV^* \leq e_B$, $V^*V \in A' \cap \langle N, e_B \rangle$, and*

$$\|e_B - V\|_{2,Tr}^2 \leq (3/2)\|e_B - h\|_{2,Tr}^2. \tag{8.3.7}$$

Proof. As we mentioned in Section 8.1, the commutant of $\langle N, e_B \rangle$ is the abelian algebra JBJ and so $\langle N, e_B \rangle$ is type I. Moreover, since N can have no finite dimensional representations, $\langle N, e_B \rangle$ is type I_∞ and thus is isomorphic to $L^\infty[0,1]\bar{\otimes}B(H)$. The projection e_B satisfies $e_B\langle N, e_B \rangle e_B = e_B B = e_B JBJ$, by Theorem 4.2.2, and is consequently abelian. Identifying e_B with a measurable function $e_B(t) \in L^\infty[0,1]\bar{\otimes}B(H)$, we see that $e_B(t)$ is a projection of rank at most 1 almost everywhere. If $z \in L^\infty[0,1]$ is the characteristic function of the set where rank $(e_B(t)) \neq 1$, then $e_B z = 0$, and $z = 0$, by Theorem 4.2.2 (v). Thus, altering $e_B(t)$ on a null set if necessary, we may assume that the rank of $e_B(t)$ is 1 for all $t \in [0,1]$.

Since the trace Tr on $L^\infty[0,1]\bar{\otimes}B(H)$ is normal and $Tr(e_B) = 1$, there exists a non-negative function $k \in L^1[0,1]$ such that

$$Tr(x) = \int_0^1 k(t)\, tr(x(t))\, dt \tag{8.3.8}$$

for those $x \in \langle N, e_B \rangle$ satisfying $Tr(|x|) < \infty$. We now apply Theorem 8.3.4, with $p(\omega)$ replaced by $e_B(t)$, to obtain a partial isometry $V(t)$ satisfying the relations $V(t)V(t)^* = e_B(t)$, $V(t)^*V(t) \in A(t)'$ and

$$\|e_B(t) - V(t)\|_{2,tr}^2 \leq (3/2)\|e_B(t) - h(t)\|_{2,tr}^2. \tag{8.3.9}$$

Then $VV^* \leq e_B, V^*V \in A' \cap \langle N, e_B \rangle$, and

$$\|e_B - V\|_{2,Tr}^2 = \int_0^1 k(t)\|e_B(t) - V(t)\|_{2,tr}^2 \, dt$$

$$\leq (3/2) \int_0^1 k(t)\|e_B(t) - h(t)\|_{2,tr}^2 \, dt$$

$$= (3/2)\|e_B - h\|_{2,Tr}^2, \qquad\qquad (8.3.10)$$

proving the result. □

Chapter 9

Perturbations

9.1 Introduction

The topic for this and the succeeding chapter is the theory of perturbations. Before giving a detailed description, we discuss the ideas in general terms without reference to norms or metrics. If we have a von Neumann subalgebra A of a II_1 factor N, and $u \in N$ is a unitary close to 1, then the algebras A and uAu^* are close and we think of uAu^* as a small perturbation of A. Conversely, if we have two algebras A and B which are close to one another, then we might expect to find a unitary $u \in N$ close to 1 so that $B = uAu^*$. This is too much to ask for in general. In these two chapters we explore whether suitable modifications can be made so that results of this type hold true. Although there are circumstances where unitary equivalence is possible, it is usually necessary to cut the algebras by projections and ask only for a partial isometry which implements a spatial isomorphism of the compressions. As we will see, the strength of the results will depend on the norms and metrics selected to define the notion of close operators and close algebras. Some theorems in this chapter are formulated for general subalgebras and thus apply in the next chapter. The main focus here is on masas, and some of these results are only valid in that case.

The easiest way to think of the distance between subalgebras of a type II_1 factor is by considering the distance between their conditional expectations as this makes available all the convergence properties of operators. The first perturbation results for masas in type II_1 factors were in the norm $\| \cdot \|$. These were developed by Christensen in a series of papers [20, 21, 22], who turned to perturbations in the $\| \cdot \|_{\infty,2}$-norm, or in $\subseteq_\delta$ in the important paper [24]. The two crucial methods Christensen introduced in the latter paper have been used in all subsequent perturbation results in this area. He used the basic construction to handle the perturbation of A to B in the algebra $\langle N, e_B \rangle = \langle N, B \rangle$ and used the fact that the element of minimal norm in a closed convex subset K of a Hilbert space is a fixed point for any unitary group leaving K invariant [24]. Popa, in an important series of papers, introduced the use of the metric and

148

algebraic properties of the 'pulldown' map from part of $\langle N, e_B \rangle$ to N and certain fine analysis of projections to strengthen the original methods of Christensen considerably (see [144, 145, 147, 148]). The crucial rigidity/perturbation results that Popa obtained are one of the foundations of the important deep theorems he has proved on II_1 factors [146, 147, 148, 149, 150].

The most important perturbation estimates for type II_1 factors are in the infinity-two norm $\|\cdot\|_{\infty,2}$ acting on the difference of the conditional expectations. Both the infinity-two norm $\|\cdot\|_{\infty,2}$ and $\subset_\delta$, *delta contained in*, used by Murray von Neumann, McDuff and Christensen are defined in Section 9.2 and the basic properties are studied.

The general method in the perturbation results in this chapter follows the same general technique developed by Christensen [24] and Popa [147]. This basic outline covers several of Popa's papers. These are the methods of this chapter, which are summarised here before we proceed further.

Summary of the techniques in perturbations of von Neumann subalgebras of II_1 factors.

Let A and B be von Neumann subalgebras of a II_1 factor N. At each stage it is only the final part that is carried forward.

(1)   If $\|(I - \mathbb{E}_B)\mathbb{E}_A\|_{\infty,2} < 1$, then 'average e_B over the unitary group $\mathcal{U}(A)$ of A' to obtain the element h of minimal $\|\cdot\|_{2,Tr}$-norm in

$$\overline{\operatorname{conv}}^w \{ u e_B u^* : u \in \mathcal{U}(A) \}$$

with the properties

$$h \in A' \cap \langle N, e_B \rangle \quad \text{and} \quad 0 < Tr(h), Tr(h^2) \leq 1.$$

(2)   From h obtain a non-zero spectral projection f_1 in $A' \cap \langle N, e_B \rangle$ with f_1 equivalent to a subprojection e_1 of e_B. There are various ways to choose such a projection, but we use the results of Chapter 8, which seem to give the best norm estimates.

(3)   From a partial isometry $v_1 \in \langle N, e_B \rangle$ with

$$v_1^* v_1 = f_1 \quad \text{and} \quad v_1 v_1^* = e_1 \leq e_B,$$

construct a non-zero *-homomorphism ϕ from A into B with

$$\phi(a) v_1 = v_1 a \quad \text{for all} \quad a \in A.$$

(4a) *EITHER*   With $\Phi : N e_B N \to N$ the pull down map $\Phi(x e_B y) = xy$ ($x, y \in N$), let $w = \Phi(v_1) \in L^2(N)$. The equation $\phi(a) v_1 = v_1 a$ ($a \in A$) of (3) gives $\phi(a) w = wa$ ($a \in A$). The polar decomposition $w = v|w|$ of the unbounded operator w affiliated to N gives a non-zero partial isometry $v \in N$ with $\phi(a) v = va$ ($a \in A$).

(4b) *OR*   By averaging $\phi(u) u^*$ over the unitary group $\mathcal{U}(A)$, the element w_0 of minimal $\|\cdot\|_2$-norm in $\overline{\operatorname{conv}}^w \{ \phi(u) u^* : u \in \mathcal{U}(A) \}$ satisfies

$$\phi(a) w_0 = w_0 a \quad \text{for all} \quad a \in A.$$

If $w_0 = v_0|w_0|$ is the polar decomposition of w_0, then

$$\phi(a)v_0 = v_0 a \quad \text{for all} \quad a \in A.$$

(5)   With additional properties on A and B cut the partial isometries and projections down and modify ϕ to be a *-isomorphism from a cutdown of A to a cutdown of B. In the case of close algebras this is done using bounds on the partial isometries and projections with a finite index bound. For two masas we use the simple observation that if one masa is contained inside another, then they are equal.

Additional information and estimates are required to ensure that the maps are non-zero, that ϕ is one-to-one on a large piece of A and to obtain $\| \cdot \|_2$-norm estimates on the final partial isometry and its associated projections. The partial isometries v and v_0 obtained in (4a) and (4b) could be different; the current estimates obtained using the pull-down map in (4a) are better than those averaging the map ϕ in (4b) as the pull-down has universal properties .

In Section 9.2, we explain the method of averaging an operator over a unitary group $\mathcal{U}(A)$. If A were hyperfinite, then this could be accomplished by using an invariant mean on a weakly dense amenable subgroup generated by the unitary groups of finite dimensional subalgebras. In general, we obtain an operator in $A' \cap \langle N, e_B \rangle$ by taking the element of minimal norm in a certain norm closed convex subset of $L^2(\langle N, e_B \rangle, Tr)$.

Section 9.3 deals with subalgebras of a II_1 factor which are close in the metric induced by the operator norm. If the two subalgebras are sufficiently close, then we show that they are unitarily conjugate by a unitary close to 1. Sections 9.4 and 9.5 address the same type of problem for the metric derived from the $\| \cdot \|_{\infty,2}$-norm, the main result being Theorem 9.5.1. Here we show that two close masas A and B have large cut downs Ap and Bq which are spatially equivalent by a partial isometry v close to 1. We give an example to demonstrate that this result is optimal. The various estimates that relate $\|1 - p\|_{2,\tau}$, $\|1 - q\|_{2,\tau}$ and $\|1 - v\|_{2,\tau}$ to $\|\mathbb{E}_A - \mathbb{E}_B\|_{\infty,2}$ are given in Section 9.6. The final section of this chapter gives a different view of the same material. This uses techniques that are simpler, but which yield estimates that are not quite as good as the earlier ones.

9.2   Averaging e_B over A

Throughout this section N will be a II_1 factor with von Neumann subalgebras A and B. If the algebras are masas, then one can average using an invariant mean on the abelian unitary group of the algebra A in place of the convex set averaging used here. This however does not simplify things as we require Hilbert norm estimates, which follow immediately from the methods employed below. Recall that e_B is the projection from $L^2(N)$ onto $L^2(B)$ and that the trace preserving conditional expectation $\mathbb{E}_B$ is the restriction of e_B to N.

The first lemma discusses some facts about convex sets and their weak closures and will be applied subsequently with $M = \langle N, e_B \rangle$. We will use $\| \cdot \|_2$ to denote the $\| \cdot \|_2$-norm, but if more than one trace is present, as will occur in subsequent results, then this will be indicated by $\| \cdot \|_{2,Tr}$.

Lemma 9.2.1. *Let M be a von Neumann algebra with a normal faithful semifinite trace Tr, and let P denote the set of projections in M of finite trace. Let $K \subseteq M^+$ be a convex set and let Λ_∞, Λ_1 and Λ_2 denote respectively the suprema over $x \in K$ of $\|x\|$, $\|x\|_1$, and $\|x\|_2$. Let $\overline{K}^w$ be the weak closure of K in M.*

(i) *For each $x \in M^+$,*

$$Tr(x) = \sup\{Tr(xp) \colon p \in P\}. \tag{9.2.1}$$

(ii) *If $\Lambda_\infty < \infty$, then $\sup\{\|x\| \colon x \in \overline{K}^w\} = \Lambda_\infty < \infty$.*

(iii) *If $\Lambda_1 < \infty$ then $\sup\{\|x\|_1 \colon x \in \overline{K}^w\} = \Lambda_1 < \infty$. Moreover, if a net (x_α) in $\overline{K}^w$ converges weakly to x, then*

$$Tr(x) \leq \limsup_\alpha Tr(x_\alpha). \tag{9.2.2}$$

(iv) *If $\Lambda_\infty, \Lambda_2 < \infty$, then $\overline{K}^w$ is a $\| \cdot \|_2$-closed convex subset of $L^2(M)$, and*

$$\sup\{\|x\|_2 \colon x \in \overline{K}^w\} = \Lambda_2. \tag{9.2.3}$$

Moreover, if $x_\alpha \to x$ weakly in $\overline{K}^w$, then $x_\alpha \to x$ in the weak topology of $L^2(M)$.

(v) *The conclusions of (iv) remain valid when the hypothesis $K \subseteq M^+$ is replaced by $K \subseteq M$.*

Proof. (i) Let $x \in M^+$ be fixed but arbitrary. The inequality

$$Tr(x) \geq \sup\{Tr(xp) \colon p \in P\} \tag{9.2.4}$$

is immediate from the relation

$$Tr(xp) = Tr(x^{1/2}px^{1/2}) \leq Tr(x), \quad p \in P. \tag{9.2.5}$$

To establish the reverse inequality, first note that the semifiniteness hypothesis on the trace gives an increasing net (p_β) from P converging weakly to 1. Then normality of the trace implies that

$$Tr(x) = \lim_\beta Tr(x^{1/2}p_\beta x^{1/2}) = \lim_\beta Tr(xp_\beta) \leq \sup\{Tr(xp) \colon p \in P\}. \tag{9.2.6}$$

(ii) This is just the w^*-compactness of norm closed balls in M.

(iii) Consider a net (x_α) in K converging weakly to $x \in \overline{K}^w \subseteq M^+$. If $p \in P$ then the linear functional $Tr(p \cdot p)$ is bounded and normal, and consequently

$$Tr(xp) = \lim_\alpha Tr(x_\alpha p). \tag{9.2.7}$$

If $\lambda < Tr(x)$ then, by (i), we may choose $p \in P$ such that $Tr(xp) > \lambda$. Then there exists α_0 so that $Tr(x_\alpha p) > \lambda$ for $\alpha \geq \alpha_0$. Then

$$\limsup_\alpha Tr(x_\alpha) \geq \sup\{\lambda : \lambda < Tr(x)\} = Tr(x). \tag{9.2.8}$$

This proves the desired inequality.

If $\Lambda_1 < \infty$ and $x \in \overline{K}^w$, choose a net (x_α) from K converging weakly to x. Then, from (9.2.8),

$$Tr(x) \leq \limsup_\alpha Tr(x_\alpha) \leq \Lambda_1, \tag{9.2.9}$$

so $x \in L^1(M)$ and $\|x\|_1 = Tr(x) \leq \Lambda_1$. This establishes that $\sup\{\|x\|_1 : x \in \overline{K}^w\} \leq \Lambda_1$, and the reverse inequality is obvious from the definition of Λ_1.

(iv) Suppose now that both Λ_∞ and Λ_2 are finite. By scaling, we may assume that both constants are no greater than 1. Let $(y_\alpha)_{\alpha \in A}$ be a net from K converging weakly to $y \in \overline{K}^w$. We will show that $y \in L^2(M)$ and, to this end, we fix an arbitrary projection $p \in P$. Then $p \in L^2(M)$, and so $y_\beta p \in L^2(M)$ for $\beta \in A$ and satisfies $\|y_\beta p\|_2^2 \leq Tr(y_\beta^2) \leq 1$. Moreover, $yp \in L^2(M)$ with $\|yp\|_2^2 = Tr(py^*yp) \leq Tr(p)$, where the last inequality is valid because $\|y\| \leq 1$. Then, for any fixed $\alpha \in A$,

$$|Tr(y_\alpha yp)| = |Tr(py_\alpha yp)| = \lim_\beta |Tr(py_\alpha y_\beta p)|$$

$$\leq \limsup_\beta \|py_\alpha\|_2 \|y_\beta p\|_2 \leq 1, \tag{9.2.10}$$

using the Cauchy–Schwarz inequality and the fact that $Tr(p \cdot p)$ is a normal and bounded linear functional. In the same way

$$Tr(y^2) = \sup_{p \in P} Tr(py^2 p) = \sup_{p \in P} \lim_\alpha |Tr(py_\alpha yp)| \leq 1, \tag{9.2.11}$$

from (9.2.10). Thus $y \in L^2(M)$. This shows that each element of $\overline{K}^w$ lies in the unit ball of $L^2(M)$.

Now consider a net (x_α) from $\overline{K}^w$ with weak limit $x \in \overline{K}^w$. From above, these elements lie in the unit ball of $L^2(M)$. In order to show that $x = \lim_\alpha x_\alpha$ in the weak topology of $L^2(M)$, it suffices to consider inner products with elements from a total subset. Since span$\{P\}$ is $\|\cdot\|_2$-dense in $L^2(M)$, it is enough to show that $\lim_\alpha Tr(x_\alpha p) = Tr(xp)$ for $p \in P$. Once again, we use boundedness and normality of the linear functional $Tr(p \cdot p)$ to conclude that

$$\lim_\alpha Tr(x_\alpha p) = \lim_\alpha Tr(px_\alpha p) = Tr(pxp) = Tr(xp), \tag{9.2.12}$$

as required. We have now shown that the identity map of $\overline{K}^w$ into $L^2(M)$ is continuous for the respective weak topologies. The image in $L^2(M)$ is then weakly compact and thus $\|\cdot\|_2$-closed. Equation (9.2.11) and the Cauchy–Schwarz inequality show that

$$\|y\|_2^2 \leq \sup_{p \in P} \limsup_\alpha \|py_\alpha\|_2 \|yp\|_2 \leq \Lambda_2 \|y\|_2, \tag{9.2.13}$$

and this implies (9.2.3).

(v) We begin with two observations. If $x \in M$ is written as $y + iz$ with y, z self-adjoint, then $y = (x + x^*)/2$, so $\|y\|_2 \leq \|x\|_2$ with a similar estimate for $\|z\|_2$. Now if $h \in M$ is self-adjoint, then the decomposition $h = h^+ - h^-$ satisfies $h^2 = (h^+)^2 + (h^-)^2$, so that $\|h^\pm\|_2 \leq \|h\|_2$.

Let L denote the convex hull of the positive and negative parts arising from the self-adjoint decompositions of all elements $k \in K$. Then $\sup_{\ell \in L} \|\ell\|_2 \leq \Lambda_2$. Since $K \subseteq (L-L)+i(L-L)$, the same containment holds with K and L replaced by their respective weak closures. Part (iv) applies to $L \subseteq M^+$, showing that $\overline{L}^w$, and hence $\overline{K}^w$, is $\| \cdot \|_2$-bounded in $L^2(M)$. The required verifications are now identical to the last paragraph of the proof of part (iv). $\square$

A *Haar unitary* in a von Neumann algebra with trace τ is a unitary u such that $\tau(u^j) = 0$ for all $j \in \mathbb{Z} \setminus \{0\}$. If A contains a masa C of N, then there is a Haar unitary u in C, since C is isomorphic to $L^\infty[0,1]$ via a *-isomorphism that sends the trace to the integral (Theorem 3.5.2).

A stronger version of Lemma 9.2.2 is given as Lemma 9.2.3; the first version is what is required in the perturbation results subsequently.

The following result is essentially in [24], and is also used in [145, 147, 148]. The operator h below will be important at several points subsequently and we will refer below to the procedure for obtaining it as *averaging* e_B *over* A. The $\| \cdot \|_{\infty,2}$-norm of a bounded map $\phi \colon N \to N$ is defined by

$$\|\phi\|_{\infty,2} = \sup\{\|\phi(x)\|_2 \colon x \in N, \ \|x\| \leq 1\}.$$

Lemma 9.2.2. *Let A and B be von Neumann subalgebras of a type II_1 factor N, and let $\overline{K}_A^w(e_B)$ be the weak closure of the set*

$$K_A(e_B) = \mathrm{conv}\,\{ue_Bu^* \colon \ u \ \text{is a unitary in } A\} \qquad (9.2.14)$$

in $\langle N, e_B \rangle$. Then $\overline{K}_A^w(e_B)$ contains a unique element h of minimal $\| \cdot \|_{2,Tr}$-norm, and this element satisfies

(i) $\|h\|, \ Tr(h), \ Tr(h^2), \ Tr(e_Bh) \leq 1;$

(ii) $h \in A' \cap \langle N, e_B \rangle, \qquad 0 \leq h \leq 1;$

(iii) $1 - Tr(e_Bh) \leq \|(I - \mathbb{E}_B)\mathbb{E}_A\|_{\infty,2}^2;$

(iv) $Tr(e_Bh) = Tr(h^2);$

(v) $\|h - e_B\|_{2,Tr} \leq \|(I - \mathbb{E}_B)\mathbb{E}_A\|_{\infty,2};$

(vi) $\|h - e_B\| \leq \|(I - \mathbb{E}_B)\mathbb{E}_A\|.$

If A is orthogonal to B and if A contains a Haar unitary, then $h = 0$.

Proof. Since $\|e_B\| = Tr(e_B) = 1$, we may apply Lemma 9.2.1 (iv) to the convex set $K = K_A(e_B)$ to see that $\overline{K}_A^w(e_B)$ is $\| \cdot \|_{2,Tr}$-closed, and thus contains a unique element h of minimal $\| \cdot \|_{2,Tr}$-norm. Moreover, this lemma shows that

$\|h\|$, $Tr(h)$, $Tr(h^2) \leq 1$, and the Cauchy–Schwarz inequality gives $Tr(e_B h) \leq \|h\|_{2,Tr}\|e_B\|_{2,Tr} \leq 1$. This proves (i).

Each $x \in K_A(e_B)$ satisfies $0 \leq x \leq 1$, and so the same is true for elements of $\overline{K}_A^w(e_B)$. In particular, $0 \leq h \leq 1$. For each unitary $u \in A$, the map $x \mapsto uxu^*$ is a $\|\cdot\|_{2,Tr}$-norm isometry which leaves $\overline{K}_A^w(e_B)$ invariant. Thus

$$uhu^* = h, \quad u \text{ unitary in } A, \tag{9.2.15}$$

by minimality of $\|h\|_{2,Tr}$, so $h \in A' \cap \langle N, e_B \rangle$. This proves (ii).

Consider a unitary $u \in A$. Since Tr is a trace,

$$
\begin{aligned}
1 - Tr(e_B u e_B u^*) &= 1 - Tr(e_B \mathbb{E}_B(u)u^*) = 1 - \tau(\mathbb{E}_B(u)u^*) \\
&= 1 - \tau(\mathbb{E}_B(u)\mathbb{E}_B(u)^*) = 1 - \|\mathbb{E}_B(u)\|_{2,\tau}^2 \\
&= \|(I - \mathbb{E}_B)(u)\|_{2,\tau}^2 \leq \|(I - \mathbb{E}_B)\mathbb{E}_A\|_{\infty,2}^2.
\end{aligned}
\tag{9.2.16}
$$

This inequality persists when $u e_B u^*$ is replaced by elements of $K_A(e_B)$, so it follows from the normality of the state $Tr(e_B \cdot e_B)$ that

$$1 - Tr(e_B h) \leq \|(I - \mathbb{E}_B)\mathbb{E}_A\|_{\infty,2}^2, \tag{9.2.17}$$

proving (iii).

Since $h \in A' \cap \langle N, e_B \rangle$,

$$Tr(u e_B u^* h) = Tr(e_B u^* h u) = Tr(e_B h) \tag{9.2.18}$$

for all unitaries $u \in A$. Part (iv) follows from this by taking suitable convex combinations and a weak limit to replace $u e_B u^*$ by h on the left-hand side of (9.2.18), again using normality of $Tr(e_B \cdot e_B)$. Now, using (iii) and (iv),

$$
\begin{aligned}
\|h - e_B\|_{2,Tr}^2 &= Tr(h^2 - 2h e_B + e_B) = Tr(e_B - h e_B) \\
&= 1 - Tr(h e_B) \leq \|(I - \mathbb{E}_B)\mathbb{E}_A\|_{\infty,2}^2,
\end{aligned}
\tag{9.2.19}
$$

proving (v).

If u is a unitary in A, then

$$
\begin{aligned}
\|e_B - u e_B u^*\| &= \|e_B u - u e_B\| = \|e_B u(1 - e_B) - (1 - e_B)u e_B\| \\
&\leq \max\{\|e_B u(1 - e_B)\|, \|(1 - e_B)u e_B\|\} \tag{9.2.20} \\
&= \max\{\|e_B(u - e_B u e_B)\|, \|(u - e_B u e_B)e_B\|\} \\
&= \max\{\|e_B(u - \mathbb{E}_B(u))\|, \|(u - \mathbb{E}_B(u))e_B\|\} \\
&\leq \|u - \mathbb{E}_B(u)\| \leq \|(1 - \mathbb{E}_B)\mathbb{E}_A\|.
\end{aligned}
$$

The inequality (9.2.20) above occurs because of the orthogonality of $e_B u(1 - e_B)$ and $(1 - e_B)u e_B$ in both range and domain spaces. Taking convex combinations in this inequality allows us to replace $u e_B u^*$ by any element of $K_A(e_B)$, from which it follows that $\|e_B - h\| \leq \|(1 - \mathbb{E}_B)\mathbb{E}_A\|$, proving (vi).

Finally, suppose that A is orthogonal to B and that u is a Haar unitary in A. Then

$$\| \sum_{j=1}^{n} u^j e_B u^{-j} \|_{2,Tr}^2 = \sum_{j,k=1}^{n} Tr(u^j e_B u^{-j+k} e_B u^{-k})$$

$$= \sum_{j,k=1}^{n} Tr(u^{j-k} \mathbb{E}_B(u^{-j+k}) e_B)$$

$$= \sum_{j=1}^{n} Tr(e_B) = n, \qquad (9.2.21)$$

since the orthogonality of A and B implies that $\mathbb{E}_B(u^r) = 0$ for all $r \in \mathbb{Z} \setminus \{0\}$ by Lemma 6.3.1. The element $x = n^{-1} \sum_{j=1}^{n} u^j e_B u^{-j}$ is in $K_A(e_B) = \text{conv}\{u e_B u^* : u \text{ is a unitary in } A\}$ and has $\|x\|_{2,Tr}^2 = 1/n$. The integer n was arbitrary, so $K_A(e_B)$ contains elements of arbitrarily small $\| \cdot \|_{2,Tr}$-norm. Thus $h = 0$. $\qquad \square$

Note that Lemma 9.2.2 (iv) can be worded

$$Tr(e_B x) = Tr(hx) \text{ for all } x \in A' \cap \langle N, e_B \rangle \text{ with } Tr(|x|) < \infty.$$

Here is a slight variation on the above Lemma 9.2.2 that may be deduced from it or proved by the same technique.

Lemma 9.2.3. *Let A and B be von Neumann subalgebras of a type II_1 factor N, let w be a unitary in N and let $\overline{K}_A^w(w^* e_B w)$ be the weak closure of the set*

$$K_A(w^* e_B w) = \text{conv}\{u w^* e_B w u^* : u \text{ is a unitary in } A\} \qquad (9.2.22)$$

in $\langle N, e_B \rangle$. Then there is a unique element h of minimal $\| \cdot \|_{2,Tr}$-norm in $\overline{K}_A^w(w^* e_B w)$, and which satisfies

(i) $\|h\|, Tr(h), Tr(h^2), Tr(e_B h) \leq 1$;

(ii) $h \in A' \cap \langle N, e_B \rangle, \qquad 0 \leq h \leq 1$;

(iii) $1 - Tr(w^* e_B w h) \leq \|(I - \mathbb{E}_B) \mathbb{E}_{wAw^*}\|_{\infty,2}^2$;

(iv) $Tr(w^* e_B w h) = Tr(h^2)$;

(v) $\|h - w^* e_B w\|_{2,Tr} \leq \|(I - \mathbb{E}_B) \mathbb{E}_{wAw^*}\|_{\infty,2}$;

(vi) $\|h - w^* e_B w\| \leq \|(I - \mathbb{E}_B) \mathbb{E}_{wAw^*}\|$.

If wAw^* is orthogonal to B and if A contains a Haar unitary, then $h = 0$. If $A = B$, then $Tr(h e_B) = \|\mathbb{E}_B(w)\|_{2,Tr}^2$.

Proof. The proof of this is similar to that of Lemma 9.2.2 except that e_B is replaced by w^*e_Bw in most places. Only the final part is new, which is proved as follows.

If $A = B$ and u is a unitary in A, then $e_Bu = ue_B$ and

$$Tr(u^*w^*e_Bwue_B) = Tr(w^*e_Bwe_B) = Tr(w^*\mathbb{E}_B(w)e_B)$$
$$= \tau(w^*\mathbb{E}_B(w)) = \|\mathbb{E}_B(w)\|_{2,\tau}^2. \qquad (9.2.23)$$

Boundedness and normality of $Tr(e_B \cdot e_B)$ allow us to take weak limits of convex combinations in equation (9.2.23), giving $Tr(he_B) = \|\mathbb{E}_B(w)\|_2^2$ as required. $\square$

Analogously to the observation made after Lemma 9.2.2, we can word Lemma 9.2.3 (iv) as

$$Tr(w^*e_Bwx) = Tr(hx) \quad \text{for all} \quad x \in A' \cap \langle N, e_B \rangle \quad \text{with} \quad Tr(|x|) < \infty.$$

9.3   Perturbing subalgebras in the uniform norm

Perturbations in the metric $d(A, B) = \|\mathbb{E}_B - \mathbb{E}_A\|$ on the conditional expectations follow directly from the basic technical results and Lemma 9.2.3.

Before presenting the main theorem of this section we discuss the functions G and H that have occurred before. We recall their definitions from (5.3.1) and (5.3.2):

$$G(\delta) = \delta + 2^{-1}(1 - (1 - 4\delta)^{1/2}) \quad \text{for} \quad 0 \leq \delta \leq 1/4 \qquad (9.3.1)$$

and

$$H(k) = 2^{1/2}(1 - (1 - k^2)^{1/2})^{1/2} \quad \text{for} \quad 0 \leq k \leq 1. \qquad (9.3.2)$$

Observe that G is increasing and convex, since

$$G'(\delta) = 1 + (1 - 4\delta)^{-1/2} \geq 0 \quad \text{and}$$
$$G''(\delta) = 2(1 - 4\delta)^{-3/2} \geq 0 \quad \text{for} \quad 0 \leq \delta < 1/4. \qquad (9.3.3)$$

The function H is also increasing and convex, as a plot would easily indicate, but this is less obvious analytically and requires the following calculation. The function $F(x) = 2(1 - (1 - x)^{1/2})$, for $0 \leq x \leq 1$, has the property that $H(k)^2 = F(k^2)$. Differentiation gives $H(k)H'(k) = kF'(k^2) \geq 0$, leading to $H'(k)^2 = k^2F'(k^2)^2/F(k^2)$ for $0 < k < 1$. Replacing k^2 by x, we see that $H'(k)$ is increasing if and only if $xF'(x)^2/F(x)$ is increasing on the interval $(0, 1)$. The further substitution $x = 1 - t^2$ makes this equivalent to the function

$$h(t) = \frac{(1 - t^2)F'(1 - t^2)^2}{F(1 - t^2)} = \frac{1 + t}{2t^2} = \frac{1}{2t^2} + \frac{1}{2t}$$

being decreasing on $(0, 1)$, and this is clearly true. Thus $H(k)$ is convex on $[0, 1]$.

The function $K(\delta) = H(2H(G(\delta)) + 2\delta)$ will appear in subsequent proofs. It is defined for any $\delta \geq 0$ satisfying $G(\delta) \leq 1$ and $2H(G(\delta)) + 2\delta \leq 1$. This function is increasing and convex since a composition of increasing convex functions vanishing at 0 is of the same type. Direct calculation shows that these constraints are met by $\delta = 1/8$ and so $K(\delta)$ is defined at least on the interval $[0, 1/8]$. Moreover

$$K(1/8) = 0.8915\ldots < 9/10. \tag{9.3.4}$$

For each $\beta \in (0, 1/8]$, the graph of $K(\delta)$ lies below the line joining the origin to the point $(\beta, K(\beta))$ since $K(\delta)$ is a convex function, and so the inequality

$$K(\delta) \leq \delta K(\beta)/\beta \quad \text{for} \quad 0 \leq \delta \leq \beta \leq 1/8 \tag{9.3.5}$$

holds. The choice of $\beta = 1/8$ in (9.3.5) gives

$$K(\delta) \leq \delta 8 K(1/8) \leq 7.2\delta \leq 8\delta \quad \text{for} \quad 0 \leq \delta \leq 1/8, \tag{9.3.6}$$

using the estimate of (9.3.4).

Theorem 9.3.1. *Let A and B be von Neumann subalgebras of a II_1 factor N. If $\|\mathbb{E}_A - \mathbb{E}_B\| < 1/8$, then there is a unitary $u \in N$ such that $uAu^* = B$ and $\|1 - u\| \leq 8\|\mathbb{E}_A - \mathbb{E}_B\|$.*

Proof. Let $\delta = \|\mathbb{E}_A - \mathbb{E}_B\|$. Average e_B over A to obtain the element $h \in A' \cap \langle N, e_B \rangle$ with $0 \leq h \leq 1$ and $\|h - e_B\| \leq \delta$ by Lemma 9.2.2 (vi). Let $G(\delta) = \delta + 2^{-1}(1 - (1 - 4\delta)^{1/2})$ be the increasing function on $[0, 1/4]$ defined in equation (9.3.1). By Lemma 5.3.1 (ii), the spectral projection f for h corresponding to the interval $[1/2, 1]$ satisfies $\|f - e_B\| \leq G(\|h - e_B\|) \leq G(\delta)$. Let $H(k) = 2^{1/2}(1 - (1 - k^2)^{1/2})^{1/2}$ be defined for $k \in [0, 1)$ as in equation (9.3.2). By Lemma 5.2.10, if $k = G(\delta) < 1$, then there is a unitary $w \in \langle N, e_B \rangle$ such that $we_B w^* = f \in A' \cap \langle N, e_B \rangle$ and $\|1 - w\| \leq H(k) = H(G(\delta))$.

From properties of e_B and B, (Lemma 4.2.2),

$$A w e_B w^* = Af = fAf \subseteq f\langle N, e_B \rangle f$$
$$= we_B \langle N, e_B \rangle e_B w^* = we_B B e_B w^* \tag{9.3.7}$$

so that $w^* A w e_B \subseteq e_B B e_B = B e_B$. Hence $w^* a w e_B = e_B w^* a w e_B = w^* a w e_B$ for all $a \in A$. The map from $B e_B$ to B given by $b e_B \mapsto b$ is an isometry in the $\|\cdot\|$-norm, since the map $b \mapsto b e_B$ is a *-isomorphism from B into $\langle N, e_B \rangle$, so is isometric. If $\|e_B b\| = 1$, then $\|b\| = 1$ and $a = \mathbb{E}_A(b) \in A$ satisfies $\|b - a\| \leq \|\mathbb{E}_B - \mathbb{E}_A\| \leq \delta$. Thus

$$\|e_B b - e_B w^* a w\| \leq \|b - w^* a w\| = \|wb - aw\|$$
$$\leq \|(w - 1)b - b(w - 1)\| + \|b - a\|$$
$$\leq 2\|w - 1\| + \delta \leq 2H(G(\delta)) + 2\delta < 1 \tag{9.3.8}$$

since $2H(G(\delta)) + 2\delta$ is increasing and $0 \leq \delta < 1/8$. Thus the quotient map from $e_B B$ onto $e_B B / e_B w^* A w$ has norm strictly less than 1 implying that $e_B w^* A w =$

$e_B B$. Since $b = 0$ if, and only if, $e_B b = 0$ for $b \in B$, the map $\phi \colon A \to B$ given by

$$w^* a w e_B = \phi(a) e_B \quad \text{for all} \quad a \in A \tag{9.3.9}$$

is well defined, injective, and surjective. Note that

$$\phi(a^*) e_B = w^* a^* w e_B = (e_B w^* a w)^* = (w^* a w e_B)^*$$
$$= (\phi(a) e_B)^* = e_B \phi(a)^* = \phi(a)^* e_B, \tag{9.3.10}$$

by equation (9.3.9). In addition,

$$\phi(a_1 a_2) e_B = w^* a_1 a_2 w e_B = w^* a_1 a_2 f w$$
$$= w^* a_1 f a_2 f w = w^* a_1 w e_B w^* a_2 w w^* f w$$
$$= \phi(a_1) e_B \phi(a_2) e_B = \phi(a_1) \phi(a_2) e_B, \tag{9.3.11}$$

for all $a_1, a_2 \in A$, using the relation $w e_B = f w$. Hence ϕ is an isometric *-isomorphism from A onto B.

If $a \in A$ with $\|a\| \leq 1$, then

$$\|\mathbb{E}_B(a) - w^* a w\| = \|w(\mathbb{E}_B(a) - a) + (w - 1)a - a(w - 1)\|$$
$$\leq \|\mathbb{E}_B(a) - a\| + 2\|w - 1\|$$
$$\leq \delta + 2H(G(\delta)) \tag{9.3.12}$$

by equation (9.3.8). Hence, by the definition of ϕ in equation (9.3.9) and by the inequality (9.3.12),

$$\|a - \phi(a)\| \leq \|a - \mathbb{E}_B(a)\| + \|\mathbb{E}_B(a) - \phi(a)\|$$
$$\leq \delta + \|(\mathbb{E}_B(a) - \phi(a)) e_B\|$$
$$= \delta + \|(\mathbb{E}_B(a) - w a w^*) e_B\|$$
$$\leq 2(\delta + H(G(\delta))) \tag{9.3.13}$$

for all $a \in A$ with $\|a\| \leq 1$. Now $2(\delta + H(G(\delta))) < 1$ since $\delta < 1/8$, and hence Theorem 5.2.11 implies that there is a unitary $u \in N$ such that $\phi(a) = uau^*$ for all $a \in A$, and u satisfies the estimate $\|1 - u\| \leq H(2\delta + 2H(G(\delta))) = K(\delta)$. The proof is completed by noting that $K(\delta) \leq 8\delta$ when $\delta \leq 1/8$, from (9.3.6). $\square$

When A and B are masas in N in Theorem 9.3.1, then it is sufficient to assume that $\|(I - \mathbb{E}_B)\mathbb{E}_A\| \leq \delta$ in the hypotheses as ϕ being an isomorphism from A onto B may be deduced from the masa property. The proof of the above theorem gives the estimate $\|1 - u\| \leq (K(\beta)/\beta)\delta$ for $\delta \in [0, \beta]$, so better constants appear from choices of $\beta < 1/8$. However, it is simple to compute that $\lim_{\beta \to 0+} K(\beta)/\beta = 6$, so the current methods will not yield any significant improvement.

9.4 Lemmas on close subalgebras

This section contains some of the implications and lemmas that are common to the main theorems of subsequent sections. Looking ahead, Theorem 9.5.1 establishes the equivalence of five statements. Some of the implications between them are proved in the lemmas of this section, and we indicate this by including the phrase

$$(i) \implies (j) \text{ in Theorem 9.5.1}$$

at the start of the appropriate proofs.

Let p and q be two abelian projections in a von Neumann algebra M with equal central supports $c_p = c_q$. Then there is a unitary $u \in M$ such that $p = uqu^*$, by a standard little lemma concerning abelian projections ([105, Prop. 6.4.6]).

The following lemma depends on results on abelian projections in masas due to Kadison [99, 100], although we will only need the more specialised version presented in Corollary 2.4.6. The first part of it is a preparatory result for use in the proof of Theorem 9.5.1.

Lemma 9.4.1. *Let N be a finite von Neumann algebra with a faithful normal trace τ, and let A and B be masas in N.*

(i) *If $h \neq 0$ is in $A' \cap \langle N, e_B \rangle$, $h \geq 0$, and if $Tr(h)$ is finite, then there is a non-zero abelian projection $f \in A' \cap \langle N, e_B \rangle$ such that f is equivalent to a projection $e_1 \leq e_B$.*

(ii) *If, in addition, z is the central support of h in $\langle N, e_B \rangle$, then the abelian projection f may be chosen in $A' \cap \langle N, e_B \rangle$ with f equivalent to ze_B in $\langle N, e_B \rangle$. If $0 \leq h \leq 1$ and e is an abelian projection in $\langle N, e_B \rangle$ with central support 1, then the choice may be made with*

$$Tr(he) \leq Tr(f) = Tr(ze) = Tr(ze_B) \leq 1. \tag{9.4.1}$$

Proof. (i)   [(2) $\implies$ (3) of Theorem 9.5.1].

If $0 < t < \|h\|$, let p_t be the spectral projection of h corresponding to the interval $[t, \|h\|]$ so that $p_t \in A' \cap \langle N, e_B \rangle$. From the inequality $tp_t \leq h$, if follows that $Tr(p_t) \leq t^{-1}Tr(h) < \infty$. Hence $p_t \langle N, e_B \rangle p_t$ is a finite type I von Neumann algebra with faithful normal trace Tr, because $\langle N, e_B \rangle = (JBJ)'$ is of type I as it is the commutant of the abelian von Neumann algebra JBJ (see Theorem 4.2.2 (i)). Thus $p_t \langle N, e_B \rangle p_t$ has no central direct summand of type I_∞ (since $p_t \langle N, e_B \rangle p_t$ has a finite trace). Then $p_t A$ is an abelian subalgebra of $p_t \langle N, e_B \rangle p_t$. Let A_1 be a masa in $p_t \langle N, e_B \rangle p_t$ containing the abelian subalgebra $p_t A$. By Corollary 2.4.6, there is a projection $q_t \in A_1$ that is abelian in $p_t \langle N, e_B \rangle p_t$ and has central support p_t there. Hence $q_t p_t = q_t$ and $q_t \langle N, e_B \rangle q_t$ is abelian so that q_t is a non-zero abelian projection in $\langle N, e_B \rangle$ with central support $c_{q_t} = c_{p_t}$ equal to the central support of p_t. This proves (i) with $f = q_t$ for some small t, say, $t = \|h\|/2$.

Note that if $0 < s < t < \|h\|$, then $p_s \geq p_t$ so that $c_{p_s} \geq c_{p_t}$.

(ii) Inductively define a sequence f_n of abelian projections in $A' \cap \langle N, e_B \rangle$ with central support $c_n = c_{f_n}$ such that

$$c_n = c_{f_n} = c_{q_{2^{-n}}},\tag{9.4.2}$$

$$f_1 = q_{2^{-1}} \leq f_n \leq f_{n+1} \quad \text{and}\tag{9.4.3}$$

$$f_n = c_n f_{n+1} \quad \text{for all} \quad n \in \mathbb{N}.\tag{9.4.4}$$

Suppose that $f_1, \ldots, f_n$ have been chosen. The central support c_{n+1} of $q_{2^{-n-1}}$ satisfies $c_{n+1} \geq c_n = c_{q_{2^{-n}}}$, since

$$c_{n+1} = c_{q_{2^{-n-1}}} = c_{p_{2^{-n-1}}} \geq c_{p_{2^{-n}}} = c_{q_{2^{-n}}} = c_n.\tag{9.4.5}$$

The abelian projections $f_n = c_n f_n$ and $c_n q_{2^{-n-1}}$ both have central support c_n in $\langle N, e_B \rangle c_n = c_n \langle N, e_B \rangle c_n$. Hence there is a unitary $v_n \in \langle N, e_B \rangle c_n$ with $v_n c_n q_{2^{-n-1}} v_n^* = f_n$ by [105, Proposition 6.4.6]. Extend the unitary v_n to a unitary u_n in $\langle N, e_B \rangle$ by $u_n = v_n + 1 - c_n$. Let $f_{n+1} = u_n q_{2^{-n-1}} u_n^*$. Then $f_n \leq f_{n+1}$ and $f_n = c_n f_{n+1}$ as required.

The abelian projections f_n, $c_n e$ and $c_n e_B$ all have central support c_n in the von Neumann algebra $\langle N, e_B \rangle c_n$ and so are unitarily equivalent there by [105, Proposition 6.4.6] and

$$Tr(f_n) = Tr(c_n e) = Tr(c_n e_B) \leq Tr(e_B) = 1.\tag{9.4.6}$$

Let f be the weak limit of f_n in $\langle N, e_B \rangle$.

The support projection p of h is also the spectral projection of this operator corresponding to the interval $(0, \|h\|]$. By definition of p_t for $0 < t < 1$, $p = \lim_{t \to 0} p_t$ weakly. Also the central support c_p of p equals the central support z of h. By [104, Proposition 5.5.3],

$$\begin{aligned} c_n = c_{f_n} &= \text{ central support of } f_n \\ &= c_{q_{2^{-n}}} = \text{ central support of } q_{2^{-n}} \\ &= c_{p_{2^{-n}}} = \text{ central support of } p_{2^{-n}} \end{aligned}\tag{9.4.7}$$

and so $\lim_{n \to \infty} c_n = z$ weakly.

Let $f_0 = c_0 = 0$. The abelian projections $f_{n+1} - f_n$ and $(c_{n+1} - c_n)e$ have central support $c_{n+1} - c_n$ in $(c_{n+1} - c_n)\langle N, e_B \rangle$ so there is a unitary w_n in $(c_{n+1} - c_n)\langle N, e_B \rangle$ with $w_n(f_{n+1} - f_n)w_n^* = (c_{n+1} - c_n)e$. Let

$$u = \sum_{n \geq 0} w_n + (1 - z).\tag{9.4.8}$$

Then u is a unitary in $\langle N, e_B \rangle$ and

$$\begin{aligned} ufu^* = \sum_{n \geq 0} u(f_{n+1} - f_n)u^* &= \sum_{n \geq 0} w_n(f_{n+1} - f_n)w_n^* \\ &= \sum_{n \geq 0} (c_{n+1} - c_n)e = ze, \end{aligned}\tag{9.4.9}$$

which proves the equivalence of f and ze in $\langle N, e_B \rangle$.

If $0 \leq h \leq 1$, then $h \leq z$ so

$$Tr(he) = Tr(ehe) \leq Tr(eze) = Tr(ze)$$
$$= Tr(ze_B) \leq Tr(e_B) = 1, \tag{9.4.10}$$

which completes the proof. $\qquad\square$

Note that if A and B are masas in N and v is a partial isometry in N with $v^*v \in A$ and $vv^* \in B$, then the statements $vAv^* = Bvv^*$ and $vA = Bv$ are equivalent since

$$vA = v(v^*v)A = vA(v^*v) = (vAv^*)v = (Bvv^*)v = Bv. \tag{9.4.11}$$

The following lemma, with or without the norm estimates, will be useful later. Its purpose is to replace a partial isometry $V \in \langle N, e_B \rangle$, implementing a homomorphism from one masa to another, by a partial isometry $v \in N$. The first step is to apply the pull down map Φ to V, which in general gives an element of $L^1(N)$, but here gives one in $L^2(N)$. This can then be viewed as a closed unbounded operator affiliated to N, and the theory developed at the end of Appendix B comes into play. Following the definitions in Section B.4, any vector $\eta \in L^2(N)$ induces a densely defined (unbounded) operator ℓ_η with closure L_η, and the domains of these operators contain $N\xi$. It will be convenient to adopt the notation $S \doteq T$ to mean that S and T agree on $N\xi$. This frees us from having to identify precisely the domain of each particular operator. We note that all unbounded operators arising in the next result are affiliated with N, and thus any bounded operators obtained from the functional calculus will lie in N. For reasons of technical simplicity, we state this lemma for the adjoint of the equation that we will encounter later. Thus, we replace the equations

$$Va = \phi(a)V, \quad e_B V = V, \tag{9.4.12}$$

by

$$aW = W\phi(a), \quad We_B = W. \tag{9.4.13}$$

We will prove the following lemma using unbounded operators (see Appendix B, particularly Sections B.4 and B.5) because it is useful to know such techniques for reading the literature in this area. However, we will then give a second proof using only bounded operators for which we will need Lemmas 9.4.3 and 9.4.4.

Lemma 9.4.2. *Let B_0 and B be von Neumann subalgebras of N, and let A be a von Neumann subalgebra of B_0 whose identity is not assumed to be that of B_0. Suppose that there exist $W \in \langle N, e_B \rangle$ and a *-homomorphism $\phi : A \to B$ such that $aW = W\phi(a)$ for $a \in A$ and such that $We_B = W$. Then there exists a non-zero partial isometry $w \in N$ with the following properties:*

(i) $w^*aw = \phi(a)w^*w, \quad a \in A;$

(ii) $\|1 - w\|_{2,\tau} \le 2\|e_B - W\|_{2,Tr}$;

(iii) $\|1 - p'\|_{2,\tau} \le \|e_B - W\|_{2,Tr}$, where $p' = w^*w \in \phi(A)' \cap N$;

(iv) if $q_1 \in N$ and $q_2 \in B$ are projections such that $q_1W = W = Wq_2$ then $q_1w = w = wq_2$;

(v) $ww^* \in A' \cap N$, $aw = w\phi(a)$ for $a \in A$, and $ww^* \in A$ when A is a masa.

Proof. Let $\eta = W\xi \in L^2(N)$. We will first show that $aL_\eta \doteq L_\eta\phi(a)$ for $a \in A$. Since Ne_BN is weakly dense in $\langle N, e_B \rangle$, we may choose, by the Kaplansky density theorem (Theorem 2.2.3), a sequence $\{y_n\}_{n=1}^\infty$ from Ne_BN converging to W in the strong* topology. Since $W = We_B$, we also have that $y_ne_B \to W$ in this topology, and each y_ne_B has the form w_ne_B for $w_n \in N$, since $e_BNe_B = Be_B$ by Lemma 4.2.3 (iii). We also note that e_B commutes with each $\phi(a) \in B$, and that $e_B\xi = \xi$. Then, for $a \in A$, $x \in N$,

$$
\begin{aligned}
aL_\eta x\xi &= aJx^*J\eta = Jx^*JaWe_B\xi \\
&= Jx^*JWe_B\phi(a)\xi = \lim_{n\to\infty} Jx^*Jw_ne_B\phi(a)\xi \\
&= \lim_{n\to\infty} Jx^*Jw_nJ\phi(a^*)J\xi = \lim_{n\to\infty} Jx^*JJ\phi(a^*)Jw_ne_B\xi \\
&= Jx^*JJ\phi(a^*)JW\xi = L_\eta\phi(a)x\xi, \qquad\qquad (9.4.14)
\end{aligned}
$$

establishing that $aL_\eta \doteq L_\eta\phi(a)$. Let $T = |L_\eta|$, and let $L_\eta = wT$ be the polar decomposition of L_η, where w is a partial isometry mapping the closure of the range of T to the closure of the range of L_η. Then

$$ awT \doteq wT\phi(a), \quad a \in A. \qquad\qquad (9.4.15) $$

Let $p' = w^*w$, the projection onto the closure of the range of T. Then $p'T \doteq T$ and (9.4.15) becomes

$$ w^*awT \doteq T\phi(a), \quad a \in A. \qquad\qquad (9.4.16) $$

For each $n \in \mathbb{N}$, let $e_n \in N$ be the spectral projection of T for the interval $[0, n]$. Then each e_n commutes with T and so we may multiply on both sides of (9.4.16) by e_n to obtain

$$ e_nw^*awe_ne_nTe_n = e_nTe_ne_n\phi(a)e_n, \quad a \in A, \qquad\qquad (9.4.17) $$

where $e_nTe_n \in N$ is now a bounded operator. When $a \ge 0$, (9.4.17) implies that $(e_nTe_n)^2$ commutes with $e_n\phi(a)e_n$, and thus so also does e_nTe_n. It follows that

$$ e_nw^*awTe_n = e_n\phi(a)Te_n, \quad a \in A, \ n \ge 1. \qquad\qquad (9.4.18) $$

For $m \le n$, we can multiply on the left by e_m and then let $n \to \infty$ to obtain that

$$ e_mw^*aw\zeta = e_m\phi(a)\zeta, \quad a \in A, \ \zeta \in \operatorname{Ran} T. \qquad\qquad (9.4.19) $$

Now let $m \to \infty$ to deduce that w^*aw and $\phi(a)$ agree on $\text{Ran}\,T$, and consequently that

$$w^*aw = w^*aww^*w = \phi(a)w^*w = \phi(a)p', \quad a \in A, \qquad (9.4.20)$$

since $p' = 1$ on $\text{Ran}\,T$. This establishes (i), and by taking $a \geq 0$ in (9.4.20), it is clear that $p' \in \phi(A)' \cap N$. We now turn to the norm estimates of (ii) and (iii).

On $[0, \infty)$, define continuous functions f_n for $n \geq 1$ by $f_n(t) = \chi_{[0,n]}(t) + nt^{-1}\chi_{(n,\infty)}(t)$, and let $h_n = f_n(T) \in N$ be the associated operators arising from the functional calculus. These functions were chosen to have the following properties: they form an increasing sequence with pointwise limit 1, each f_n dominates a positive multiple of each $\chi_{[0,m]}$, and each $tf_n(t)$ is a bounded function. Thus $\{h_n\}_{n=1}^{\infty}$ increases strongly to 1 and each Th_n is a bounded operator and thus in N. The range of each h_n contains the range of each e_m, and so $L_\eta h_n$ and L_η have identical closures of ranges for every $n \geq 1$. Thus w is also partial isometry in the polar decomposition $L_\eta h_n = wTh_n$, $n \geq 1$. The point of introducing the h_n's is to reduce to the case of bounded operators where we can now apply Lemma 5.2.7. This gives

$$\|1 - w\|_{2,\tau} \leq 2\|1 - L_\eta h_n\|_{2,\tau}, \quad \|1 - p'\|_{2,\tau} \leq \|1 - L_\eta h_n\|_{2,\tau}, \qquad (9.4.21)$$

for all $n \geq 1$, noting that $1 - p'$ and $1 - ww^*$ have the same $\|\cdot\|_{2,\tau}$-norm. For each $b \in B$,

$$L_\eta b\xi = Jb^*JW\xi = \lim_{n \to \infty} Jb^*Jw_n e_B\xi = \lim_{n \to \infty} w_n e_B Jb^*J\xi = Wb\xi, \qquad (9.4.22)$$

using that e_B commutes with both b and J. Thus $L_\eta e_B = We_B = W$ and, since $ww^*L_\eta \doteq L_\eta$, we also have $ww^*W = W$. Returning to (9.4.21), we obtain

$$\begin{aligned}
\|1 - L_\eta h_n\|_{2,\tau} &= \|e_B - L_\eta h_n e_B\|_{2,Tr} = \|e_B - wTh_n e_B\|_{2,Tr} \\
&= \|e_B - wh_n Te_B\|_{2,Tr} = \|e_B - wh_n w^* L_\eta e_B\|_{2,Tr} \\
&= \|e_B - wh_n w^* We_B\|_{2,Tr}. \qquad (9.4.23)
\end{aligned}$$

Since Tr is normal we may let $n \to \infty$ in this last equation, giving $\lim_{n\to\infty} \|1 - L_\eta h_n\|_{2,\tau} = \|e_B - W\|_{2,Tr}$. The inequalities of (ii) and (iii) now follow by letting $n \to \infty$ in (9.4.21). We now establish (iv).

Let $q_1 \in N$ be such that $q_1 W = W$. For each $x \in N$,

$$L_\eta(x\xi) = Jx^*JW\xi = Jx^*Jq_1W\xi = q_1Jx^*JW\xi \qquad (9.4.24)$$

and so q_1 is the identity on the closure of the range of L_η which is also the range of w. Thus $q_1 w = w$.

Now suppose that $q_2 \in B$ is such that $Wq_2 = W$. By replacing q_2 by $1 - q_2$, we may prove the equivalent statement that $wq_2 = 0$ follows from $Wq_2 = 0$. Then

$$0 = Wq_2 = We_Bq_2 = Wq_2e_B = L_\eta q_2e_B = wTq_2e_B. \qquad (9.4.25)$$

Multiply by $e_n w^*$ to obtain $e_n Tq_2 e_B = 0$. Since $e_n T \in N$, it follows from Lemma 4.2.3 (v) that $e_n Tq_2 = 0$ for all $n \geq 1$. Then $q_2 Te_n = 0$ for all $n \geq 1$, and letting n increase, we find that q_2 annihilates the closure of the range of T which is also the range of w^*. Thus $q_2 w^* = 0$ and so $wq_2 = 0$, completing the proof of (iv).

We now consider (v). Let ww^* be denoted by p, and observe that $pap = w\phi(a)w^*$, for $a \in A$, by multiplying on the left in (i) by w and on the right by w^*. The map $a \mapsto w\phi(a)w^*$ is a *-homomorphism on A because $w^*w = p'$ lies in $\phi(A)' \cap N$ by (iii), and so the map $\psi(a) = pap$ is also a *-homomorphism on A. Let a be an arbitrary self-adjoint element of A, and write $a = pap + pa(1 - p) + (1 - p)ap + (1 - p)a(1 - p)$. The relation $\psi(a^2) = \psi(a)^2$ gives

$$papap + pa(1 - p)ap = papap, \tag{9.4.26}$$

showing that $pa(1 - p)ap = 0$. Thus $(1 - p)ap = 0$, so $ap = pap$. Taking the adjoint shows that $ap = pa$, and so p must lie in $A' \cap N$. Using (i), we see that

$$\begin{aligned} aw = apw = paw = ww^*aw = w\phi(a)w^*w \\ = w\phi(a)p' = wp'\phi(a) = w\phi(a), \end{aligned} \tag{9.4.27}$$

for $a \in A$, proving the first two statements in (v). If, in addition, A is a masa, then $A' \cap N = A$, showing that $p \in A$. $\qquad\square$

The next two lemmas are required in order to give a bounded operator proof of Lemma 9.4.2.

Lemma 9.4.3. *Let N be a finite von Neumann algebra with a normalised faithful normal trace τ, and let $v \in N$. Let p be the spectral projection of vv^* for the interval $[0, 4]$. Then*

$$\|pv\| \leq 2 \quad \text{and} \quad \|1 - pv\|_2 \leq \|1 - v\|_2. \tag{9.4.28}$$

Proof. The first inequality is immediate from the choice of p, and so we consider the second. The functional calculus gives $4(1 - p) \leq (1 - p)vv^*(1 - p)$, and so

$$4\tau(1 - p) \leq \tau((1 - p)vv^*(1 - p)), \tag{9.4.29}$$

which is the inequality

$$2\|1 - p\|_2 \leq \|(1 - p)v\|_2. \tag{9.4.30}$$

Multiplying (9.4.30) by $\|(1 - p)v\|_2$ gives

$$2\|1 - p\|_2\|(1 - p)v\|_2 \leq \|(1 - p)v\|_2^2. \tag{9.4.31}$$

From the Cauchy–Schwarz inequality and (9.4.31),

$$\begin{aligned} 2|\tau((1 - p)v)| = 2|\langle(1 - p)v, 1 - p\rangle| \\ \leq 2\|1 - p\|_2\|(1 - p)v\|_2 \leq \|(1 - p)v\|_2^2, \end{aligned} \tag{9.4.32}$$

and so

$$\tau((1-p)v + v^*(1-p)) \leq 2|\tau((1-p)v)|$$
$$\leq \|(1-p)v\|_2^2 = \tau((1-p)vv^*), \qquad (9.4.33)$$

which can be rearranged to

$$\tau(-pv - pv^* + pvv^*p) \leq \tau(-v - v^* + vv^*). \qquad (9.4.34)$$

Add 1 to both sides of (9.4.34) to conclude that

$$\|1 - pv\|_2^2 = \tau(1 - pv - pv^* + pvv^*p)$$
$$\leq \tau(1 - v - v^* + vv^*) = \|1 - v\|_2^2, \qquad (9.4.35)$$

and then take the square root to complete the proof. $\qquad \square$

We now present a different type of averaging result, again based on the technique of choosing the element of minimal norm in a closed convex subset of a Hilbert space. We will then use this to give a second proof of Lemma 9.4.2 which does not use unbounded operators.

For $x \in N$ and $\eta \in L^2(N)$, it will be convenient to write ηx for the vector $Jx^*J\eta$. Below, $\mathcal{U}(A)$ will denote the unitary group of A.

Lemma 9.4.4. *Let A be a von Neumann subalgebra of a finite von Neumann algebra N with normalised faithful normal trace τ, and let $\phi\colon A \to N$ be a *-homomorphism, not assumed to be unital. Let $\eta \in L^2(N)$ be given and let*

$$K_A(\eta) = \overline{\mathrm{conv}}\{u\eta\phi(u^*)\colon u \in \mathcal{U}(A)\},$$

where the closure is taken in the $\|\cdot\|_2$-norm. Let ζ be the unique vector of minimal $\|\cdot\|_2$-norm in $K_A(\eta)$.

(i) $v\zeta\phi(v^*) = \zeta$ for all $v \in \mathcal{U}(A)$;

(ii) *if η_1 satisfies $v\eta_1\phi(v^*) = \eta_1$ for all $v \in \mathcal{U}(A)$, then $\|\zeta - \eta_1\|_2 \leq \|\eta - \eta_1\|_2$;*

(iii) *if $\eta \in N\xi \subseteq L^2(N)$, then $\zeta \in N\xi$.*

Proof. (i) For each $v \in \mathcal{U}(A)$, the map $v \cdot \phi(v^*)$ is contractive on $L^2(N)$ and maps $K_A(\eta)$ to itself. Thus ζ and $v\zeta\phi(v^*)$ are both elements of minimal norm in $K_A(\eta)$, so uniqueness gives $v\zeta\phi(v^*) = \zeta$.

(ii) If η_1 satisfies $v\eta_1\phi(v^*) = \eta_1$ for all $v \in \mathcal{U}(A)$, then

$$\|\eta_1 - v\eta\phi(v^*)\|_2 = \|v(\eta_1 - \eta)\phi(v^*)\|_2 \leq \|\eta_1 - \eta\|_2. \qquad (9.4.36)$$

The inequality $\|\eta_1 - \zeta\|_2 \leq \|\eta_1 - \eta\|_2$ follows by taking convex combinations and norm limits in (9.4.36).

(iii) Suppose that $\eta = x\xi$ for some $x \in N$. The weak closure in N of $\mathrm{conv}\{ux\phi(u^*)\colon u \in \mathcal{U}(A)\}$ is compact, and so its embedding into $L^2(N)$ is weakly compact and thus equal to $K_A(\eta)$. It follows that $\zeta \in N\xi$. $\qquad \square$

We now refer the reader back to the statement of Lemma 9.4.2.

Second proof of Lemma 9.4.2. We have a *-homomorphism $\phi\colon A \to B$ and an operator $W \in \langle N, e_B \rangle$ such that $We_B = W$ and $aW = W\phi(a)$ for $a \in A$. We will first consider the special case when $1 \in A$.

Let $\eta_1 = W\xi$. By the Kaplansky density theorem, Theorem 2.2.3, and Lemma 4.2.3 (iv), there is a net $(w_\alpha e_B)$, $w_\alpha \in N$, converging *-strongly to We_B. Since e_B commutes with B, and hence with each $\phi(a)$ for $a \in A$, we see that

$$a\eta_1 = aWe_B\xi = We_B\phi(a)\xi = \lim_\alpha w_\alpha \phi(a)\xi$$

$$= \lim_\alpha w_\alpha J\phi(a)^* J\xi = \lim_\alpha J\phi(a)^* Jw_\alpha e_B\xi$$

$$= J\phi(a)^* JW\xi = \eta_1\phi(a), \quad a \in A. \tag{9.4.37}$$

Noting that the choice $a = 1$ gives $\eta_1 = \eta_1\phi(1)$, we conclude that $u\eta_1\phi(u^*) = \eta_1$ for all $u \in \mathcal{U}(A)$. If we apply Lemma 9.4.4 to the vectors $w_\alpha\xi$, then the result is a net $(v_\alpha) \in N$ satisfying

$$av_\alpha = v_\alpha\phi(a) \text{ for } a \in A, \tag{9.4.38}$$

and

$$\|v_\alpha\xi - \eta_1\|_2 \le \|w_\alpha\xi - \eta_1\|_2. \tag{9.4.39}$$

Now let p_α be the spectral projection of $v_\alpha v_\alpha^*$ for the interval $[0, 4]$. By Lemma 9.4.3 we have

$$\|p_\alpha v_\alpha\| \le 2 \text{ and } \|1 - p_\alpha v_\alpha\|_2 \le \|1 - v_\alpha\|_2. \tag{9.4.40}$$

Moreover, from (9.4.38),

$$av_\alpha v_\alpha^* = v_\alpha\phi(a)v_\alpha^*, \quad a \in A, \tag{9.4.41}$$

and so $v_\alpha v_\alpha^* \in A' \cap N$, as is the spectral projection p_α. Thus

$$ap_\alpha v_\alpha = p_\alpha av_\alpha = p_\alpha v_\alpha\phi(a), \quad a \in A. \tag{9.4.42}$$

By dropping to a subnet and using the uniform bound of (9.4.40), we may assume that the net $(p_\alpha v_\alpha)$ converges weakly to an operator $v \in N$ satisfying $\|v\| \le 2$. Taking weak limits over α in (9.4.42) leads to

$$av = v\phi(a), \quad a \in A. \tag{9.4.43}$$

Now let $v = w|v|$ be the polar decomposition of v, where w is a partial isometry. If we substitute this into (9.4.43) and multiply on the left by $|v|w^*$, then the resulting equation is

$$|v|w^*aw|v| = |v|^2\phi(a), \quad a \in A. \tag{9.4.44}$$

Thus $|v|^2 \in \phi(A)'$, and the same is true for $|v|$. Returning to (9.4.43) and multiplying on the left by w^*, we have

$$w^*aw|v| = |v|\phi(a) = \phi(a)|v|, \quad a \in A, \tag{9.4.45}$$

and so w^*aw and $\phi(a)$ are equal on the closure of the range of $|v|$, which is also the range of the projection w^*w. Thus $w^*aw = \phi(a)w^*w$ for $a \in A$, proving the first part of the lemma.

We now consider the $\|\cdot\|_2$-norm estimates. We have

$$\|e_B - W\|_{2,Tr}^2 = \|(1-W)e_B\|_{2,Tr}^2 = Tr(e_B(1-W^*)(1-W)e_B)$$
$$= \langle e_B(1-W^*)(1-W)e_B\xi, \xi \rangle = \|\xi - \eta_1\|_{2,\tau}^2, \qquad (9.4.46)$$

where the third equality is justified by noting that $e_B(1-W^*)(1-W)e_B$ has the form be_B for some $b \in B$. Strong convergence of the net $(w_\alpha e_B)$ to We_B then gives

$$\lim_\alpha \|1 - w_\alpha\|_{2,\tau}^2 = \lim_\alpha \|\xi - w_\alpha\xi\|_{2,\tau}^2 = \|\xi - \eta_1\|_{2,\tau}^2, \qquad (9.4.47)$$

since $We_B\xi = W\xi = \eta_1$. Now

$$\|1 - v_\alpha\|_{2,\tau} = \|\xi - v_\alpha\xi\|_{2,\tau} \leq \|\xi - w_\alpha\xi\|_{2,\tau} + \|v_\alpha\xi - w_\alpha\xi\|_{2,\tau}$$
$$\leq \|\xi - w_\alpha\xi\|_{2,\tau} + \|v_\alpha\xi - \eta_1\|_{2,\tau} + \|w_\alpha\xi - \eta_1\|_{2,\tau}$$
$$\leq \|\xi - w_\alpha\xi\|_{2,\tau} + 2\|w_\alpha\xi - \eta_1\|_{2,\tau}, \qquad (9.4.48)$$

where we have used the inequality (9.4.39) for the last step. Then (9.4.47) and (9.4.48) combine to yield

$$\limsup_\alpha \|1 - v_\alpha\|_{2,\tau} \leq \limsup_\alpha \|\xi - w_\alpha\xi\|_{2,\tau} = \|\xi - \eta_1\|_{2,\tau}, \qquad (9.4.49)$$

using $\lim_\alpha(w_\alpha\xi - \eta_1) = \lim_\alpha(w_\alpha e_B\xi - We_B\xi) = 0$ in $\|\cdot\|_{2,\tau}$-norm. By equations (9.4.40) and (9.4.49),

$$\limsup_\alpha \|1 - p_\alpha v_\alpha\|_{2,\tau} \leq \limsup_\alpha \|1 - v_\alpha\|_{2,\tau}$$
$$\leq \|\xi - \eta_1\|_{2,\tau} = \|e_B - W\|_{2,Tr}. \qquad (9.4.50)$$

Since $p_\alpha v_\alpha \to v$ weakly, we conclude that $\|1 - v\|_{2,\tau} \leq \|e_B - W\|_{2,Tr}$, and so $\|1 - w\|_{2,\tau} \leq 2\|e_B - W\|_{2,Tr}$, by Lemma 5.2.7. The same lemma also establishes

$$\|1 - p'\|_{2,\tau} \leq \|1 - v\|_{2,\tau} \leq \|e_B - W\|_{2,Tr}, \qquad (9.4.51)$$

and we have proved the second and third parts.

Now suppose that $q_1 \in N$ is a projection such that $q_1 W = W$. By working with $1 - q_1$, we may instead suppose that $q_1 W = 0$ and prove that $q_1 w = 0$. Since $w_\alpha e_B \to We_B$ strongly, this implies that $\lim_\alpha \|w_\alpha\xi - \eta_1\|_{2,\tau} = 0$ and $\lim_\alpha \|q_1 w_\alpha\xi\|_{2,\tau} = 0$, using $q_1\eta_1 = q_1 W\xi = 0$. From (9.4.39), we also obtain $\lim_\alpha \|v_\alpha\xi - \eta_1\|_{2,\tau} = 0$ and so $\lim_\alpha \|q_1 v_\alpha\xi\|_{2,\tau} = 0$. Thus $\lim_\alpha \|q_1 p_\alpha v_\alpha\xi\|_{2,\tau} = 0$ since

$$\tau(q_1 p_\alpha v_\alpha v_\alpha^* p_\alpha q_1) \leq \tau(q_1 v_\alpha v_\alpha^* q_1). \qquad (9.4.52)$$

For each $x \in N$,

$$|\langle q_1 v\xi, x\xi \rangle| = \lim_\alpha |\langle q_1 p_\alpha v_\alpha \xi, x\xi \rangle|$$

$$\leq \lim_\alpha \|q_1 p_\alpha v_\alpha \xi\|_{2,\tau} \|x\|_{2,\tau} = 0, \qquad (9.4.53)$$

and so $q_1 v\xi = 0$. Thus $q_1 v = 0$, implying that $q_1 w = 0$ since the closures of the ranges of v and w are equal.

Now consider a projection $q_2 \in B$ such that $W q_2 = 0$. Then

$$0 = W q_2 \xi = W e_B q_2 \xi = \lim_\alpha w_\alpha e_B q_2 \xi$$

$$= \lim_\alpha w_\alpha q_2 e_B \xi = \lim_\alpha w_\alpha q_2 \xi \qquad (9.4.54)$$

in $\|\cdot\|_{2,\tau}$-norm. Thus

$$\lim_\alpha v_\alpha q_2 \xi = \lim_\alpha J q_2 J v_\alpha \xi$$

$$= \lim_\alpha J q_2 J w_\alpha \xi + \lim_\alpha J q_2 J(v_\alpha - w_\alpha)\xi = 0, \qquad (9.4.55)$$

using (9.4.39) to conclude that $\lim_\alpha v_\alpha \xi = \lim_\alpha w_\alpha \xi = \eta_1$. Consequently, $\lim_\alpha p_\alpha v_\alpha q_2 \xi = 0$ in $\|\cdot\|_{2,\tau}$-norm. Since $p_\alpha v_\alpha \to v$ weakly, we obtain $v q_2 \xi = 0$, and so $v q_2 = 0$ and $q_2 v^* = 0$. Thus $q_2 v^* v = 0$. Since the closures of the ranges of $v^* v$ and w^* are equal, it follows that $q_2 w^* = 0$ and $w q_2 = 0$, as required, proving the fourth part.

Finally, suppose that the identity of A is a projection $p \neq 1$. We let $\tilde{A} = A \oplus \mathbb{C}(1 - p)$, and extend ϕ to $\tilde{\phi} \colon \tilde{A} \to B$ by

$$\tilde{\phi}(a + \lambda(1 - p)) = \phi(a) + \lambda(1 - \phi(p))$$

for $a \in A$ and $\lambda \in \mathbb{C}$. Since $pW = W\phi(p)$, it is easy to check that $\tilde{a}W = W\tilde{\phi}(\tilde{a})$ for $\tilde{a} \in \tilde{A}$. By the previous special case, the partial isometry w such that $w^* \tilde{a} w = \tilde{\phi}(\tilde{a}) w^* w$ has the requisite properties.

Since we have already given a bounded operator version of (v) in the original proof, we do not repeat it here. $\qquad \square$

Corollary 9.4.5. Let A and B be masas in a II_1 factor N. If there are a *-homomorphism $\phi_1 \colon A \to B$ and a partial isometry $v_1 \in \langle N, e_B \rangle$ such that

$$\phi_1(a)v_1 = v_1 a, \quad a \in A, \qquad (9.4.56)$$

then there exist a non-zero partial isometry $v_0 \in N$ and a projection $q_0 = v_0 v_0^* \in \phi_1(A)' \cap N$ satisfying

$$\phi_1(a)v_0 = v_0 a, \quad \phi_1(a)q_0 = v_0 a v_0^*, \quad for\ a \in A. \qquad (9.4.57)$$

Proof. In Lemma 9.4.2 (v), take $\phi = \phi_1$ and $W = v_1^*$. Then there exists a partial isometry $w \in N$ such that

$$w\phi_1(a) = aw, \quad a \in A. \qquad (9.4.58)$$

If we take the adjoint of this equation and define $v_0 = w^*$, then we obtain

$$\phi_1(a)v_0 = v_0 a, \quad a \in A. \tag{9.4.59}$$

Now set $p_0 = v_0^* v_0$ and $q_0 = v_0 v_0^*$. The required properties $p_0 \in A' \cap N = A$ and $q_0 \in \phi_1(A)' \cap N$ are consequences of the equations

$$v_0^* \phi_1(a) v_0 = v_0^* v_0 a \quad \text{and} \quad \phi_1(a) v_0 v_0^* = v_0 a v_0^* \quad \text{for } a \in A. \tag{9.4.60}$$

$\square$

Lemma 9.4.6. *Let N be a finite von Neumann algebra with faithful normal trace τ and let A and B be masas in N. If f is a non-zero projection in $A' \cap \langle N, e_B \rangle$ equivalent to a subprojection e_1 of e_B, then there are non-zero projections $p \in A$ and $q \in B$, and a partial isometry $v \in N$ such that*

(i) $v^* v = p$, $vv^* = q$, $vAv^* = Bq$;

(ii) *if $\phi(x) = vxv^*$ for all $x \in N$, then ϕ is a *-isomorphism from pNp onto qNq, and a *-isomorphism from pA onto qB.*

Proof. [(3) $\implies$ (4) of Theorem 9.5.1.]
Step 1. *We define a map ϕ_1 into B.*
Since f is equivalent to $e_1 \leq e_B$, there is a partial isometry v_1 in $\langle N, e_B \rangle$ such that

$$v_1^* v_1 = f, \quad v_1 v_1^* = e_1 \leq e_B \tag{9.4.61}$$

and

$$\|v_1 - f\|_{2,Tr}, \quad \|v_1 - e_1\|_{2,Tr} \leq 2^{1/2} \|e_1 - f\|_{2,Tr}, \tag{9.4.62}$$

where the norm estimates in (9.4.62) may be found in Lemma 5.2.9 (ii).
    Define $\theta : \langle N, e_B \rangle \to e_1 \langle N, e_B \rangle e_1$ by

$$\theta(x) = v_1 x v_1^* \quad \text{for all} \quad x \in \langle N, e_B \rangle. \tag{9.4.63}$$

Then θ is a norm-reducing normal *-map. Since $e_B e_1 = e_1$ and $e_1 v_1 = v_1$, the range of θ is contained in $e_B \langle N, e_B \rangle e_B = e_B B$, this last equality being Theorem 4.2.2 (iii). Part (v) of the same theorem shows that there is a well defined norm-reducing normal *-map $\phi_1 : \langle N, e_B \rangle \to B$ given by

$$\phi_1(x)e_B = e_B \phi_1(x) = \theta(x) = v_1 x v_1^* \quad \text{for} \quad x \in \langle N, e_B \rangle. \tag{9.4.64}$$

This equation leads to

$$\phi_1(a)v_1 = v_1 a v_1^* v_1 = v_1 a f = v_1 f a = v_1 a \tag{9.4.65}$$

for $a \in A$, and taking the adjoint of (9.4.65) gives

$$v_1^* \phi_1(a) = a v_1^* \quad \text{for all} \quad a \in A. \tag{9.4.66}$$

Let

$$\mathcal{F} = \{f\}' \cap \langle N, e_B \rangle, \tag{9.4.67}$$

which contains A since $f \in A' \cap \langle N, e_B \rangle$. If $x \in \mathcal{F} = \{f\}' \cap \langle N, e_B \rangle$ and $y \in \langle N, e_B \rangle$, then

$$e_B \phi_1(xy) = v_1 xy v_1^* = v_1 fxy v_1^* = v_1 x f y v_1^* = v_1 x v_1^* (v_1 y v_1^*)$$
$$= e_B \phi_1(x) e_B \phi_1(y) = e_B \phi_1(x) \phi_1(y) \tag{9.4.68}$$

so that $\phi_1(xy) = \phi_1(x)\phi_1(y)$, using Lemma 4.2.3 (v). Similarly, or using the *-preserving property of ϕ_1, $\phi_1(xy) = \phi_1(x)\phi_1(y)$, where at least one of x or y lies in $\mathcal{F}$. Let $e_0 = \phi_1(f)$. Then e_0 is a projection in B with $e_0 e_B = e_1$, from (9.4.64). Observe that $y \in \langle N, e_B \rangle \cap \mathrm{Ker}(\phi_1)$ is equivalent to $v_1 y v_1^* = 0 = fyf$. Hence

$$\mathrm{Ker}(\phi_1) = (1 - f)\langle N, e_B \rangle + \langle N, e_B \rangle (1 - f). \tag{9.4.69}$$

If $b \in B$, then $v_1^* b v_1 \in \mathcal{F}$, since $v_1 f = v_1$, and

$$\phi_1(v_1^* b v_1) e_B = v_1 v_1^* b v_1 v_1^* = e_1 b e_1 = e_0 e_B b e_0 e_B = b e_0 e_B. \tag{9.4.70}$$

It follows that

$$\phi_1(v_1^* b v_1) = b e_0 \tag{9.4.71}$$

since $b \mapsto b e_B$ is one-to-one. Thus $\phi_1(\mathcal{F}) = B e_0$ and ϕ_1 is a *-isomorphism from $f\mathcal{F}$ onto $B e_0$.

If $x \in \mathcal{F}$, then

$$\phi_1(x) v_1 = \phi_1(x) e_B v_1 = v_1 x v_1^* v_1 = v_1 x f = v_1 f x = v_1 x. \tag{9.4.72}$$

Step 2. v_1 *is replaced by a partial isometry in* N.
If we restrict (9.4.72) to $A \subseteq \mathcal{F}$, then we are in the situation of Corollary 9.4.5, and we have a partial isometry $v_0 \in N$ with source and range projections p_0 and q_0 satisfying

$$p_0 \in A' \cap N = A, \quad q_0 \in \phi_1(A)' \cap N, \quad \text{and} \tag{9.4.73}$$
$$\phi_1(a) v_0 = v_0 a, \quad \phi_1(a) q_0 = v_0 a v_0^* \quad \text{for all} \quad a \in A. \tag{9.4.74}$$

Step 3. *Cut downs of* $\phi_1(A)$ *and* B *are equal.*
Define the map $\psi : p_0 N p_0 \to q_0 N q_0$ by

$$\psi(x) = v_0 x v_0^* , \quad x \in p_0 N p_0. \tag{9.4.75}$$

Since $p_0 = v_0^* v_0$ and $q_0 = v_0 v_0^*$, it is easy to see that ψ is a normal *-isomorphism from $p_0 N p_0$ onto $q_0 N q_0$. The von Neumann algebra $p_0 N p_0$ contains the masa $A p_0$ and thus $\psi(A p_0)$ is a masa in $q_0 N q_0$. From equations (9.4.73) and (9.4.74), it follows that

$$\psi(a p_0) = v_0 a p_0 v_0^* = v_0 a v_0^* = \phi_1(a) q_0 = q_0 \phi_1(a) \quad \text{for all} \quad a \in A, \tag{9.4.76}$$

so that $\phi_1(A)q_0$ is a masa in $q_0 N q_0$ and

$$(\phi_1(A)q_0)' \cap q_0 N q_0 = \phi_1(A)q_0. \tag{9.4.77}$$

If $a \in A$ and $b \in B$, then

$$(\phi_1(a)q_0)(q_0 b q_0) = q_0 \phi_1(a) b q_0 = q_0 b \phi_1(a) q_0 = (q_0 b q_0)(\phi_1(a)q_0), \tag{9.4.78}$$

since $\phi_1(A) \subseteq B$ and $q_0 \in \phi_1(A)' \cap q_0 N q_0$. Thus the elements of $q_0 B q_0$ commute with the masa $\phi_1(A)q_0$ in $q_0 N q_0$ so that $q_0 B q_0 \subseteq \phi_1(A)q_0$. On the other hand $\psi(Ap_0) = \psi(p_0 A p_0) = q_0 \phi_1(A) q_0 \subseteq q_0 B q_0$ and so

$$\phi_1(A)q_0 = q_0 B q_0. \tag{9.4.79}$$

Note that

$$\psi(p_0) = q_0 = \phi_1(1)q_0 = e_0 q_0, \tag{9.4.80}$$

and $q_0 \leq e_0$ follows from (9.4.80).

Step 4.   *Twisting ϕ_1 onto qB.*

Let P be the von Neumann algebra $(\phi_1(A)e_0)' \cap e_0 N e_0$. Then $Be_0 \subseteq P$ because $\phi_1(A)e_0 \subseteq Be_0$ is an abelian algebra. Also $q_0 \in P$, since $q_0 \leq e_0$ and $q_0 \in \phi_1(A)'$, by equation (9.4.73). By Theorem 5.4.1 (ii),

$$\begin{aligned} q_0 P q_0 &= q_0((\phi_1(A)e_0)' \cap e_0 N e_0)q_0 \\ &= (\phi_1(A)q_0)' \cap q_0 N q_0 = \phi_1(A)q_0, \end{aligned} \tag{9.4.81}$$

so that q_0 is an abelian projection in P as $\phi_1(A)q_0$ is abelian. Since B is a masa in N, Be_0 is a masa in $e_0 N e_0$, and hence Be_0 is a masa in P. By Corollary 2.4.6 applied to the masa $Be_0 \subseteq P$, there is a projection $q_2 \in Be_0$ with central support in $e_0 P e_0$ equal to e_0 such that q_2 and q_0 are equivalent in P. Let v_2 be a partial isometry in P such that

$$v_2^* v_2 = q_0 \quad \text{and} \quad v_2 v_2^* = q_2. \tag{9.4.82}$$

Let

$$v = v_2 v_0, \quad p = v^* v \quad \text{and} \quad q = vv^*. \tag{9.4.83}$$

From this definition of p and q, and equation (9.4.76)

$$p = v_0^* v_2^* v_2 v_0 = v_0^* q_0 v_0 = p_0 \in A \quad \text{and} \tag{9.4.84}$$

$$q = v_2 v_0 v_0^* v_2^* = v_2 q_0 v_2^* = q_2 \in B. \tag{9.4.85}$$

From equations (9.4.82) and (9.4.74), $q_0 = v_2^* v_2$ and $\phi_1(a)q_0 = v_0 a v_0^*$ $(a \in A)$, we obtain $v_2^* \phi_1(a) v_2 = \phi_1(a) v_2^* v_2 = v_0 a v_0^*$ and hence

$$\phi_1(a)q = \phi_1(a)q_2 = v_2 v_2^* \phi_1(a) v_2 v_2^* = v_2 v_0 a v_0^* v_2^* = vav^* \tag{9.4.86}$$

for all $a \in A$.

Step 5.   *Defining ϕ as a twist of ϕ_1*

Let $\phi : N \to N$ be defined by $\phi(x) = vxv^*$ for all $x \in N$. Then ϕ is a *-isomorphism from pNp onto qNq, since v is a partial isometry with $p = v^*v$ and $q = vv^*$. The algebra pA is a masa in pNp, since $p \in A$ by equation (9.4.84), so that $\phi(pA) = vAv^*$ is a masa in qNq, and Bq is a masa in qNq, since $q \in B$ by equation (9.4.85). From the definitions of ϕ and v, and equation (9.4.74)

$$\phi(a) = vav^* = v_2v_0av_0^*v_2^* = v_2\phi_1(a)q_0v_2^* = v_2\phi_1(a)v_2^* \tag{9.4.87}$$

for all $a \in A$, since $q_0v_2^* = v_2^*$. The choice of $v_2 \in P$ means that v_2 commutes with $\phi_1(A)q_0$ so that

$$\phi(a) = \phi_1(a)v_2v_2^* = \phi_1(a)q_2 = \phi_1(a)q \in Bq \tag{9.4.88}$$

for all $a \in A$, by equation (9.4.85). Hence $\phi(A) \subseteq qB$ so these two algebras are equal, since $\phi(A)$ is a masa in qNq and qB is abelian.   □

The technique used in the first two parts of the previous proof becomes an important lemma for A and B non-abelian (see Chapter 10).

Lemma 9.4.7. *Let N be a type II_1 factor, and let A and B be masas in N. If the non-zero projections $p \in A$ and $q \in B$, and the partial isometry $v \in N$ satisfy $v^*v = p$, $vv^* = q$ and $vAv^* = Bq$, then there is a unitary element $w \in N$ such that*

$$\|(I - \mathbb{E}_B)\mathbb{E}_{wAw^*}\|_{\infty,2} \leq \|1 - p\|_2 < 1. \tag{9.4.89}$$

Proof. [(4) $\implies$ (1) of Theorem 9.5.1]

Let v_0 be a partial isometry in N such that

$$v_0^*v_0 = 1 - p = 1 - v^*v, \tag{9.4.90}$$
$$v_0v_0^* = 1 - q = 1 - vv^*; \tag{9.4.91}$$

such a partial isometry exists since the projections have the same trace. Then $w = v + v_0$ is a unitary in N with $qw = wp = qwp = v$. Thus

$$wAw^*q = qwAw^*q = qwpApw^*q = vAv^* = Bq. \tag{9.4.92}$$

It follows that $xq \in Bq$, $\mathbb{E}_B(xq) = xq$, and

$$x - \mathbb{E}_B(x) = (x - \mathbb{E}_B(x))(1 - q) \tag{9.4.93}$$

for all $x \in wAw^*$. Hence if $x \in wAw^*$ with $\|x\| \leq 1$, then

$$\|(I - \mathbb{E}_B)(x)\|_2 = \|(I - \mathbb{E}_B)(x - xq)\|_2$$
$$= \|(I - \mathbb{E}_B)((1 - q)x)\|_2 \leq \|1 - q\|_2 \tag{9.4.94}$$

so that

$$\|(I - \mathbb{E}_B)\mathbb{E}_{wAw^*}\|_{\infty,2} \leq \|1 - q\|_2 = \|1 - p\|_2. \tag{9.4.95}$$

□

Lemma 9.4.8. *Let N be a type II_1 factor, let A and B be von Neumann sub-algebras of N. If $\delta > 0$ and $y_1, \ldots, y_n$ are in N with*

$$\sum_{i,j=1}^{n} \|\mathbb{E}_B(y_i^* u y_j)\|_2^2 \geq \delta \qquad (9.4.96)$$

for all unitaries $u \in A$, then there is a non-zero $h \in A' \cap \langle N, e_B \rangle$ such that

$$Tr(h^2) < \infty, \qquad (9.4.97)$$

$$\delta \leq Tr(h^2) = Tr\left(h \left(\sum_{i=1}^{n} y_i e_B y_i^* \right) \right) \leq \sum_{i=1}^{n} \|y_i\|^2 \quad \text{and} \qquad (9.4.98)$$

$$0 \leq h \leq \left\| \sum_{i=1}^{n} y_i e_B y_i^* \right\| 1. \qquad (9.4.99)$$

Proof. [(5) $\implies$ (2) of Theorem 9.5.1]
If $y_1, \ldots, y_n \in N$ and $\delta > 0$ are given satisfying

$$\sum_{i,j=1}^{n} \|\mathbb{E}_B(y_i^* u y_j)\|_2^2 \geq \delta \qquad (9.4.100)$$

for all unitaries $u \in A$, let

$$x = \sum_{i=1}^{n} y_i e_B y_i^* \in \langle N, e_B \rangle. \qquad (9.4.101)$$

Using

$$Tr(y^* e_B y e_B) = Tr(y^* \mathbb{E}_B(y) e_B) = \tau(y^* \mathbb{E}_B(y)) = \|\mathbb{E}_B(y)\|_2^2 \qquad (9.4.102)$$

yields

$$Tr(x^* u x u^*) = \sum_{i,j=1}^{n} Tr(y_i e_B y_i^* u y_j e_B y_j^* u^*)$$

$$= \sum_{i,j=1}^{n} Tr(y_j^* u^* y_i e_B (y_i^* u y_j) e_B)$$

$$= \sum_{i,j=1}^{n} \|\mathbb{E}_B(y_i^* u y_j)\|_2^2 \geq \delta \qquad (9.4.103)$$

for all unitaries $u \in A$ by inequality (9.4.100). Let

$$K = \text{conv}\{u x u^* : u \in \mathcal{U}(A)\}. \qquad (9.4.104)$$

With the same technique as Lemma 9.2.1, $uKu^* = K$ and $u\overline{K}^w u^* = \overline{K}^w$ for all $u \in \mathcal{U}(A)$, where $\overline{K}^w$ is the weak closure of K in $\langle N, e_B \rangle$. For each $u \in \mathcal{U}(A)$,

$$\|u x u^*\|_{2,Tr} \leq \sum_{i,j=1}^{n} \|y_i^* e_B y_j\|_{2,Tr} \leq \sum_{i,j=1}^{n} \|y_i\|^2 \qquad (9.4.105)$$

since $\|e_B\|_{2,Tr} = 1$. Thus K is a bounded set in the $\|\cdot\|_{2,Tr}$- and $\|\cdot\|$-norms so $\overline{K}^w$ is a weakly, and hence norm, closed subset of $L^2(\langle N, e_B \rangle, Tr)$. Moreover $z \mapsto uzu^*$ is an isometry on $L^2(\langle N, e_B \rangle, Tr)$ so that $u\overline{K}^w u^* = \overline{K}^w$ for all $u \in \mathcal{U}(A)$. Let h be the unique element of minimal $\|\cdot\|_{2,Tr}$-norm in $\overline{K}^w$. By the uniqueness of h, $uhu^* = h$ for all $u \in \mathcal{U}(A)$ so that $h \in A' \cap \langle N, e_B \rangle$.

The inequality $Tr(x) \le \sum_{i=1}^n \|y_i\|^2$ follows from (9.4.101) and implies that

$$\delta \le Tr(xh) \le \sum_{i=1}^n \|y_i\|^2 < \infty, \qquad (9.4.106)$$

using the normality of Tr. From this we see that, for all $u \in \mathcal{U}(A)$,

$$Tr(uxu^*h) = Tr(xu^*hu) = Tr(xh) \ge \delta \qquad (9.4.107)$$

so that $Tr(h^2) \ge \delta$ by another application of convex combinations and the normality of Tr. Thus $h^2 \in \langle N, e_B \rangle$ with $0 \le h^2$ and $0 < Tr(h^2) < \infty$ proving (9.4.99). Since $\|uxu^*\| = \|\sum_{i=1}^n y_i e_B y_i^*\|$ for all unitaries u in A, $\|h\| \le \|\sum_{i=1}^n y_i e_B y_i\| \le \sum_{i=1}^n \|y_i\|^2$ by another application of weak limits. $\square$

9.5 Distances and groupoid normalisers

In this section we consider two masas A and B in a II$_1$ factor N, and we relate the distance between A and B in the $\|\cdot\|_{\infty,2}$-norm (or more generally between B and a unitary conjugate of A) to the existence of a partial isometry that intertwines Ap and Bq for projections $p \in A$ and $q \in B$. This is a crucial tool in recent work of Popa [145, 146, 147, 149, 150] and is also the inspiration for the general perturbation work in Chapter 10.

Theorem 9.5.1 is the most important one for perturbations, and we note that the long proof that it requires has been divided into a sequence of lemmas in Section 9.4.

Theorem 9.5.1. *Let N be a II$_1$ factor, and let A and B be masas in N. The following conditions are equivalent:*

(1) *there is a unitary element $w \in N$ such that $\|(I - \mathbb{E}_B)\mathbb{E}_{wAw^*}\|_{\infty,2} < 1$;*

(2) *there is a non-zero $h \in A' \cap \langle N, e_B \rangle$ with $h \ge 0$ and $Tr(h), Tr(h^2) < \infty$;*

(3) *there exists a non-zero abelian projection $f \in A' \cap \langle N, e_B \rangle$;*

(4) *there exist non-zero projections $p \in A$ and $q \in B$, and a partial isometry $v \in N$ such that $v^*v = p$, $vv^* = q$ and $vAv^* = Bq$;*

(5) *there are elements $y_1, \ldots, y_n \in N$ and a constant $\delta > 0$ such that*

$$\sum_{i,j=1}^n \|\mathbb{E}_B(y_i^* u y_j)\|_2^2 \ge \delta \qquad (9.5.1)$$

for all unitaries $u \in A$.

Proof. (1) $\implies$ (2)

Lemma 9.2.3 (ii) gives the existence of h and that $h \geq 0$, while parts (i) and (v) of the same lemma ensure respectively that $Tr(h), Tr(h^2) < \infty$, and that $h \neq 0$.

(2) $\implies$ (3)

This is Lemma 9.4.1 (i).

(3) $\implies$ (4)

Since e_B is an abelian projection in $\langle N, e_B \rangle$ whose central support is 1, we have that f is equivalent to a subprojection of e_B (see [105, Proposition 6.4.6]). This implication now follows from Lemma 9.4.6.

(4) $\implies$ (1)

This is Lemma 9.4.7.

(4) $\implies$ (5)

Using p, q and v of (4), let $n = 1$, $y_1 = v$ and $\delta = \tau(p) > 0$. For each unitary u in A, we have $v^*uv = v^*puv \in B$ and $vv^* \in A$ so that

$$\|\mathbb{E}_B(y_1^* u y_1)\|_2^2 = \|\mathbb{E}_B(v^* uv)\|_2^2 = \|v^* uv\|_2^2$$
$$= \tau(v^* uvv^* u^* v) = \tau(p) = \delta, \qquad (9.5.2)$$

which proves (5).

(5) $\implies$ (2)

This is Lemma 9.4.8. $\qquad\square$

Remark 9.5.2. The implication (3) $\implies$ (2) follows directly as in [147, Theorem A.1]. Let p, q and v be as in (3). If $a \in A$, then $v^* e_B va = v^* e_B vpa = v^* e_B (vav^*) v = v^* (vav^*) e_B v = pav^* e_B v = av^* e_B v$, since $v^* v = p \in A$ and $vav^* \in B$. Hence $v^* e_B v \in A' \cap \langle N, e_B \rangle$ with $0 < Tr(v^* e_B v) = Tr(e_B vv^* e_B) \leq Tr(e_B) = 1$. $\qquad\square$

Adding the hypothesis *Cartan* or *semi-regular* enables one to strengthen the conclusion of Theorem 9.5.1 considerably as follows [147].

Corollary 9.5.3. *Let N be a separable* II_1 *factor, and let A and B be either Cartan masas or semi-regular masas in N. The following conditions are equivalent:*

(1) *there is a unitary element $w \in N$ such that $\|(I - \mathbb{E}_B) \mathbb{E}_{wAw^*}\|_{\infty,2} < 1$;*

(2) *there is a non-zero $h \in A' \cap \langle N, e_B \rangle$ with $h \geq 0$ and $Tr(h), Tr(h^2) < \infty$;*

(3) *there is a unitary $u \in N$ with $uAu^* = B$.*

Proof. The implication (1) $\implies$ (2) has already been proved in Theorem 9.5.1, and the implication (3) $\implies$ (1) is trivial. Thus we only need to show that (2) $\implies$ (3). Assume that condition (2) holds, and note that it is the same as the second condition in Theorem 9.5.1.

By the equivalence of parts (2) and (4) in Theorem 9.5.1, there are non-zero projections $p \in A$ and $q \in B$, and a partial isometry $v \in N$ such that $v^* v = p$, $vv^* = q$ and $vAv^* = Bq$. This is the hypothesis for Theorem 6.2.7, from which we conclude that $uAu^* = B$ for some unitary $u \in N$. $\qquad\square$

The following example is included to show that masas can have spatially isomorphic cutdowns but need not be unitarily equivalent.

Example 9.5.4. Let R be the hyperfinite II_1 factor and let $p \in R$ be an arbitrary non-trivial projection. Then pRp and $(1-p)R(1-p)$ are both isomorphic to R by *-isomorphisms θ and ϕ respectively. From Examples 3.4.3 and 7.5.3, we may choose a Cartan masa A_1 and a singular masa A_2 in R. Then define $A = \theta(A_1) \oplus \phi(A_1)$ and $B = \theta(A_1) \oplus \phi(A_2)$. The first is Cartan, the second is not, and so they cannot be unitarily equivalent. However, $Ap = Bp$, and are spatially isomorphic by the partial isometry $v = p$. $\qquad\square$

9.6 Numerical constants for perturbations

As we have already seen, closeness of masas A and B implies that they have spatially isomorphic cutdowns Ap and Bq. One expects that, as the distance between the masas shrinks, the corresponding projections p and q should increase, and this is indeed the case. The purpose of this section is to quantify these statements by providing estimates for various distances. We also include a result that shows that unitaries u for which A is close to uAu^* must themselves be close to normalising unitaries. We begin with a preparatory result.

Proposition 9.6.1. *Let A be an abelian von Neumann subalgebra of N where the identity of A is not required to be that of N. Let $\varepsilon > 0$ and let $q \in N$ be a projection satisfying*

(i) $\tau(q) \geq 1 - \varepsilon$;

(ii) $q \in A' \cap N$, and Aq is a masa in qNq;

(iii) *if $a \in A$ and $aq = 0$, then $a = 0$.*

Let B be a masa containing A. Then there exists a projection $\tilde{q} \in B$, $\tilde{q} \leq q$, and $\tau(\tilde{q}) \geq 1 - 2\varepsilon$. Moreover, we may choose $\tilde{q}$ to have the form pq for some projection $p \in A$.

Proof. We will first assume that A contains the identity element of N, and remove this restriction at the end. By (ii), we may define a masa in qNq by $C = Aq$. By (iii), the map

$$aq \mapsto a \mapsto a(1-q), \qquad a \in A, \tag{9.6.1}$$

is a well-defined *-homomorphism θ of C into $(1-q)N(1-q)$. Writing elements of N as matrices relative to the decomposition $1 = q + (1-q)$, we see that

$$A = \left\{ \begin{pmatrix} c & 0 \\ 0 & \theta(c) \end{pmatrix} : c \in C \right\}, \tag{9.6.2}$$

and $q = \begin{pmatrix} 1 & 0 \\ 0 & 0 \end{pmatrix}$.

If

$$b = \begin{pmatrix} b_1 & b_2 \\ b_3 & b_4 \end{pmatrix} \in B, \tag{9.6.3}$$

then b commutes with A, so $b_1 \in C' = C$, from (9.6.2). Applying this to b, b^* and bb^* gives $b_1, b_1^*, b_2 b_2^* \in C$. Thus any partial isometry $v \in qB(1-q)$ (which is then the $(1,2)$ entry of a matrix in B) has the property that $vv^* \in C$.

Let $\{v_i\}_{i=1}^\infty$ be a set of partial isometries in $qB(1-q)$ which is maximal with respect to having orthogonal initial projections and orthogonal final projections. Let q_1 denote the projection $\sum_{i=1}^\infty v_i v_i^* \in C$. Normality of the trace implies that

$$\tau(q_1) = \sum_{i=1}^\infty \tau(v_i v_i^*) = \tau\left(\sum_{i=1}^\infty v_i^* v_i\right) \le \varepsilon, \tag{9.6.4}$$

since each projection $v_i^* v_i$ lies under $1-q$. Let $q_2 = q - q_1$, $q_3 = \theta(q_1)$, $q_4 = \theta(q_2)$. Then $\{q_i\}_{i=1}^4$ is a set of orthogonal projections which sum to 1, and we will subsequently write elements of N as 4×4 matrices relative to the decomposition $1 = \sum_{i=1}^4 q_i$. Define two $*$-homomorphisms α and β as the restrictions of θ to Cq_1 and Cq_2 respectively. Then

$$A = \left\{ \begin{pmatrix} f & 0 & 0 & 0 \\ 0 & g & 0 & 0 \\ 0 & 0 & \alpha(f) & 0 \\ 0 & 0 & 0 & \beta(g) \end{pmatrix} : f \in Cq_1, g \in Cq_2 \right\}, \tag{9.6.5}$$

and $q = \begin{pmatrix} 1&0&0&0 \\ 0&1&0&0 \\ 0&0&0&0 \\ 0&0&0&0 \end{pmatrix}$. Then $q_1 + q_3 \in A$ and so commutes with B. This forces each $b \in B$ to have the form

$$\begin{pmatrix} * & 0 & * & 0 \\ 0 & * & 0 & * \\ * & 0 & * & 0 \\ 0 & * & 0 & * \end{pmatrix}, \tag{9.6.6}$$

so each element of $qB(1-q)$ has the form

$$\begin{pmatrix} 0 & 0 & * & 0 \\ 0 & 0 & 0 & * \\ 0 & 0 & 0 & 0 \\ 0 & 0 & 0 & 0 \end{pmatrix}. \tag{9.6.7}$$

Since $q_1 v_i = v_i$ for all i, it follows that each v_i has the form

$$\begin{pmatrix} 0 & 0 & * & 0 \\ 0 & 0 & 0 & 0 \\ 0 & 0 & 0 & 0 \\ 0 & 0 & 0 & 0 \end{pmatrix}, \tag{9.6.8}$$

and so

$$v_i v_i^* \le q_1, \quad v_i^* v_i \le q_3, \quad i \ge 1. \tag{9.6.9}$$

Suppose that there is an element of B with a non-zero $(2,4)$ or $(4,2)$ entry. By taking adjoints if necessary, we may assume that the $(2,4)$ entry is non-zero. Multiplication on the left by $q_2 + q_4 \in A$ gives an element

$$b = \begin{pmatrix} 0 & 0 & 0 & 0 \\ 0 & g & 0 & x \\ 0 & 0 & 0 & 0 \\ 0 & y & 0 & z \end{pmatrix} \in B \qquad (9.6.10)$$

with $g \in Cq_2$ and $x \neq 0$. The condition $B \subseteq A' \cap N$ leads, as before, to the conclusion that xx^* is a non-zero element of Cq_2 which, being a masa in $q_2 N q_2$, is isomorphic to $L^\infty[0,1]$ by Theorem 3.5.2. We may then pick a non-negative $h \in Cq_2$ so that hxx^* is a non-zero projection in Cq_2. Let $k = h^{1/2}$. Then $v = kx$ is a non-zero partial isometry. Multiplication of b on the left by

$$\begin{pmatrix} 0 & 0 & 0 & 0 \\ 0 & k & 0 & 0 \\ 0 & 0 & 0 & 0 \\ 0 & 0 & 0 & \beta(k) \end{pmatrix} \in A \qquad (9.6.11)$$

gives an element

$$\begin{pmatrix} 0 & 0 & 0 & 0 \\ 0 & * & 0 & v \\ 0 & 0 & 0 & 0 \\ 0 & * & 0 & * \end{pmatrix} \in B, \qquad (9.6.12)$$

so

$$\begin{pmatrix} 0 & 0 & 0 & 0 \\ 0 & 0 & 0 & v \\ 0 & 0 & 0 & 0 \\ 0 & 0 & 0 & 0 \end{pmatrix} \in qB(1-q) \qquad (9.6.13)$$

is a non-zero partial isometry. The initial and final projections are under q_4 and q_2 respectively, and this contradicts the maximality of $\{v_i\}_{i=1}^\infty$. Thus no $(2,4)$ or $(4,2)$ entry can be non-zero, so each element of B has the form

$$\begin{pmatrix} * & 0 & * & 0 \\ 0 & * & 0 & 0 \\ * & 0 & * & 0 \\ 0 & 0 & 0 & * \end{pmatrix}. \qquad (9.6.14)$$

This implies that $q_2 \in B' \cap N = B$. Define $p = q_2 + q_4 \in A$, and let $\tilde{q} = q_2 = pq$. Then

$$\tau(\tilde{q}) = \tau(q - q_1) \geq 1 - 2\varepsilon \qquad (9.6.15)$$

from (i) and (9.6.4).

We now turn to the general case. Let e be the unit of A and suppose that $e \neq 1$. By (ii), $eq = q$ so $e \geq q$. Since A is isomorphic to Aq, by (iii), we may define a normal $*$-isomorphism γ from A to $(1-e)B(1-e)$. We may then define

$$A_1 = \{a + \gamma(a) \colon a \in A\}, \qquad (9.6.16)$$

which also satisfies the hypotheses. From the first part, there is a projection $p \in A$ such that

$$\tau((p + \gamma(p))q) \geq 1 - 2\varepsilon \tag{9.6.17}$$

and $(p + \gamma(p))q \in B$. Since $\gamma(p)q = 0$, the result follows. $\hspace{1cm} \square$

Lemma 9.6.2. *Let A be a masa in a type II_1 factor N and let $\delta \geq 0$. If $u \in N$ is a unitary for which there exists $\tilde{u} \in \mathcal{N}(A)$ with $\|u - \tilde{u}\|_{2,\tau} \leq \delta$, then*

$$\|\mathbb{E}_A - \mathbb{E}_{uAu^*}\|_{\infty,2} \leq 4\delta. \tag{9.6.18}$$

Proof. Define w to be $u\tilde{u}^*$. Then $wAw^* = uAu^*$, so it suffices to estimate $\|\mathbb{E}_A - \mathbb{E}_{wAw^*}\|_{\infty,2}$. Let $h = 1 - w$. Then, for $x \in N$, $\|x\| \leq 1$,

$$
\begin{aligned}
\|\mathbb{E}_A(x) - \mathbb{E}_{wAw^*}(x)\|_2 &= \|\mathbb{E}_A(x) - w\mathbb{E}_A(w^*xw)w^*\|_2 \\
&= \|w^*\mathbb{E}_A(x)w - \mathbb{E}_A(w^*xw)\|_2 \\
&\leq \|w^*\mathbb{E}_A(x)w - \mathbb{E}_A(x)\|_2 + \|\mathbb{E}_A(x) - \mathbb{E}_A(w^*xw)\|_2 \\
&\leq \|\mathbb{E}_A(x)w - w\mathbb{E}_A(x)\|_2 + \|x - w^*xw\|_2 \\
&= \|\mathbb{E}_A(x)h - h\mathbb{E}_A(x)\|_2 + \|hx - xh\|_2 \\
&\leq 4\|h\|_2 = 4\|1 - u\tilde{u}^*\|_2 = 4\|\tilde{u} - u\|_2 \leq 4\delta, \tag{9.6.19}
\end{aligned}
$$

which gives (9.6.18). $\hspace{1cm} \square$

Theorem 9.6.3. *Let A be a masa in a separable type II_1 factor N and let $\delta > 0$.*

  (i) *If u is a unitary such that $\|\mathbb{E}_A - \mathbb{E}_{uAu^*}\|_{\infty,2} \leq \delta$, then there exists a unitary $\tilde{u} \in \mathcal{N}(A)$ such that $\|u - \tilde{u}\|_{2,\tau} \leq 5\delta$.*

  (ii) *The inequality*

$$5\|\mathbb{E}_A - \mathbb{E}_{uAu^*}\|_{\infty,2} \geq \|u - \mathbb{E}_{\mathcal{N}(A)''}(u)\|_2$$

holds for all unitaries $u \in N$.

Proof. We assume that $\delta \leq 2/5$, otherwise the theorem is vacuously true by taking $\tilde{u} = 1$.

Let B be the masa uAu^*. By averaging e_B over A, we find an element $h \in A' \cap \langle N, e_B \rangle^+$ such that $\|h - e_B\|_{2,Tr} \leq \delta$ and $Tr(h), Tr(h^2) < \infty$. By Theorem 8.3.5 there exists a partial isometry $V \in \langle N, e_B \rangle$ such that $VV^* \leq e_B$, $V^*V \in A' \cap \langle N, e_B \rangle$, and

$$\|e_B - V\|_{2,Tr}^2 \leq (3/2)\|h - e_B\|_{2,Tr}^2 \leq 3\delta^2/2. \tag{9.6.20}$$

Then $a \mapsto VaV^*$, $a \in A$, defines a *-homomorphism of A into $e_B\langle N, e_B \rangle e_B = Be_B$, and so there is a *-homomorphism $\phi: A \to B$ such that $VaV^* = \phi(a)e_B$. Thus

$$Va = V(V^*V)a = VaV^*V = \phi(a)e_BV = \phi(a)V, \quad a \in A, \tag{9.6.21}$$

using the fact that $V^*V \in A' \cap \langle N, e_B \rangle$. Lemma 9.4.2, applied to the adjoint of this last equation, gives a partial isometry $v \in N$ such that

$$vav^* = \phi(a)vv^*, \quad a \in A, \tag{9.6.22}$$

and also gives the estimate

$$\|1 - v\|_{2,\tau} \leq 2\|e_B - V\|_{2,Tr} \leq 2\sqrt{3/2}\delta = \sqrt{6}\delta. \tag{9.6.23}$$

The projection $q = vv^*$ lies in $\phi(A)' \cap N$ and satisfies, from Lemma 9.4.2, the inequality

$$\|1 - q\|_{2,\tau} \leq \|e_B - V\|_{2,Tr} \leq \sqrt{3/2}\delta, \tag{9.6.24}$$

which is equivalent to $\tau(q) \geq 1 - 3\delta^2/2$. Let $p = v^*v \in N$. Then, for $a \in A$,

$$\begin{aligned} pa(1-p)a^*p &= v^*(vaa^*v^* - vav^*va^*v^*)v \\ &= v^*(\phi(aa^*)q - \phi(a)q\phi(a^*)q)v = 0, \end{aligned} \tag{9.6.25}$$

from which it follows that $pa(1-p) = 0$, and $p \in A' \cap N = A$. Then v implements a *-isomorphism of pNp onto qNq sending the masa $Ap \subseteq pNp$ onto $\phi(A)q$, where $\phi(A)$ is contained in the masa B. If $\phi(ap)q = 0$ then $vapv^* = 0$ and it follows that $ap = 0$, multiplying on the left and right by v^* and v respectively. Then $\phi(ap) = 0$, and the hypotheses of Proposition 9.6.1 are satisfied with $\varepsilon = 3\delta^2/2$ and A replaced by $\phi(Ap)$. Thus there is a projection $\tilde{q} \in B$ with $\tilde{q} \leq q$, $\tau(\tilde{q}) \geq 1 - 3\delta^2$, and $\tilde{q} = \phi(p_1)q$ for some projection $p_1 \in Ap$. Now define a partial isometry $v_1 \in N$ to be vp_1. Then

$$v_1v_1^* = vp_1v^* = \phi(p_1)q = \tilde{q} \tag{9.6.26}$$

while

$$v_1^*v_1 = p_1v^*vp_1 = p_1pp_1 = p_1. \tag{9.6.27}$$

Thus v_1 implements an isomorphism of Ap_1 onto $B\tilde{q} = uAu^*\tilde{q}$. This shows that u^*v_1 is in the normalising groupoid $\mathcal{GN}(A)$ of A, and so has the form wp_2 where p_2 is a projection in A and $w \in \mathcal{N}(A)$, by Lemma 6.2.3 (iii). We now show that u is close to w^*.

Since $\|u - w^*\|_{2,\tau} = \|1 - w^*u^*\|_{2,\tau}$, we define w_1 to be $w^*u^* \in \mathcal{U}(N)$ and estimate $\|1 - w_1\|_{2,\tau}$.

We begin from the relation $w_1v_1 = p_2$. Then

$$\tau(p_2) = \tau(v_1^*v_1) = \tau(v_1v_1^*) = \tau(\tilde{q}) \geq 1 - 3\delta^2, \tag{9.6.28}$$

and so

$$\|1 - p_2\|_{2,\tau}^2 = \tau(1 - p_2) \leq 3\delta^2. \tag{9.6.29}$$

Now

$$p_2 = p_2v^*v = w_1v_1v^*v = w_1vp_1v^*v = w_1\tilde{q}v \tag{9.6.30}$$

and so $w_1^*p_2 = \tilde{q}v$. Thus

$$w_1^* = \tilde{q}v + w_1^*(1 - p_2) \tag{9.6.31}$$

and

$$\begin{aligned}
\|w_1^* - 1\|_{2,\tau} &= \|\tilde{q}v - 1 + w_1^*(1 - p_2)\|_{2,\tau} \\
&= \|\tilde{q}(v - 1) - (1 - \tilde{q}) + w_1^*(1 - p_2)\|_{2,\tau} \\
&\leq \|\tilde{q}(v - 1) - (1 - \tilde{q})\|_{2,\tau} + \|1 - p_2\|_{2,\tau} \\
&\leq \|\tilde{q}(v - 1) - (1 - \tilde{q})\|_{2,\tau} + \sqrt{3}\delta.
\end{aligned} \tag{9.6.32}$$

Orthogonality gives

$$\begin{aligned}
\|\tilde{q}(v - 1) - (1 - \tilde{q})\|_{2,\tau}^2 &= \|\tilde{q}(v - 1)\|_{2,\tau}^2 + \|1 - \tilde{q}\|_{2,\tau}^2 \\
&\leq \|v - 1\|_{2,\tau}^2 + \tau(1 - \tilde{q}) \\
&\leq 6\delta^2 + 3\delta^2 = 9\delta^2.
\end{aligned} \tag{9.6.33}$$

Putting this estimate into (9.6.32) leads to

$$\|1 - w_1\|_{2,\tau} = \|1 - w_1^*\|_{2,\tau} \leq (3 + \sqrt{3})\delta \leq 5\delta, \tag{9.6.34}$$

and so $\|u - w^*\|_{2,\tau} \leq 5\delta$. Set $\tilde{u} = w^* \in \mathcal{N}(A)$ to complete the proof of (i).
The inequality of (ii) is an immediate consequence of (i). $\qquad\square$

In recent work, Chifan [12] has obtained an improved version of Theorem 9.6.3 (ii) by removing the constant 5. However, his more sophisticated methods do not seem to improve the estimate of part (i).

The following result summarises the previous two theorems, giving inequalities in both directions. Let $d_2(x, \mathcal{N}(A))$ denote the quantity

$$\inf\{\|x - w\|_{2,\tau} : w \in \mathcal{N}(A)\}.$$

Theorem 9.6.4. *Let N be a separable II_1 factor with a masa A and let $u \in N$ be a unitary. Then*

$$d_2(u, \mathcal{N}(A))/5 \leq \|\mathbb{E}_A - \mathbb{E}_{uAu^*}\|_{\infty,2} \leq 4d_2(u, \mathcal{N}(A)). \tag{9.6.35}$$

The next result shows that if two masas are close in $\|\cdot\|_{\infty,2}$-norm, then large compressions are spatially isomorphic by a partial isometry close to 1. This is similar to several results that precede it in this chapter, but the objective is to present good constants that quantify the notion of closeness.

Theorem 9.6.5. *Let A and B be masas in a separable II_1 factor N and suppose that $\|\mathbb{E}_A - \mathbb{E}_B\|_{\infty,2} \leq \delta$. Then there exist projections $p \in A$ and $q \in B$ and a partial isometry $v \in N$ such that*

$$vAv^* = Bq, \quad vv^* = q, \quad v^*v = p. \tag{9.6.36}$$

and

$$\|1 - p\|_{2,\tau} = \|1 - q\|_{2,\tau} \leq \sqrt{3}\delta, \quad \|1 - v\|_{2,\tau} \leq 3\delta. \tag{9.6.37}$$

Proof. If we follow the proof of Theorem 9.6.3 with changes of notation, then we obtain a partial isometry $v_1 \in N$ and a projection $p \in A$ such that

$$\|1 - v_1\|_{2,\tau} \leq \sqrt{6}\delta, \quad \tau(p) \geq 1 - 3\delta^2, \tag{9.6.38}$$

and the partial isometry $v = v_1 p$ has the properties that $v^* v = p$, vv^* is a projection $q \in B$ and v implements a spatial isomorphism of Ap onto Bq (v_1, v, p and q are respectively v, v_1, p_1 and $\tilde{q}$ in Theorem 9.6.3). Then

$$\|1 - p\|_{2,\tau} = (1 - \tau(p))^{1/2} \leq \sqrt{3}\delta, \tag{9.6.39}$$

with an identical estimate for $\|1 - q\|_{2,\tau}$. Thus

$$\|1 - v\|_{2,\tau}^2 = \|p + (1 - p) - v_1 p\|_{2,\tau}^2 = \|1 - p\|_{2,\tau}^2 + \|p(1 - v_1)\|_{2,\tau}^2$$
$$\leq 3\delta^2 + 6\delta^2 = 9\delta^2, \tag{9.6.40}$$

using orthogonality of $1 - p$ and $p(1 - v_1)$. The estimate $\|1 - v\|_{2,\tau} \leq 3\delta$ follows. $\qquad\qquad\square$

Corollary 9.6.6. *Let $\delta > 0$ and suppose that A and B are masas in a separable II_1 factor N with $\|\mathbb{E}_A - \mathbb{E}_B\|_{\infty,2} \leq \delta$. Then there are a masa B_1 in N, a unitary $u \in N$ and a projection $q \in B_1 \cap B$ such that*

   (i) $B_1 = uAu^*$;

   (ii) $B_1 q = Bq$;

   (iii) $\|1 - q\|_{2,\tau} \leq \sqrt{3}\delta$ and $\|1 - u\|_{2,\tau} \leq 4\delta$.

Proof. We will assume that $\delta \leq 1/\sqrt{3}$, otherwise the conclusions are obtained vacuously by taking $q = 0$ and $u = 1$. The hypotheses of Theorem 9.6.5 are met, and so there exist projections $p \in A$ and $q \in B$ and a partial isometry $v \in N$ such that

$$vAv^* = Bq, \quad vv^* = q, \quad v^* v = p, \tag{9.6.41}$$

and

$$\|1 - p\|_{2,\tau} = \|1 - q\|_{2,\tau} \leq \sqrt{3}\delta, \quad \|1 - v\|_{2,\tau} \leq 3\delta. \tag{9.6.42}$$

The projections $1 - p$ and $1 - q$ have the same trace and so are equivalent by a partial isometry $w \in N$ satisfying $ww^* = 1 - q$ and $w^* w = 1 - p$. Then $v + e^{it}w$ is a unitary, denoted u, where t is chosen so that $\tau(e^{it}w) \in i\mathbb{R}$ and thus $\operatorname{Re}\tau(e^{it}w) = 0$. Using orthogonality of q and $1 - q$, we have

$$\|1 - u\|_{2,\tau}^2 = \|q + (1 - q) - qv - (1 - q)e^{it}w\|_{2,\tau}^2$$
$$= \|q(1 - v)\|_{2,\tau}^2 + \|(1 - q)(1 - e^{it}w)\|_{2,\tau}^2$$
$$\leq 9\delta^2 + \tau((1 - q)((1 - e^{it}w)(1 - e^{-it}w^*)))$$
$$= 9\delta^2 + \tau(1 - q) + \tau(ww^*) - 2\operatorname{Re}\tau(e^{it}w)$$
$$\leq 9\delta^2 + 3\delta^2 + \|w\|_{2,\tau}^2 \leq 15\delta^2. \tag{9.6.43}$$

This gives the estimate $\|1 - u\|_{2,\tau} \leq 4\delta$, and (iii) is proved.

If we let $B_1 = uAu^*$, then the relations $vp = v$ and $wp = 0$ imply that $upu^* = q$ and so $up = qu$. Thus $q \in B_1' = B_1$, and

$$qB_1 = quAu^* = upAu^* = upApu^* = vAv^* = Bq. \qquad (9.6.44)$$

This establishes (ii), and (i) is true by definition of B_1. $\qquad \square$

9.7 Perturbations of masas by averaging

In this section we will give an alternative approach to the perturbation results that we have already presented, based on an averaging technique due to Christensen [24]. The constants in the estimates are not quite as good, but this method is of independent interest, and so we include it here. In the first lemma, due to Christensen [24], note that we do not assume that ϕ is unital.

Lemma 9.7.1. *Let N be a II_1 factor, and let M be a unital von Neumann subalgebra. Let $\varepsilon > 0$ and let $\phi \colon M \to N$ be a *-homomorphism with the property that*

$$\|\phi(x) - x\|_2 \leq \varepsilon \|x\|, \quad x \in M. \qquad (9.7.1)$$

Then there exist a partial isometry $v_1 \in N$ and projections $p_1 = v_1^* v_1 \in M' \cap N$ and $q_1 = v_1 v_1^* \in \phi(M)' \cap N$ satisfying

$$\phi(x)v_1 = v_1 x, \quad v_1 x v_1^* = \phi(x)q_1, \quad x \in M, \qquad (9.7.2)$$

and

$$\|1 - p_1\|_2 = \|1 - q_1\|_2 \leq \varepsilon, \quad \|1 - v_1\|_2 \leq 2\varepsilon. \qquad (9.7.3)$$

Proof. Consider

$$K = \mathrm{conv}\,\{\phi(u)u^* \colon u \in \mathcal{U}(M)\} \subseteq N. \qquad (9.7.4)$$

By Lemma 9.2.1 (v), the weak closure $\overline{K}^w$ is a $\|\cdot\|_2$-closed convex subset of $L^2(N)$. Moreover K is invariant for the $\|\cdot\|_2$-reducing maps $k \mapsto \phi(u)ku^*$, $k \in K$, $u \in \mathcal{U}(M)$, and the same is then true for $\overline{K}^w$. Let w be the unique element of minimal $\|\cdot\|_2$-norm in $\overline{K}^w$. Then

$$\phi(u)wu^* = w, \quad u \in \mathcal{U}(M), \qquad (9.7.5)$$

by uniqueness of w. It follows that

$$\phi(u)w = wu, \quad u \in \mathcal{U}(M), \qquad (9.7.6)$$

and thus this relation also holds with a general element of M replacing u. Multiplication on the left by w^* shows that $w^*w \in M' \cap N$, and so this algebra also contains $|w|$ and its spectral projections. Let $w = v_1|w|$ be the polar decomposition of w. Then

$$\phi(x)v_1|w| = v_1|w|x = v_1 x|w|, \quad x \in M. \qquad (9.7.7)$$

Let f_t be the spectral projection of $|w|$ for the interval $(t, \|w\|]$, for any $t \in [0, \|w\|]$. Then, for $t > 0$, $|w| f_t$ is invertible in $f_t N f_t$ with inverse denoted y_t. Multiplication on the right by $f_t y_t$ in equation (9.7.7) gives

$$\phi(x) v_1 f_t = v_1 x f_t = v_1 f_t x, \quad x \in M, \tag{9.7.8}$$

and then let $t \to 0+$ to obtain

$$\phi(x) v_1 = v_1 x, \quad x \in M, \tag{9.7.9}$$

since $v_1 f_0 = v_1$ and $\lim_{t \to 0+} f_t = f_0$ strongly. Now

$$\|\phi(u) u^* - 1\|_2 = \|\phi(u) - u\|_2 \leq \varepsilon, \quad u \in \mathcal{U}(M), \tag{9.7.10}$$

and so $\overline{K}^w$ is contained in the $L^2(N)$-ball of radius ε centred at 1. Thus $\|1 - w\|_2 \leq \varepsilon$.

Let $p_1 = v_1^* v_1$ and $q = v_1 v_1^*$. The estimates $\|1 - p_1\|_2 = \|1 - q_1\|_2 \leq \varepsilon$ and $\|1 - v_1\|_2 \leq 2\varepsilon$ come from Lemma 5.2.7, while the properties $p_1 \in M' \cap N$ and $q_1 \in \phi(M)' \cap N$ follow from the equations

$$v_1^* v_1 x = v_1^* x v_1, \quad \phi(x) v_1 v_1^* = v_1 x v_1^*, \quad x \in M. \tag{9.7.11}$$

The latter equation also yields $v_1 x v_1^* = \phi(x) q_1$ for $x \in M$. $\square$

The last result of this section is a restatement of Theorem 9.6.5 but with slightly worse constants. As already mentioned, the objective is to give a proof which illustrates another set of useful techniques.

Theorem 9.7.2. *Let A and B be masas in a separable II_1 factor N, let $\delta > 0$ and suppose that $\|\mathbb{E}_A - \mathbb{E}_B\|_{\infty,2} \leq \delta$. Then there exist a partial isometry $v \in N$ and projections $p \in A$, $q \in B$ such that*

$$v^* v = p, \quad v v^* = q, \quad v A v^* = Bq. \tag{9.7.12}$$

Moreover, these operators may be chosen to satisfy

$$\|1 - p\|_{2,\tau} = \|1 - q\|_{2,\tau} \leq \sqrt{12}\delta, \quad \|1 - v\|_{2,\tau} \leq 6\delta. \tag{9.7.13}$$

Proof. Following the proof of Theorem 9.6.5, we obtain a partial isometry $V \in \langle N, e_B \rangle$ and a *-homomorphism $\phi \colon A \to B$ such that

$$e_B V = V, \quad V a = \phi(a) V, \quad a \in A, \tag{9.7.14}$$

and $\|e_B - V\|_{2,Tr} \leq \sqrt{3/2}\delta$. We now estimate $\|\phi(x) - x\|_{2,\tau}$ for $x \in A$. Since $\phi(x)$ commutes with e_B, these inequalities give

$$
\begin{aligned}
\|\phi(x) - x\|_{2,\tau} &= \|\phi(x) e_B - e_B x\|_{2,Tr} \\
&= \|\phi(x) V + \phi(x)(e_B - V) - V x - (e_B - V) x\|_{2,Tr} \\
&= \|\phi(x)(e_B - V) - (e_B - V) x\|_{2,Tr} \\
&\leq 2\|e_B - V\|_{2,Tr} \|x\| \leq \sqrt{6}\delta\|x\|,
\end{aligned}
\tag{9.7.15}
$$

for $x \in A$. Now apply the previous lemma with $\varepsilon = \sqrt{6}\delta$ and $M = A$, to conclude that there exist a partial isometry $v_1 \in N$ and projections $p_1 = v_1^* v_1 \in A' \cap N = A$, $q_1 = v_1 v_1^* \in \phi(A)' \cap N$ such that

$$v_1 a v_1^* = \phi(a) q_1, \quad a \in A, \tag{9.7.16}$$

and

$$\|1 - p_1\|_{2,\tau} = \|1 - q_1\|_{2,\tau} \le \sqrt{6}\delta, \quad \|1 - v_1\|_{2,\tau} \le 2\sqrt{6}\delta. \tag{9.7.17}$$

If $\phi(a p_1) q_1 = 0$ then $v_1 a p_1 v_1^* = 0$. Multiply on the left and right respectively by v_1^* and v_1 to obtain $p_1 a p_1 = 0$, and so $a p_1 = 0$ and $\phi(a p_1) = 0$. Thus we can apply Proposition 9.6.1 to $\phi(A p_1)$ to obtain a projection $p_2 \in A$ so that $q = \phi(p_2 p_1) q_1 \in B$ with $\tau(q) \ge 1 - 12\delta^2$, since $\tau(q_1) \ge 1 - 6\delta^2$. Now define $p = p_1 p_2 \in A$ and $v = v_1 p$. Then

$$v v^* = v_1 p_1 p_2 v_1^* = \phi(p_1 p_2) q_1 = q, \tag{9.7.18}$$

and

$$v v^* = p_1 p_2 v_1 v_1^* p_1 p_2 = p_1 p_2 = p. \tag{9.7.19}$$

Thus

$$\|1 - p\|_{2,\tau} = \|1 - q\|_{2,\tau} \le \sqrt{12}\delta. \tag{9.7.20}$$

Since $v(Ap)v^*$ is a masa in qNq and is contained in the masa Bq, we must have equality. It remains to estimate $\|1 - v\|_{2,\tau}$, for which we follow the calculation of equation (9.6.40). We obtain

$$\|1 - v\|_{2,\tau}^2 = \|p + (1 - p) - v_1 p\|_{2,\tau}^2 = \|1 - p\|_{2,\tau}^2 + \|p(1 - v_1)\|_{2,\tau}^2$$
$$\le 12\delta^2 + 24\delta^2 = 36\delta^2, \tag{9.7.21}$$

and so $\|1 - v\|_{2,\tau} \le 6\delta$, completing the proof. $\qquad\square$

Chapter 10

General perturbations

10.1 Introduction

In chapter 9, we developed the theory of perturbations of masas in type II_1 factors making use of the special structure of such subalgebras. Here we turn to the general theory. The results are essentially the same since close subalgebras will be shown to have spatially isomorphic cutdowns by projections, and the two chapters could have been combined into this one. However, the techniques of the previous chapter do produce significantly better numerical estimates and also give Theorem 9.6.3 on normalising unitaries for which we know no general counterpart.

In Section 10.2, we give a very brief survey of the theory of subfactors, just those parts that we will use subsequently. Section 10.3 considers the situation of a containment $M \subseteq N$ where these two algebras are close in an appropriate sense. The main result is Theorem 10.3.5, which shows that there is a large projection p in the relative commutant $M' \cap N$ so that $Mp = pNp$. This is the crucial result for the perturbation theorems of Section 10.4, the most general one being Theorem 10.4.1.

Much of the material of this chapter is taken from [152], which was based on earlier results from [147].

10.2 The Jones index

In this section we will briefly describe those parts of subfactor theory that we will use in this chapter. There are several good accounts of the theory in [95, 97, 144] and so we will only state the relevant results.

Let N be a II_1 factor in standard form, represented by left multiplication on $L^2(N)$, with a normalised normal trace τ (also denoted τ_N). Let M be a subfactor with trace τ_M, the restriction of τ_N. Then M' is a factor which is either of type II_1 or II_∞. In the latter case the index $[N : M]$ is defined to be ∞. In the former case there is a normalised trace $\tau_{M'}$ on M', and τ_M and $\tau_{M'}$

are used to define the index as follows. Let η be any non-zero vector in $L^2(N)$. The projections onto the closed cyclic subspaces $\overline{M\eta}$ and $\overline{M'\eta}$ lie respectively in M' and M, and we denote them respectively by p' and p. Then $[N:M]$ is defined to be $\tau_M(p)/\tau_{M'}(p')$, and this quantity is independent of the vector η [94]. If ξ is the vector corresponding to $1 \in N$, then the choice $\eta = \xi$ gives $p' = e_M$ and $p = 1$. The von Neumann algebra $\langle N, e_M \rangle$ has canonical trace Tr and commutant JMJ (see Chapter 4) and so $M' = J\langle N, e_M \rangle J$. The uniqueness of the trace up to scaling then shows that $\tau_{M'}(e_M) = Tr(e_M)/Tr(1) = Tr(1)^{-1}$, and so $[N:M] = Tr(1)$. In particular, $[N:M] \in [1, \infty]$, and equals 1 precisely when $M = N$. The following theorem was proved by Jones [94] and initiated the theory of subfactors.

Theorem 10.2.1. *If $M \subseteq N$ is an inclusion of II_1 factors, then the value of the index $[N:M]$ is contained in the set*

$$\{4\cos^2(\pi/n) : n \geq 3\} \cup [4, \infty],$$

and every such value can occur. Moreover, when $[N:M] < 4$, the relative commutant $M' \cap N$ is automatically trivial.

Remark 10.2.2. The discrete values of the index begin with 1 and 2, and it is worth noting that $Tr(1) < 2$ must then imply $Tr(1) = 1$. It follows from this that $M = N$. $\qquad\square$

We will exploit the gaps in the set of index values in the next result. Recall from Section 9.2 the procedure for obtaining a positive operator h by taking the element of minimal $\|\cdot\|_{2,Tr}$-norm in the closed convex hull of $\{ue_Bu^* : u \in U(A)\}$ for $A, B \subseteq N \subseteq \langle N, e_B \rangle$. Under additional hypotheses we obtain information on the spectrum $\sigma(h)$ of h.

Theorem 10.2.3. *Let $Q_0 \subseteq Q_1$ be a containment of finite von Neumann algebras and let τ be a unital faithful normal trace on Q_1. Suppose that*

$$Q_0' \cap Q_1 = Z(Q_0) = Z(Q_1), \tag{10.2.1}$$

and let $h \in \langle Q_1, e_{Q_0} \rangle$ be the operator obtained from averaging e_{Q_0} over Q_1. Then

$$h \in Z(Q_0) = Q_1' \cap \langle Q_1, e_{Q_0} \rangle = Z(Q_1), \tag{10.2.2}$$

and the spectrum of h lies in the set

$$S = \{(4\cos^2(\pi/n))^{-1} : n \geq 3\} \cup [0, 1/4]. \tag{10.2.3}$$

In particular, the spectrum of h lies in $\{1\} \cup [0, 1/2]$, and the spectral projection q_1 corresponding to $\{1\}$ is the largest central projection for which $Q_0q_1 = Q_1q_1$.

Proof. Let Q_2 denote $\langle Q_1, e_{Q_0} \rangle$. Then $Q_2' = JQ_0J$, so $Q_1' \cap Q_2 = JQ_1J \cap (JQ_0J)'$, which is $J(Q_0' \cap Q_1)J = J(Z(Q_1))J = Z(Q_1)$. In addition, $Z(Q_2) = Z(Q_2') = Z(JQ_0J) = Z(Q_1)$, so the algebras $Z(Q_0), Z(Q_1), Z(Q_2), Q_0' \cap Q_1$

and $Q_1' \cap Q_2$ coincide under these hypotheses. Since $h \in Q_1' \cap Q_2$, by Lemma 9.2.2 (ii), we have thus established (10.2.2).

The set S consists of an interval and a decreasing sequence of points. If the spectrum of h is not contained in S, then we may find a closed interval $[a,b] \subseteq S^c \cap (1/4,1)$ so that the corresponding spectral projection z of h is non-zero and also lies in $Z(Q_1)$. By cutting the algebras by z, we may assume that $a1 \leq h \leq b1$ and $[a,b] \cap S = \emptyset$ so $a \geq 1/4$. The trace Tr on Q_2 coming from the basic construction satisfies $Tr(1) \leq a^{-1}Tr(h) = a^{-1}Tr(e_{Q_0}) \leq 4$ and is thus finite. Let $\mathbb{T}_Z$ denote the centre-valued trace on Q_2, whose restrictions to Q_0 and Q_1 are also the centre-valued traces on these subalgebras. Then $\mathbb{T}_Z(e_{Q_0}) \geq a1$. If Q_2 has a central summand of type I_n with corresponding central projection p, then cutting by p gives containment of type I_n algebras with equal centres and are thus equal to each other. This would show that $hp = p$ and 1 would lie in the spectrum of h, contrary to assumption. Thus Q_2 is a type II_1 von Neumann algebra. Then 1 may be expressed as a sum of four equivalent projections $\{p_i\}_{i=1}^4$ each having central trace $(4^{-1})1$. Thus each p_i is equivalent to a subprojection of e_{Q_0} and so there exist partial isometries $v_i \in Q_2$ such that $1 = \sum_{i=1}^4 v_i e_{Q_0} v_i^*$. Since $e_{Q_0}Q_1 = e_{Q_0}Q_2$, we may replace each v_i by an operator $w_i \in Q_1$, yielding $1 = \sum_{i=1}^4 w_i e_{Q_0} w_i^*$. For each $x \in Q_1$, multiply on the right by $x e_{Q_0}$ to obtain

$$x e_{Q_0} = \sum_{i=1}^4 w_i \mathbb{E}_{Q_0}(w_i^* x) e_{Q_0}, \quad x \in Q_1, \tag{10.2.4}$$

so $x = \sum_{i=1}^4 w_i \mathbb{E}_{Q_0}(w_i^* x)$, and Q_1 is a finitely generated right Q_0-module. In a similar fashion $Q_2 = \sum_{i=1}^4 w_i e_{Q_0} Q_2 = \sum_{i=1}^4 w_i e_{Q_0} Q_1$, and so Q_2 is finitely generated over Q_1. This is a standard argument in subfactor theory (see [133]) which we include for the reader's convenience.

Let Ω be the spectrum of $Z(Q_2)$, and fix $\omega \in \Omega$. Then

$$I_2 = \{x \in Q_2 : \mathbb{T}_Z(x^*x)(\omega) = 0\} \tag{10.2.5}$$

is a maximal norm closed ideal in Q_2 and Q_2/I_2 is a type II_1 factor, denoted M_2, with trace $\tau_\omega = \omega \circ \mathbb{T}_Z$, by Theorem A.3.1. Similar constructions yield maximal ideals $I_k \subseteq Q_k$, $k = 0, 1$, and factors $M_k = Q_k/I_k$. Equality of the centres gives $Q_k \cap I_2 = I_k$ for $k = 0, 1$, and so $M_0 \subseteq M_1 \subseteq M_2$ is an inclusion of factors. Let $\pi \colon Q_2 \to M_2$ denote the quotient map, and let $e = \pi(e_{Q_0})$. From above we note that M_2 is a finitely generated M_1-module.

Consider $x \in I_1$. By uniqueness of the centre-valued trace, the composition of $\mathbb{E}_{Q_0}$ with the restriction of $\mathbb{T}_Z$ to Q_0 is $\mathbb{T}_Z$. Thus

$$\mathbb{T}_Z(\mathbb{E}_{Q_0}(x^*x)))(\omega) = \mathbb{T}_Z(x^*x)(\omega), \quad x \in Q_1. \tag{10.2.6}$$

Conditional expectations are completely positive unital maps [193, 194, 195] and so the inequality (3.6.9) gives

$$\mathbb{E}_{Q_0}(x^*)\mathbb{E}_{Q_0}(x) \leq \mathbb{E}_{Q_0}(x^*x), \quad x \in Q_1, \tag{10.2.7}$$

showing that $\mathbb{E}_{Q_0}$ maps I_1 to I_0. Thus there is a well defined τ_ω-preserving conditional expectation $\mathbb{E}\colon M_1 \to M_0$ given by $\mathbb{E}(x + I_1) = \mathbb{E}_{Q_0}(x) + I_0$ for $x \in Q_1$. From above, e commutes with M_0 and M_2 is generated by M_1 and e. Moreover, $exe = \mathbb{E}(x)e$ for $x \in M_1$ by applying π to the equation $e_{Q_0} x e_{Q_0} = \mathbb{E}_{Q_0}(x)e_{Q_0}$ for $x \in Q_1$. Thus M_2 is the extension of M_1 by M_0 with Jones projection e, by Theorem 4.3.15. Now $\mathbb{T}_Z(e_{Q_0})(\omega) = h(\omega) \in [a, b]$, so $\tau_\omega(e) = h(\omega)$ while $\tau_\omega(1) = 1$. It follows that $[M_1\colon M_0]^{-1} \in [a, b]$, contradicting the theorem of Jones, Theorem 10.2.1, on the possible values of the index.

Now let q_1 be the spectral projection of h corresponding to $\{1\}$. If we cut by q_1 then we may assume that $h = 1$. But then $e_{Q_0} = 1$ and $Q_0 = Q_1$. We conclude that $Q_0 q_1 = Q_1 q_1$. On the other hand, let $z \in Z(Q_1)$ be a projection such that $Q_0 z = Q_1 z$. Then $e_{Q_0} z = z$, so $hz = z$, showing that $z \leq q_1$. Thus q_1 is the largest central projection with the stated property. $\qquad\square$

10.3 Containment of finite algebras

In this section we consider an inclusion $M \subseteq N$ of finite von Neumann algebras where N has a faithful normal unital trace τ. Recall that $N \subset_\delta M$ means that $\|I - \mathbb{E}_M\|_{\infty,2} \leq \delta$, where we are interested in small positive values of δ. Our objective is to show that, by cutting the algebras by a suitable projection p in the centre of the relative commutant $M' \cap N$ of large trace dependent on δ, we may arrive at $Mp = pNp$. This is achieved in the following lemmas which are independent of one another. However, we have chosen the notation so that they may be applied sequentially to our original inclusion $M \subseteq N$. The definition of $N \subset_\delta M$ depends implicitly on the trace τ assigned to N. Since we will be rescaling traces at various points, we will make this explicit by adopting the notation $N \subset_{\delta,\tau} M$. If $\tau_1 = \lambda\tau$ for some $\lambda > 0$, then

$$\|x\|_{2,\tau_1} = \sqrt{\lambda}\|x\|_{2,\tau}, \qquad x \in N. \tag{10.3.1}$$

Consequently $N \subset_{\delta,\tau} M$ becomes $N \subset_{\sqrt{\lambda}\,\delta,\tau_1} M$ for this change of trace.

It is worth noting that the beginning of the proof of the next lemma shows that if $M_0 \subseteq N_0$, $N_0 \subset_{\delta_0,\tau_0} M_0$ for a von Neumann algebra N_0 with a faithful normal unital trace τ_0, then $M_0' \cap N_0 \subset_{\delta_0,\tau_0} Z(M_0)$.

Lemma 10.3.1. *Let $\delta_1 \in (0, 1)$ and consider an inclusion $M_1 \subseteq N_1$, where N_1 has a unital faithful normal trace τ_1 relative to which $N_1 \subset_{\delta_1,\tau_1} M_1$. Then there exists a projection $p_1 \in Z(M_1' \cap N_1)$ such that $\tau_1(p_1) \geq 1 - \delta_1^2$ and $(M_1 p_1)' \cap p_1 N_1 p_1$ is abelian.*

Setting $M_2 = M_1 p_1$, $N_2 = p_1 N_1 p_1$, and $\tau_2 = \tau_1(p_1)^{-1}\tau_1$, we have $M_2' \cap N_2$ is abelian and $N_2 \subset_{\delta_2,\tau_2} M_2$, where δ_2 is defined by $\delta_2^2 = \delta_1^2(1 - \delta_1^2)^{-1}$.

Proof. Let $C = M_1' \cap N_1$, which contains $Z(M_1)$. If $c \in C$, $\|c\| \leq 1$, we may choose $m \in M_1$ to satisfy $\|c - m\|_{2,\tau_1} \leq \delta_1$. Conjugation by unitaries from M_1 leaves c invariant, so Dixmier's approximation theorem, Theorem 2.2.2, shows that there is an element $z \in Z(M_1)$ such that $\|c - z\|_{2,\tau_1} \leq \delta_1$. Thus

$C \subset_{\delta_1} Z(M_1)$. Let A be a maximal abelian subalgebra of C which contains $Z(M_1)$, and note that $Z(C) \subseteq A$. Choose a projection $p_1 \in Z(C)$, maximal with respect to the property that Cp_1 is abelian. We now construct a unitary $u \in C(1 - p_1)$ such that $\mathbb{E}_{A(1-p_1)}(u) = 0$.

The algebra $C(1 - p_1)$ may be decomposed as a direct sum

$$C(1 - p_1) = \bigoplus_{k \geq 0} C_k, \qquad (10.3.2)$$

where C_0 is type II_1 and each C_k for $k \geq 1$ has the form $\mathbb{M}_{n_k} \otimes A_k$ for an abelian subalgebra A_k of A. Let q_k, $k \geq 0$, be the identity element of C_k. Then Aq_k, $k \geq 1$, contains A_k and is maximal abelian in C_k. By Theorem 2.4.3, Aq_k is, up to unitary equivalence, $\mathbb{D}_k \otimes A_k$ for the diagonal subalgebra $\mathbb{D}_k \subseteq \mathbb{M}_{n_k}$. Note that the choice of p_1 implies $n_k \geq 2$. For $k \geq 1$, let u_k be a unitary in $\mathbb{M}_{n_k}$ ($\cong \mathbb{M}_{n_k} \otimes 1$) which cyclically permutes the basis for $\mathbb{D}_k$. For $k = 0$, choose two equivalent orthogonal projections in Aq_0 which sum to q_0, let $v \in C_0$ be an implementing partial isometry, and let $u_0 = v + v^*$. Then $u = \sum_{k=0}^{\infty} u_k$ is a unitary in $C(1 - p_1)$ for which $\mathbb{E}_{A(1-p_1)}(u) = 0$. Thus $w = p_1 + u$ is a unitary in C. Then

$$\delta_1 \geq \|w - \mathbb{E}_{Z(M_1)}(w)\|_{2,\tau_1} \geq \|w - \mathbb{E}_A(w)\|_{2,\tau_1}$$
$$= \|u - \mathbb{E}_{A(1-p_1)}(u)\|_{2,\tau_1} = \|u\|_{2,\tau_1} = \|1 - p_1\|_{2,\tau_1}, \qquad (10.3.3)$$

and the inequality $\tau_1(p_1) \geq 1 - \delta_1^2$ follows.

Now let $M_2 = M_1 p_1$, $N_2 = p_1 N_1 p_1$ and $\tau_2 = \tau_1(p_1)^{-1}\tau_1$. Then $M_2' \cap N_2 = Cp_1$, which is abelian, and $N_2 \subset_{\delta_2,\tau_2} M_2$, where $\delta_2 = \delta_1(1 - \delta_1^2)^{-1/2}$.   □

Lemma 10.3.2. *Let* $\delta_2 \in (0, 2^{-1})$ *and consider an inclusion* $M_2 \subseteq N_2$, *where* N_2 *has a unital faithful normal trace* τ_2 *relative to which* $N_2 \subset_{\delta_2,\tau_2} M_2$. *Further suppose that* $M_2' \cap N_2$ *is abelian. Then there exists a projection* $p_2 \in M_2' \cap N_2$ *such that* $\tau_2(p_2) \geq 1 - 4\delta_2^2$ *and* $(M_2 p_2)' \cap (p_2 N_2 p_2) = Z(M_2 p_2)$. *In particular, when* N_2 *is abelian we have* $N_2 p_2 = M_2 p_2$.

Setting $M_3 = M_2 p_2$, $N_3 = p_2 N_2 p_2$ *and* $\tau_3 = \tau_2(p_2)^{-1}\tau_2$, *we have* $M_3' \cap N_3 = Z(M_3)$ *and* $N_3 \subset_{\delta_3,\tau_3} M_3$, *where* δ_3 *is defined by* $\delta_3^2 = \delta_2^2(1 - 4\delta_2^2)^{-1}$.

Proof. Let $A = Z(M_2)$ and let $C = M_2' \cap N_2$, which is abelian by hypothesis. It is easy to see that $C \subset_{\delta_2,\tau_2} A$, by applying Dixmier's approximation theorem, Theorem 2.2.2. Then A and C are abelian, and $\|c - \mathbb{E}_A(c)\|_{2,\tau} \leq \delta_2$ whenever $\|c\| \leq 1$. Thus the hypotheses of Theorem 4.4.4 are met and the result follows.   □

Lemma 10.3.3. *Let* $\delta_3 \in (0, 4^{-1})$ *and consider an inclusion* $M_3 \subseteq N_3$, *where* N_3 *has a unital faithful normal trace* τ_3 *relative to which* $N_3 \subset_{\delta_3,\tau_3} M_3$. *Further suppose that* $Z(M_3) = M_3' \cap N_3$. *Then there exists a projection* $p_3 \in Z(M_3)$ *such that* $\tau_3(p_3) \geq 1 - 16\delta_3^2$ *and*

$$Z(M_3 p_3) = (M_3 p_3)' \cap (p_3 N_3 p_3) = Z(p_3 N_3 p_3). \qquad (10.3.4)$$

Setting $M_4 = M_3 p_3$, $N_4 = p_3 N_3 p_3$ and $\tau_4 = \tau_3(p_3)^{-1}\tau_3$, we have

$$Z(M_4) = M_4' \cap N_4 = Z(N_4) \tag{10.3.5}$$

and $N_4 \subset_{\delta_4,\tau_4} M_4$, where δ_4 is defined by $\delta_4^2 = \delta_3^2(1 - 16\delta_3^2)^{-1}$.

Proof. Since $Z(N_3) \subseteq M_3' \cap N_3$ we have, by hypothesis, that $Z(N_3) \subseteq Z(M_3)$. If $x \in Z(M_3)$, $\|x\| \leq 1$, and u is a unitary in N_3 then choose $m \in M_3$ such that $\|u - m\|_{2,\tau_3} \leq \delta_3$. It follows that

$$\|ux - xu\|_{2,\tau_3} = \|(u - m)x - x(u - m)\|_{2,\tau_3} \leq 2\delta_3, \tag{10.3.6}$$

and so $\|uxu^* - x\|_{2,\tau_3} \leq 2\delta_3$. Suitable convex combinations of terms of the form uxu^* converge in norm to an element of $Z(N_3)$, showing that $Z(M_3) \subset_{2\delta_3,\tau_3} Z(N_3)$. Now apply Lemma 10.3.2 to the inclusion $Z(N_3) \subseteq Z(M_3)$, taking $\delta_2 = 2\delta_3$. We conclude that there is a projection $p_3 \in Z(M_3)$ such that $\tau_3(p_3) \geq 1 - 16\delta_3^2$ and $Z(N_3)p_3 = Z(M_3)p_3$.

Now let $M_4 = M_3 p_3$, $N_4 = N_3 p_3$ and $\tau_4 = \tau_3(p_3)^{-1}\tau_3$. Then (10.3.5) is satisfied and $N_4 \subset_{\delta_4,\tau_4} M_4$, where $\delta_4 = \delta_3(1 - 16\delta_3^2)^{-1/2}$. $\qquad\square$

Lemma 10.3.4. *Let $\delta_4 \in (0, 2^{-1/2})$ and consider an inclusion $M_4 \subseteq N_4$, where N_4 has a unital faithful normal trace τ_4 relative to which $N_4 \subset_{\delta_4,\tau_4} M_4$. Further suppose that*

$$Z(M_4) = M_4' \cap N_4 = Z(N_4). \tag{10.3.7}$$

Then there exists a projection $p_4 \in Z(M_4)$ such that $\tau_4(p_4) \geq 1 - 2\delta_4^2$ and $M_4 p_4 = N_4 p_4$.

Proof. Consider the basic construction $M_4 \subseteq N_4 \subseteq \langle N_4, e_{M_4} \rangle$ with associated projection e_{M_4}, and let $h \in N_4' \cap \langle N_4, e_{M_4} \rangle$ be the operator obtained from e_{M_4} by averaging over the unitary group of N_4. By hypothesis, the conditions of Theorem 10.2.3 are met, and so $h \in Z(N_4)$ and has spectrum contained in $\{1\} \cup [0, 2^{-1}]$. Let $q \in Z(N_4) = Z(M_4)$ be the spectral projection of h for the eigenvalue 1, and note that $h(1 - q) \leq (1 - q)/2$. Fix an arbitrary $\varepsilon > 0$ and suppose that

$$Tr(e_{M_4} u(e_{M_4}(1 - q))u^*) \geq (2^{-1} + \varepsilon)Tr(e_{M_4}(1 - q)) \tag{10.3.8}$$

for all unitaries $u \in M_4$. Taking the average leads to

$$Tr(e_{M_4} h(1 - q)) \geq (2^{-1} + \varepsilon)Tr(e_{M_4}(1 - q)), \tag{10.3.9}$$

and so $\tau_4(h(1-q)) \geq (2^{-1}+\varepsilon)\tau_4(1-q)$. If $q = 1$ then we already have $N_4 = M_4$; otherwise the last inequality gives a contradiction and so (10.3.8) fails for every $\varepsilon > 0$. The presence of $(1 - q)$ in (10.3.8) ensures that this inequality fails for a unitary $u_\varepsilon \in N_4(1 - q)$. Thus

$$Tr(e_{M_4} u_\varepsilon(e_{M_4}(1 - q))u_\varepsilon^*) < (2^{-1} + \varepsilon)Tr(e_{M_4}(1 - q)) \tag{10.3.10}$$

for each $\varepsilon > 0$. Define a unitary in N_4 by $v_\varepsilon = q + u_\varepsilon$. By hypothesis,

$$
\begin{aligned}
\delta_4^2 &\geq \|q + u_\varepsilon - \mathbb{E}_{M_4}(q + u_\varepsilon)\|_{2,\tau_4}^2 = \|(I - \mathbb{E}_{M_4})(u_\varepsilon)\|_{2,\tau_4}^2 \\
&= \|u_\varepsilon\|_{2,\tau_4}^2 - \|\mathbb{E}_{M_4}(u_\varepsilon)\|_{2,\tau_4}^2 = \tau_4(1-q) - \tau_4(\mathbb{E}_{M_4}(u_\varepsilon)u_\varepsilon^*) \\
&= \tau_4(1-q) - Tr(e_{M_4}u_\varepsilon e_{M_4}(1-q)u_\varepsilon^*) \\
&\geq \tau_4(1-q) - (2^{-1} + \varepsilon)\tau_4(1-q), \qquad\qquad (10.3.11)
\end{aligned}
$$

where we have used (10.3.10) and the fact that $q \in Z(M_4) = Z(\langle N_4, e_{M_4} \rangle)$. Letting $\varepsilon \to 0$ in (10.3.11), we obtain $\tau_4(q) \geq 1 - 2\delta_4^2$.

Define $p_4 = q \in Z(M_4)$. The basic construction for $M_4 p_4 \subseteq N_4 p_4$ is obtained from the basic construction for $M_4 \subseteq N_4$ by cutting by the central projection q. Since $hq = q$, it follows that $N_4 p_4 = M_4 p_4$, completing the proof. $\qquad\square$

We now summarise these lemmas.

Theorem 10.3.5. *Let N be a von Neumann algebra with a unital faithful normal trace τ, let M be a von Neumann subalgebra, and let δ be a positive number in the interval $(0, (23)^{-1/2})$. If $N \subset_{\delta,\tau} M$, then there exists a projection $p \in Z(M' \cap N)$ such that $\tau(p) \geq 1 - 23\delta^2$ and $Mp = pNp$.*

Proof. We apply the previous four lemmas successively to cut by projections until the desired conclusion is reached. Each projection has trace at least a fixed proportion of the trace of the previous one, so the estimates in these lemmas combine to give

$$
\tau(p) \geq (1 - \delta_1^2)(1 - 4\delta_2^2)(1 - 16\delta_3^2)(1 - 2\delta_4^2), \qquad\qquad (10.3.12)
$$

where the δ_i's satisfy the relations

$$
\delta_1^2 = \delta^2, \quad \delta_2^2 = \frac{\delta_1^2}{1 - \delta_1^2}, \quad \delta_3^2 = \frac{\delta_2^2}{1 - 4\delta_2^2}, \quad \delta_4^2 = \frac{\delta_3^2}{1 - 16\delta_3^2}. \qquad (10.3.13)
$$

Substitution of (10.3.13) into (10.3.12) gives $\tau(p) \geq 1 - 23\delta^2$. $\qquad\square$

Remark 10.3.6. The assumption that $\delta < (23)^{-1/2}$ in Theorem 10.3.5 guarantees that the δ_i's in the lemmas fall in the correct ranges. This theorem is still true, but vacuous, for $\delta \geq (23)^{-1/2}$. The constant 23 can be improved under additional hypotheses by joining the sequence of lemmas at a later point. If the inclusion $M \subseteq N$, $N \subset_\delta M$ also satisfies the hypotheses of Lemmas 10.3.2, 10.3.3 or 10.3.4 then 23 can be replaced respectively by 22, 18 or 2. $\qquad\square$

For the case when the larger algebra is a factor, the estimate in Theorem 10.3.5 can be considerably improved.

Theorem 10.3.7. *Let N be a type II$_1$ factor with a unital faithful normal trace τ, let M be a von Neumann subalgebra, and let δ be a positive number in the interval $(0, (2/5)^{1/2})$. If $N \subset_{\delta,\tau} M$, then there exists a projection $p \in M' \cap N$ with $\tau(p) \geq 1 - \delta^2/2$ such that $Mp = pNp$.*

Proof. Let $q \in M' \cap N$ be a projection with $\tau(q) \geq 1/2$. Then $1 - q$ is equivalent in N to a projection $e \leq q$. Let $v \in N$ be a partial isometry such that $vv^* = e$, $v^*v = 1 - q$. Let $w = v + v^* \in N$ and note that $\|w\| = 1$ and $\mathbb{E}_M(w) = 0$. Then

$$\|w - \mathbb{E}_M(w)\|_{2,\tau}^2 = \|v + v^*\|_2^2 = 2\tau(1 - q), \qquad (10.3.14)$$

so we must have $2\tau(1 - q) \leq \delta^2$, or $\tau(q) \geq 1 - \delta^2/2$. On the other hand, if $\tau(q) \leq 1/2$ then this argument applies to $1 - q$, giving $\tau(q) \leq \delta^2/2$. Thus the range of the trace on projections in $M' \cap N$ is contained in $[0, \delta^2/2] \cup [1 - \delta^2/2, 1]$.

By Zorn's lemma and the normality of the trace, there is a projection $p \in M' \cap N$ which is minimal with respect to the property of having trace at least $1 - \delta^2/2$. We now show that p is a minimal projection in $M' \cap N$. If not, then p can be written $p_1 + p_2$ with $\tau(p_1)$, $\tau(p_2) > 0$. By choice of p, we see that $\tau(p_1)$, $\tau(p_2) \leq \delta^2/2$. It follows that

$$1 - \delta^2/2 \leq \tau(p) = \tau(p_1) + \tau(p_2) \leq \delta^2, \qquad (10.3.15)$$

which contradicts $\delta^2 < 2/5$. Thus p is minimal in $M' \cap N$, so Mp has trivial relative commutant in pNp, since any nontrivial projection in $(Mp)' \cap pNp$ would lie strictly below p in $M' \cap N$. Let τ_1 be the normalised trace $\tau(p)^{-1}\tau$ on pNp. Then $pNp \subset_{\delta_1, \tau_1} Mp$, where $\delta_1 = \delta(1 - \delta^2/2)^{-1/2} < 2^{-1/2}$.

We have now reached the situation of a subfactor inclusion $P \subseteq Q$, $Q \subset_\delta P$ for a fixed $\delta < 2^{-1/2}$ and $P' \cap Q = \mathbb{C}1$. Since

$$Q' \cap \langle Q, e_P \rangle = J(P' \cap Q)J = \mathbb{C}1, \qquad (10.3.16)$$

the operator h obtained from averaging e_P over unitaries in Q is $\lambda 1$ for some $\lambda > 0$. By Lemma 9.2.2 (iii) and (iv), we have $1 - \lambda \leq \delta^2 < 1/2$ and $\lambda = \lambda^2 Tr(1)$, yielding $Tr(1) = 1/\lambda < 2$. Thus $[Q : P] < 2$, so $Q = P$ from Remark 10.2.2. Applying this to $Mp \subseteq pNp$, we conclude equality as desired. $\qquad \square$

10.4 Close von Neumann algebras

This section contains the main result of the chapter, dealing with the relationship between close subalgebras of a type II_1 factor. The idea is to cut the algebras by large projections so that one cutdown is contained in and close to the other. We can then invoke the results of the previous section. The details are contained in Theorem 10.4.1 below. The last step of the proof is handled by the results on commutants of compressions from Theorem 5.4.1.

Theorem 10.4.1. *Let $\delta > 0$, let B_0 and B be von Neumann subalgebras of a type II_1 factor N with unital faithful normal trace τ, and suppose that $\|\mathbb{E}_B - \mathbb{E}_{B_0}\|_{\infty,2} \leq \delta$. Then there exist projections $q_0 \in B_0$, $q \in B$, $q_0' \in B_0' \cap N$, $q' \in B' \cap N$, $p_0' = q_0 q_0'$, $p' = q q'$, and a partial isometry $v \in N$ such that $vp_0' B_0 p_0' v^* = p' B p'$, $vv^* = p'$, $v^*v = p_0'$. Moreover, v can be chosen to satisfy $\|1 - v\|_{2,\tau} \leq 69\delta$, $\|1 - p'\|_{2,\tau} \leq 35\delta$ and $\|1 - p_0'\|_{2,\tau} \leq 35\delta$.*

Under the additional hypothesis that the relative commutants of B_0 and B are respectively their centres, the projections may be chosen so that $p_0' \in B_0$ and $p' \in B$.

Proof. We assume that $\delta < (35)^{-1}$, otherwise we may take $v = 0$. Let e_B be the Jones projection for the basic construction $B \subseteq N \subseteq \langle N, e_B \rangle$ and let $h \in B_0' \cap \langle N, e_B \rangle$ be the operator obtained from averaging e_B over the unitary group of B_0 (see Lemma 9.2.2). If we denote by e the spectral projection of h for the interval $[1/2, 1]$, then $e \in B_0' \cap \langle N, e_B \rangle$ and $\|e_B - e\|_{2,Tr} \leq 2\delta$, by Lemma 5.3.1 (i). Then

$$eB_0 = eB_0 e \subseteq e\langle N, e_B \rangle e. \tag{10.4.1}$$

Consider $x \in e\langle N, e_B \rangle e$ with $\|x\| \leq 1$. Since $e_B \langle N, e_B \rangle e_B = Be_B$, there exists $b \in B$, $\|b\| \leq 1$, such that $e_B x e_B = b e_B$. Then

$$\|x - \mathbb{E}_{B_0}(b)e\|_{2,Tr}^2 = \|(x - \mathbb{E}_{B_0}(b))e\|_{2,Tr}^2$$
$$= \|(x - \mathbb{E}_{B_0}(b))ee_B\|_{2,Tr}^2 + \|(x - \mathbb{E}_{B_0}(b))e(1 - e_B)\|_{2,Tr}^2 \tag{10.4.2}$$

and we estimate these terms separately. For the first, we have

$$\|(x - \mathbb{E}_{B_0}(b))ee_B\|_{2,Tr} = \|e(x - \mathbb{E}_{B_0}(b))e_B\|_{2,Tr}$$
$$\leq \|e(x - \mathbb{E}_B(b))e_B\|_{2,Tr} + \|e(\mathbb{E}_B(b) - \mathbb{E}_{B_0}(b))e_B\|_{2,Tr}$$
$$\leq \|e(x - e_B x e_B)e_B\|_{2,Tr} + \delta$$
$$= \|e(e - e_B)(xe_B)\|_{2,Tr} + \delta$$
$$\leq \|e - e_B\|_{2,Tr} + \delta \leq 3\delta. \tag{10.4.3}$$

For the second term in (10.4.2), we have

$$\|(x - \mathbb{E}_{B_0}(b))e(1 - e_B)\|_{2,Tr}^2 = \|(x - \mathbb{E}_{B_0}(b))e(e - e_B)\|_{2,Tr}^2$$
$$\leq \|x - \mathbb{E}_{B_0}(b)\|^2 \|e - e_B\|_{2,Tr}^2$$
$$\leq 16\delta^2. \tag{10.4.4}$$

Substituting (10.4.3) and (10.4.4) into (10.4.2) gives

$$\|x - \mathbb{E}_{B_0}(b)e\|_{2,Tr}^2 \leq 25\delta^2. \tag{10.4.5}$$

Hence $e\langle N, e_B \rangle e \subset_{5\delta,Tr} B_0 e$. Since $\|e - e_B\|_{2,Tr}^2 \leq 4\delta^2$, it follows that

$$1 + 4\delta^2 \geq Tr(e) \geq 1 - 4\delta^2. \tag{10.4.6}$$

If we now define a unital trace on $e\langle N, e_B \rangle e$ by $\tau_1 = Tr(e)^{-1} Tr$, then $e\langle N, e_B \rangle e \subset_{\varepsilon, \tau_1} B_0 e$ where $\varepsilon = 5\delta(1 - 4\delta^2)^{-1/2}$. By Theorem 10.3.5, there exists a projection $f \in (B_0 e)' \cap e\langle N, e_B \rangle e$ with $\tau_1(f) \geq 1 - 23\varepsilon^2$ such that $B_0 f = f\langle N, e_B \rangle f$ (since $ef = f$).

Let $V \in \langle N, e_B \rangle$ be the partial isometry in the polar decomposition of $e_B f$, so that $e_B f = (e_B f e_B)^{1/2} V$. The inequality $\|V - e_B\|_{2,Tr} \leq \sqrt{2} \, \|e_B - f\|_{2,Tr}$

is obtained from Lemma 5.2.9 (ii), and we estimate this last quantity. We have

$$
\begin{aligned}
\|e_B - f\|_{2,Tr}^2 &= Tr(e_B + f - 2e_B f) \\
&= Tr(e_B + e - 2e_B e - (e - f) + 2e_B(e - f)) \\
&= \|e_B - e\|_{2,Tr}^2 + Tr(2e_B(e - f) - (e - f)) \\
&\leq \|e_B - e\|_{2,Tr}^2 + Tr(e - f) \leq 4\delta^2 + 23\varepsilon^2 Tr(e) \\
&\leq 4\delta^2 + (23)(25)\delta^2(1 + 4\delta^2)/(1 - 4\delta^2)
\end{aligned}
\tag{10.4.7}
$$

so $\|V - e_B\|_{2,Tr} \leq \sqrt{2}\,\delta_1$ where δ_1^2 is the last quantity above. Then $VV^* \in e_B\langle N, e_B\rangle e_B = Be_B$, and $V^*V \in f\langle N, e_B\rangle f = B_0 f$, by the choice of f. Thus there exist projections $p_0 \in B_0$, $p \in B$ so that

$$
V^*V = p_0 f, \qquad VV^* = pe_B. \tag{10.4.8}
$$

If $z \in Z(B_0)$ is the central projection corresponding to the kernel of the homomorphism $b_0 \mapsto b_0 f$ on B_0, then by replacing p_0 by $p_0(1 - z)$, we may assume that $p_0 b_0 p_0 = 0$ whenever $p_0 b_0 p_0 f = 0$. Since $p_0(1-z)f = p_0 f$, (10.4.8) remains valid and we note that the following relations (and their adjoints) hold:

$$
V = Vp_0 = Vf = pV = e_B V. \tag{10.4.9}
$$

Define $\Theta \colon p_0 B_0 p_0 \to \langle N, e_B\rangle$ by

$$
\Theta(b_0) = Vb_0 V^*, \quad b_0 \in p_0 B_0 p_0. \tag{10.4.10}
$$

We will show that Θ is a *-isomorphism onto $pBpe_B$. Now $pe_B V = V$ from (10.4.9), so the range of Θ is contained in $pe_B\langle N, e_B\rangle e_B p = pBpe_B$. Since

$$
V^*\Theta(b_0)V = p_0 f b_0 p_0 f, \quad b_0 \in p_0 B_0 p_0, \tag{10.4.11}
$$

from (10.4.8), the choice of p_0 shows that Θ has trivial kernel. The map is clearly self-adjoint, and we check that it is a homomorphism. For the elements $p_0 b_0 p_0$, $p_0 b_1 p_0 \in p_0 B_0 p_0$,

$$
\begin{aligned}
\Theta(p_0 b_0 p_0)\Theta(p_0 b_1 p_0) &= Vp_0 b_0 p_0 V^* Vp_0 b_1 p_0 V^* = Vp_0 b_0 p_0 f p_0 b_1 p_0 V^* \\
&= Vf p_0 b_0 p_0 b_1 p_0 V^* = \Theta(p_0 b_0 p_0 b_1 p_0),
\end{aligned}
\tag{10.4.12}
$$

using $Vf = V$. Finally we show that Θ maps onto $pBpe_B$. Given $b \in B$, let

$$
x = V^* pbpe_B V = p_0 V^* pbpe_B Vp_0 \in p_0 f\langle N, e_B\rangle f p_0 = p_0 B_0 p_0 f. \tag{10.4.13}
$$

Then x has the form $p_0 b_0 p_0 f$ for some $b_0 \in B_0$. Thus

$$
\begin{aligned}
\Theta(p_0 b_0 p_0) &= Vp_0 b_0 p_0 V^* = Vp_0 b_0 p_0 fV^* \\
&= VxV^* = VV^* pbpe_B VV^* \\
&= pe_B pbpe_B pe_B = pbpe_B,
\end{aligned}
\tag{10.4.14}
$$

and this shows surjectivity. Thus the map $\Theta\colon p_0B_0p_0 \to pBpe_B$ is a surjective *-isomorphism, and so can be expressed as $\Theta(p_0b_0p_0) = \theta(p_0b_0p_0)e_B$ where $\theta\colon p_0B_0p_0 \to pBp$ is a surjective *-isomorphism. From the definitions of these maps,

$$Vb_0 = \theta(b_0)V, \qquad b_0 \in p_0B_0p_0. \tag{10.4.15}$$

If we let $A = p_0B_0p_0 \subseteq B_0$, $\phi = \theta$, and $W = V^*$, then the adjoints of (10.4.9) and (10.4.15) are exactly the hypotheses of Lemma 9.4.2, so this lemma gives a partial isometry $w \in N$ such that

$$b_0w = w\theta(b_0), \qquad b_0 \in p_0B_0p_0. \tag{10.4.16}$$

Defining v to be w^* and taking adjoints in (10.4.16) gives

$$vb_0 = \theta(b_0)v, \qquad b_0 \in p_0B_0p_0, \tag{10.4.17}$$

and so vv^* is a projection $p' \in N$ which commutes with $\theta(p_0B_0p_0) = pBp$. By Lemma 9.4.2 (iv) and (10.4.9), $pv = v$, and so $p' = vv^* \in (pBp)' \cap pNp$, and similarly v^*v is a projection $p_0' \in (p_0B_0p_0)' \cap p_0Np_0$. Moreover

$$vB_0v^* = vp_0B_0p_0v^* = pBpp' = p'Bp'. \tag{10.4.18}$$

It remains to estimate $\|1 - v\|_{2,\tau}$. Since

$$\|V - e_B\|_{2,Tr} \leq \sqrt{2}\,\|e_B - f\|_{2,Tr} \leq \sqrt{2}\,\delta_1, \tag{10.4.19}$$

from (10.4.7), we obtain

$$\|1 - v\|_{2,\tau} \leq 2\sqrt{2}\,\delta_1 \tag{10.4.20}$$

from Lemma 9.4.2 (ii). From the definition of δ_1 and the requirement that $\delta < (35)^{-1}$, we see that

$$8\delta_1^2/\delta^2 \leq (69)^2, \tag{10.4.21}$$

by evaluating the term $(1 + 4\delta^2)(1 - 4\delta^2)^{-1}$ at $\delta = 1/35$. The estimate $\|1 - v\|_{2,\tau} \leq 69\delta$ follows. Lemma 9.4.2 (iii) and (10.4.19) give

$$\|1 - p'\|_{2,\tau} \leq \|e_B - V\|_{2,\tau} \leq \sqrt{2}\,\delta_1 \leq (69/2)\delta \leq 35\delta \tag{10.4.22}$$

with a similar estimate for $\|1 - p_0'\|_{2,\tau}$. The fact that each projection is a product of a projection from the algebra and one from the relative commutant follows from Theorem 5.4.1. The last statement of the theorem is an immediate consequence of the first part, because now the relative commutants are contained in the algebras. $\qquad\square$

This theorem is the most general of its kind, and shows that suitable cutdowns of close algebras are spatially isomorphic. Under more restrictive hypotheses the constants can be considerably improved, as the next result for close subfactors shows. The proof follows that of Theorem 10.4.1 so we only point out the numerical differences.

Theorem 10.4.2. *Let $\delta > 0$, let B_0 and B be subfactors of N and suppose that $\|\mathbb{E}_B - \mathbb{E}_{B_0}\|_{\infty,2} \leq \delta$. Then there exist projections $q_0 \in B_0$, $q \in B$, $q_0' \in B_0' \cap N$, $q' \in B' \cap N$, $p_0 = q_0 q_0'$, $p = qq'$, and a partial isometry $v \in N$ such that $vp_0 B_0 p_0 v^* = pBp$, $vv^* = p$, $v^* v = p_0$, and*

$$\|1 - v\|_{2,\tau} \leq 13\delta, \qquad \tau(p) = \tau(p_0) \geq 1 - 67\delta^2. \tag{10.4.23}$$

If, in addition, the relative commutants of B_0 and B are both trivial and $\delta < 67^{-1/2}$, then B and B_0 are unitarily conjugate in N.

Proof. We assume that $\delta < 67^{-1/2}$, otherwise take $v = 0$. The proof is identical to that of Theorem 10.4.1 except that we now have an inclusion $eB_0e \subseteq e\langle N, B\rangle e$ of factors. Our choice of δ allows us a strict upper bound of $(2/5)^{-1/2}$ on the ε which appears immediately after (10.4.6). Thus the estimate of Theorem 10.3.7 applies, which allows us to replace 23 by $1/2$ in (10.4.7). This gives

$$8\delta_1^2/\delta^2 \leq 145 < 169 \tag{10.4.24}$$

and the estimates of (10.4.23) follow.

If the relative commutants are trivial then $p \in B$ and $p_0 \in B_0$, so v implements an isomorphism between pBp and $p_0 B_0 p_0$ which then easily extends to unitary conjugacy between B and B_0. $\qquad\square$

We conclude this chapter by remarking that the projections from the relative commutant in Theorems 10.4.1 and 10.4.2 are essential. Let R be the hyperfinite type II_1 factor, choose a projection $p \in R$ with $\tau(p) = 1 - \delta$, where δ is small, and let θ be an isomorphism of pRp onto $(1 - p)R(1 - p)$. Let $B_0 = \{x + \theta(x) : x \in pRp\}$ and let B have a similar definition but using an isomorphism ϕ such that $\theta^{-1}\phi$ is a properly outer automorphism of pRp. Such an example shows that the projections from the relative commutants in the previous two theorems cannot be avoided.

Chapter 11

Singular masas

11.1 Introduction

In this chapter we prove a theorem (Theorem 11.1.2) for singular masas in a separable II_1 factor showing that they are relatively rigid with respect to perturbations by unitaries, or suitable nilpotent partial isometries. We do this by using a technical rigidity-perturbation result of Popa (Theorem 9.5.1). Conditions on a subgroup H of a discrete I.C.C. group G are given that ensure that $L(H)$ satisfies strong singularity conditions in $L(G)$. We discuss the Laplacian algebra $\{h\}''$ in the free group factor $L(\mathbb{F}_2)$. If a and b denote the generators of $\mathbb{F}_2$, then h is defined to be the self-adjoint element $a + a^{-1} + b + b^{-1} \in L(\mathbb{F}_2)$, and $\{h\}''$ is shown to also satisfy these strong singularity conditions. We begin with a brief discussion of the various metric invariants for singular masas that have played a role in the theory, but which have been superseded by the results in Theorem 11.1.2.

Dixmier [47] introduced singular masas in 1954 and Pukánszky [154] defined his invariant in 1960, which enabled him to show that there was a countable family of pairwise non-conjugate singular masas in the hyperfinite II_1 factor R (see Chapter 7). The next few advances concerning singular masas were due to Popa [139, 141], in that he related singularity to the Pukánszky invariant (see Corollary 7.2.2) and showed that there were sufficiently many singular masas in a separable II_1 factor N to differentiate subfactors M of N with $M' \cap N = \mathbb{C}1$ (see Theorem 12.4.3 and Corollary 12.4.5). Popa showed that singular masas exist by introducing an invariant, called the δ-invariant, for a masa in a II_1 factor (see [139, p. 164] and the proof of Theorem 12.4.3). In the notation of these notes the δ-invariant of A is defined for a masa A in a II_1 factor N by

$$\delta(A) = \inf\{\|(I - \mathbb{E}_A)\mathbb{E}_{vAv^*}\|_{\infty,2}\|v\|_2^{-1} : v \text{ a partial isometry in } N$$
$$\text{with } vv^* \text{ and } v^*v \text{ orthogonal projections in } A\}. \quad (11.1.1)$$

Here

$$\|(I - \mathbb{E}_A)\mathbb{E}_{vAv^*}\|_{\infty,2} = \sup\{\|(I - \mathbb{E}_A)(vxv^*)\|_2 : x \in A, \|x\| \leq 1\}. \quad (11.1.2)$$

Popa [139] showed that a masa A is singular in N if, and only if, $\delta(A) > 0$. Subsequently, he proved in [145] that this potentially continuous invariant for singular masas is actually discrete with $\delta(A) = 0$ or 1. Moreover $\delta(A) = 1$ if, and only if, A is singular in N. His introduction of the δ-invariant was one of the motivations for the introduction of strong singularity, and the α-invariant, by the authors in [174]. White, Wiggins and the authors showed in [177] that the α-invariant takes only the values 0 and 1 like the δ-invariant. Related to these invariants were various ideas of the conditional expectation $\mathbb{E}_A$ being 'an asymptotic homomorphism' with respect to some sequence of unitaries in A (see [160, 161, 174, 175]). In [174], a masa A in a II_1 factor N was defined to have the *asymptotic homomorphism property* (AHP) if there exists a unitary $v \in A$ such that

$$\lim_{|n| \to \infty} \|\mathbb{E}_A(xv^n y) - \mathbb{E}_A(x)v^n \mathbb{E}_A(y)\|_2 = 0 \qquad (11.1.3)$$

for all $x, y \in N$. In that paper it was shown that strong singularity is a consequence of this property. Subsequently it was observed in [161, Lemma 2.1] that a weaker property, which we will call the *weak asymptotic homomorphism property*, $(WAHP)$, suffices to imply strong singularity.

Motivated by ergodic actions of groups, Jolissaint and Stalder [92] introduced the condition that an abelian von Neumann algebra is *strongly mixing*, which is related to the asymptotic homomorphism condition. They prove that there are singular masas that are not strongly mixing, and hence do not have the asymptotic homomorphism property with respect to any Haar unitary (see their paper for related conditions, results and discussion [92]).

In 1965, Tauer [190] constructed families of masas in the hyperfinite II_1 factor R using matrix methods and a twist argument on half of the diagonal projections to yield classes of masas including singular ones. White developed and extended her results, providing a continuous family of singular masas in R and relating these to centralising sequences and the Pukánszky invariant [208]. The Laplacian masa $B = \{h\}''$ in $L(\mathbb{F}_k)$ $(k \geq 2)$ was shown to be a masa by Pytlik [155] and to be singular by Rădulescu [156]. Rădulescu's proof involves a fine analysis of various words and elements in $L(\mathbb{F}_k)$ to deduce that $(1 - e_B)\mathcal{B}'$ is a type I_∞ von Neumann algebra. He and Boca used similar techniques to study singular masas in a free product of finite groups [8]. The spectrum of h had been studied by Kesten [108], Cohen [32] and with Trenholme [33]. There is further discussion of masas in Chapter 7 on the Pukánszky invariant.

Before giving the main result of this chapter here are the definitions of strong singularity and the weak asymptotic homomorphism property.

Definition 11.1.1. Let A be a masa in a II_1 factor N.

(1) The masa A is *strongly singular* if

$$\|\mathbb{E}_A - \mathbb{E}_{uAu^*}\|_{\infty,2} \geq \|u - \mathbb{E}_A(u)\|_2 \quad \text{for all unitaries} \quad u \in N. \qquad (11.1.4)$$

(2) The masa A has *the weak asymptotic homomorphism property* if, for each $\varepsilon > 0$ and each finite subset $x_1, \ldots, x_n \in N$, there is a unitary $u \in A$ such

that the inequalities

$$\|\mathbb{E}_A(x_i u x_j^*) - \mathbb{E}_A(x_i)u\mathbb{E}_A(x_j^*)\|_2 < \varepsilon \quad \text{for} \quad 1 \le i, j \le n \qquad (11.1.5)$$

hold.   □

Note that inequality (11.1.5) may be replaced by the inequality

$$\|\mathbb{E}_A(x_i u y_j) - \mathbb{E}_A(x_i)u\mathbb{E}_A(y_j)\|_2 < \varepsilon \quad \text{for} \quad 1 \le i \le m,\ 1 \le j \le n \qquad (11.1.6)$$

by changing the set in N to $\{x_1, \ldots, x_m, y_1^*, \ldots, y_n^*\}$. Moreover, in checking the weak asymptotic homomorphism property for a masa A in N, it is sufficient to prove inequality (11.1.6) for the x_i and y_j in a set whose linear span is dense in N in the $\|\cdot\|_2$-norm. This observation is used in the proof of Corollary 11.3.4.

Here is the main result of this chapter. We note that the equivalence of (1) and (2) in the statement of the theorem is just Popa's earlier result that his δ-invariant takes only the value 1 on singular masas [145].

Theorem 11.1.2. *Let A be a masa in a separable II_1 factor N. The following conditions are equivalent:*

(1) *A is singular;*

(2) *for each partial isometry $v \in N$ with vv^* and v^*v orthogonal projections in A,*

$$\sup\{\|(I - \mathbb{E}_A)(vxv^*)\|_2 : x \in A, \|x\| \le 1\} = \tau(vv^*) = \|v\|_2^2; \qquad (11.1.7)$$

(3) *A is strongly singular;*

(4) *A has the weak asymptotic homomorphism property.*

This theorem will follow from various lemmas in the subsequent sections of this chapter, as outlined in the following comments.

The implication (3) $\implies$ (1) of Theorem 11.1.2 follows directly from the definition of strong singularity in Definition 11.1.1. If $u \in \mathcal{N}(A)$, then $uAu^* = A$, and the defining inequality (11.1.4) implies that $u = \mathbb{E}_A(u) \in A$. This implication provided the motivation to introduce strong singularity in [174].

Lemma 11.2.1 is (2) $\implies$ (1), Lemma 11.2.3 is both implications (4) $\implies$ (2) and (4) $\implies$ (3), while Lemma 11.3.2 is (1) $\implies$ (4).

11.2   Basic lemmas

The implications between the various equivalent statements in Theorem 11.1.2 are indicated in the proofs of the following lemmas.

Lemma 11.2.1. *Let A be a masa in a separable II_1 factor N. If*

$$\sup\{\|(I - \mathbb{E}_A)(vxv^*)\|_2 : x \in A, \|x\| \le 1\} = \tau(vv^*) = \|v\|_2^2 \qquad (11.2.1)$$

for each partial isometry $v \in N$ with vv^* and v^*v orthogonal projections in A, then A is singular.

Proof. [(2) $\implies$ (1) in Theorem 11.1.2]

Suppose that there exists a unitary $u \in N \setminus A$ with $uAu^* = A$. By Theorem 3.5.2, there is an increasing sequence of finite dimensional abelian von Neumann subalgebras A_n of A, with each A_n having all of its minimal projections of equal trace and with A the weak closure of $\bigcup_n A_n$. By Lemma 3.6.8

$$\lim_n \|\mathbb{E}_A(u) - \mathbb{E}_{A'_n \cap N}(u)\|_2 = 0. \tag{11.2.2}$$

Since $u \neq \mathbb{E}_A(u)$, there is an A_n having minimal projections $p_1, \cdots, p_k$ such that

$$u - \mathbb{E}_{A'_n \cap N}(u) = \sum_{j=1}^k (up_j - p_j up_j) = \sum_{j=1}^k (1 - p_j) up_j \neq 0 \tag{11.2.3}$$

by Remark 3.6.7 and equality (11.2.2). Thus there is a j with $1 \leq j \leq k$ such that $v = (1 - p_j)up_j \neq 0$. Since $u^* p_j u$ and $up_j u^*$ are in A,

$$vv^* = (1 - p_j)up_j u^*(1 - p_j) = (1 - p_j)up_j u^* \quad \text{and} \tag{11.2.4}$$
$$v^* v = p_j(1 - up_j u^*) \tag{11.2.5}$$

are non-zero orthogonal projections in A. For all $x \in A$, $(I - \mathbb{E}_A)(vxv^*) = 0$ because $vAv^* = (1 - p_j)up_j u^* A = vv^* A$. Hence

$$\sup\{\|(I - \mathbb{E}_A(vxv^*)\|_2 : x \in A, \|x\| \leq 1\} = 0, \tag{11.2.6}$$

and so $v = 0$, by hypothesis. This contradiction proves that no such u can exist, and A is singular. $\square$

The above proof shows that if A is not singular, then $\delta(A) = 0$.

Remark 11.2.2. The following observation is useful in applying Lemma 11.2.3. If A and A_0 are masas in a II_1 factor N and w is a partial isometry in N with $w^* Aw \subseteq A_0$ and $wA_0 w^* \subseteq A$, then $w^* w \in A_0$, $ww^* \in A$ and $w^* Aw = A_0 w^* w$. Let $p = ww^* \in A$ and $q = w^* w \in A_0$. The map $pNp \to qNq : pxp \mapsto w^* pxpw = w^* xw$ is a *-isomorphism from pNp onto qNq sending the masa pA into the masa qA_0. Hence

$$w^* Aw = w^* pAw = A_0 q = A_0 w^* w, \tag{11.2.7}$$

as required. $\square$

Note that separability is not required in the proof of the following lemma and that the non-separable case of this is used in the proof of Corollary 11.3.5.

Lemma 11.2.3. *Let A be a masa in a II_1 factor N. If A has the weak asymptotic homomorphism property, then A is strongly singular, and*

$$\sup\{\|(I - \mathbb{E}_A)(vxv^*)\|_2 : x \in A, \|x\| \leq 1\} = \tau(vv^*) = \|v\|_2^2 \tag{11.2.8}$$

for each partial isometry $v \in N$ with vv^* and $v^* v$ orthogonal projections in A.

Proof. Let $x \in N$ and let $\varepsilon > 0$. By the weak asymptotic homomorphism property there is a unitary w in A such that

$$\|\mathbb{E}_A(xwx^*) - \mathbb{E}_A(x)w\mathbb{E}_A(x^*)\|_2 < \varepsilon. \qquad (11.2.9)$$

We now consider two implications in the equivalences of Theorem 11.1.2.
$(4) \implies (3)$ Let $x = u^*$ with u a unitary in N. Using the two relations

$$\|\mathbb{E}_A(u^*)w\mathbb{E}_A(u)\|_2 \leq \|\mathbb{E}_A(u)\|_2 \leq 1, \quad \|\mathbb{E}_{uAu^*}(w)\|_2 = \|\mathbb{E}_A(u^*wu)\|_2, \quad (11.2.10)$$

it follows from (11.2.9) that

$$\begin{aligned}
\|(I - \mathbb{E}_{uAu^*})\mathbb{E}_A\|_{\infty,2}^2 &\geq \|(I - \mathbb{E}_{uAu^*})(w)\|_2^2 \\
&= 1 - \|\mathbb{E}_A(u^*wu)\|_2^2 \\
&\geq 1 - (\|\mathbb{E}_A(u^*)w\mathbb{E}_A(u)\|_2 + \varepsilon)^2 \\
&\geq 1 - \|\mathbb{E}_A(u)\|_2^2 - 2\varepsilon - \varepsilon^2 \\
&= \|u - \mathbb{E}_A(u)\|_2^2 - 2\varepsilon - \varepsilon^2. \qquad (11.2.11)
\end{aligned}$$

Thus $\|(I - \mathbb{E}_{uAu^*})\mathbb{E}_A\|_{\infty,2} \geq \|u - \mathbb{E}_A(u)\|_2^2$.
$(4) \implies (2)$ Let $x = v$ be a partial isometry in N with v^*v and vv^* orthogonal projections in A. By the A-module property of $\mathbb{E}_A$, we have

$$\mathbb{E}_A(v) = \mathbb{E}_A((vv^*)v(v^*v)) = vv^*\mathbb{E}_A(v)v^*v = 0, \qquad (11.2.12)$$

so that

$$\|\mathbb{E}_A(vwv^*)\|_2 < \|\mathbb{E}_A(v)w\mathbb{E}_A(v^*)\|_2 + \varepsilon = \varepsilon \qquad (11.2.13)$$

by (11.2.9). Thus

$$\begin{aligned}
\|(I - \mathbb{E}_A)(vwv^*)\|_2^2 &> \|vwv^*\|_2^2 - \varepsilon^2 \\
&= \tau(w(v^*v)w^*(v^*v)) - \varepsilon^2 \\
&= \tau((v^*v)v^*v) - \varepsilon^2 \\
&= \|v\|_2^2 - \varepsilon^2 \qquad (11.2.14)
\end{aligned}$$

since $w(v^*v) = (v^*v)w$. This proves (2). $\square$

Lemma 11.2.3 is useful when considering masas in group factors arising from abelian subgroups, a situation already discussed in Section 3.3. Let G be a discrete I.C.C. group with an abelian subgroup H, and recall from Section 3.3 the following condition on the inclusion $H \subseteq G$:

(C2) for each finite subset B of $G \setminus H$, there exists $h \in H$ such that $b_1 h b_2 \notin H$ for all $b_1, b_2 \in B$.

If (C2) is satisfied, then $L(H)$ is a singular masa in $L(G)$, by Lemma 3.3.2. Lemma 11.2.3 gives another way of seeing this. The following is an apparently stronger statement, but we will soon prove that singularity, strong singularity, and the weak asymptotic homomorphism property are all equivalent.

Lemma 11.2.4. *Let G be a discrete I.C.C. group with an abelian subgroup H satisfying condition (C2). Then $L(H)$ has the weak asymptotic homomorphism property and is a strongly singular masa in $L(G)$.*

Proof. We will verify the equivalent formulation (11.1.6) of the weak asymptotic homomorphism property. As noted after that equation, it suffices to consider a total subset of elements in the orthogonal complement of $L(H)$, and we take $S = \{\lambda_g : g \in G \setminus H\}$. Given a finite subset $B \subseteq G \setminus H$, condition (C2) allows us to choose $h \in H$ such that $b_1 h b_2 \notin H$ for all $b_1, b_2 \in B$. But then $\mathbb{E}_{L(H)}(\lambda_{b_1} \lambda_h \lambda_{b_2})$, $\mathbb{E}_{L(H)}(\lambda_{b_1})$, and $\mathbb{E}_{L(H)}(\lambda_{b_2})$ are all equal to 0, and thus (11.1.6) holds for any choice of $\varepsilon > 0$. Thus $L(H)$ has the WAHP, and strong singularity follows from Lemma 11.2.3. $\qquad\square$

11.3 Singular to WAHP

The implication (1) $\implies$ (4) of Theorem 11.1.2 is the deep implication depending on Popa's rigidity-perturbation result of Theorem 9.5.1. The perturbation ideas are turned into the correct form for the application in the next lemma.

Lemma 11.3.1. *Let A and A_0 be masas in a separable II$_1$ factor N and let $e \in A$ and $f \in A_0$ be non-zero projections with the property that no non-zero partial isometry $w \in N$ satisfies the conditions $ww^* \in A$, $ww^* \leq e$, $w^*w \in A_0$, $w^*w \leq f$, and $w^*Aw = A_0 w^*w$. If $\varepsilon > 0$ and $x_1, \ldots, x_k \in N$, then there exists a unitary $u \in A$ such that*

$$\|\mathbb{E}_{A_0}(f x_i e u x_j^* f)\|_2 < \varepsilon \qquad (11.3.1)$$

for $1 \leq i, j \leq k$.

Proof. By Theorem 3.5.2, a masa in a separable II$_1$ factor is isomorphic to $L^\infty[0, 1]$ with the trace corresponding to $\int_0^1 \cdot \, dt$. Hence, in such a masa there are finite sets of orthogonal projections summing to 1 and with pre-assigned traces. Using this observation, choose two sets of orthogonal projections

$$\{e_1, \ldots, e_m\} \subseteq A \text{ and } \{f_1, \ldots, f_n\} \subseteq A_0,$$

each finite set summing to 1, and satisfying the relations

(i) $e_1 = e$, $f_1 = f$;

(ii) $\tau(e_i) = \tau(e)$ for $1 \leq i \leq m - 1$;

(iii) $\tau(f_j) = \tau(f)$ for $1 \leq j \leq n - 1$;

(iv) $\tau(e_m) \leq \tau(e)$, $\tau(f_n) \leq \tau(f)$.

The equations in (i) are required in the final construction of $u \in A$. Again, we use the isomorphisms of A and A_0 with $L^\infty[0, 1]$ (Theorem 3.5.2) to choose two projections $p_m \in A$ and $q_m \in A_0$ with $p_m \leq e$, $q_n \leq f$, $\tau(p_m) = \tau(e_m)$ and $\tau(q_n) = \tau(f_n)$. Then select partial isometries $v_i, w_j \in N$ satisfying the relations

(v) $v_i^* v_i = e_i$, $v_i v_i^* = e$, for $1 \le i \le m - 1$;

(vi) $w_j^* w_j = f_j$, $w_j w_j^* = f$ for $1 \le j \le n - 1$;

(vii) $v_m^* v_m = e_m$, $v_m v_m^* = p_m$;

(viii) $w_n^* w_n = f_n$, $w_n w_n^* = q_n$.

These partial isometries exist because the projections in (i)–(iv) have been chosen with appropriate traces.

Define two subalgebras of N by

$$B = \sum_{i=1}^{m} v_i^* A v_i \quad \text{and} \quad B_0 = \sum_{j=1}^{n} w_j^* A_0 w_j. \tag{11.3.2}$$

These two algebras B and B_0 are masas in N, because A and A_0 are masas in N. We now show that there cannot exist a non-zero partial isometry $v \in N$ satisfying $vv^* \in B$, $v^* v \in B_0$, and $v^* B v = B_0 v^* v$. Suppose that such a partial isometry exists. Then, for a suitable choice of i and j, the element $w = v_i v w_j^*$ is non-zero. This element w is a partial isometry, because $v_i^* v_i$ and vv^* are projections in B, and $w_j^* w_j$ and $v^* v$ are projections in B_0 with both B and B_0 abelian. Hence w is a non-zero partial isometry satisfying $ww^* \le e$, $w^* w \le f$, and

$$w^* A w = w_j v^* v_i^* A v_i v w_j^* \subseteq w_j v^* B v w_j^* = w_j B_0 v^* v w_j^*$$
$$\subseteq w_j B_0 w_j^* = w_j w_j^* A_0 w_j w_j^* = f_1 A_0 f_1 = A_0 f_1 \subseteq A_0. \tag{11.3.3}$$

Interchanging the roles of A and A_0 in (11.3.3) gives $w A_0 w^* \subseteq A e_1 \subseteq A$ in the same way. Then $ww^* \in A$, $w^* w \in A_0$ and $w^* A w = A_0 w^* w$, by Remark 11.2.2. Moreover,

$$ww^* = v_i v w_j^* w_j v^* v_i^* \le v_i v_i^* = e_1 \le e \tag{11.3.4}$$

and

$$w^* w = w_j v^* v_i^* v_i v w_j^* \le w_j w_j^* = f_1 \le f. \tag{11.3.5}$$

This contradicts the hypothesis, and so no such non-zero partial isometry v can exist. This shows that condition (4) of Theorem 9.5.1 fails to hold for the masas B and B_0. Then the equivalent formulation (5) of that theorem must fail for these masas. Thus, given $\varepsilon > 0$ and the elements $y_i = f x_i e$, $1 \le i \le k$, we obtain a unitary $u_1 \in B$ such that

$$\| \mathbb{E}_{B_0}(f x_i e u_1 e x_j^* f) \|_2 < \varepsilon, \quad 1 \le i, j \le k. \tag{11.3.6}$$

By construction $e = e_1 \in B$ and so commutes with u_1. Thus $u = e u_1 + (1 - e)$ defines a unitary in A such that $eu = e u_1 \in A$, since $eB = eA$. We may then replace $e u_1 e = eu$ by eu in (11.3.6). Moreover, $f B_0 = f A_0$, and so (11.3.6) may be rewritten as

$$\| \mathbb{E}_{A_0}(f x_i e u x_j^* f) \|_2 < \varepsilon, \quad 1 \le i, j \le k, \tag{11.3.7}$$

reaching the desired conclusion.   □

The next result is (1) $\implies$ (4) in Theorem 11.1.2.

Lemma 11.3.2. *Let A be a singular masa in a separable type II_1 factor N. If $x_1, \ldots, x_k \in N$ and $\varepsilon > 0$ are given, then there is a unitary $u \in A$ such that*

$$\|\mathbb{E}_A(x_i u x_j^*) - \mathbb{E}_A(x_i) u \mathbb{E}_A(x_j^*)\|_2 < \varepsilon \qquad (11.3.8)$$

for $1 \leq i, j \leq k$.

Proof. [(1) $\implies$ (4) in Theorem 11.1.2]
If $x, y \in N$ and $u \in A$, then

$$\mathbb{E}_A(xuy^*) - \mathbb{E}_A(x)u\mathbb{E}_A(y^*) = \mathbb{E}_A((x - \mathbb{E}_A(x))u\mathbb{E}_A(y - \mathbb{E}_A(y))^*) \qquad (11.3.9)$$

by the module properties of $\mathbb{E}_A$. Thus inequality (11.3.8) follows if we can establish that

$$\|\mathbb{E}_A(x_i u x_j^*)\|_2 < \varepsilon, \quad 1 \leq i, j \leq k, \qquad (11.3.10)$$

when the x_i's also satisfy $\mathbb{E}_A(x_i) = 0$ for $1 \leq i \leq k$. We assume this extra condition, and prove inequality (11.3.10). By scaling, there is no loss of generality in assuming $\|x_i\| \leq 1$ for $1 \leq i \leq k$.

Let $\delta = \varepsilon/4$. From Lemma 3.6.8 and Corollary 3.6.9, there is a finite dimensional abelian subalgebra $A_0 \subseteq A$ with minimal projections $e_1, \ldots, e_n$ such that

$$\|\sum_{m=1}^{n} e_m x_i e_m\|_2 = \|\mathbb{E}_A(x_i) - \mathbb{E}_{A_0' \cap N}(x_i)\|_2 < \delta \qquad (11.3.11)$$

for $1 \leq i \leq k$. Any partial isometry $v \in N$ satisfying $vAv^* = Avv^*$ has the form pu for a projection $p \in A$ and a normalising unitary $u \in N$ by Lemma 6.2.3 (iii). The singularity of A then shows that $v \in A$, making it impossible to satisfy the two inequalities $vv^* \leq e_m$ and $v^*v \leq (1 - e_m)$ simultaneously unless $v = 0$. Thus no non-zero partial isometry $v \in N$ satisfies $vv^* \leq e_m$, $v^*v \leq (1 - e_m)$, and $vAv^* = Avv^*$. The hypotheses of Lemma 11.3.1 are satisfied, and applying this result with ε replaced by δ/n gives unitaries $u_m \in A$ such that

$$\|\mathbb{E}_A((1 - e_m)x_i e_m u_m x_j^* (1 - e_m))\|_2 < \delta/n \qquad (11.3.12)$$

for $1 \leq m \leq n$ and $1 \leq i, j \leq k$. Define a unitary $u \in A$ by $u = \sum_{m=1}^{n} u_m e_m$, and let $y_i = \sum_{m=1}^{n} (1 - e_m)x_i e_m$, for $1 \leq i \leq k$. We have

$$x_i - y_i = x_i - \sum_{m=1}^{n} (1 - e_m)x_i e_m = \sum_{m=1}^{n} e_m x_i e_m = \mathbb{E}_{A_0' \cap N}(x_i) \qquad (11.3.13)$$

for $1 \leq i \leq k$. The inequalities

$$\|x_i - y_i\|_2 \leq \|x_i\|_2 \leq 1, \quad \|x_i - y_i\| \leq \|x_i\| \leq 1, \quad \|y_i\| \leq 2, \qquad (11.3.14)$$

for $1 \leq i \leq k$, follow immediately.

If we apply $\mathbb{E}_A$ to the identity

$$x_i u x_j^* = (x_i - y_i)u x_j^* + y_i u(x_j^* - y_j^*) + y_i u y_j^*, \tag{11.3.15}$$

then (11.3.14) gives

$$\|\mathbb{E}_A(x_i u x_j^*)\|_2 \leq \|x_i - y_i\|_2 + \|y_i\|\|x_j - y_j\|_2 + \|\mathbb{E}_A(y_i u y_j^*)\|_2$$
$$< 3\delta + \|\mathbb{E}_A(y_i u y_j^*)\|_2, \tag{11.3.16}$$

and we estimate the last term. The identity

$$y_i u y_j^* = \sum_{m,s=1}^{n} (1 - e_m) x_i e_m u e_s x_j^* (1 - e_s)$$
$$= \sum_{m=1}^{n} (1 - e_m) x_i e_m u_m x_j^* (1 - e_m) \tag{11.3.17}$$

holds because each e_s commutes with u and $e_m e_s = 0$ for $m \neq s$. The last sum has n terms, so the inequalities

$$\|\mathbb{E}_A(y_i u y_j^*)\|_2 < \delta, \quad 1 \leq i, j \leq k, \tag{11.3.18}$$

are immediate from (11.3.12). Together (11.3.16) and (11.3.18) yield

$$\|\mathbb{E}_A(x_i u x_j^*)\|_2 < 3\delta + \delta = \varepsilon, \quad 1 \leq i, j \leq k, \tag{11.3.19}$$

as required.   □

Since the weak asymptotic homomorphism property is a consequence of applying Lemma 11.3.2 to the set of elements $x_1, \ldots, x_k, y_1^*, \ldots, y_k^* \in N$, we immediately obtain the following corollary, which is the final step in proving Theorem 11.1.2.

Corollary 11.3.3. *Let A be a singular masa in a separable II_1 factor N. Then A has the weak asymptotic homomorphism property.*

Here are some corollaries for singular masas of Theorem 11.1.2. The first depends on the equation $\mathbb{E}_{A \overline{\otimes} B} = \mathbb{E}_A \otimes \mathbb{E}_B$.

Corollary 11.3.4. *Let $A \subseteq M$ and $B \subseteq N$ be masas in separable II_1 factors M and N. If A and B are both singular, then $A \overline{\otimes} B$ is a singular masa in $M \overline{\otimes} N$.*

Proof. Theorem 11.1.2 shows that singularity and the weak asymptotic homomorphism property are equivalent for masas in separable II_1 factors. By Tomita's commutant theorem (Theorem 2.2.4 or [105, Theorem 11.2.16]), $A \overline{\otimes} B$ is a masa in $M \overline{\otimes} N$, since

$$(A \overline{\otimes} B)' \cap (M \overline{\otimes} N) = (A' \cap M) \overline{\otimes} (B' \cap N) = A \overline{\otimes} B. \tag{11.3.20}$$

To verify that the weak asymptotic homomorphism property carries over to tensor products, it suffices to check this property on a set whose linear span is $\|\cdot\|_2$ - norm dense in $M\overline{\otimes}N$, as was observed following Definition 11.1.1. In this case the appropriate set is

$$\{x \otimes y \colon x \in M,\ y \in N\}.$$

Let $\varepsilon > 0$, let $x_1,\ldots,x_n \in M$ and $y_1,\ldots,y_n \in N$ and assume that $\|x_i\| \le 1$ and $\|y_i\| \le 1$ for $1 \le i,j \le n$. By Theorem 11.1.2, both A and B satisfy the weak asymptotic homomorphism property so there are unitaries $u \in A$ and $v \in B$ such that the relations

$$\|\mathbb{E}_A(x_i u x_j^*) - \mathbb{E}_A(x_i)u\mathbb{E}_A(x_j^*)\|_2 < \varepsilon \tag{11.3.21}$$

and

$$\|\mathbb{E}_B(y_i v y_j^*) - \mathbb{E}_B(y_i)v\mathbb{E}_B(y_j^*)\|_2 < \varepsilon, \quad \text{for} \quad 1 \le i,j \le n, \tag{11.3.22}$$

hold. From the equation $\mathbb{E}_{A\overline{\otimes}B} = \mathbb{E}_A \otimes \mathbb{E}_B$ and (11.3.20), we obtain

$$\|\mathbb{E}_{A\overline{\otimes}B}((x_i \otimes y_i)(u \otimes v)(x_j^* \otimes y_j^*)) - \mathbb{E}_{A\overline{\otimes}B}(x_i \otimes y_i)(u \otimes v)\mathbb{E}_{A\overline{\otimes}B}(x_j^* \otimes y_j^*)\|_2$$
$$= \|\mathbb{E}_A(x_i u x_j^*) \otimes \mathbb{E}_B(y_i v y_j^*) - \mathbb{E}_A(x_i)u\mathbb{E}_A(x_j^*) \otimes \mathbb{E}_B(y_i)v\mathbb{E}_B(y_j^*)\|_2$$
$$\le \|\mathbb{E}_A(x_i u x_j^*) - \mathbb{E}_A(x_i)u\mathbb{E}_A(x_j^*)\|_2 + \|\mathbb{E}_B(y_i v y_j^*) - \mathbb{E}_B(y_i)v\mathbb{E}_B(y_j^*)\|_2$$
$$< 2\varepsilon \quad \text{for} \quad 1 \le i,j \le n. \tag{11.3.23}$$

Here we have added and subtracted a suitable term, used the triangle inequality and the estimates $\|x_i\| \le 1$ and $\|y_i\| \le 1$. $\qquad\square$

The previous corollary is in [177] and the following corollary was proved originally by Popa using his results on the δ-invariant for singular masas in [139]. For the definition of N^ω and A^ω, see Appendix A, Section A.4.

Corollary 11.3.5. *Let A be a singular masa in a separable II_1 factor N. Then A^ω has the weak asymptotic homomorphism property in N^ω and is strongly singular.*

Proof. Let $\varepsilon > 0$ and let $x_1,\ldots,x_k \in N^\omega$ have representatives $(x_{n\,1}),\ldots,(x_{n\,k})$ in $\ell^\infty(N)$. By Theorem 11.1.2 the weak asymptotic homomorphism property holds for A in N so there are unitaries u_n in A such that

$$\|\mathbb{E}_A(x_{n\,i}u_n x_{n\,j}^*) - \mathbb{E}_A(x_{n\,i})u_n\mathbb{E}_A(x_{n\,j}^*)\|_2 < \varepsilon \tag{11.3.24}$$

for all $1 \le i,j \le k$ and all $n \in \mathbb{N}$. Let the unitary $u = \pi(u_n) \in A^\omega$ correspond to (u_n). Then

$$\|\mathbb{E}_A^\omega(x_i u x_j^*) - \mathbb{E}_A^\omega(x_i)u\mathbb{E}_A^\omega(x_j^*)\|_{2,\omega} \le \varepsilon \tag{11.3.25}$$

by inequality (11.3.24), so A^ω has the weak asymptotic property. The strong singularity follows from Lemma 11.2.3, which does not depend on separability of the factor. $\qquad\square$

Remark 11.3.6. By refining the methods presented in this section, Chifan [12] has strengthened Corollary 11.3.4 by showing that

$$\left(\mathcal{N}(A\overline{\otimes}B)\right)'' = \left(\mathcal{N}(A)''\right)\overline{\otimes}\left(\mathcal{N}(B)''\right)$$

holds for all masas A and B. For singular masas this reduces to Corollary 11.3.4.   □

11.4   A basis condition for singularity

In this section a condition on an orthonormal basis of a subalgebra of N is given to ensure that certain subalgebras are singular. The hypotheses of the next theorem are satisfied by a diffuse abelian von Neumann subalgebra B of a separable II_1 factor, as in that case the algebra is isomorphic to $L^\infty[0,1]$, (Theorem 3.5.2), so contains a Haar unitary u with $\{u^n\colon n \in \mathbb{Z}\}$ an orthonormal basis of $L^2(B, \tau)$. In our subsequent applications N will be a II_1 factor; note that B is not assumed to be abelian in the next theorem.

Theorem 11.4.1. *Let N be a separable finite von Neumann algebra with faithful normal trace τ and let $Y \subseteq N$ be a subset whose $\|\cdot\|_2$-closed linear span is $L^2(N)$. Let B be a von Neumann subalgebra of N that contains an infinite set $\mathcal{W} = \{w_n\colon n \in \mathbb{N}\}$ of mutually orthogonal unitaries. If $\{v_n\colon n \in \mathbb{N}\} \subset B$ is an orthonormal basis for $L^2(B)$ such that*

$$\sum_{n=1}^{\infty} \|\mathbb{E}_B(xv_n y) - \mathbb{E}_B(x)v_n\mathbb{E}_B(y)\|_2^2 < \infty \qquad (11.4.1)$$

for all $x, y \in Y$, then

$$\lim_n \|\mathbb{E}_B(xw_n y) - \mathbb{E}_B(x)w_n\mathbb{E}_B(y)\|_2 = 0 \qquad (11.4.2)$$

for all $x, y \in N$.

Proof. Since $\mathcal{W}$ is an infinite orthonormal subset of $L^2(B)$,

$$\lim_{n\to\infty} \tau(w_n^* v_j) = 0 \quad \text{for all} \quad j \in \mathbb{N}. \qquad (11.4.3)$$

Let F be a finite subset of Y and let $\varepsilon > 0$. By the hypothesis (inequality (11.4.1)), choose $k \in \mathbb{N}$ such that

$$\sum_{j=k+1}^{\infty} \|\mathbb{E}_B(xv_j y) - \mathbb{E}_B(x)v_j\mathbb{E}_B(y)\|_2^2 < \varepsilon \qquad (11.4.4)$$

for all $x, y \in F$. Let $\beta = \max\{\|x\|\colon x \in F\}$. By (11.4.3) there is an $n_0 \in \mathbb{N}$ such that

$$|\tau(w_n^* v_j)| \le \varepsilon k^{-1}\beta^{-2} \qquad (11.4.5)$$

for $1 \le j \le k$ and all $n \ge n_0$. Since $\{v_j : j \in \mathbb{N}\}$ is an orthonormal basis of $L^2(B)$, we obtain

$$w_n = \sum_{j=1}^{\infty} \tau(w_n v_j^*) v_j \quad \text{for all } n \in \mathbb{N}, \tag{11.4.6}$$

which leads to

$$\|\mathbb{E}_B(x w_n y) - \mathbb{E}_B(x) w_n \mathbb{E}_B(y)\|_2$$
$$= \left\| \sum_{j=1}^{\infty} \tau(w_n v_j^*) \big(\mathbb{E}_B(x v_j y) - \mathbb{E}_B(x) v_j \mathbb{E}_B(y) \big) \right\|_2 \tag{11.4.7}$$

for all $x, y \in F$. Split this sum at $j = k$ and estimate each part separately.

For all $n \ge n_0$, the inequality (11.4.5) gives

$$\left\| \sum_{j=1}^{k} \tau(w_n v_j^*) \big(\mathbb{E}_B(x v_j y) - \mathbb{E}_B(x) v_j \mathbb{E}_B(y) \big) \right\|_2$$
$$\le \sum_{j=1}^{k} \varepsilon k^{-1} \beta^{-2} \|\mathbb{E}_B(x v_j y) - \mathbb{E}_B(x) v_j \mathbb{E}_B(y)\|_2$$
$$\le \sum_{j=1}^{k} \varepsilon k^{-1} \beta^{-2} 2\|x\|\|y\|$$
$$\le 2\varepsilon \tag{11.4.8}$$

for all $x, y \in F$, since $\|v_j\|_2 = 1$ and $\beta = \max\{\|x\| : x \in F\}$. The orthonormality of $\{v_j : j \in \mathbb{N}\}$ implies that

$$1 = \|w_n\|_2^2 = \sum_{j=1}^{\infty} |\tau(w_n v_j^*)|^2, \tag{11.4.9}$$

and combining this relation with (11.4.4) and the Cauchy–Schwarz inequality leads to

$$\left\| \sum_{j=k+1}^{\infty} \tau(w_n v_j^*) \big(\mathbb{E}_B(x v_j y) - \mathbb{E}_B(x) v_j \mathbb{E}_B(y) \big) \right\|_2$$
$$\le \sum_{j=k+1}^{\infty} |\tau(w_n v_j^*)| \, \|\mathbb{E}_B(x v_j y) - \mathbb{E}_B(x) v_j \mathbb{E}_B(y)\|_2$$
$$\le \left(\sum_{j=k+1}^{\infty} |\tau(w_n v_j^*)|^2 \right)^{1/2} \left(\sum_{j=k+1}^{\infty} \|\mathbb{E}_B(x v_j y) - \mathbb{E}_B(x) v_j \mathbb{E}_B(y)\|_2^2 \right)^{1/2}$$
$$< \varepsilon \tag{11.4.10}$$

for all $x, y \in F$. The right-hand side of (11.4.7) can be estimated by using (11.4.8) and (11.4.10), resulting in the inequality

$$\|\mathbb{E}_B(xw_ny) - \mathbb{E}_B(x)w_n\mathbb{E}_B(y)\|_2 < 3\varepsilon \qquad (11.4.11)$$

for all $x, y \in F$ and all $n \geq n_0$. Since $\varepsilon > 0$ was arbitrary, (11.4.11) implies that

$$\lim_{n \to \infty} \|\mathbb{E}_B(xw_ny) - \mathbb{E}_B(x)w_n\mathbb{E}_B(y)\|_2 = 0 \qquad (11.4.12)$$

for all $x, y \in Y$. The lemma follows, because $\|w_n\| = 1$ for all $n \in \mathbb{N}$ and the linear span of Y is $\|\cdot\|_2$-dense in N. $\square$

Remark 11.4.2. (i)   If $x, y \in N$ and $w \in B$, then the identity

$$\mathbb{E}_B(xwy) - \mathbb{E}_B(x)w\mathbb{E}_B(y) = \mathbb{E}_B((x - \mathbb{E}_B(x))wy)$$
$$= \mathbb{E}_B((x - \mathbb{E}_B(x))w(y - \mathbb{E}_B(y))) \qquad (11.4.13)$$

reduces the general case of Lemma 11.4.1 to that where $\mathbb{E}_B(x) = 0 = \mathbb{E}_B(y)$.
(ii)   If B is an abelian von Neumann subalgebra of the II_1 factor N, and if the conclusion of Theorem 11.4.1 holds so that

$$\lim_n \|\mathbb{E}_B(xw_ny) - \mathbb{E}_B(x)w_n\mathbb{E}_B(y)\|_2 = 0 \qquad (11.4.14)$$

for all $x, y \in N$, then B is diffuse. To see this, we argue as follows.

Let p be a non-zero minimal projection in B. Then $pB = \mathbb{C}p$, so each unitary $w \in B$ has the form $w = \lambda p + (1 - p)w$ for a scalar λ of modulus 1. In pNp, decompose p as a sum $p_1 + p_2$ of orthogonal equivalent projections, and choose a partial isometry $v \in N$ such that $vv^* = p_1$, $v^*v = p_2$. Then $v = p_1vp_2$, so the module properties of $\mathbb{E}_B$ give $\mathbb{E}_B(v) = 0$ since B is abelian and $p_1p_2 = 0$. Now let $u = v + v^*$, a unitary in pNp satisfying $\mathbb{E}_B(u) = 0$. For any unitary $w = \lambda p + (1 - p)w \in B$,

$$uwu^* = upwu^* = \lambda uu^* = \lambda p,$$

so $\|\mathbb{E}_B(uwu^*)\|_2^2 = \tau(p)$, while $\mathbb{E}_B(u)w\mathbb{E}_B(u^*) = 0$. Thus (11.4.14) can never be satisfied by any sequence of unitaries from B, showing that no such minimal projection p can exist. Thus B is diffuse.
(iii)   If B is a diffuse abelian von Neumann subalgebra of N, then $(B, \tau) \cong (L^\infty[0, 1], \int_0^1 \cdot \, dt)$ by Theorem 3.5.2, so there is a Haar unitary $u \in B$ that generates B and $\{u^n : n \in \mathbb{Z}\}$ is an orthonormal basis of $L^2(B)$.
(iv)   Note that the conclusions of Theorem 11.4.1 are closely related to conditions of Jolissaint and Stalder [92].
(v)   If $\{v_n : n \in \mathbb{N}\}$ and $\{w_n : n \in \mathbb{N}\}$ are orthonormal bases of $L^2(B)$ in B, then

$$\sum_{n=1}^{\infty} \|\mathbb{E}_B(xv_ny) - \mathbb{E}_B(x)v_n\mathbb{E}_B(y)\|_2^2$$

$$= \sum_{n=1}^{\infty} \|\mathbb{E}_B(xw_ny) - \mathbb{E}_B(x)w_n\mathbb{E}_B(y)\|_2^2 \qquad (11.4.15)$$

for all $x, y \in N$. This follows by standard properties of orthonormal bases by expanding each v_n in terms of the w_j and is easily seen in the case where $\mathbb{E}_B(x) = \mathbb{E}_B(y) = 0$. $\square$

Corollary 11.4.3. *Let N be a separable II_1 factor, let B be a diffuse von Neumann subalgebra of N and let Y be a subset of N whose linear span is $\|\cdot\|_2$-dense in N. If there is an orthonormal basis $\{v_n \colon n \in \mathbb{N}\} \subseteq B$ of $L^2(B)$ with*

$$\sum_{n=1}^{\infty} \|\mathbb{E}_B(xv_ny) - \mathbb{E}_B(x)v_n\mathbb{E}_B(y)\|_2^2 < \infty \qquad (11.4.16)$$

for all $x, y \in Y$, then B is a singular masa in N.

Proof. Let A be a masa in B. Then A is a diffuse separable abelian von Neumann algebra so, by Theorem 3.5.2, (A, τ) is isomorphic to $\left(L^\infty[0,1], \int_0^1 \cdot \, dt\right)$. From this we obtain that $L^2(A)$ has an orthonormal basis $\{u^n \colon n \in \mathbb{Z}\}$, where u is a Haar unitary in A, and thus also in B. Then the conditions of Theorem 11.4.1 are satisfied and

$$\lim_{n \to \infty} \|\mathbb{E}_B(xu^ny) - \mathbb{E}_B(x)u^n\mathbb{E}_B(y)\|_2^2 = 0 \qquad (11.4.17)$$

for all $x, y \in N$.

If w lies in the unitary normaliser $\mathcal{N}(B)$, then the map $b \mapsto wbw^*$ defines a $*$-automorphism of B, and thus wu^nw^* is a unitary in B for all $n \in \mathbb{N}$. Then (11.4.17) implies that

$$\lim_{n \to \infty} \|\mathbb{E}_B(w)u^n\mathbb{E}_B(w^*)\|_2 = 1, \qquad (11.4.18)$$

where we have taken x and y to be respectively w and w^*. This is impossible unless $\|\mathbb{E}_B(w)\|_2 = 1$, in which case $w = \mathbb{E}_B(w) \in B$. Thus B is singular. $\square$

Remark 11.4.4. A class of examples of von Neumann algebras satisfying the conditions of the above theorem may be constructed using a condition of Jolissant and Stalder [92, Theorem 1.2(2)].

Let H be an infinite subgroup of a countable discrete group G and assume the extra condition that $L(H)$ is diffuse (automatically satisfied when $L(H)$ is a masa or a subfactor in $L(G)$). If for each pair $g_1, g_2 \in G \setminus H$, there is a finite subset $E \subset H$ such that

$$g_1(H \setminus E)g_2 \cap H = \emptyset, \qquad (11.4.19)$$

then $B = L(H)$ will be shown to satisfy the hypotheses of Corollary 11.4.3 as a subalgebra of the von Neumann algebra $(L(G), \tau)$.

Take H as the orthonormal basis of $L^2(H)$ and let $Y = G \subseteq L(G)$ with $Y'' = L(G)$. The term $\mathbb{E}_B(g_1hg_2) - \mathbb{E}_B(g_1)h\mathbb{E}_B(g_2)$ is zero if either g_1 or g_2 is in H, since $\mathbb{E}_B$ is a B-module map. If $g_1, g_2 \in G \setminus H$, choose a finite subset $E \subseteq H$ such that $g_1hg_2 \notin H$ for all $h \in H \setminus E$. Hence $\mathbb{E}_B(g_1hg_2) - \mathbb{E}_B(g_1)h\mathbb{E}_B(g_2) = 0$

for all $h \in H \setminus E$ as the three conditional expectations are zero. Thus all but a finite number of terms in the series

$$\sum_{h \in H} \|\mathbb{E}_B(g_1 h g_2) - \mathbb{E}_B(g_1) h \mathbb{E}_B(g_2)\|_2^2$$

are zero so the series converges.

There are many groups and subgroups with these properties. Here are two examples related to the free groups used in this chapter:

(1) $H = \langle a \rangle \subseteq G = \mathbb{F}_2 = \langle a, b \rangle$;

(2) $H = \mathbb{F}_2 = \langle a, b \rangle \subseteq G = \mathbb{F}_3 = \langle a, b, c \rangle$. $\square$

11.5 Enumeration of words in $\mathbb{F}_2$

In order to show that the Laplacian masa $\{h\}''$ satisfies the properties of Corollary 11.4.3, estimates on certain classes of words in $\mathbb{F}_2$ are required. The following definition establishes the notation that we will use.

Definition 11.5.1. Let $\mathbb{F}_2$ denote the free group on 2 generators a, b and let $S = \{a, b, a^{-1}, b^{-1}\}$. Let W_n be the set of all reduced words in $\mathbb{F}_2$ of length n (in particular, $W_1 = S$). We let w_n denote the sum in $\mathbb{C}\mathbb{F}_2$ of all reduced words from $\mathbb{F}_2$ of length n for $n \geq 1$, and we adopt the convention that $W_0 = \{e\}$ and that $w_0 = e$. $\square$

We list here the basic relations amongst the w_n's, which are all self-adjoint elements of $L(\mathbb{F}_2)$. The first two are just definitions and the second two are proved by simple algebra in $\mathbb{F}_2$.

(i) $w_0 = e$;

(ii) $w_1 = a + b + a^{-1} + b^{-1}$;

(iii) $w_1^2 = w_2 + 4w_0 = w_2 + 4e$;

(iv) $w_1 w_n = w_n w_1 = w_{n+1} + 3w_{n-1}$ for $n \geq 2$.

These equations show that the algebra $\mathrm{Alg}(w_1)$ generated by w_1 is equal to the linear span of the set $\{w_n : n \geq 0\}$. Two words of different lengths are orthogonal in $\ell^2(\mathbb{F}_2)$, and so $\{w_n : n \geq 0\}$ is an orthogonal set of vectors in this space. The quantity $\|w_n\|_2^2$ is equal to the number of terms in the sum for w_n, and this is the number of words in $\mathbb{F}_2$ of length n. There are four choices for the first letter of such a word and three choices for each subsequent letter, giving $\|w_n\|_2^2 = 4 \cdot 3^{n-1}$ when $n \geq 1$, and $\|w_0\|_2^2 = 1$ is clear in the exceptional case $n = 0$.

Let B denote the abelian von Neumann subalgebra of $L(\mathbb{F}_2)$ generated by w_1. This algebra is called the *radial*, or *Laplacian*, algebra (or masa) of $L(\mathbb{F}_2)$. The remarks above show that

$$\{v_n = w_n/\|w_n\|_2 : n \geq 0\} \tag{11.5.1}$$

is an orthonormal basis for $L^2(B)$. From the equation $\mathbb{E}_B = e_B|_N$, if follows that

$$\mathbb{E}_B(x) = \sum_{n \geq 0} \tau(xw_n)w_n/\|w_n\|_2^2 = \sum_{n \geq 0} \tau(xv_n)v_n. \qquad (11.5.2)$$

The quantities in the next definition will be needed subsequently in Corollary 11.5.6.

Definition 11.5.2.   (i) For each pair $x, y \in S = \{a, a^{-1}, b, b^{-1}\}$ and for each $n \geq 2$, let $w_n(x,y)$ be the sum of all reduced words of length n which begin with x and end with y. We denote the number of such reduced words by $\nu_n(x,y)$, and this is equal to $\|w_n(x,y)\|_2^2$.

(ii) More generally, if σ and τ are subsets of S, then $w_n(\sigma, \tau)$ will denote the sum of all words beginning with a letter from σ and ending with a letter from τ. We will abuse notation slightly by continuing to use $w_n(x,y)$ rather than $w_n(\{x\}, \{y\})$ for one point sets.

(iii) The numbers α_n, β_n, and γ_n are defined by

$$\alpha_n = \nu_n(a, b), \quad \beta_n = \nu_n(a, a), \quad \gamma_n = \nu_n(a, a^{-1}),$$

for $n \geq 2$.   □

For any pair $x, y \in S$, there is an automorphism derived from permuting the generators of $\mathbb{F}_2$ and their inverses such that the pair (x, y) is sent to one of (a, b), (a, a) or (a, a^{-1}). Hence $w_n(x, y)$ corresponds, under this automorphism, to one of $w_n(a, b)$, $w_n(a, a)$ or $w_n(a, a^{-1})$ and thus $\nu_n(x, y)$ is one of α_n, β_n or γ_n for each $n \geq 2$.

Lemma 11.5.3. *For* $n \geq 2$,

(i) $w_{n+1}(a, b) = a\big(w_n(a, b) + w_n(b, b) + w_n(b^{-1}, b)\big);$

(ii) $w_{n+1}(a, a) = a\big(w_n(a, a) + w_n(b, a) + w_n(b^{-1}, a)\big);$

(iii) $w_{n+1}(a, a^{-1}) = a\big(w_n(a, a^{-1}) + w_n(b, a^{-1}) + w_n(b^{-1}, a^{-1})\big).$

Proof. This is a simple direct application of the definition of $w_n(x, y)$ and the unique definition of each word in terms of S.   □

Lemma 11.5.4. *The following relations hold for* α_n, β_n *and* γ_n:

(i) $\alpha_2 = 1$, $\beta_2 = 1$, *and* $\gamma_2 = 0$;

(ii) *for* $n \geq 2$,

(1) $\alpha_{n+1} = \alpha_n + \beta_n + \gamma_n;$

(2) $\beta_{n+1} = \beta_n + 2\alpha_n;$

(3) $\gamma_{n+1} = \gamma_n + 2\alpha_n;$

(iii) *for $n \geq 2$,*

$$|t_n - 4^{-1}3^{n-1}| \leq 3, \qquad (11.5.3)$$

where t_n denotes any one of the three quantities α_n, β_n, or γ_n.

Proof. (i)   These are clear since $w_2(a,b) = ab$, $w_2(a,a) = a^2$, and $w_2(a,a^{-1}) = 0$ since there are no reduced words of length 2 which begin with a and end with a^{-1}.

(ii)   These three equations follow from the corresponding three parts of Lemma 11.5.3 using $\alpha_n = \|w_n(a,b)\|_2^2$, $\beta_n = \|w_n(a,a)\|_2^2$ and $\gamma_n = \|w_n(a,a^{-1})\|_2^2$, orthogonality and the observation above on $\nu_n(x,y)$.

(iii)   The equations of part (ii) show that

$$\beta_{n+1} - \gamma_{n+1} = \beta_n - \gamma_n \qquad (11.5.4)$$

and so this difference is independent of n. Since $\beta_2 - \gamma_2 = 1$, it follows that

$$\beta_n = 1 + \gamma_n \quad \text{for} \quad n \geq 2. \qquad (11.5.5)$$

Subtracting (3) from (1) shows that

$$\alpha_{n+1} - \gamma_{n+1} = \beta_n - \alpha_n = 1 + (\gamma_n - \alpha_n) \qquad (11.5.6)$$

by (11.5.5). Since $\alpha_2 - \gamma_2 = 1$, an induction argument using equation (11.5.6) shows that

$$\alpha_n = \gamma_n + (1 + (-1)^n)/2 \quad \text{for} \quad n \geq 2. \qquad (11.5.7)$$

From equations (11.5.7) and (11.5.5) if follows that

$$|\alpha_n - \gamma_n| \leq 1 \quad \text{and} \quad |\alpha_n - \beta_n| \leq 2 \qquad (11.5.8)$$

for all $n \geq 2$. There are $4 \cdot 3^{n-1}$ words of length n in $\mathbb{F}_2$, which may be split into three classes: those beginning and ending with different symbols, those with the same symbol and those with a symbol and its inverse. These three correspond to $w_n(a,b)$, $w_n(a,a)$ and $w_n(a,a^{-1})$. Enumeration of the possibilities (for example, $w_n(a,a)$, $w_n(b,b)$, $w_n(a^{-1},a^{-1})$, and $w_n(b^{-1},b^{-1})$ lead to the $4\beta_n$ term below) leads to

$$8\alpha_n + 4\beta_n + 4\gamma_n = 4.3^{n-1} \qquad (11.5.9)$$

or

$$2\alpha_n + \beta_n + \gamma_n = 3^{n-1} \qquad (11.5.10)$$

for all $n \geq 2$. Substituting the estimates from inequality (11.5.8) into (11.5.10) yields

$$|4\alpha_n - 3^{n-1}| \leq 3 \quad \text{for all} \quad n \geq 2. \qquad (11.5.11)$$

If we combine the inequalities (11.5.8) and (11.5.11), then the estimates

$$|\gamma_n - 4^{-1}3^{n-1}| \leq 7/4 \quad \text{and} \quad |\beta_n - 4^{-1}3^{n-1}| \leq 11/4, \qquad (11.5.12)$$

for all $n \geq 2$, are immediate and (11.5.3) is established. $\square$

Remark 11.5.5. Well known properties of linear recurrence relations enable one to obtain closed forms for α_n, β_n and γ_n from Lemma 11.5.4. Induction, generating functions or the observation that

$$\begin{pmatrix} \alpha_{n+1} \\ \beta_{n+1} \\ \gamma_{n+1} \end{pmatrix} = \begin{pmatrix} 1 & 1 & 1 \\ 2 & 1 & 0 \\ 2 & 0 & 1 \end{pmatrix} \begin{pmatrix} \alpha_n \\ \beta_n \\ \gamma_n \end{pmatrix}$$

would all derive exact formulae for these quantities. In the latter case, the eigenvalues and eigenvectors of this 3×3 matrix enable one to give the closed form. □

The estimates in the next corollary could be improved by a more detailed analysis, but are sufficient for subsequent use. The quantities in the statement of the next result were defined in Definition 11.5.2, and we use the notation $|\sigma|$ for the cardinality of a set σ. The result says, in essence, that up to a bounded error, $\nu_n(\sigma, \tau)$ depends only on the sizes of the two sets.

Corollary 11.5.6. *If σ_1, σ_2, τ_1 and τ_2 are non-empty subsets of S with*

$$|\sigma_1| = |\sigma_2| \quad \text{and} \quad |\tau_1| = |\tau_2|, \tag{11.5.13}$$

then

$$|\nu_n(\sigma_1, \tau_1) - \nu_n(\sigma_2, \tau_2)| \le 96 \tag{11.5.14}$$

for all $n \ge 2$.

Proof. By the definition of $\nu_n(\sigma, \tau)$ in Definition 11.5.2,

$$\begin{aligned} |\nu_n(\sigma_1, \tau_1) - &\nu_n(\sigma_2, \tau_2)| \\ &= | \sum_{x_1 \in \sigma_1, y_1 \in \tau_1} \nu_n(x_1, y_1) - \sum_{x_2 \in \sigma_2, y_2 \in \tau_2} \nu_n(x_2, y_2)|. \end{aligned} \tag{11.5.15}$$

Each $\nu_n(x, y)$ satisfies

$$|\nu_n(x, y) - 4.3^{n-1}| \le 3 \tag{11.5.16}$$

by Lemma 11.5.4 (iii), so that adding and subtracting $|\sigma_1| \cdot |\tau_1| = |\sigma_2| \cdot |\tau_2|$ times $4 \cdot 3^{n-1}$ and using the triangle inequality show that

$$|\nu_n(\sigma_1, \tau_1) - \nu_n(\sigma_2, \tau_2)| \le 2|\sigma_1| \cdot |\tau_1| \cdot 3 \le 96, \tag{11.5.17}$$

since $|\sigma| \le 4$ for any subset σ of S. □

Lemma 11.5.7. *Let $x = x_j \ldots x_1$ and $y = y_1 \ldots y_k$ be words in $\mathbb{F}_2$ written in reduced form. If $n \ge j + k + 2$, $0 \le r \le j$, and $0 \le s \le k$, then there are subsets $\sigma_r(x)$ and $\tau_s(y)$ of S for $0 \le r \le j$ and $0 \le s \le k$ such that the number of reduced words in $x w_n y$ resulting from r cancellations on the left and s cancellations on the right is equal to $\nu_{n-r-s}(\sigma_r(x), \tau_s(y))$.*

Moreover

$$|\sigma_0(x)| = |\sigma_j(x)| = |\tau_0(y)| = |\tau_k(y)| = 3 \qquad (11.5.18)$$

and

$$|\sigma_r(x)| = |\tau_s(y)| = 2 \qquad (11.5.19)$$

for $0 < r < j$ *and* $0 < s < k$.

Proof. Let

$$\sigma_0(x) = S \setminus \{x_1\}, \quad \sigma_j(x) = S \setminus \{x_j\}, \qquad (11.5.20)$$

$$\tau_0(y) = S \setminus \{y_1\}, \quad \tau_k(y) = S \setminus \{y_k\}. \qquad (11.5.21)$$

For $0 < r < j$ and $0 < s < k$, let

$$\sigma_r(x) = S \setminus \{x_{r+1}^{-1}, x_r\}, \quad \tau_s(x) = S \setminus \{y_{s+1}^{-1}, y_s\}. \qquad (11.5.22)$$

Note that the pairs $\{x_{r+1}^{-1}, x_r\}$ and $\{y_{s+1}^{-1}, y_s\}$ consist of two elements otherwise cancellation would occur in x and y. The cardinalities of the sets in equations (11.5.20) and (11.5.21) are all 3, while those in (11.5.22) are 2. If the reduced word $z = x_j \cdots x_{r+1} v y_{s+1} \cdots y_k$ results from r cancellations in x and s cancellations in y, then v must not begin with x_{r+1}^{-1} and must not end with y_{s+1}^{-1}. These restrictions do not apply if $r = j - 1$ or $s = k - 1$. The original word of w_n that cancelled to z is $x_1^{-1} \cdots x_r^{-1} v y_s^{-1} \cdots y_1^{-1}$, which requires that v does not begin with x_r or end with x_s. Thus the first symbol of v must lie in $\sigma_r(x)$ and the last symbol in $\tau_s(y)$ for exactly r and s cancellations. The word v clearly has length $n - r - s$ so the lemma follows. $\qquad \square$

The constants in the next lemma are not best possible but are sufficient for subsequent use.

Lemma 11.5.8. *There is a function* $K \colon \mathbb{N} \to \mathbb{N}$ *such that if* $x, y \in \mathbb{F}_2$ *and* $n \geq |x| + |y| + 2$, *then*

$$\|\mathbb{E}_B(xw_n y) - \mathbb{E}_B(x)w_n\mathbb{E}_B(y)\|_2 \leq K(|x|)K(|y|)\|w_n\|_2^{-1}. \qquad (11.5.23)$$

Proof. Let x and y be fixed words in $\mathbb{F}_2$ of lengths j and k, respectively, with $j, k \geq 1$. Let z be an arbitrary word of length j and let $n \geq j + k + 2$.

If a word v has length p, then it is orthogonal to w_n for all $n \neq p$ and hence $\mathbb{E}_B(v) = w_p \|w_p\|_2^{-2}$, since $\tau(vw_p) = 1$. Now consider $xw_n y$. This is a sum of reduced words with various lengths. If v is a reduced word of length n, then the unreduced word xvy has $j + k + n$ letters. If it is possible to carry out r cancellations on the left and s cancellations on the right, then the resulting reduced word has length $j + k + n - 2(r + s)$, and Lemma 11.5.7 shows that $\nu_{n-r-s}(\sigma_r(x), \tau_s(y))$ distinct words arise in this way. If we let r and s vary and take into account all the possibilities, the result is

$$\mathbb{E}_B(xw_n y)$$

$$= \sum_{r=0}^{j} \sum_{s=0}^{k} \nu_{n-r-s}(\sigma_r(x), \tau_s(y)) \|w_{j+k+n-2(r+s)}\|_2^{-2} w_{j+k+n-2(r+s)}. \qquad (11.5.24)$$

If $z \in \mathbb{F}_2$ is another word, then by Corollary 11.5.6 and Lemma 11.5.7, the sets $\sigma_r(x)$ and $\sigma_r(z)$ arising from xw_ny and zw_ny are equal (see (11.5.19) and (11.5.20)). Moreover,

$$|\nu_p(\sigma_r(x), \tau_s(y)) - \nu_p(\sigma_r(z), \tau_s(y))| \leq 96 \text{ for all } p \geq 2, \tag{11.5.25}$$

from (11.5.17). Hence there are integers $\lambda_{r,s}$ depending on x, y, z with $|\lambda_{r,s}| \leq 96$ for all r, s such that

$$\mathbb{E}_B(xw_ny) - \mathbb{E}_B(zw_ny)$$
$$= \sum_{r=0}^{j} \sum_{s=0}^{k} \lambda_{r,s} w_{n+j+k-2(r+s)} \|w_{n+j+k-2(r+s)}\|_2^{-2}. \tag{11.5.26}$$

Because $\|w_p\|_2^2 = 4 \cdot 3^{p-1} = 3^{p-n}\|w_n\|_2^2$ is the number of words in $\mathbb{F}_2$ of length p, the choice $p = n + j + k - 2(r + s)$ gives

$$\|w_{n+j+k-2(r+s)}\|_2^2 = 3^{j+k-2(r+s)}\|w_n\|_2^2 \geq 3^{(j+k)}\|w_n\|_2^2 \tag{11.5.27}$$

for $0 \leq r \leq j$ and $0 \leq s \leq k$. By the triangle inequality and (11.5.26), we obtain the estimate

$$\|\mathbb{E}_B(xw_ny) - \mathbb{E}_B(zw_ny)\|_2$$
$$\leq 96 \sum_{r=0}^{j} \sum_{s=0}^{k} \|w_{n+j+k-2(r+s)}\|_2^{-1}$$
$$\leq 96(j+1)(k+1)3^{(j+k)/2}\|w_n\|_2^{-1}. \tag{11.5.28}$$

If this inequality (11.5.28) is summed over all words z of length j (there are $\|w_j\|_2^2$ of these), then

$$\|\|w_j\|_2^2\mathbb{E}_B(xw_ny) - \mathbb{E}_B(w_jw_ny)\|_2$$
$$\leq \sum_{|z|=j} \|\mathbb{E}_B(xw_ny) - \mathbb{E}_B(zw_ny)\|_2$$
$$\leq 96(j+1)(k+1)3^{(j+k)/2}\|w_n\|_2^{-1}\|w_j\|_2^2. \tag{11.5.29}$$

Since $\mathbb{E}_B(x) = w_j\|w_j\|_2^{-2}$ and $\mathbb{E}_B(w_jw_ny) = w_jw_n\mathbb{E}_B(y)$, we have

$$\mathbb{E}_B(x)w_n\mathbb{E}_B(y) = w_jw_n\mathbb{E}_B(y)\|w_j\|_2^{-2} = \mathbb{E}_B(w_jw_ny)\|w_j\|_2^{-2}. \tag{11.5.30}$$

Then (11.5.29) and (11.5.30) imply that

$$\|\mathbb{E}_B(xw_ny) - \mathbb{E}_B(zw_ny)\|_2 \leq 96(j+1)(k+1)3^{(j+k)/2}\|w_n\|_2^{-1}$$
$$\leq 96(j+1)(k+1)3^{(j+k)/2}\|w_n\|_2^{-1}. \tag{11.5.31}$$

If we define $K(j)$ to be $(96)^{1/2}3^{j/2}(j+1)$ for all $j \in \mathbb{N}$, then the lemma follows.

$\square$

Theorem 11.5.9. *Let $B = \{h\}''$ be the Laplacian algebra in $L(\mathbb{F}_2)$. If $v_n = w_n\|w_n\|_2^{-1}$ for all $n \geq 0$, then $\{v_n : n \geq 0\}$ is an orthonormal basis for $L^2(B)$ satisfying*

$$\sum_{n=1}^{\infty} \|\mathbb{E}_B(xv_ny) - \mathbb{E}_B(x)v_n\mathbb{E}_B(y)\|_2^2 < \infty \qquad (11.5.32)$$

for all $x, y \in \mathbb{F}_2$.

Proof. If $x, y \in \mathbb{F}_2$, then define k to be $|x| + |y| + 2$ and let C be $K(|x|)^2 K(|y|)^2$, where $K(\cdot)$ is the bound of Lemma 11.5.8. Then Lemma 11.5.8 gives the estimate

$$\|\mathbb{E}_B(xv_ny) - \mathbb{E}_B(x)v_n\mathbb{E}_B(y)\|_2^2 \leq C\|w_n\|_2^{-4} = C.4^{-2}3^{-2n+2} \qquad (11.5.33)$$

for all $n \geq k$. The desired inequality (11.5.32) follows immediately from (11.5.33). $\qquad\square$

11.6   The Laplacian masa

Recall that in $L(\mathbb{F}_2)$, the self-adjoint element h is defined as the sum of the generators and their inverses. This section is devoted to the proof that the Laplacian algebra $B = \{h\}'' \subseteq L(\mathbb{F}_2)$ is a masa in $L(\mathbb{F}_2)$ using some nice commutator inequalities of Pytlik [155] and then deducing singularity from the previous sections. An alternative method is to show that B is diffuse by calculating that the spectrum of h is $[-2\sqrt{3}, 2\sqrt{3}]$ so that $B \cong L^\infty[-2\sqrt{3}, 2\sqrt{3}]$. From this it follows that B contains a Haar unitary, so B is a singular masa by applying the results of Sections 11.4 and 11.5. The reader is referred to the papers of Cohen [32] and Pytlik [155] or to the book by Figà-Talamanca and M. Picardello [69] for the spectrum of h. If $W \subseteq \mathbb{F}_2$ and if a vector $\eta \in \ell^2(\mathbb{F}_2)$ has orthonormal expansion $\eta = \sum_{x \in \mathbb{F}_2} \eta_x x$, then we will use the notation $\eta|_W$ for the restricted sum $\sum_{x \in W} \eta_x x$.

In the following lemma, c could be a or a^{-1} if $n > 2$ and could be a if $n = 2$ but cannot be a^{-1} in this case. If $n = 1$, then $c = b$ or b^{-1} and $z = e$.

Lemma 11.6.1. *Let $x, y \in \mathbb{F}_2$ with $|x| = |y| = n$, where $x = za$ and $y = cz$ with $|z| = n - 1$, and a and c are generators of $\mathbb{F}_2$ or their inverses. If $\varepsilon > 0$, then there is a vector $\eta \in \ell^2(\mathbb{F}_2)$ satisfying the inequalities*

$$\|(h\eta - \eta h) - (x - y)\|_2 < \varepsilon, \qquad \|\eta\|_2 \leq 6\varepsilon^{-2}. \qquad (11.6.1)$$

Proof. Before beginning the proof, we refer back to Definition 11.5.1 for the W_k's appearing below.

For $m \geq 0$, let ψ_m be the sum of all reduced words in $\mathbb{F}_2$ of the form $uczav$ with $|u| = |v| = m$. This is also the sum of all words $uczav$ with $|u| = |v| = m$ with the restrictions that u does not end in c^{-1} and v does not begin with a^{-1}. There are 3^{2m} words in the sum of ψ_m, as may be seen by counting from z outwards, so $\|\psi_m\|_2 = 3^m$ for all m, since $\psi_0 = cza$. The following properties hold from the definition of ψ_m:

(1) $h\psi_m$ and $\psi_m h$ are supported on $W_{2m+n} \cup W_{2m+n+2}$;

(2) $3h\psi_m|_{W_{2m+n+2}} = \psi_{m+1}h|_{W_{2m+n+2}}$;

(3) $3\psi_m|_{W_{2m+n+2}}h = h\psi_{m+1}|_{W_{2m+n+2}}$.

The first is clear. The second is valid because there is only one way of achieving each word in the sum of the left-hand side but three cancellations which will give it on the right-hand side. The same argument establishes (3).

Let $0 < \delta < 1$ and let

$$\eta_\delta = \sum_{m=0}^{\infty} (1 - \delta)^m 3^{-m} \psi_m. \tag{11.6.2}$$

By the pairwise orthogonality of ψ_m ($m \geq 0$) and $\|\psi_m\|_2 = 3^m$,

$$\|\eta_\delta\|_2^2 = \sum_{m=0}^{\infty} (1 - \delta)^{2m} 3^{-2m} \|\psi_m\|_2^2$$
$$= (1 - (1 - \delta)^2)^{-1} = (2\delta - \delta^2)^{-1} < \delta^{-1}, \tag{11.6.3}$$

so that $\|\eta_\delta\|_2 < \delta^{1/2}$. By the restrictions on the support of ψ_m ($m \geq 0$) in (1), $h\eta_\delta - \eta_\delta h$ is supported on $\bigcup_m W_{2m+n}$.

If $0 \leq r < n$, then

$$(h\eta_\delta - \eta_\delta h)|_{W_r} = 0. \tag{11.6.4}$$

If $r = n$ or $m = 0$, then

$$(h\eta_\delta - \eta_\delta h)|_{W_r} = h\psi_0 - \psi_0 h|_{W_n} = za - cz = x - y. \tag{11.6.5}$$

If $r = 2m + n + 2$ with $m = 0, 1, \ldots$, then

$$(h\eta_\delta - \eta_\delta h)|_{W_r}$$
$$= ((1 - \delta)^m 3^{-m}(h\psi_m - \psi_m h) + (1 - \delta)^{m+1} 3^{-m-1}(h\psi_{m+1} - \psi_{m+1}h))|_{W_r}$$
$$= ((1 - \delta)^m 3^{-m}(h\psi_m - \psi_m h) - (1 - \delta)^{m+1} 3^{-m-1}3(h\psi_m - \psi_m h))|_{W_r}$$
$$= (1 - \delta)^m 3^{-m} \delta (h\psi_m - \psi_m h)|_{W_r} \tag{11.6.6}$$

by equations (2) and (3).

These equations yield

$$(h\eta_\delta - \eta_\delta h) - (x - y)$$
$$= \sum_{m=0}^{\infty} (1 - \delta)^m 3^{-m} \delta (h\psi_m - \psi_m h)|_{W_{2m+n+2}}. \tag{11.6.7}$$

By the orthogonality of the supports and this equation (11.6.7),

$$\|(h\eta_\delta - \eta_\delta h) - (x - y)\|_2^2$$

$$= \left\|\sum_{m=0}^{\infty}(h\eta_\delta - \eta_\delta h)|_{W_{2m+n+2}}\right\|_2^2$$

$$= \sum_{m=0}^{\infty}(1 - \delta)^{2m}3^{-2m}\delta^2\|(h\psi_m - \psi_m h)|_{W_{2m+n+2}}\|_2^2. \qquad (11.6.8)$$

In $(h\psi_m - \psi_m h)|_{W_{2m+n+2}}$, any words appearing twice will cancel, so this expression consists of a sum of distinct words with coefficients ± 1. Counting the number of possibilities leads to the upper estimate

$$\|(h\psi_m - \psi_m h)|_{W_{2m+n+2}}\|_2^2 \leq 2 \cdot 3 \cdot 3^{2m}. \qquad (11.6.9)$$

By inequalities (11.6.9) and (11.6.8),

$$\|(h\eta_\delta - \eta_\delta h) - (x - y)\|_2^2$$

$$\leq \sum_{m=0}^{\infty}(1 - \delta)^{2m}3^{-2m}\delta^2 2.3^{2m+1}$$

$$\leq 6\delta^2(1 - (1 - \delta)^2)^{-1} \leq 6\delta(2 - \delta)^{-1} < 6\delta. \qquad (11.6.10)$$

If we now choose δ to satisfy $(6\delta)^{1/2} = \varepsilon$, then (11.6.1) follows from (11.6.2) and (11.6.10). $\qquad\qquad\qquad\qquad\qquad\qquad\qquad\qquad\qquad\qquad\qquad\qquad\qquad\qquad\qquad \square$

The hypotheses and conclusions of Lemma 11.6.1 are invariant under an automorphism of $L(\mathbb{F}_2)$ induced by an automorphism of $\mathbb{F}_2$ that permutes the generators and their inverses. Thus a and c in the lemma can be any elements of W_1 subject to the restrictions when $n = 1$ or $n = 2$.

Lemma 11.6.2. *If x, $y \in \mathbb{F}_2$ with $|x| = |y|$ and if $\varepsilon > 0$, then there is an $\eta \in \ell^2(\mathbb{F}_2)$ such that*

$$\|(h\eta - \eta h) - (x - y)\|_2 < \varepsilon. \qquad (11.6.11)$$

Proof. Let $n > 2$ and let $x = c_1 \cdots c_n$ and $y = d_n \cdots d_1$ with $c_j, d_j \in W_1$. If $d_1 \neq c_1^{-1}$, let $x_0 = x$, and define x_j for $j \geq 1$ by

$$x_j = d_j \cdots d_1 x c_n^{-1} \cdots c_{n-j+1}^{-1} = d_j \cdots d_1 c_1 \cdots c_{n-j}. \qquad (11.6.12)$$

Then $x_n = y$ and each pair (x_j, x_{j+1}) satisfies the conditions of Lemma 11.6.1. For the case $d_1 = c_1^{-1}$, choose a generator c_0 with $c_0 \neq d_1$ and $c_0 \neq c_1$, and define x_j by

$$x_j = \begin{cases} x, & j = 0, \\ c_0 x c_n^{-1}, & j = 1, \\ d_{j-1} \cdots d_1 c_0 x c_n^{-1} \cdots c_{n-j+1}^{-1}, & 2 \leq j \leq n, \\ d_n \cdots d_1 = y, & j = n+1. \end{cases} \qquad (11.6.13)$$

Each pair (x_j, x_{j+1}) satisfies the conditions of Lemma 11.6.1, since $x_n = d_{n-1} \cdots d_1 c_0$. By Lemma 11.6.1, there are vectors $\eta_j \in \ell^2(\mathbb{F}_2)$ with

$$\|(h\eta_j - \eta_j h) - (x_j - x_{j+1})\|_2 < \varepsilon/(n+1) \tag{11.6.14}$$

for $0 \le j \le n - 1$ or $0 \le j \le n$ in the two cases. Let $\eta = \sum_j \eta_j$. Then

$$\|(h\eta - \eta h) - (x - y)\|_2 \le \sum_j \|(h\eta_j - \eta_j h) - (x_j - x_{j+1})\|_2$$

$$< \varepsilon. \tag{11.6.15}$$

If $n = 1$ with $x = a$ or $y = a$ or a^{-1}, then use $x_0 = a$, $x_1 = b$ and $x_2 = y$ in Lemma 11.6.1 with η the sum of two of the η_j's. This completes the proof. $\square$

We note that if $|x| = |y| = n$, then the choice in Lemma 11.6.2 gives $\|\eta\|_2 \le 6(n+1)\varepsilon^{-2}$.

Theorem 11.6.3. *If $h = a + a^{-1} + b + b^{-1}$ and $B = \{h\}''$ is the Laplacian algebra in $L(\mathbb{F}_2)$, then B is a masa in $L(\mathbb{F}_2)$.*

Proof. Let $z \in L(\mathbb{F}_2)$ with $zh = hz$ and $\|z\|_2 = 1$. Let $x, y \in \mathbb{F}_2$ with $|x| = |y| = n$ and let $\varepsilon > 0$. By Lemma 11.6.2, there is a vector $\eta \in \ell^2(\mathbb{F}_2)$ such that

$$\|(h\eta - \eta h) - (x - y)\|_2 < \varepsilon. \tag{11.6.16}$$

If we apply the Cauchy–Schwarz inequality and the self-adjointness of h, then (11.6.16) yields

$$|\tau(z(x^{-1} - y^{-1}))| = |\langle zh - hz, \eta \rangle_\tau - \langle z, x - y \rangle_\tau|$$
$$= |\langle z, (h\eta - \eta h - x + y) \rangle_\tau| < \varepsilon. \tag{11.6.17}$$

Hence $\tau(zx^{-1}) = \tau(zy^{-1})$ for all $x, y \in \mathbb{F}_2$ with $|x| = |y|$ so that $\|w_n\|_2^2 \tau(zx^{-1}) = \tau(zw_n)$, because $\|w_n\|_2^2$ is the number of words in $\mathbb{F}_2$ of length n. This implies that

$$z = \sum_{n=0}^{\infty} \left(\sum_{|x|=n} \tau(zx^{-1})x \right)$$

$$= \sum_{n=0}^{\infty} \|w_n\|_2^{-2} \tau(zw_n) w_n$$

$$= \mathbb{E}_B(z). \tag{11.6.18}$$

This shows that B' lies in B, so B is a masa. $\square$

Theorem 11.6.4. *The Laplacian masa $B = \{h\}''$ is singular in $L(\mathbb{F}_2)$.*

Proof. By Theorem 11.6.3, B is a masa in $L(\mathbb{F}_2)$ so is a diffuse abelian subalgebra of $L(\mathbb{F}_2)$ by Corollary 3.5.3. By Theorem 3.5.2, the algebra B is isomorphic to $L^\infty[0, 1]$ so contains a Haar unitary. Thus the algebra B satisfies the hypotheses of Corollary 11.4.3 with respect to the basis $\|w_n\|_2^{-1} w_n$ $(n \ge 0)$ of B, and hence is singular. $\square$

Just as the previous theorems may be extended to the Laplacian algebra in the free group factor $L(\mathbb{F}_k)$ for $k \geq 2$, so the theorems may be extended by replacing the generators by positive powers of generators (see [123, 175]).

The treatment of the singularity of the Laplacian masa B given here is taken from [175]. At that time, this was the only way to prove that this masa is strongly singular. After the equivalence of singularity and strong singularity (Theorem 11.1.2) was shown in [177], an alternative route is then to appeal to the singularity of B as proved by Rădulescu in [156].

Chapter 12

Existence of special masas

12.1 Introduction

This chapter is devoted to results of Popa [136, 139, 140, 141] showing that masas with certain additional properties exist in subfactors with trivial relative commutant in separable II_1 factors. His results are more general, and his methods apply more widely than to the II_1 factor case presented here. The proofs are neater for the II_1 factors; the inductive nature of his methods restricts them to the case of separable predual. Note that although Zorn's Lemma may easily be used to show that masas exist, no additional structure is provided by this method.

If N is a separable type II_1 factor and if $A \subseteq M \subseteq N$ are von Neumann algebras with A a masa in N, then

$$M' \cap N \subseteq A' \cap N = A \subseteq N.$$

This condition is thus necessary if masas in N that are contained in M are to exist, and Theorem 12.2.4 (which comes from [136]) shows that it is sufficient. Clearly more cannot be said about A, because if M is a masa in N then $A = M$. The condition that M is a subfactor of N with trivial relative commutant, $M' \cap N = \mathbb{C}1$, ensures that M is large enough to have many masas of N in it. Theorems 12.3.1 and 12.4.3 (first proved in [139, 140]) shows that in this situation M contains both singular and semiregular masas of N. Voiculescu's result [199] that there is no Cartan masa in a free group factor $L(\mathbb{F}_n)$ shows that the algebra cannot be taken to be Cartan, and Dykema's extension of this result [59] shows that the masa cannot have Pukánszky invariant a finite subset of $\mathbb{N}$. The main results of this chapter are summarised for a subfactor with trivial relative commutant in the next theorem.

Theorem 12.1.1. *Let N be a separable type II_1 factor and let M be a subfactor with $M' \cap N = \mathbb{C}1$.*

(i) *There is a singular masa A of N contained in M.*

(ii) *There is a masa A of N and a hyperfinite subfactor R of M with A a Cartan masa in R.*

(iii) *If M is a hyperfinite subfactor of N, then there is a masa A of N with A a Cartan masa in M.*

Note that in case (ii) we have $A \subseteq R \subseteq M$ so $R' \cap N \subseteq A' \cap N = A$ so that $R' \cap N = \mathbb{C}1$ because $A \subseteq R$.

The basic idea is to construct the masa A as the weak closure of an increasing sequence of finite dimensional abelian von Neumann subalgebras A_n of N with various good $\|\cdot\|_2$-norm properties. The basic $\|\cdot\|_2$-norm approximation property of Lemma 12.2.3 ensures that A is a masa and the other approximations lead to semiregularity (Theorem 12.3.1) and singularity (Theorem 12.4.3). The whole process starts with an approximation based on the Dixmier style process arising from averaging using minimal $\|\cdot\|_2$-norm techniques and $M' \cap N \subseteq M$. The following elementary observation will be used many times in this chapter with no subsequent comment. If e_j ($1 \leq j \leq n$) is a finite set of elements in N with $e_k^* e_j = 0$ for all $j \neq k$, then orthogonality implies that

$$\left\| \sum_{j=1}^{n} e_j x_j \right\|_2^2 = \sum_{j=1}^{n} \| e_j x_j \|_2^2 \qquad (12.1.1)$$

for all $x_j \in N$, $1 \leq j \leq n$.

An important result of Popa [139] is proved in Theorem 12.5.6 concerning masas in N associated with a countable group of freely acting properly outer automorphisms of N. We briefly mention another result from the same paper which we will not prove:

> *Let N be a separable type II_1 factor and let θ be an aperiodic automorphism of N. If $\varepsilon > 0$, there is a unitary u in N, a masa A in N and a hyperfinite subfactor R of N with A a Cartan masa in R such that $Ad(u)\theta(R) = R$ and $\|u - 1\|_2 < \varepsilon$.*

12.2   Approximations in subalgebras

This section contains the technical approximation lemmas on which the rest of this chapter is built. Lemma 12.2.1 is one of the typical averaging arguments discussed in section 3.6. Iterating this lemma as in the proof of the Dixmier Approximation Theorem (see Theorem 2.2.2 for the statement and [105, Section 8.3] for the proof) starts a contraction argument for a single element orthogonal to $M' \cap N$. Subsequent lemmas extend it to finite sets and remove the orthogonality requirement. In Theorem 12.2.4, we prove that masas exist in von Neumann subalgebras M of N with $M' \cap N \subseteq M$. Note that if $M \subseteq N$, then $M' \cap N \subseteq M$ is equivalent to $M' \cap N$ is the centre of M.

In several of the lemmas in this section the minimal projections are all chosen to have equal trace 2^{-k} for some positive integer k. This power 2^{-k} may be replaced throughout by m^{-k}, where $m \geq 2$ is a positive integer.

Lemma 12.2.1. *Let M be a von Neumann subalgebra of a II_1 factor N, let $\mathcal{U}(M)$ be the unitary group of M and let x be in N. If $\mathbb{E}_{M' \cap N}(x) = 0$ and $\varepsilon > 0$, then there is a unitary u in M such that*

$$\|uxu^* - x\|_2^2 \geq (2 - \varepsilon)\|x\|_2^2. \tag{12.2.1}$$

Proof. As in Lemma 3.6.5, the set

$$K_M(x) = \overline{\mathrm{conv}}^w\{uxu^* : u \in \mathcal{U}(M)\}$$

has a unique element h of minimal $\| \cdot \|_2$-norm, and $h = \mathbb{E}_{M' \cap N}(x)$.

Suppose that

$$2\|x\|_2^2 - 2\mathrm{Re}\,\tau(uxu^*x^*) = \|uxu^* - x\|_2^2 \leq (2 - \varepsilon)\|x\|_2^2 \tag{12.2.2}$$

for all $u \in \mathcal{U}(M)$, where Re is the real part. Then $2\mathrm{Re}\,\tau(uxu^*x^*) \geq \varepsilon\|x\|_2^2$ for all $u \in \mathcal{U}(M)$ so by convexity and continuity,

$$2\mathrm{Re}\,\tau(\mathbb{E}_{M' \cap N}(x)x^*) \geq \varepsilon\|x\|_2^2, \tag{12.2.3}$$

which contradicts $\mathbb{E}_{M' \cap N}(x) = 0$. $\qquad\square$

Lemma 12.2.2. *Let N be a II_1 factor and let M be a von Neumann subalgebra of N with $M' \cap N \subseteq M$. If x is in N with $\mathbb{E}_{M' \cap N}(x) = 0$, then there is a finite dimensional abelian *-subalgebra A of M such that*

$$\|\mathbb{E}_{A' \cap N}(x)\|_2^2 \leq 3/4\|x\|_2^2 \tag{12.2.4}$$

and all the minimal projections of A may be chosen to have the same trace 2^{-k} for some positive integer k.

Proof. Assume that $\|x\| = 1$. By Lemma 12.2.1 (ii) there is a unitary $u \in \mathcal{U}(M)$ such that

$$\|uxu^* - x\|_2 > 5/4\|x\|_2, \tag{12.2.5}$$

since $\|x\|_2 > 0$. By the spectral theorem for the unitary u there are spectral projections $f_1, \ldots, f_t$ in M and complex numbers $\mu_1, \ldots, \mu_t$ of modulus 1 such that $\sum f_j = 1$ and $\|u - \sum \mu_j f_j\| < 1/16$. By approximating the finite dimensional algebra $\sum \mathbb{C}f_j$ by a suitable finite dimensional abelian von Neumann algebra A all of whose minimal projections e_j, $1 \leq j \leq 2^k$, have equal trace, there are complex numbers λ_j of modulus 1 such that

$$\left\| u - \sum_{j=1}^{2^k} \lambda_j e_j \right\|_2 < 1/8. \tag{12.2.6}$$

Each λ_j may be chosen to be a suitable μ_j and each f_j will be approximately a finite sum of some e_i's in $\| \cdot \|_2$-norm. Inequalities (12.2.5) and (12.2.6), and $\|x\| \leq 1$ imply that

$$\left\| \left(\sum_{j=1}^{2^k} \lambda_j e_j \right) x \left(\sum_{j=1}^{2^k} \lambda_j e_j \right)^* - x \right\|_2 > \|x\|_2. \tag{12.2.7}$$

The elements $e_i x e_j$, $1 \leq i, j \leq 2^k$, are pairwise orthogonal in $L^2(N)$ since $e_i e_j = 0$ if $i \neq j$. Using this orthogonality, in addition to $\lambda_j \overline{\lambda}_j = 1$ and the inequality $|\lambda_i \overline{\lambda}_j - 1| \leq 2$ for all i, j, gives

$$4\|x\|_2^2 - 4\|\sum_j e_j x e_j\|_2^2 = 4\|\sum_{i,j} e_i x e_j\|_2^2 - 4\|\sum_j e_j x e_j\|_2^2$$

$$= 4\|\sum_{i \neq j} e_i x e_j\|_2^2 \geq \sum_{i \neq j} \|(\lambda_i \overline{\lambda}_j - 1) e_i x e_j\|_2^2$$

$$= \|\sum_{i,j}(\lambda_i \overline{\lambda}_j - 1) e_i x e_j\|_2^2$$

$$= \|(\sum_i \lambda_i e_i) x (\sum_j \lambda_j e_j)^* - x\|_2^2 > \|x\|_2^2 \qquad (12.2.8)$$

by (12.2.7). Thus

$$\|\mathbb{E}_{A' \cap N}(x)\|_2^2 = \left\|\sum_{j=1}^{2^k} e_j x e_j\right\|_2^2 < 3/4\|x\|_2^2. \qquad (12.2.9)$$

□

Note that any finite dimensional abelian von Neumann algebra B containing A in the above lemma has

$$\|\mathbb{E}_{B' \cap N}(x)\|_2 \leq 3/4\|x\|_2, \qquad (12.2.10)$$

since $B' \cap N \subseteq A' \cap N$.

Lemma 12.2.3. *Let N be II_1 factor, let M be a von Neumann subalgebra of N with $M' \cap N \subseteq M$, let B be a finite dimensional abelian von Neumann subalgebra of M, and let $\varepsilon > 0$. If $x_1, \ldots, x_n$ are in N, then there is a finite dimensional abelian von Neumann subalgebra A of M with $B \subseteq A$ such that*

$$\|\mathbb{E}_{A' \cap N}(x_j) - \mathbb{E}_A(x_j)\|_2 < \varepsilon \qquad (12.2.11)$$

for $1 \leq j \leq n$.

If all the minimal projections of B have equal trace (an integer power of 2), then A may be chosen with all it minimal projections of equal trace (an integer power of 2).

Proof. The proof is in three parts. The algebra A is first constructed for a single element x satisfying $\mathbb{E}_{M' \cap N}(x) = 0$ so $\|\mathbb{E}_{A' \cap N}(x)\|_2 < \varepsilon$. This is then extended to a finite number of elements with this property and the final result follows with the aid of the basic masa approximation in Corollary 3.6.9.

An increasing family of finite dimensional abelian von Neumann subalgebras $B = C_0 \subseteq C_1 \subseteq \ldots \subseteq M$ is constructed such that

$$\|\mathbb{E}_{C_k' \cap N}(x)\|_2^2 \leq (3/4)^k \|x\|_2^2 \qquad (12.2.12)$$

and the minimal projections f in C_k all have equal trace (an integer power of 2) if all the projections of B have equal trace (an integer power of 2).

Note that inequality (12.2.12) holds for $k = 0$ because $\mathbb{E}_{B' \cap N}$ is contractive for the $\|\cdot\|_2$-norm. Suppose that the algebras up to and including C_n have been constructed satisfying the required conditions and that the minimal projections of C_n are f_j ($1 \leq j \leq r$). Apply Lemma 12.2.2 to the element $f_j x f_j$ in the factor $f_j N f_j$ with von Neumann subalgebra $f_j M f_j$ for $1 \leq j \leq r$. Observe that

$$(f_j M f_j)' \cap f_j N f_j = f_j(M' \cap N)f_j \subseteq f_j M f_j \qquad (12.2.13)$$

with these algebras acting on $L^2(f_j N f_j)$ and that

$$\mathbb{E}_{f_j M f_j}(f_j x f_j) = f_j \mathbb{E}_M(x) f_j = 0 \qquad (12.2.14)$$

since $f_j \in M$ for $1 \leq j \leq r$. For each j there is a family of pairwise orthogonal projections in $f_j M f_j$ summing to f_j such that if B_j is the abelian algebra they generate, then

$$\|\mathbb{E}_{B_j' \cap f_j N f_j}(f_j x f_j)\|_2^2 \leq 3/4 \|f_j x f_j\|_2^2. \qquad (12.2.15)$$

The $\|\cdot\|_2$-norm in this inequality is calculated initially in the algebra $f_j N f_j$ and then in N as the $\|\cdot\|_2$-norm in $f_j N f_j$ is just a scalar multiple of the $\|\cdot\|_2$-norm in N. Let C_{k+1} be the abelian von Neumann subalgebra of M generated by the family B_j ($1 \leq j \leq r$) so that

$$C_{k+1} = \bigoplus_{j=1}^{r} B_j. \qquad (12.2.16)$$

By orthogonality, calculating all $\|\cdot\|_2$-norms in N,

$$\|\mathbb{E}_{C_{k+1}' \cap N}(x)\|_2^2 = \sum_{j=1}^{r} \|\mathbb{E}_{B_j' \cap f_j N f_j}(f_j x f_j)\|_2^2 \leq 3/4 \sum \|f_j x f_j\|_2^2$$
$$= 3/4 \|\mathbb{E}_{C_k' \cap N}(x)\|_2^2 \leq (3/4)^{k+1} \|x\|_2^2. \qquad (12.2.17)$$

by inequalities (12.2.15) and (12.2.12). The minimal projections in C_{k+1} can be constructed to all have the same trace that is an integer power of 2 if this applies to A_k though the choice has to be made all at one time to get a high enough power of 2^{-1} to deal with all the algebras B_j ($1 \leq j \leq r$). This proves the existence of the family C_j for a single element of N with $\mathbb{E}_{M' \cap N}(x) = 0$.

Now let $\{x_1, \ldots, x_n\}$ be a finite subset of N with $\mathbb{E}_M(x_j) = 0$ for $1 \leq j \leq n$. By induction on j one constructs a sequence of finite dimensional abelian von Neumann subalgebras A_j ($0 \leq j \leq n$) of M such that $A_0 = B$, $A_j \subseteq A_{j+1}$ and $\|\mathbb{E}_{A_j' \cap N}(x_j)\|_2 < \varepsilon \|x_j\|_2$ for $0 \leq j \leq n$. Once A_j is constructed this follows from the first part by taking k large enough that $(3/4)^k < \varepsilon$ and choosing $B = A_j$ and $x = x_{j+1}$, and letting A_{j+1} be the C_k of the first part of this proof. Once the sequence A_j ($0 \leq j \leq n$) is constructed, then

$$\|\mathbb{E}_{A_n' \cap N}(x_j)\|_2 \leq \|\mathbb{E}_{A_j' \cap N}(x_j)\|_2 \leq \varepsilon \|x_j\|_2 \qquad (12.2.18)$$

for $1 \leq j \leq n$ since $A'_n \cap N \subseteq A'_j \cap N$.

The condition $\mathbb{E}_{M' \cap N}(x_j) = 0$ is now removed. Let $\varepsilon > 0$ and let $x_1, \ldots, x_n$ be in N. Let

$$\varepsilon' = 2^{-1}\varepsilon(1 + \sum_{j=1}^{n} \|x_j\|_2)^{-1}, \qquad (12.2.19)$$

let $x'_j = \mathbb{E}_M(x_j)$ and let $x''_j = x_j - x'_j$ for $1 \leq j \leq n$. Then $\mathbb{E}_M(x''_j) = 0$ for $1 \leq j \leq n$. By the proof above there is a finite dimensional abelian *-subalgebra C of M containing B such that

$$\|\mathbb{E}_{C' \cap N}(x''_j)\|_2 \leq \varepsilon' \|x''_j\|_2 \qquad (12.2.20)$$

for $1 \leq j \leq n$. Let e_i, $1 \leq i \leq t$, be the set of minimal projections in C. Let $A_{0,i}$ be a masa in the von Neumann algebra $e_i M e_i$ for $1 \leq i \leq t$. By Corollary 3.6.9, there is a finite dimensional von Neumann subalgebra $A_{1,i}$ of $A_{0,i}$ such that

$$\|\mathbb{E}_{A'_{1,i} \cap e_i M e_i}(e_i x'_j e_i) - \mathbb{E}_{A_{1,i}}(e_i x'_j e_i)\|_2 \leq \varepsilon' \|e_i x_j e_i\|_2 \qquad (12.2.21)$$

for $1 \leq j \leq n$ and $1 \leq i \leq t$. This first holds in the $\| \cdot \|_2$-norm of $e_i N e_i$ and then in N, as these norms are scalar multiples of one another.

Let $A = \bigoplus_{i=1}^{t} A_{1,i}$. Then A is a finite dimensional von Neumann subalgebra of M containing B and C; if the choice of all the $A_{1,i}$'s is carefully made together then all the minimal projections of all the $A_{1,i}$'s have the same trace, which is an integer power of 2. Since x'_j is in M,

$$\|\mathbb{E}_{A' \cap N}(x'_j) - \mathbb{E}_A(x'_j)\|_2^2 = \|\mathbb{E}_{A' \cap M}(x'_j) - \mathbb{E}_A(x'_j)\|_2$$

$$= \sum_{i=1}^{t} \|e_i \mathbb{E}_{A'_{1,i} \cap e_i M e_i}(x'_j)e_i - e_i \mathbb{E}_{A_{1,i}}(x'_j)e_i\|_2^2$$

$$= \sum_{i=1}^{t} \|\mathbb{E}_{A'_{1,i} \cap e_i M e_i}(e_i x'_j e_i) - \mathbb{E}_{A_{1,i}}(e_i x'_j e_i)\|_2^2$$

$$\leq (\varepsilon')^2 \sum_{i=1}^{t} \|e_i x'_j e_i\|_2^2 = (\varepsilon')^2 \|x'_j\|_2^2 \qquad (12.2.22)$$

by orthogonality and inequality (12.2.21).

Hence

$$\|\mathbb{E}_{A' \cap N}(x_j) - \mathbb{E}_A(x_j)\|_2$$
$$\leq \|\mathbb{E}_{A' \cap N}(x''_j) - \mathbb{E}_A(x''_j)\|_2 + \|\mathbb{E}_{A' \cap N}(x'_j) - \mathbb{E}_A(x'_j)\|_2$$
$$\leq \varepsilon' \|x'_j\|_2 + \|\mathbb{E}_{C' \cap N}(x''_j)\|_2 \qquad (12.2.23)$$

since $A' \cap N \subseteq C' \cap N$, $A \supseteq C$ and $\mathbb{E}_A(x''_j) = 0$, because $A \subseteq M$ and $\mathbb{E}_M(x''_j) = 0$.

By (12.2.22) and (12.2.23) it follows that

$$\|\mathbb{E}_{A' \cap N}(x_j) - \mathbb{E}_A(x_j)\|_2 \leq \varepsilon'(\|x'_j\|_2 + \|x''_j\|_2)$$
$$\leq 2\varepsilon' \|x_j\|_2 < \varepsilon \qquad (12.2.24)$$

for $1 \leq j \leq n$ by the choice of ε' (12.2.19). $\square$

At the 1967 operator algebras conference at Baton Rouge, an influential list of important problems was informally circulated by Kadison (and published many years later by Ge in [79]). One of these problems (fully solved in [136]) is:

> *if M is a subfactor of a separable factor N with $M' \cap N = \mathbb{C}1$, must there exist a masa A of M that is a masa of N?*

Note that for a subfactor M of N the conditions $M' \cap N \subseteq M$ and $M' \cap N = \mathbb{C}1$ are equivalent. We will describe the positive solution for II_1 factors in Section 12.3. Theorem 12.2.4 is a precursor to Theorems 12.3.1 and 12.4.3 in which additional properties are built into A in the inductive construction.

Theorem 12.2.4. *Let M be a von Neumann subalgebra of a separable type II_1 factor N with $M' \cap N \subseteq M$. If B is a finite dimensional abelian *-subalgebra of M, then there is a masa A of N with $B \subseteq A \subseteq M$. In particular, this holds for any subfactor with trivial relative commutant.*

Proof. Let $\{x_j\}_{j=0}^{\infty}$ be a sequence in the unit ball of N with $x_0 = 1$ that is $\| \cdot \|_2$ dense in this ball. By induction an increasing sequence $\{A_n\}_{n=0}^{\infty}$ of finite dimensional von Neumann subalgebras of M is constructed so that $A_0 = B$ and

$$\|\mathbb{E}_{A_n' \cap N}(x_j) - \mathbb{E}_{A_n}(x_j)\|_2 \leq 2^{-n} \quad \text{for} \quad 1 \leq j \leq n. \tag{12.2.25}$$

Once A_n has been chosen, select A_{n+1} by Lemma 12.2.3 (with B there equal to A_n) so that $A_n \subseteq A_{n+1}$ and $\|\mathbb{E}_{A_{n+1}' \cap N}(x_j) - \mathbb{E}_{A_{n+1}(x_j)}\|_2 \leq 2^{-n-1}$ for $1 \leq j \leq n$. If A is the weak closure of $\bigcup_{n \geq 0} A_n$, then $B \subseteq A \subseteq N$ and A is an abelian von Neumann algebra. By (12.2.25) and the $\| \cdot \|_2$-density of $\{x_j : j \geq 0\}$ in the unit ball of N, it follows that $\lim_{n \to \infty} \|\mathbb{E}_{A_n' \cap N}(x) - \mathbb{E}_A(x)\|_2 = 0$ for each $x \in N$. Hence A is a masa in N by Corollary 3.6.9. $\qquad\square$

12.3 Constructing semiregular masas

The main result of this section (Theorem 12.3.1) is that there is an abundance of semiregular masas in a separable II_1 factor. As we will see in Corollary 12.3.2, there are sufficiently many to separate the subfactors with trivial relative commutant.

Note that if A is a masa in N which is Cartan in a subfactor $M \subseteq N$ of trivial relative commutant, then the algebra $\mathcal{N}(A, N)''$ generated by the normaliser of A in N lies between M and N, and is consequently a subfactor of N. Thus the masas constructed in Theorem 12.3.1 are automatically semiregular.

Theorem 12.3.1. *Let N be a separable II_1 factor.*

(i) *If M is a subfactor of N with $M' \cap N = \mathbb{C}1$ and C is a finite dimensional abelian *-subalgebra of M with all its minimal projections of equal trace, then there is a masa A in N containing C and a hyperfinite subfactor R of M with A a Cartan masa in R and $R' \cap N = \mathbb{C}1$.*

(ii) *If M is a hyperfinite subfactor of N with $M' \cap N = \mathbb{C}1$, then there is a masa A in N with A a Cartan masa in M.*

Proof. Part (i) is proved and then the additional inductive condition needed for (ii) is indicated.

Let $\{x_j : j \geq 0\}$ be a $\| \cdot \|_2$-norm dense sequence in the unit ball of N with $x_0 = 1$. This exists by the separability of N in $\| \cdot \|_2$-norm. Let C have minimal projections e_{ii}^0 for $1 \leq i \leq k_0$ and let M_0 be a matrix subalgebra of M with diagonal C and matrix units e_{ij}^0 $(1 \leq i, j \leq k_0)$. By induction choose increasing sequences of

(1) finite dimensional abelian *-subalgebras $A_0 = C \subseteq A_1 \subseteq A_2 \ldots$ and

(2) matrix algebras $M_0 \subseteq M_1 \subseteq M_2 \ldots$ of M such that, if e_{ij}^n $(1 \leq i, j \leq k_n)$ are the matrix units of M_n, then

(3) $\{e_{ii} : 1 \leq i \leq k_n\}$ is the set of minimal projections of A_n, and

(4) $\|\mathbb{E}_{A_n' \cap N}(x_j) - \mathbb{E}_{A_n}(x_j)\|_2 < 2^{-n}$ for $1 \leq j \leq n$.

Observe that (3) says that A_n is the diagonal of M_n with respect to the matrix units e_{ij}^n. The algebras $A_0 \subseteq M_0$ have been constructed. Suppose that A_n, M_n, $e_{i,j}^n$ $(1 \leq i, j \leq k_n)$ have been constructed to satisfy (1) to (4). To construct M_{n+1} we do the construction of $e_{11}M_{n+1}e_{11}$ in $e_{11}Me_{11}$ and then shift this to form M_{n+1} by the matrix units e_{ij}^n. This is the reason that the elements x_j $(1 \leq j \leq n+1)$ are pulled back to $e_{11}^n N e_{11}^n$ in (12.3.1). Note that the subfactor $e_{11}^n M e_{11}^n$ has trivial relative commutant $\mathbb{C}e_{11}$ in $e_{11}^n N e_{11}^n$ by Theorem 5.4.1. By Lemma 12.2.3 there is a finite dimensional abelian *-subalgebra B of $e_{11}^n M e_{11}^n$ with all the minimal projections of B having the same trace such that

$$\|(\mathbb{E}_{B' \cap e_{11}^n N e_{11}^n} - \mathbb{E}_B)(e_{1j}^n x_t e_{j1}^n)\|_2 < 2^{-(n+1)} k_n^{-1/2} \qquad (12.3.1)$$

for $1 \leq t \leq n+1$ and $1 \leq j \leq k_n$. Let e_{ii}^{n+1} $(1 \leq i \leq m_n)$ denote the set of minimal projections of B, which are all equivalent in $e_{11}^n M e_{11}^n$. Thus there are partial isometries v_j in M with

$$v_j^* v_j = e_{11}^{n+1} \quad \text{and} \quad v_j v_j^* = e_{jj}^{n+1} \quad \text{for} \quad 1 \leq j \leq m_n. \qquad (12.3.2)$$

Let $e_{ij}^{n+1} = v_i v_j^*$ for $1 \leq i, j \leq m_n$. Then $\{e_{ij}^{n+1} : 1 \leq i, j \leq m_n\}$ form a set of matrix units for a subalgebra of $e_{11}^n M e_{11}^n$, which are now moved to form the matrix algebra M_{n+1} using e_{st}^n.

If $k_{n+1} = k_n m_n$, then

$$\{e_{i1}^n e_{st}^{n+1} e_{1j}^n : 1 \leq s, t \leq m_n, \ 1 \leq i, j \leq k_n\} \qquad (12.3.3)$$

is a set of $k_{n+1} \times k_{n+1}$ matrix units for a *-subalgebra M_{n+1} of M. The abelian algebra spanned by

$$\{e_{j1}^n e_{tt}^{n+1} e_{1j}^n : 1 \leq t \leq m_n, \ 1 \leq j \leq k_n\}, \qquad (12.3.4)$$

is denoted A_{n+1} and is the diagonal of $M_{n+1} \cong \mathbb{M}_{k_{n+1}}$. Then $A_n \subseteq A_{n+1}$ and $M_n \subseteq M_{n+1}$.

It is necessary to confirm (4) at $n+1$. The projections e_{jj}^n are in A_{n+1}, so

$$\mathbb{E}_{A'_{n+1} \cap N}(x) - \mathbb{E}_{A_{n+1}}(x) = \sum_{j=1}^{k_n} e_{jj}^n \big(\mathbb{E}_{A'_{n+1} \cap N}(x) - \mathbb{E}_{A_{n+1}}(x)\big) e_{jj}^n \qquad (12.3.5)$$

for all $x \in N$. The pairwise orthogonality of $e_{jj}^n N e_{jj}^n$ for $1 \le j \le k_n$ implies that

$$\|\mathbb{E}_{A'_{n+1} \cap N}(x) - \mathbb{E}_{A_{n+1}}(x)\|_2^2 = \sum_{j=1}^{k_n} \|e_{jj}^n \big(\mathbb{E}_{A'_{n+1} \cap N}(x) - \mathbb{E}_{A_{n+1}}(x)\big) e_{jj}^n\|_2^2$$

$$= \sum_{j=1}^{k_n} \|\big(\mathbb{E}_{A'_{n+1} \cap N} - \mathbb{E}_{A_{n+1}}\big)\big(e_{jj}^n x e_{jj}^n\big)\|_2^2, \qquad (12.3.6)$$

since $e_{jj}^n \in A_n \subseteq A_{n+1} \subseteq A'_{n+1} \cap N$.

The map $e_{jj}^n x e_{jj}^n \mapsto e_{1j}^n x e_{j1}^n$ is a $\| \cdot \|_2$-isometric isomorphism from $e_{jj}^n N e_{jj}^n$ onto $e_{1j}^n N e_{j1}^n$ that carries $e_{jj}^n \big(A'_{n+1} \cap N\big) e_{jj}^n$ onto $e_{1j}^n \big(A'_{n+1} \cap N\big) e_{j1}^n$. The definition of A_{n+1} in terms of the minimal projections of B in (12.3.4) implies that $e_{11}^n \big(A'_{n+1} \cap N\big) e_{11}^n = e_{11}^n \big(B' \cap N\big) e_{11}^n$ and $e_{11}^n A_{n+1} e_{11}^n = e_{11}^n B e_{11}^n$. Thus

$$\|\mathbb{E}_{A'_{n+1} \cap N}(x) - \mathbb{E}_{A_{n+1}}(x)\|_2^2$$

$$= \sum_{j=1}^{k_n} \|\big(\mathbb{E}_{B' \cap e_{11}^n N e_{11}^n} - \mathbb{E}_B\big)\big(e_{1j}^n x e_{j1}^n\big)\|_2^2 \qquad (12.3.7)$$

for all $x \in N$ by (12.3.6). Replacing x in (12.3.7) by x_t for $1 \le t \le n+1$ and using (12.3.1) leads to

$$\|\mathbb{E}_{A'_{n+1} \cap N}(x_t) - \mathbb{E}_{A_{n+1}}(x_t)\|_2^2 \le \sum_{j=1}^{k_n} 4^{-(n+1)} k_n^{-1} = 4^{-(n+1)}, \qquad (12.3.8)$$

which proves (4) for $n+1$.

Let R be the weak closure of $\cup M_n$ and let A be the weak closure of $\cup A_n$. Clearly R is a hyperfinite subfactor of M and A is contained in R. By (4) and the $\| \cdot \|_2$-density of $\{x_j : j \ge 0\}$ in the closed unit ball of N, it follows that $\lim_{n \to \infty} \|\mathbb{E}_{A'_n \cap N}(x) - \mathbb{E}_{A_n}(x)\|_2 = 0$ for all $x \in N$. By Corollary 3.6.9, A is a masa in N. Hence $R' \cap N \subseteq A' \cap N = A$, and so $R' \cap N \subseteq R' \cap A \subseteq R' \cap R = \mathbb{C}1$ as required. Finally note that if u is a unitary in M_n with $uA_nu^* = A_n$, then $uA_ku^* = A_k$ for all $k \ge n$ so that $uAu^* = A$. Each M_n is generated by its unitaries that normalise A_n, and thus R is generated by the nomaliser $\mathcal{N}(A, R)$. Hence A is a Cartan masa in R so is semiregular in N.

(ii)   Now additionally suppose that M is hyperfinite and let $\{y_n : n \ge 0\}$ be a $\| \cdot \|_2$-dense sequence in the unit ball of M with $y_0 = 1$. Add to the inductive conditions (1) to (4) of part (i) the inductive hypothesis

(5) $\|y_j - \mathbb{E}_{M_n}(y_j)\|_2 \leq 2^{-n}$ for $1 \leq j \leq n$.

Suppose that A_n, M_n and e_{ij}^n have been constructed to satisfy (1) to (5). Then proceed as in part (i) to construct algebras and matrix units as before but denoting the $n+1$ versions just constructed by A_{n+1}^0, M_{n+1}^0 and $e_{ij}^{0,n+1}$, which satisfy (1) to (4) for $n+1$.

Since M is hyperfinite, it is the weak closure of an increasing sequence of matrix subalgebras, and so there is a matrix algebra M_{n+1} with $M_{n+1}^0 \subseteq M_{n+1} \subseteq M$ such that

$$\|y_j - \mathbb{E}_{M_{n+1}}(y_j)\|_2 \leq 2^{-n-1} \quad \text{for} \quad 1 \leq j \leq n+1. \tag{12.3.9}$$

Let A_{n+1} be a masa (= diagonal) in M_{n+1} containing A_{n+1}^0 and let e_{ij}^{n+1} be a set of matrix units of M_{n+1} with $e_{st}^{0,n+1}$ sums of suitable e_{ij}^{n+1} for all s,t. By Corollary 3.6.9,

$$\|\mathbb{E}_{A_{n+1}' \cap N}(x_j) - \mathbb{E}_{A_{n+1}}(x_j)\|_2 \leq \|\mathbb{E}_{A_{n+1}^{0'} \cap N}(x_j) - \mathbb{E}_{A_{n+1}^0}(x_j)\|_2$$
$$\leq 2^{-n-1} \tag{12.3.10}$$

for $1 \leq j \leq n+1$. The properties (1) to (4) now follow easily for A_{n+1} and M_{n+1}. Thus A is a masa in N as before and is a Cartan masa in $R = \overline{\cup M_n}^w$. Here condition (5) implies that $R = M$ completing the proof. $\qquad \square$

Corollary 12.3.2. *Let N be a separable II_1 factor with subfactors M_1 and M_2 both with trivial relative commutant. If $M_1 \not\subseteq M_2$, then there is a semiregular masa A in N with $A \subseteq M_1$ and $A \not\subseteq M_2$.*

Proof. Let $h = h^*$ be in M_1 but not in M_2. By approximating h using the spectral theorem, there is a finite dimensional abelian *-subalgebra C of M_1 with $C \not\subseteq M_2$ with all the minimal projections of C having the same trace. By Theorem 12.3.1 there is a semiregular masa A of N with $C \subseteq A \subseteq M_1$. $\qquad \square$

Remark 12.3.3. The results in this section are from [136, Theorem 3.2] (see also [189]) where a more general result for a semifinite subfactor M of a separable factor N is proved. In the case where $M = N$ the results of section 12.2 are not required and the situation is much simpler. See [172] for an elementary account of this, which has been used in calculations of the continuous Hochschild cohomology groups $H^n(N, N) = 0$, $n \geq 1$. $\qquad \square$

12.4 Constructing singular masas

In this section we show that there are enough singular masas in a separable type II_1 factor to separate subfactors with trivial relative commutant. The method is similar to that used to show that there are many semiregular masas: an additional condition is added to those in the inductive construction of the basic masa (Theorem 12.3.1). A metric condition is required to pass from a countable dense set of partial isometries, or unitaries, having a singularity property to their closure to force singularity. The method used here is due to Popa [139] and involves showing that his delta invariant $\delta(A)$ of the constructed masa A has $\delta(A) > 1/49$. This is equivalent to the statement:

> if A is a masa in N and if v is a partial isometry in N with vv^* and v^*v orthogonal projections in A, then
>
> $$\|(I - \mathbb{E}_A)\mathbb{E}_{vAv^*}\|_{\infty,2} \geq \|vv^*\|_2/49.$$

The advantage of this inequality is that its validity on a countable dense set of nilpotent partial isometries in N implies that it holds for all nilpotent isometries in N. A partial isometry v is nilpotent if, and only if, vv^* and v^*v are orthogonal projections.

Lemma 12.4.1 is the core of the proof of Theorem 12.4.3.

Lemma 12.4.1. *Let N be a II_1 factor, let M be a subfactor with $M' \cap N = \mathbb{C}1$, let B be a finite dimensional abelian *-subalgebra of M with $1 \in B$ and let $\varepsilon > 0$. If v is a partial isometry in N with $v^2 = 0$ and F is a finite subset of N, then there is a finite dimensional abelian *-subalgebra A of M with $B \subseteq A$ such that*

(1) $\|\mathbb{E}_A(x) - \mathbb{E}_{A'\cap N}(x)\|_2 \leq \varepsilon$ *for all $x \in F$,*

and at least one of the following statements holds:

(2) *there is an element a in A with $\|a\| = 1$ such that*

$$\|vav^* - \mathbb{E}_{A'\cap N}(vav^*)\| > \|vv^*\|_2/49; \qquad (12.4.1)$$

(3) *there is an element a in A with $\|a\| = 1$ such that*

$$\|v^*av - \mathbb{E}_{A'\cap N}(v^*av)\| > \|vv^*\|_2/49. \qquad (12.4.2)$$

Proof. By Lemma 12.2.3 there is a finite dimensional abelian *-subalgebra C of M with $B \subseteq C$ such that

$$\|\mathbb{E}_C(x) - \mathbb{E}_{C'\cap N}(x)\|_2 < \varepsilon \text{ for all } x \in F. \qquad (12.4.3)$$

Each finite dimensional abelian *-subalgebra A of M with $C \subseteq A$ will also satisfy this inequality (12.4.5) by Corollary 3.6.9, since

$$\|\mathbb{E}_A(x) - \mathbb{E}_{A'\cap N}(x)\|_2 \leq \|\mathbb{E}_C(x) - \mathbb{E}_{C'\cap N}(x)\|_2, \qquad (12.4.4)$$

for $x \in N$.

Let $\alpha > 0$, to be chosen later. Let $e_1 = v^*v$ and $e_2 = vv^*$. If

$$\|e_j - \mathbb{E}_{C'\cap N}(e_j)\|_2 > \alpha\|e_j\|_2 \qquad (12.4.5)$$

for either $j = 1$ or 2, then take $A = C$ and $a = 1$ so giving (2) or (3).

Suppose that

$$\|e_j - \mathbb{E}_{C' \cap N}(e_j)\|_2 \le \alpha \|e_j\|_2 \text{ for } j = 1, 2. \tag{12.4.6}$$

Note that $\|e_1\|_2 = \|e_2\|_2$. Since $e_1 e_2 = v^* v v v^* = 0$, there are projections f_1 and f_2 in $C' \cap N$ such that $f_1 f_2 = 0$ and

$$\|e_j - f_j\|_2 \le (2^{3/2} + 2)\|e_j - \mathbb{E}_{C' \cap N}(e_j)\|_2 \le 5\alpha \|e_j\|_2, \tag{12.4.7}$$

by Lemma 5.3.3 and inequality (12.4.6).

Let $g_1, \ldots, g_n$ be the minimal projections of the finite dimensional *-subalgebra $f_1 C$. The algebra $g_j M g_j$ is a II_1 factor for each j. By splitting the projections g_j in half, there are two projections g_{1j} and g_{2j} and a partial isometry w_j in $g_j M g_j$ such that

$$g_j = g_{1j} + g_{2j}, \ g_{1j} = w_j^* w_j \text{ and } g_{2j} = w_j w_j^* \text{ for } 1 \le j \le n. \tag{12.4.8}$$

Let

$$x_1 = \sum_{j=1}^{n} (g_{1j} - g_{2j}) \tag{12.4.9}$$

and

$$x_2 = f_2 \mathbb{E}_{C' \cap N}(v x_1 v^*) f_2. \tag{12.4.10}$$

Since g_{ij} are pairwise orthogonal projections and the g_j are the minimal projections of $f_1 C$, we have

$$x_1 x_1^* = x_1^2 = \sum_{j=1}^{n} (g_{1j} + g_{2j}) = \sum_{j=1}^{n} g_j = f_1 \tag{12.4.11}$$

so

$$\|x_1\|_2 = \|f_1\|_2. \tag{12.4.12}$$

There are now two possibilities each leading to a different construction of A and $a \in A$ with the choice depending on $\|v^* x_1 v - x_2\|_2$. Let $\beta > 0$ to be chosen subsequently together with α and with both choices being independent of N, M, C, F and ε.

First possibility: Suppose that

$$\|v^* x_1 v - x_2\|_2 \ge \beta \|e_1\|_2. \tag{12.4.13}$$

Let A be the finite dimensional abelian *-subalgebra of M generated by the pairwise orthogonal projections g_{ij} ($i = 1, 2; \ j = 1, \ldots, n$) and by the abelian algebra $(1 - f_1)C$, so $A = (1 - f_1)C + \bigoplus_{ij} \mathbb{C} g_{ij}$. Let $a = x_1$ so $\|a\| = 1$ by equation (12.4.9). Since $f_1 = \sum_j g_j = \sum_{i,j} g_{ij}$ is in A, we have $(1 - f_1)C = (1 - f_1)A$. Thus the two sets of minimal projections q in A and C with $q f_2 \ne 0$ are the same. If $x \in N$, then

$$
\begin{aligned}
\mathbb{E}_{A' \cap N}(f_2 x f_2) &= \sum \{q f_2 x f_2 q : q \text{ a minimal projection in } A\} \\
&= \sum \{q f_2 x f_2 q : q f_2 \neq 0, \ q \text{ a minimal projection in } A\} \\
&= \sum \{q f_2 x f_2 q : q f_2 \neq 0, \ q \text{ a minimal projection in } C\} \\
&= \sum \{q f_2 x f_2 q : q f_2 \neq 0, \ q \text{ a minimal projection in } C\} \\
&= \mathbb{E}_{C' \cap N}(f_2 x f_2) \qquad\qquad\qquad\qquad (12.4.14)
\end{aligned}
$$

by the equation (3.6.42) for the conditional expectation of $\mathbb{E}_{A' \cap N}$ for a finite dimensional abelian *-subalgebra A. Since $f_2 \in C' \cap N$, equation (12.4.10) gives

$$
\begin{aligned}
x_2 &= f_2 \mathbb{E}_{C' \cap N}(v x_1 v^*) f_2 = \mathbb{E}_{C' \cap N}(f_2 v x_1 v^* f_2) \\
&= \mathbb{E}_{A' \cap N}(f_2 v x_1 v^* f_2). \qquad\qquad\qquad\qquad (12.4.15)
\end{aligned}
$$

Using $e_2 v = v$, (12.4.15) implies that

$$
\begin{aligned}
\|x_2 - \mathbb{E}_{A' \cap N}(v x_1 v^*)\|_2 &= \|\mathbb{E}_{A' \cap N}(f_2 v x_1 v^* f_2 - v x_1 v^*)\|_2 \\
&\leq \|f_2 v x_1 v^* f_2 - v x_1 v^*\|_2 \\
&\leq \|(f_2 - e_2) v x_1 v^* f_2\|_2 + \|e_2 v x_1 v^* (f_2 - e_2)\|_2 \\
&\leq 2 \|f_2 - e_2\|_2. \qquad\qquad\qquad\qquad (12.4.16)
\end{aligned}
$$

If $a = x_1$, then by inequalities (12.4.7), (12.4.13) and (12.4.16)

$$
\begin{aligned}
\|v a v^* - \mathbb{E}_{A' \cap N}(v a v^*)\|_2 &\geq \|v x_1 v^* - x_2\|_2 - \|x_2 - \mathbb{E}_{A' \cap N}(v x_1 v^*)\|_2 \\
&\geq \|v x_1 v^* - x_2\|_2 - 2 \|f_2 - e_2\|_2 \\
&\geq (\beta - 10\alpha) \|e_2\|_2. \qquad\qquad\qquad\qquad (12.4.17)
\end{aligned}
$$

This proves (2) if $\beta = 11\alpha$ provided that $1/49 < \alpha < 1/48$, where the choice of the constant α lies in the little optimisation at the end of the second possibility.
Second possibility: Suppose that

$$
\|v^* x_1 v - x_2\|_2 < \beta \|e_1\|_2. \qquad\qquad\qquad\qquad (12.4.18)
$$

The method here is to rotate each of the projections g_{1j}, g_{2j} through an angle of $\pi/4$ using the w_j of (12.4.8), and to define A in terms of the resulting projections. The rotated projections and their properties are handled first.

Let

$$
\begin{aligned}
h_{1j} &= 2^{-1}(g_j + w_j + w_j^*), \\
h_{2j} &= g_j - h_{1j} \text{ for } 1 \leq j \leq n. \qquad\qquad\qquad\qquad (12.4.19)
\end{aligned}
$$

The *-algebra generated by the partial isometry w_j is isomorphic to the 2×2 matrix algebra $\mathbb{M}_2$ with w_j, $h_{1j}4$, and h_{2j} corresponding respectively to the matrices

$$
\begin{pmatrix} 0 & 1 \\ 0 & 0 \end{pmatrix}, \quad 2^{-1} \begin{pmatrix} 1 & 1 \\ 1 & 1 \end{pmatrix}, \quad 2^{-1} \begin{pmatrix} 1 & -1 \\ -1 & 1 \end{pmatrix}
$$

since $w_j w_j^* + w_j^* w_j = g_j$ and $g_j w_j = w_j = w_j g_j$ by equality (12.4.8). From this isomorphism, it follows that

$$h_{kj} g_{ij} h_{kj} = 2^{-1} h_{kj} \text{ for } i, k = 1, 2 \text{ and } j = 1, \ldots, n. \qquad (12.4.20)$$

By the definition of x_1 in equation (12.4.9) and by equation (12.4.20),

$$\sum_{j,k} h_{kj} x_1 h_{kj} = \sum_{i,j,k} h_{kj} g_{1i} h_{kj} - h_{kj} g_{1i} h_{kj} = 0. \qquad (12.4.21)$$

Clearly

$$\sum_{j,k} h_{kj} = \sum_j g_j = f_1. \qquad (12.4.22)$$

Let A be a finite dimensional abelian *-subalgebra of M containing C such that

(i) $f_1, f_2 \in A$;

(ii) $f_1 A$ has minimal projections h_{kj} for $k = 1, 2$ and $j = 1, \ldots, n$;

(iii) $f_2 A$ is such that $\|x_2 - \mathbb{E}_{f_2 A}(x_2)\|_2 \leq \delta \|e_1\|_2$;

where δ is chosen so that $0 < 2\delta < \alpha$. Note that the choice of $f_2 A$ to satisfy (iii) is possible, because x_2 is a self-adjoint element of the algebra $f_2(C' \cap N) f_2$ by equation (12.4.10) so can be approximated by an element in a finite dimensional abelian *-subalgebra of $f_2(C' \cap N) f_2$ by the spectral theorem.

Observe that the definition of $\mathbb{E}_{A' \cap N}$, (i) and (ii) imply that

$$\mathbb{E}_{A' \cap N}(x_1) = 0 \qquad (12.4.23)$$

as this is equation (12.4.21). Let

$$a = \mathbb{E}_A(x_2). \qquad (12.4.24)$$

By definition of x_2 in equation (12.4.10)

$$\begin{aligned} a &= \mathbb{E}_A(f_2 \mathbb{E}_{C' \cap N}(v x_1 v^*) f_2) = f_2 \mathbb{E}_A \mathbb{E}_{C' \cap N}(v x_1 v^*) f_2 \\ &= f_2 \mathbb{E}_A(v x_1 v^*) f_2 \end{aligned} \qquad (12.4.25)$$

since $f_2 \in A$ and $A \subseteq C' \cap N$. Using equation (12.4.24) and the fact that $\mathbb{E}_{A' \cap N}$ is $\| \cdot \|_2$-norm reducing, we obtain

$$\begin{aligned} \|v^* a v - &\mathbb{E}_{A' \cap N}(v^* a v)\|_2 \\ &\geq \|x_1 - \mathbb{E}_{A' \cap N}(x_1)\|_2 - \|\mathbb{E}_{A' \cap N}(x_1 - v^* a v)\|_2 - \|x_1 - v^* a v\|_2 \\ &\geq \|x_1\|_2 - 2\|x_1 - v^* a v\|_2. \end{aligned} \qquad (12.4.26)$$

We now estimate the last two terms in (12.4.26). By inequality (12.4.7) and equality (12.4.12),

$$\|x_1\|_2 = \|f_1\|_2 \geq \|e_1\|_2 - \|e_1 - f_1\|_2 \geq (1 - 5\alpha) \|e_1\|_2. \qquad (12.4.27)$$

By definition of a in equation (12.4.24) and using inequality (iii),

$$
\begin{aligned}
\|x_1 - v^* a v\|_2 &\leq \|x_1 - v^* x_2 v\|_2 + \|v^*(x_2 - \mathbb{E}_A(x_2))v\|_2 \\
&\leq \|x_1 - v^* x_2 v\|_2 + \|x_2 - \mathbb{E}_A(x_2)\|_2 \\
&= \|x_1 - v^* x_2 v\|_2 + \|x_2 - \mathbb{E}_{f_2 A}(x_2)\|_2 \\
&\leq \|x_1 - v^* x_2 v\|_2 + \delta\|e_1\|_2 \quad\quad\quad\quad\quad (12.4.28)
\end{aligned}
$$

since $\mathbb{E}_A(x_2) = \mathbb{E}_{f_2 A}(x_2)$ by equation (12.4.10). An estimate for $\|x_1 - v^* x_2 v\|_2$ is required. By equation (12.4.9), since $f_1 x_1 f_1 = x_1$,

$$
\begin{aligned}
\|x_1 - v^* x_2 v\|_2 &= \|f_1 x_1 f_1 - v^* x_2 v\|_2 \\
&\leq \|f_1 x_1 (f_1 - e_1)\|_2 + \|(f_1 - e_1)x_1 e_1\|_2 + \|e_1 x_1 e_1 - v^* x_2 v\|_2 \\
&\leq 2\|f_1 - e_1\|_2 + \|v^*(vx_1 v^* - x_2)v\|_2 \\
&\leq 2\|f_1 - e_1\|_2 + \|vx_1 v^* - x_2\|_2 \\
&< (10\alpha + \beta)\|e_1\|_2 \quad\quad\quad\quad\quad (12.4.29)
\end{aligned}
$$

by inequalities (12.4.7) and (12.4.18) using $e_1 = v^* v$. Assembling inequalities (12.4.26), (12.4.27), (12.4.27) and (12.4.29) together yields

$$
\begin{aligned}
\|v^* a v - \mathbb{E}_{A' \cap N}(v^* a v)\|_2 &\geq (1 - 5\alpha - 2(10\alpha + \beta) - 2\delta)\|e_1\|_2 \\
&= (1 - 25\alpha - 2\beta - 2\delta)\|e_1\|_2. \quad\quad (12.4.30)
\end{aligned}
$$

To obtain good estimates here in inequalities (12.4.17) and (12.4.30), and to satisfy inequality (12.4.5), we require the maximum value of

$$
\min\{\alpha, \beta - 10\alpha, 1 - 25\alpha - 2\beta - 2\delta\}
$$

subject to $\alpha, \beta, \delta > 0$. The minimum occurs when

$$
\alpha = \beta - 10\alpha = 1 - 25\alpha - 2\beta - 2\delta
$$

so that $\beta = 11\alpha$ and $1 = 48\alpha + 2\delta$. Taking $0 < 2\delta < \alpha$ gives $1/48 < \alpha < 1/49$ as required. $\qquad\square$

Remark 12.4.2. Using $2^{3/2} + 2$ in place of 5 in inequality (12.4.7) yields $\alpha > 1/33$. However there is little point in trying to improve the inequality a little as in Theorem 11.1.2 we showed that the constant can be taken to be 1.

A little problem that arises from this proof is the following. If v is a partial isometry in N with $v^2 = 0$ and A is a masa, or an abelian von Neumann subalgebra, is $\|\mathbb{E}_A(v)\|_2 \leq 2^{-1/2}\|v\|_2$? $\qquad\square$

Theorem 12.4.3. *Let N be a separable II$_1$ factor and let M be a subfactor of N with $M' \cap N = \mathbb{C}1$. If B is a finite dimensional abelian *-subalgebra of M, then there is a singular masa A in N with $B \subseteq A \subseteq M$.*

Proof. Let $\{x_n\}_{n=1}^{\infty}$ be a sequence that is dense in the unit ball of N in the $\|\cdot\|_2$-norm. Let $\{v_n\}_{n=1}^{\infty}$ be a sequence of partial isometries in N with $v_n^2 = 0$

for all n that is dense in $\|\cdot\|_2$-norm in the set of all partial isometries v in N with $v^2 = 0$. Let $\alpha > 1/49$ be the constant of Lemma 12.4.1.

By Lemma 12.4.1 choose an increasing sequence $\{A_n\}_{n=0}^{\infty}$ of finite dimensional abelian *-subalgebras of M satisfying $A_0 = B$ and

$$\|\mathbb{E}_{A_n}(x_j) - \mathbb{E}_{A'_n \cap N}(x_j)\|_2 < 2^{-n} \quad \text{for } 1 \leq j \leq n, \tag{12.4.31}$$

and one of the following two conditions holds: either

$$\sup\{\|v_n x v_n^* - \mathbb{E}_{A'_n \cap N}(v_n x v_n^*)\|_2 : x \in A_n, \|x\| \leq 1\} > \alpha\|vv^*\|_2 \tag{12.4.32}$$

or

$$\sup\{\|v_n^* x v_n - \mathbb{E}_{A'_n \cap N}(v_n^* x v_n)\|_2 : x \in A_n, \|x\| \leq 1\} > \alpha\|vv^*\|_2. \tag{12.4.33}$$

In Lemma 12.4.1 take $F = \{x_1, \ldots, x_n\}$ and $v = v_n$.

Let A be the weak closure of $\cup A_n$. Then $B \subseteq A \subseteq N$ and A is a maximal abelian *-subalgebra of N by (12.4.31) and Corollary 3.6.9. Let v be a partial isometry in N with $v^2 = 0$ and let $\varepsilon > 0$. By the choice of the sequence $\{v_n\}_{n=1}^{\infty}$, there is an integer n such that

$$\|v - v_n\|_2 < \varepsilon. \tag{12.4.34}$$

Either inequality (12.4.32) or (12.4.33) holds. Suppose that inequality (12.4.32) holds and let $x \in A$ with $\|x\| \leq 1$ and

$$\|v_n x v_n^* - \mathbb{E}_{A'_n \cap N}(v_n x v_n^*)\|_2 > \alpha\|v_n v_n^*\|_2. \tag{12.4.35}$$

If inequality (12.4.33) holds the calculation is similar to that below but with $v_n x v_n^*$ and vxv^* replaced by $v_n^* x v_n$ and $v^* x v$. Using that $\mathbb{E}_{A'_n \cap N}$ is a projection gives

$$\|vxv^* - \mathbb{E}_{A'_n \cap N}(vxv^*)\|_2 \geq \|v_n x v_n^* - \mathbb{E}_{A'_n \cap N}(v_n x v_n^*)\|_2 - 4\|v - v_n\|_2$$
$$> \alpha\|v_n v_n^*\|_2 - 4\varepsilon \tag{12.4.36}$$

by inequalities (12.4.34) and (12.4.35). Here

$$\|v_n v_n^*\|_2 = \|v_n\|_2 \geq \|v\| - \|v - v_n\|_2 > \|v\|_2 - \varepsilon. \tag{12.4.37}$$

Since $A'_n \cap N \supseteq A$,

$$\|vxv^* - \mathbb{E}_A(vxv^*)\|_2 \geq \|vxv^* - \mathbb{E}_{A'_n \cap N}(vxv^*)\|_2 \geq \alpha\|vv^*\|_2 - 5\varepsilon. \tag{12.4.38}$$

This implies that

$$\|(I - \mathbb{E}_A)\mathbb{E}_{vAv^*}\|_{\infty,2} \geq \alpha\|vv^*\|_2. \tag{12.4.39}$$

The similar calculation with v and v^* interchanged implies that if v is a partial isometry in N with $v^2 = 0$, then either

$$\|(I - \mathbb{E}_A)\mathbb{E}_{vAv^*}\|_{\infty,2} \geq \alpha\|vv^*\|_2 \tag{12.4.40}$$

or

$$\|(I - \mathbb{E}_A)\mathbb{E}_{v^*Av}\|_{\infty,2} \geq \alpha \|v^*v\|_2. \tag{12.4.41}$$

If v is a partial isometry in N with vv^*, $v^*v \in A$, $v^2 = 0$ and $vAv^* = vv^*A$, then $v^*Av = v^*vA$ and the left-hand sides of inequalities (12.4.40) and (12.4.41) are zero so $v = 0$.

Let u be a unitary in N with $uAu^* = A$ and suppose that u is not in $A = A' \cap N$. Then there is a projection p in A with $u^*pu \neq p$. Let $q = p(1 - u^*pu)$. Then q is a non-zero projection in A with $uqu^* \leq 1 - p$ so that $ququ^* = 0$ and $quq = 0$. Let $v = uq$. Then v is a non-zero partial isometry in N with $v^2 = uquq = 0$, $vAv^* = uqAu^* \subseteq A$ and $v^*Av = qA \subseteq A$. This contradicts that such a v must be zero and completes the proof. □

Remark 12.4.4. Theorem 12.4.3 says in effect that the singular masa has Popa's delta invariant $\delta(A) \geq \alpha > 1/49$ (see Section 11.1). Note that due to a theorem of S. Popa (Theorem 11.1.2) it is now known that the invariant $\delta(A) = 1$ for all singular masas A. □

Corollary 12.4.5. *Let N be a separable II_1 factor and let $\mathcal{SM}(N)$ denote the set of all singular masas in N. If M is a subfactor of N with $M' \cap N = \mathbb{C}1$, let*

$$M^\circ = \{A \in \mathcal{SM}(N) : A \subseteq M\}.$$

If M_1 and M_2 are subfactors of N with trivial relative commutants, then $M_1 \subseteq M_2$ if, and only if, $M_1^\circ \subseteq M_2^\circ$.

Proof. Clearly $M_1 \subseteq M_2$ implies that $M_1^\circ \subseteq M_2^\circ$. If $M_1 \nsubseteq M_2$, then there is a projection p in M_1 that is not in M_2. By Theorem 12.4.3 there is a singular masa A in N with $\mathbb{C}1 + \mathbb{C}p \subseteq A \subseteq M_1$ so that $M_1^\circ \nsubseteq M_1^\circ$. □

Observe that the hypothesis that M_j are subfactors with trivial relative commutant cannot be weakened as there might be no singular masas of N in M_j.

The results in this section are from [139]. The technique of either leaving a 2×2 diagonal of a matrix alone or sometimes twisting it to achieve singularity is used in Tauer's work [190] and also in the work of White [208].

12.5 Singularity and automorphisms

Popa [137] proved that if $\mathcal{G}$ is a countable group of properly outer automorphims on a separable II_1 factor N, then there is a singular masa A in N with $\theta(A) \neq A$ for all θ in $\mathcal{G}$ (see Theorem 12.5.6). The result is deduced from Theorem 12.4.3 by taking a crossed product with $\mathcal{G}$ and using some preparatory lemmas due to Jones and Popa [96] (see Lemma 6.2.4).

Definition 12.5.1. Let N be a von Neumann algebra.

(i) An automorphism θ acts *freely* on a von Neumann algebra N if $w \in N$ and $xw = w\theta(x)$ for all $x \in N$ implies that $w = 0$.

(ii) If $\alpha \colon \mathcal{G} \to \mathrm{Aut}(N)$ is a homomorphism, then $\mathcal{G}$ acts *freely* on N if each α_g acts freely for $g \in \mathcal{G} \setminus \{e\}$.

(iii) An action is *ergodic* if an element $x \in N$ satisfying $\alpha_g(x) = x$ for all $g \in \mathcal{G}$ must lie in $\mathbb{C}1$.   □

Lemma 12.5.2. *If θ is an outer automorphism on a II_1 factor N, then θ acts freely on N.*

Proof. Let $w \in N$ be non-zero and suppose that

$$xw = w\theta(x) \quad \text{for all} \quad x \in N. \tag{12.5.1}$$

Then

$$xww^* = w\theta(x)w^* = w(w\theta(x^*))^* = w(x^*w)^* = ww^*x \tag{12.5.2}$$

for all $x \in N$ so $ww^* = \lambda 1$ for some $\lambda > 0$. Let $u = \lambda^{-1/2}w$. Then u is a unitary in N, since N is a finite factor gives $u^*u = 1$, and $\theta(x) = u^*xu$ ($x \in N$) follows from equation (12.5.1). This contradicts θ being an outer automorphism.   □

Remark 12.5.3. If N is a finite von Neumann algebra with faithful normal trace τ and if the discrete group $\mathcal{G}$ acts on N as trace preserving automorphisms, let $N \rtimes_\alpha \mathcal{G}$ be the von Neumann crossed product [198], [105, Chapter 13]. Let u_g be the unitary implementing α_g by $\alpha_g(x) = u_g x u_g^*$ for all $x \in N$ and all $g \in \mathcal{G}$. An element w in $N \rtimes_\alpha \mathcal{G}$ may be represented by $w = \sum_g x_g u_g$, where $x_g \in \mathcal{G}$ for all $g \in \mathcal{G}$. The trace τ is defined on $N \rtimes_\alpha \mathcal{G}$ by $\tau(\sum_g x_g u_g) = \tau(x_e)$ and the $\| \cdot \|_2$-norm on $N \rtimes_\alpha \mathcal{G}$ is given by $\| \sum_g x_g u_g \|_2^2 = \sum_g \| x_g \|_2^2$ for such a representation.   □

Lemma 12.5.4. *Let N be a finite von Neumann algebra with a faithful normal trace τ and let the discrete group $\mathcal{G}$ act as trace preserving automorphisms on N. If the discrete group $\mathcal{G}$ acts freely on N and ergodically on its centre $N' \cap N$, then $M = N \rtimes_\alpha \mathcal{G}$ is a type II_1 factor and $N' \cap M = N' \cap N$.*

Proof. Let $w = \sum_g x_g u_g$ be in $N' \cap M$ with $x_g \in N$ for all g. For all $y \in N$,

$$\sum_g y x_g u_g = \sum_g x_g u_g y = \sum_g x_g \alpha_g(y) u_g \tag{12.5.3}$$

so that

$$y x_g = x_g \alpha_g(y) \quad \text{for all } g \in \mathcal{G}. \tag{12.5.4}$$

If $g \in \mathcal{G} \setminus \{e\}$, then equation (12.5.4) implies that $x_g = 0$, since α_g acts freely on N. If $g = e$, then $x_e \in N'$. This shows that $N' \cap M \subseteq N' \cap N \subseteq N' \cap M$ as required.

If $x \in M' \cap M \subseteq N' \cap M$, then $x \in N$ and $\alpha_g(x) = u_g x u_g^* = x$ for all $g \in \mathcal{G}$ so that $x = \lambda 1$ for some $\lambda \in \mathbb{C}$, by the ergodicity of the action of $\mathcal{G}$ on the centre $N' \cap N$. Hence M is a II_1 factor.   □

Definition 12.5.5. An automorphism α of a II_1 factor N is *properly outer* if for each non-zero projection p in N with $\alpha(p) = p$, the restriction $\alpha|_{pNp}$ of α to pNp is outer.

A group of automorphisms $\mathcal{G}$ on N is called *properly outer* if all automorphisms $\alpha \in \mathcal{G} \setminus \{I\}$ are properly outer. $\square$

Theorem 12.5.6. *Let N be a separable II_1 factor and let $\mathcal{G}$ be a countable group of properly outer automorphisms on N. If B is a finite dimensional abelian *-subalgebra of N, there is a singular masa A in N containing B such that $\phi(pA) \nsubseteq A$ for all non-zero projections $p \in A$ and all $\phi \in \mathcal{G} \setminus \{I\}$.*

Proof. Regard $\mathcal{G}$ as a discrete group acting on N and note that $\mathcal{G}$ acts freely on N by Lemma 12.5.2 and ergodically on the centre $\mathbb{C}1$. Hence $M = N \rtimes \mathcal{G}$ is a separable II_1 factor and N is a subfactor with $N' \cap M = N' \cap N = \mathbb{C}1$ by Lemma 12.5.4 . By Theorem 12.4.3 there is a singular masa A in M with $B \subseteq A \subseteq N$. Let u be the unitary in $M = N \rtimes \mathcal{G}$ implementing the automorphism ϕ in $\mathcal{G} \setminus \{I\}$ by $\phi(x) = uxu^*$ $(x \in N)$. Suppose that there is a non-zero projection p in A with $\phi(pA) \subseteq A$ and let $v = up$. Then v is a partial isometry in M such that $v^*v = p$, $vv^* = q$ is a projection in A, $\phi(p) = upu^* = vv^* = q$ and $vAv^* = \phi(pA) = q\phi(A) \subseteq qA$. Note that $\phi|_{pMp}$ is a *-isomorphism from pMp onto qMq mapping the masa pA into the masa qA. Hence $\phi(pA) = qA$ and $vAv^* = qA$. This implies that v is in the groupoid normaliser of A in M, and so has the form $v = qu$ where q is a projection in A and $u \in \mathcal{N}(A)$, by Lemma 6.2.4. Then $u \in A$ since A is singular, giving $v \in A$. Thus $p = v^*v = vv^* = q$ since A is abelian. This implies that $\phi|_{pNp}$ is an inner automorphism on pNp implemented by the unitary v in pNp. This contradicts ϕ being a properly outer automorphism of N. $\square$

Remark 12.5.7. Theorem 12.5.6 is in [139] in a more general form. That paper presents various lemmas for properly outer automorphisms of a continuous semifinite von Neumann algebra, and deduces Theorem 12.5.6 from these.

Can we use the perturbation results of Popa on $\delta(A) = 1$ for singular masas to show that $\|(I - \mathbb{E}_A)\mathbb{E}_{\phi(A)}\|_{\infty,2} = 1$ for all $\phi \in \mathcal{G} \setminus \{I\}$? $\square$

Chapter 13

Irreducible hyperfinite subfactors

13.1 Introduction

This chapter is devoted to the construction of irreducible hyperfinite subfactors R in a separable II_1 factor N with suitable additional properties available for R in its embedding in N. All these results depend on inductive matrix methods that were developed by Popa [136]. The method has already been used extensively in Chapter 12 for the construction of singular and semiregular masas.

In Section 13.2, a basic method is presented to show that an irreducible hyperfinite subfactor exists in each separable II_1 factor. Section 13.3 shows that if A is a Cartan masa in a separable II_1 factor N, then there is an irreducible hyperfinite subfactor R in N with $A \subseteq R$ and A Cartan in R (see [141]). Section 13.4 discusses the basic theory of property Γ factors, a topic which we will revisit in greater depth in Appendix A. This is applied in Section 13.5 to prove a useful result (Theorem 13.5.4) on the existence of a masa in a Γ factor that is Cartan in an irreducible hyperfinite subfactor and that contains unitaries that can be used in the Γ condition. This theorem combines methods from [140] and from [30, Theorem 5.3] that give the Γ condition.

The following result from Chapter 12 is the basic one for hyperfinite subfactors containing masas and the main theorems of this chapter will be refinements:

> Let N be a separable II_1 factor. There is a masa A in N and a hyperfinite subfactor M of N containing A such that A is a Cartan subalgebra of M and $M' \cap N$ is $\mathbb{C}1$.

13.2 Irreducible hyperfinite subfactors exist

This section is devoted to the construction of an irreducible hyperfinite subfactor R in a separable II_1 factor N; such subfactors were called Popa subfactors in

[172]. It is easy to get a hyperfinite subfactor R in a (separable) II_1 factor by putting in an increasing chain of matrix subalgebras and taking R to be the weak closure of the union. An extra approximation is needed to ensure that it is irreducible. In other sections such subfactors are constructed with additional properties. However, the proof here is more elementary than those cases so is included to illustrate a slightly different way of handling things. We will need the following two lemmas.

Lemma 13.2.1. *Let N be a II_1 factor and let M be a matrix subalgebra of N. Then $\mathbb{E}_{M'\cap N}(x) = \tau(x)$ for all $x \in M$.*

Proof. Since M is a matrix subalgebra of N, the basic theory of factors (see [104, 105, 187]) shows that $(M' \cap N)' \cap N = M$. If $x \in M$ and $b \in M' \cap N$, then

$$b\mathbb{E}_{M'\cap N}(x) = \mathbb{E}_{M'\cap N}(bx) = \mathbb{E}_{M'\cap N}(xb) = \mathbb{E}_{M'\cap N}(x)b \qquad (13.2.1)$$

by the module property of $\mathbb{E}_{M'\cap N}$. Hence $\mathbb{E}_{M'\cap N}(x) \in M \cap M' = \mathbb{C}1$ so $\mathbb{E}_{M'\cap N}(x) = \lambda 1$ and $\lambda = \tau(x)$ by the trace preserving property of $\mathbb{E}_{M'\cap N}$. $\square$

Lemma 13.2.2. *Let N be a II_1 factor and let y be a self-adjoint element in N. If $\varepsilon > 0$, then there exists a matrix subalgebra M of N such that*

$$\|\mathbb{E}_{M'\cap N}(y) - \tau(y)1\|_2 < \varepsilon. \qquad (13.2.2)$$

Proof. By elementary spectral theory there are mutually orthogonal projections $p_1, \ldots, p_n$ in N (actually in $\{y\}''$) and $\lambda_1, \ldots, \lambda_n \in \mathbb{R}$ such that $\tau(p_i) = n^{-1}$ for $1 \le i \le n$ and

$$\left\| y - \sum_{i=1}^n \lambda_i p_i \right\|_2 < \varepsilon/2. \qquad (13.2.3)$$

Since N is a II_1 factor, there is an $n \times n$ matrix subalgebra M of N such that $\{p_1, \ldots, p_n\}$ is the set of minimal projections in the diagonal of M. Applying τ and $\mathbb{E}_{M'\cap N}$ to $y - \sum_{i=1}^n \lambda_i p_i$ gives

$$\left| \tau(y) - \sum_{i=1}^n \lambda_i n^{-1} \right| < \varepsilon/2 \qquad (13.2.4)$$

and

$$\left\| \mathbb{E}_{M'\cap N}(y) - \sum_{i=1}^n \lambda_i n^{-1} \right\|_2 = \left\| \mathbb{E}_{M'\cap N}(y) - \sum_{i=1}^n \lambda_i \mathbb{E}_{M'\cap N}(p_i) \right\|_2$$
$$< \varepsilon/2 \qquad (13.2.5)$$

by inequality (13.2.3) and Lemma 13.2.2. The lemma follows from this. $\square$

The following theorem is due to Popa [136].

Theorem 13.2.3. *Let N be a separable II_1 factor. There exists an irreducible hyperfinite subfactor R of N.*

Proof. Choose a $\|\cdot\|_2$-dense sequence $\{x_n\}_1^\infty$ in the set of all self-adjoint elements of N with $x_1 = 1$. By induction we construct an ascending sequence of matrix subalgebras M_n of N such that

$$\|\mathbb{E}_{M_n' \cap N}(x_n) - \tau(x_n)1\|_2 < 1 \quad \text{for all} \quad n \in \mathbb{N}. \tag{13.2.6}$$

Observe that $M_n \cong \mathbb{M}_k$ for a k considerably larger than n in most cases. Let $M_1 = \mathbb{C}1$. Suppose that matrix subalgebras $M_1 \subseteq \ldots \subseteq M_n$ of N have been constructed satisfying the inductive hypothesis (13.2.6). Decompose $N = M_n \otimes (M_n' \cap N)$, where $M_n' \cap N$ is a II_1 factor. Let $y = \mathbb{E}_{M_n' \cap N}(x_{n+1}) \in M_n' \cap N$ with $\tau(y) = \tau(\mathbb{E}_{M_n' \cap N}(x_{n+1})) = \tau(x_{n+1})$. In the II_1 factor $M_n' \cap N$ there is a matrix subalgebra $\widetilde{M}$ such that

$$\|\mathbb{E}_{\widetilde{M}' \cap (M_n' \cap N)}(y) - \tau(y)1\|_2 < \varepsilon, \tag{13.2.7}$$

by Lemma 13.2.2.

Let

$$M_{n+1} = \left(M_n \cup \widetilde{M}\right)'' = M_n \otimes \widetilde{M}, \tag{13.2.8}$$

which is a tensor since M_n and $\widetilde{M}$ are matrix subalgebras with $\widetilde{M} \subseteq M_n'$. The containment $M_{n+1}' \cap N \subseteq M_n' \cap N$ follows from $M_n \subseteq M_{n+1}$, and thus

$$\mathbb{E}_{M_{n+1}' \cap N}(x_{n+1}) = \mathbb{E}_{M_{n+1}' \cap N}\mathbb{E}_{M_n' \cap N}(x_{n+1}) = \mathbb{E}_{M_{n+1}' \cap N}(y). \tag{13.2.9}$$

Then

$$\mathbb{E}_{M_{n+1}' \cap N}(x_{n+1}) = \mathbb{E}_{M_{n+1}' \cap (M_n' \cap N)}(y) \tag{13.2.10}$$

because $y \in M_n' \cap N \subseteq N$.

The equation $M_{n+1} = \left(M_n \cup \widetilde{M}\right)'' = M_n \otimes \widetilde{M}$ gives $M_{n+1}' = \widetilde{M}' \cap M_n'$ so that

$$\mathbb{E}_{M_{n+1}' \cap (M_n' \cap N)} = \mathbb{E}_{\widetilde{M}' \cap M_n' \cap N}. \tag{13.2.11}$$

Hence $\mathbb{E}_{M_{n+1}' \cap N}(x_{n+1}) = \mathbb{E}_{\widetilde{M}' \cap M_n' \cap N}(y)$ by the equations (13.2.9) and (13.2.11). The trace preserving property of the conditional expectation gives

$$\|\mathbb{E}_{M_{n+1}' \cap N}(x_{n+1}) - \tau(x_{n+1})1\|_2 = \|\mathbb{E}_{M_{n+1}' \cap N}(y) - \tau(y)1\|_2 < 1. \tag{13.2.12}$$

This ends the construction of the sequence M_n.

Let $R = \left(\bigcup M_n\right)''$, and consider $x \in R' \cap N$. For each $k \in \mathbb{N}$ choose an $x_{n(k)}$ such that

$$\|kx - x_{n(k)}\|_2 < 1. \tag{13.2.13}$$

By the inequality (13.2.6) in the induction hypothesis,

$$\|\mathbb{E}_{M'_{n(k)} \cap N}(x_{n(k)}) - \tau(x_{n(k)})\|_2 < 1. \tag{13.2.14}$$

Since $M_{n(k)} \subseteq R$ and $x \in R'$, we have $x \in M'_{n(k)} \cap N$ so that $\mathbb{E}_{M'_{n(k)} \cap N}(x) = x$. Thus, from inequality (13.2.13), we obtain the inequalities

$$\|kx - \mathbb{E}_{M'_n \cap N}(x_{n(k)})\|_2 < 1 \tag{13.2.15}$$

and

$$|k\tau(x) - \tau(x_{n(k)})| < 1 \tag{13.2.16}$$

so that $\|kx - k\tau(x)1\|_2 < 3$. Dividing by k and letting $k \to \infty$ gives $x = \tau(x)1$. Thus $R' \cap N = \mathbb{C}1$. $\qquad \square$

Remark 13.2.4. One can give an elementary proof of Theorem 13.2.3 that avoids direct use of the conditional expectation. It is a standard philosophy that to study subalgebras M of a II_1 factor N one investigates $\mathbb{E}_M$, e_M and $\langle N, e_M \rangle$ as has been emphasised in these notes, so the conditional expectation version is used in Theorem 13.2.3. Here is the outline of the direct approach avoiding explicit use of conditional expectations, though the maps introduced are the conditional expectations of the commutants of matrix subalgebras.

Let N be a separable II_1 factor for the remainder of this remark. If G is a compact group of unitaries in N with normalised Haar measure μ_G, let $\phi_G : N \to N$ be defined by

$$\phi_G(x) = \int_G uxu^* \, d\mu_G(u) \quad (x \in N). \tag{13.2.17}$$

When this is used in the discussion, G is a finite group and $\int_G \cdot \, d\mu_G$ is the average over G.

The steps are as follows. Show that if $H \subseteq G$ are such subgroups of N, then

$$\phi_G \phi_H = \phi_H \phi_G = \phi_G \tag{13.2.18}$$

and

$$\|\phi_G(x) - \tau(x)1\|_2 \leq \|\phi_H(x) - \tau(x)\|_2 \quad (x \in N). \tag{13.2.19}$$

Using the technique of Lemma 13.2.1 show that if $\varepsilon > 0$ and $y = y^* \in N$, then there exists an $n \times n$ matrix subalgebra $\widetilde{M}$ of N with matrix units $e_{ii} \in \{y\}''$ such that if G is the group generated by the natural permutation matrices in $\widetilde{M}$, then $\|\phi_G(y) - \tau(y)1\|_2 < \varepsilon$. Here G is the group generated by the unitary permutation matrices u_{ij} interchanging i^{th} and j^{th} basis elements and leaving the others fixed:

$$u_{ij} = e_{ij} + e_{ji} + \sum_{s \neq i,j} e_{ss} \quad (1 \leq i, j \leq n) \tag{13.2.20}$$

in $\widetilde{M}$, where e_{ij} $(1 \leq i, j \leq n)$ are the matrix units of M. The rest of the proof roughly follows the inductive construction in Theorem 13.2.3 but with $\mathbb{E}_{\widetilde{M}' \cap N}$ replaced by ϕ_G.

This all works of course because $\phi_G = \mathbb{E}_{\widetilde{M}' \cap N}$ as may be seen by applying the uniqueness condition of the conditional expectation in Lemma 3.6.2. $\qquad \square$

13.3   Cartan masas in hyperfinite subfactors

The main result of this section is Theorem 13.3.2, which constructs a hyperfinite subfactor containing a given Cartan subalgebra. Throughout, A denotes a Cartan subalgebra of a separable II_1 factor N. Recall from Section 3.3 that a Cartan subalgebra A of N is a masa in N such that the unitary normaliser $\mathcal{N}(A) = \{u \in \mathcal{U}(N): uAu^* = A\}$ of A in N generates N as a von Neumann algebra. From Section 6.2, recall that $\mathcal{GN}(A)$ denotes the groupoid normaliser of A:

$$\mathcal{GN}(A) = \{v \in N: v \text{ is a partial isometry}, vAv^* \subseteq A \text{ and } v^*Av \subseteq A\}.$$

The following lemma ensures that the hyperfinite algebra M constructed in Theorem 13.3.2 is generated by the unitary normaliser of A in it.

Lemma 13.3.1. *If $\{f_1, \ldots, f_k\}$ is a set of orthogonal equivalent projections in a Cartan subalgebra A of N, then there are matrix units f_{ij} $(1 \le i, j \le k)$ in $\mathcal{GN}(A)$ such that $f_{ii} = f_i$ for $1 \le i \le k$.*

Proof. By Lemma 6.2.6, there are partial isometries $f_{1j} \in \mathcal{GN}(A)$, $1 \le j \le k$, such that

$$f_{1j}f_{1j}^* = f_1, \quad f_{1j}^*f_{1j} = f_j, \quad 1 \le j \le k, \tag{13.3.1}$$

where $f_{11} = f_1$. If we define $f_{ij} = f_{1i}^*f_{1j}$, then each f_{ij} lies in $\mathcal{GN}(A)$, by Lemma 6.2.3, and we now have the desired set of matrix units. $\qquad\square$

Theorem 13.3.2. *Let N be a separable II_1 factor and let A be a Cartan subalgebra of N. Then there is an irreducible hyperfinite subfactor M of N such that A is a Cartan subalgebra of M.*

Proof. The idea of the proof is to construct an increasing sequence of matrix algebras using techniques from Chapter 12 to ensure that the weak closure of the union has the required properties.

Let $\{a_n: n \in \mathbb{N}\}$ be a sequence of projections in A whose linear span is dense in A in the strong operator topology. By induction, we construct an increasing sequence

$$M_0 = \mathbb{C}1 \subseteq M_1 \subseteq M_2 \ldots$$

of finite dimensional matrix subalgebras of N with matrix units (e_{ij}^n) satisfying the induction hypotheses

(1) $e_{ij}^n \in \mathcal{GN}(A)$ for $1 \le i, j \le k(n)$, where M_n is isomorphic to $\mathbb{M}_{k(n)}$;

(2) if A_n is the finite dimensional subalgebra generated by the set

$$\{e_{ii}^n: 1 \le i \le k(n)\},$$

then $A_n \subseteq A$ and

$$\|a_j - \mathbb{E}_{A_n}(a_j)\|_2 \le 2^{-n} \quad \text{for} \quad 1 \le j \le n. \tag{13.3.2}$$

Suppose that M_n has been constructed. Since $w \in \mathcal{GN}(A)$ implies that $wAw^* \subseteq A$ by definition of $\mathcal{GN}(A)$, the finite set

$$\mathcal{F} = \{e_{1i}^n a_t e_{i1}^n : 1 \leq t \leq n+1, \ 1 \leq i \leq k(n)\}$$

is contained in A. By Corollary 3.6.9, there is a finite dimensional subalgebra B of A that contains A_n and is such that

$$\|(I - \mathbb{E}_B)(e_{1i}^n a_j e_{i1}^n)\|_2 \leq k(n)^{-1} 2^{-(n+1)} \tag{13.3.3}$$

for $1 \leq i \leq k(n)$ and $1 \leq j \leq n+1$, while the minimal projections $f_1, \ldots, f_\ell$ in B are all equivalent.

The projection e_{11}^n is in $A_n \subseteq B$, so is a sum of some of the minimal projections in B, and e_{11}^n acts as an identity for the finite set

$$\{e_{1i}^n a_t e_{i1}^n : 1 \leq i \leq k(n), \ 1 \leq t \leq n+1\}.$$

Thus those f_j that do not occur in the sum for e_{11}^n may be discarded, and we assume that

$$e_{11}^n = f_1 + \cdots + f_m. \tag{13.3.4}$$

By Lemma 13.3.1 there are matrix units f_{ij}, $1 \leq i, j \leq m$, such that $f_{ii} = f_i$ for $1 \leq i \leq m$. The matrix units $e_{k\ell}^{n+1}$ $(1 \leq k, \ell \leq mk(n))$ are defined by using a tensor product idea. Let

$$e_{i,j,s,t}^{n+1} = e_{s1}^n f_{ij} e_{1t}^n \quad \text{for} \quad 1 \leq s, t \leq k(n) \text{ and } 1 \leq i, j \leq m, \tag{13.3.5}$$

and let M_{n+1} be the linear span of these matrix units. Then M_{n+1} is isomorphic to the algebra of $k(n+1) \times k(n+1)$ matrices, where $k(n+1) = mk(n)$ and M_{n+1} contains M_n. The diagonal of M_{n+1} is contained in A since each e_{ij}^n is in $\mathcal{GN}(A)$.

To finish the induction we need to prove inequality (13.3.2) from (13.3.3). If L_x and R_x denote respectively left and right multiplication by x, observe that $L_{e_{i1}^n} R_{e_{1i}^n}$ is a $\| \cdot \|_2$-norm isometry from $e_{1i}^n A e_{i1}^n$ onto $e_{ii}^n A e_{ii}^n$. Using the A_{n+1}-module property of $\mathbb{E}_{A_{n+1}}$ and the equation

$$e_{11}^n A_{n+1} e_{11}^n = e_{11}^n B e_{11}^n, \tag{13.3.6}$$

it follows that

$$\|(I - \mathbb{E}_{A_{n+1}})(e_{ii}^n x e_{ii}^x)\|_2 = \|(I - \mathbb{E}_{A_{n+1}})(e_{1i}^n x e_{1i}^n)\|_2$$
$$= \|(I - \mathbb{E}_B)(e_{1i}^n x e_{1i}^n)\|_2 \tag{13.3.7}$$

for all x in A. Since A is abelian and e_{ii}^n $(1 \leq i \leq k(n))$ are orthogonal in A_{n+1},

$$\|x - \mathbb{E}_{A_{n+1}}(x)\|_2^2 = \|\sum e_{ii}^n (I - \mathbb{E}_{A_{n+1}})(x) e_{ii}^n\|_2^2$$
$$= \sum \|e_{ii}^n (I - \mathbb{E}_{A_{n+1}})(x) e_{ii}^n\|_2^2$$
$$= \sum \|(I - \mathbb{E}_{A_{n+1}})(e_{ii}^n x e_{ii}^n)\|_2^2$$
$$= \sum \|(I - \mathbb{E}_B)(e_{1i}^n x e_{i1}^n)\|_2^2 \tag{13.3.8}$$

for all x in A, by inequality (13.3.7), where the sums are taken over $1 \leq i \leq k(n)$. Inequalities (13.3.3) and (13.3.8) give inequality (13.3.2) directly.

Let M be the weak closure of $\bigcup M_n$, a subfactor of N. Note that the weak closure of $\bigcup A_n$ is equal to A by inequality (13.3.2) and the choice of $\{a_n : n \in \mathbb{N}\}$. Thus $A \subseteq M$, and is Cartan in this subfactor of N by Corollary 6.2.5 (ii). $\qquad\square$

The crossed product construction of Murray and von Neumann of a discrete non-amenable group G acting ergodically on $L^\infty(X, \mu)$ provides many separable non-hyperfinite II_1 factors N, to which Theorem 13.3.2 applies. There are also separable non-hyperfinite II_1 factors with non-conjugate Cartan subalgebras [41], although Cartan masas in the hyperfinite factor R are unique up to unitary conjugacy [40].

13.4   Property Γ

In seeking to give examples of non-isomorphic II_1 factors, Murray and von Neumann [118] introduced property Γ with which they were able to distinguish the hyperfinite II_1 factor R from the group factor $L(\mathbb{F}_2)$. In this section we will first present two equivalent formulations of property Γ, and then give some other characterisations which are more flexible. The one due to Dixmier [50] is stated as Theorem 13.4.6 but the proof is postponed to Appendix A since it requires some results concerning ultrafilters.

Definition 13.4.1. A II_1 factor N is said to have *property* Γ if, given $\varepsilon > 0$ and $x_1, \ldots, x_k \in N$, there exists a trace zero unitary $u \in N$ such that

$$\|ux_i - x_i u\|_2 < \varepsilon, \quad 1 \leq i \leq k. \tag{13.4.1}$$

An alternative formulation is the existence, for a fixed but arbitrary finite set $F \subseteq N$, of a sequence $\{u_n\}_{n=1}^\infty$ of trace zero unitaries in N satisfying

$$\lim_{n \to \infty} \|u_n x - x u_n\|_2 = 0, \quad x \in F. \tag{13.4.2}$$

$\qquad\square$

It will be convenient to relax the requirement that $\tau(u_n) = 0$ in the second version of Definition 13.4.1, by only asking that $\lim_{n \to \infty} \tau(u_n) = 0$. This is Corollary 13.4.4, and the following preliminary results are needed in order to see that this is possible. Theorem 13.4.2 and Corollary 13.4.3 are phrased in terms of matrices and factors respectively, but an examination of the proofs shows that they are really about properties of abelian algebras, $\ell^\infty(n)$ in the first case and $L^\infty[0, 1]$ in the second.

Theorem 13.4.2. *Let $\varepsilon > 0$ be fixed and let n be any integer satisfying $n\varepsilon > 1$. Let τ denote the normalised trace on $\mathbb{M}_n$. If $U \in \mathbb{M}_n$ is a diagonal unitary matrix with $|\tau(U)| < \varepsilon$, then there exists a diagonal unitary matrix $U_0 \in \mathbb{M}_n$ satisfying $\tau(U_0) = 0$ and $\|U - U_0\|_2 < 25\sqrt{\varepsilon}$.*

Proof. We begin by making three simple observations concerning the geometry of the unit disc $\mathbb{D}$ and the unit circle $\mathbb{T}$.

(i) If z_1, $z_2 \in \mathbb{D}$, and the angle between them is at least $\pi/4$, then

$$|(z_1 + z_2)/2| \le \cos(\pi/8) \le 0.93 < 19/20. \qquad (13.4.3)$$

This follows by examining the extremal case $z_1 = e^{i\pi/8}$ and $z_2 = e^{-i\pi/8}$.

(ii) If $0 \le R < 1$ and $|z| \le R$, then the horizontal line joining z to the unit semicircle in the left half plane has length at least $1 - R$. The minimal such distance occurs when $z = -R$ which is a distance $1 - R$ from the closest point -1 on the unit circle.

(iii) For each integer $k \ge 2$, every point $re^{i\theta} \in \mathbb{D}$ is the average of k points in $\mathbb{T}$. This is clear for $k = 2$. If the result is true for $k - 1$, then take $z_k = e^{i\theta}$ and choose $z_1, \ldots, z_{k-1}$ to average to $(kre^{i\theta} - e^{i\theta})/(k - 1) \in \mathbb{D}$. A simple computation gives $re^{i\theta}$ as the average of $z_1, \ldots, z_k$.

By multiplying U by a suitable scalar of modulus 1, we may assume that $\tau(U)$ is both real and positive. The strategy of the proof will be to alter a small proportion of the eigenvalues in such a way that the sum of their imaginary parts is unchanged while making an appropriate reduction in the sum of their real parts. The three observations will allow us to achieve this. We need to consider two ranges for ε so we begin by assuming that $\varepsilon < 1/150$.

Divide $\mathbb{T}$ into eight equal arcs $A_1, \ldots, A_8$ where A_m corresponds to the interval $(m - 1)\pi/4 \le \operatorname{Arg} z < m\pi/4$ for $1 \le m \le 8$. There must be at least one arc which contains more that $10n\varepsilon$ eigenvalues, otherwise we would have at most $80n\varepsilon < n$ eigenvalues. The interval $(10n\varepsilon, 11n\varepsilon)$ has length greater than 1, and so let k be the minimal integer in this interval. There are three cases to consider, the first two of which will be shown to be impossible. Note that (c) covers all cases outside (a) and (b) since if there are three or more arcs with at least k eigenvalues each, then at least two are non-adjacent and so separated by an angle of at least $\pi/4$.

(a) Only one arc has at least k eigenvalues.

(b) Exactly two arcs have at least k eigenvalues each and they are adjacent.

(c) There are two non-adjacent arcs which each contain at least k eigenvalues.

(a) Since we will show that $|\tau(U)| > 3\varepsilon$ in this case, we may rotate the eigenvalues so that the arc with at least k eigenvalues is now $-\pi/8 \le \operatorname{Arg} z < \pi/8$, which lies to the right of the line $x = \cos(\pi/8)$. The other seven arcs contain at most $70n\varepsilon$ eigenvalues, and so

$$\operatorname{Re} \tau(U) \ge \left(-70n\varepsilon + (n - 70n\varepsilon) \cos(\pi/8) \right)/n$$
$$\ge -70\varepsilon + (150\varepsilon - 70\varepsilon) \cos(\pi/8) > 3\varepsilon, \qquad (13.4.4)$$

by evaluation of the last term. This contradicts $|\tau(U)| < \varepsilon$.

(b)   By rotation as in (a), we may assume that the two arcs containing at least k eigenvalues are A_1 and A_8, which both lie to the right of the line $x = 1/\sqrt{2}$. The other six arcs account for at most $60n\varepsilon$ eigenvalues so a similar estimate to that of (a) gives

$$\operatorname{Re}\tau(U) \geq (-60n\varepsilon + (n - 60n\varepsilon)/\sqrt{2})/n$$
$$\geq -60\varepsilon + (150\varepsilon - 60\varepsilon)/\sqrt{2} > 3\varepsilon, \qquad (13.4.5)$$

and again we have a contradiction.

(c)   Select k eigenvalues from each of two non-adjacent arcs. Then (i) implies that the average of these $2k$ eigenvalues lies in the disc of radius $19/20$. Observations (ii) and (iii) allow us to change these $2k$ values within $\mathbb{T}$ so that the sum of the imaginary parts is unchanged but any reduction in the sum of the real parts up to $(1/20)(2k) > n\varepsilon$ is possible. Thus we can define a diagonal unitary U_0 with $\tau(U_0) = 0$ by moving $2k < 22n\varepsilon$ eigenvalues of U to other points on the circle. Each eigenvalue moves a distance at most 2, and so we have the estimate

$$\|U - U_0\|_2^2 \leq 2^2 \, 22\varepsilon = 88\varepsilon. \qquad (13.4.6)$$

Thus $\|U - U_0\|_2 \leq 10\sqrt{\varepsilon}$.

Now consider $\varepsilon \geq 1/150$. Take U_0 to be any $n \times n$ diagonal unitary matrix of zero trace, possible by (iii). Then $\|U - U_0\|_2 \leq 2 \leq 2\sqrt{150\varepsilon} < 25\sqrt{\varepsilon}$. This last estimate is then valid for all values of ε.   □

Corollary 13.4.3. *Let N be a II_1 factor and let $\varepsilon > 0$ be given. If $u \in N$ is a unitary satisfying $|\tau(u)| < \varepsilon$, then there exists a unitary $\tilde{u} \in N$ satisfying $\tau(\tilde{u}) = 0$ and $\|u - \tilde{u}\|_2 < 26\sqrt{\varepsilon}$.*

Proof. The functional calculus in the abelian von Neumann algebra generated by u yields orthogonal projections $p_1, \ldots, p_n \in N$ and scalars $\lambda_1, \ldots, \lambda_n$ of modulus 1 such that $\sum_{i=1}^n \lambda_i p_i$ approximates u as closely as we like in $\|\cdot\|_2$-norm. Further small perturbations of the projections allow us to assume that they all have rational trace, whereupon appropriate subdivision of these projections allows us to assume that they all have equal trace. The result is that we may choose the projections and scalars so that

$$\left|\tau\left(\sum_{i=1}^n \lambda_i p_i\right)\right| < \varepsilon \qquad (13.4.7)$$

and

$$\left\|u - \sum_{i=1}^n \lambda_i p_i\right\|_2 < \sqrt{\varepsilon}. \qquad (13.4.8)$$

Further subdivision of the p_i's will ensure that n is so large that $n\varepsilon > 1$. Then the p_i's are the minimal diagonal projections in a matrix subfactor $\mathbb{M}_n \subseteq N$. We can apply Theorem 13.4.2 to see that there is a trace zero unitary $\tilde{u} \in \mathbb{M}_n$ with $\|\sum_{i=1}^n \lambda_i p_i - \tilde{u}\|_2 < 25\sqrt{\varepsilon}$, and the result follows from (13.4.8) and the triangle inequality.   □

Corollary 13.4.4. *Let N be a II_1 factor. The following statements are equivalent:*

(i) *N has property Γ;*

(ii) *given $x_1, \ldots, x_k \in N$, there exists a sequence of unitaries $\{u_n\}_{n=1}^{\infty}$ in N such that $\lim_{n \to \infty} \tau(u_n) = 0$ and*

$$\lim_{n \to \infty} \|u_n x_i - x_i u_n\|_2 = 0, \quad 1 \leq i \leq k. \tag{13.4.9}$$

Proof. (i) $\implies$ (ii) is immediate from Definition 13.4.1 so we only prove the other direction. Suppose that (ii) holds. If we have a sequence of unitaries $\{u_n\}_{n=1}^{\infty}$ satisfying (ii), then Corollary 13.4.3 allows us to find a sequence of unitaries $\{\tilde{u}_n\}_{n=1}^{\infty}$ so that $\tau(\tilde{u}_n) = 0$ and $\lim_{n \to \infty} \|u_n - \tilde{u}_n\|_2 = 0$. This second condition ensures that the $\tilde{u}_n$'s asymptotically commute with the x_i's, showing that N has property Γ. $\qquad\square$

The following theorem will be used in the next section. Although it is easy to state, there does not seem to be a short proof of it. The one given here requires the previous results of this section. A more sophisticated approach would use the ultrafilters of Appendix A, which would have the advantage of extending the result to finite index inclusions of factors [133].

Since we have two factors below, we use τ_M and τ_N to distinguish their traces.

Theorem 13.4.5. *Let N and M be type II_1 factors and suppose that there exists a matrix algebra M_r such that N is isomorphic to $\mathrm{M}_r \otimes M$. Then N has property Γ if and only if the same is true of M.*

Proof. Suppose that M has property Γ, and let $x_1, \ldots, x_k \in N$ be a finite set of elements. Let $\{y_i : 1 \leq i \leq kr^2\} \subseteq M$ be a listing of the matrix entries when we view the x_i's as matrices over M. There is a sequence of trace zero unitaries $\{u_n\}_{n=1}^{\infty}$ in M which asymptotically commutes with the y_i's in $\|\cdot\|_{2,\tau_M}$-norm, and then the sequence of unitaries $\{I_r \otimes u_n\}_{n=1}^{\infty}$ in N asymptotically commutes with the x_i's in $\|\cdot\|_{2,\tau_N}$-norm and $\tau_N(I_r \otimes u_n) = 0$ for $n \geq 1$. Thus N has property Γ.

Conversely suppose that N has property Γ, and consider a finite set $\{x_1, \ldots, x_k\} \subseteq M$. Fix a finite group G of unitaries in M_r which generates this matrix algebra, and apply Definition 13.4.1 to

$$F = \{I_r \otimes x_1, \ldots, I_r \otimes x_k\} \cup \{v \otimes 1_M : v \in G\} \subseteq N.$$

We obtain a sequence $\{u_n\}_{n=1}^{\infty}$ of trace zero unitaries in N which asymptotically commutes with the elements of F in $\|\cdot\|_{2,\tau_N}$-norm. For $n \geq 1$, let

$$t_n = \frac{1}{|G|} \sum_{v \in G} (v \otimes 1_M) u_n (v^* \otimes 1_M) \in N. \tag{13.4.10}$$

The u_n's asymptotically commute with the $(v \otimes 1_M)$'s, and so $\lim_{n\to\infty} \|u_n - t_n\|_{2,\tau_N} = 0$. Moreover, $\tau_N(t_n) = 0$ and $\|t_n\| \leq 1$ for $n \geq 1$. Each t_n lies in $\mathbb{M}_r' \cap N$, and so has the form $I_r \otimes s_n$ for some $s_n \in M$ satisfying $\|s_n\| \leq 1$ and $\tau_M(s_n) = 0$. Since $\lim_{n\to\infty} \|u_n^* u_n - t_n^* t_n\|_{2,\tau_N} = 0$, it follows that $\lim_{n\to\infty} \|1_M - s_n^* s_n\|_{2,\tau_M} = 0$, which says that the s_n's are behaving asymptotically as unitaries.

Let $s_n = v_n |s_n|$ be the polar decomposition of s_n for $n \geq 1$, and let $p_n = v_n v_n^*$, $q_n = v_n^* v_n$. These projections have equal trace and so we may find partial isometries $\tilde{v}_n$ satisfying $1_M - p_n = \tilde{v}_n \tilde{v}_n^*$ and $\tilde{v}_n^* \tilde{v}_n = 1_M - q_n$. Since $p_n s_n = s_n$, we have that $\lim_{n\to\infty} \|1_M - s_n^* p_n s_n\|_{2,\tau_M} = 0$, showing that $\lim_{n\to\infty} \|p_n\|_{2,\tau_M} = 1$. Thus $\lim_{n\to\infty} \tau_M(1_M - p_n) = 0$, which is the same as $\lim_{n\to\infty} \|\tilde{v}_n\|_{2,\tau_M} = 0$. Now define unitaries by $w_n = v_n + \tilde{v}_n \in M$. Applying τ_M to the inequality

$$(1_M - s_n^* s_n)^2 = (1_M - |s_n|^2)^2 \geq (1_M - |s_n|)^2 \tag{13.4.11}$$

shows that $\lim_{n\to\infty} \|1_M - |s_n|\|_{2,\tau_M} = 0$, and so $\lim_{n\to\infty} \|s_n - v_n\|_{2,\tau_M} = 0$. This yields $\lim_{n\to\infty} \|s_n - w_n\|_{2,\tau_M} = 0$. The s_n's asymptotically commute with the x_i's, and so the same is true for the w_n's. Since $\lim_{n\to\infty} \tau_M(s_n - w_n) = 0$, we obtain $\lim_{n\to\infty} \tau_M(w_n) = 0$. From Corollary 13.4.4, we conclude that M has property Γ. □

The following characterisation of property Γ, due to Dixmier [50], has been very useful (see [30]). It will be proved in Appendix A, although we will make use of it in the next section.

Theorem 13.4.6. *A separable II_1 factor N has property Γ if, and only if, for each $m, k \in \mathbb{N}$, $x_1, \cdots, x_m \in N$ and $\varepsilon > 0$, there exist mutually orthogonal projections $p_1, \cdots, p_k$ in N each with trace k^{-1} such that*

$$\|p_i x_j - x_j p_i\|_2 < \varepsilon \quad for \quad 1 \leq j \leq m, \ 1 \leq i \leq k. \tag{13.4.12}$$

Remark 13.4.7. We mentioned at the beginning of this section that property Γ could distinguish certain factors. In the hyperfinite factor R, any finite set of elements can be simultaneously well approximated in $\|\cdot\|_2$-norm by elements in some large matrix subfactor M. Any trace zero unitary from $M' \cap R$ will then fulfill the definition of property Γ for R. Establishing that $L(\mathbb{F}_2)$ does not have property Γ is more complicated and is postponed to Appendix A (see also [105, Theorem 6.7.8]). □

In the following example, we expand this remark to include McDuff factors, infinite tensor products, and tensor products of two factors when at least one has property Γ.

Example 13.4.8. Let (N_k, τ_k), $k \geq 1$, be a sequence of II_1 factors with normal normalised traces, and let $M_k = N_1 \overline{\otimes} \cdots \overline{\otimes} N_k$ with the product trace. Then $M_1 \subseteq M_2 \subseteq \ldots$, and we let A be the C^*-algebra generated by $\bigcup_{k=1}^\infty M_k$, which has a trace τ induced by the traces on the M_k's. Let π be the GNS representation of (A, τ) and let N be the weak closure of $\pi(A)$. The trace τ on A is given by a vector state which induces a normal trace on N, also denoted by τ. Then N is a II$_1$ factor, since any two normal normalised traces on N would agree on each M_k and hence on N. This is the infinite tensor product $\overline{\otimes}_{k=1}^{\infty} N_k$, in which $\bigcup_{k=1}^{\infty} M_k$ is $\|\cdot\|_2$-dense. To check property Γ, we only have to consider elements $x_1, \ldots, x_m \in M_k$ for a sufficiently large choice of k. We may then choose a trace zero unitary $u \in N_{k+1}$, and this commutes with each x_i, verifying property Γ.

If we write the hyperfinite II$_1$ factor R as $\overline{\otimes}_{k=1}^{\infty} \mathbb{M}_2$ and group the matrix factors appropriately, then we see that $R \cong \overline{\otimes}_{k=1}^{\infty} R_k$, where each R_k is a copy of R. Consequently, for any II$_1$ factor M, the McDuff factor $M \overline{\otimes} R$ is isomorphic to an infinite tensor product and so has property Γ by the previous discussion.

We also note that $M_1 \overline{\otimes} M_2$ has property Γ whenever at least one factor, say M_1, has this property. It is easily checked that any family (u_α) of trace zero unitaries verifying property Γ for M_1 gives rise to a family $(u_\alpha \otimes 1)$ verifying property Γ for $M_1 \overline{\otimes} M_2$. $\square$

Remark 13.4.9. There is a more general form of Corollary 13.4.3 available which we will need in Theorem A.5.3: given $\varepsilon > 0$ and $\lambda \in \mathbb{D}$, there is a constant $K(|\lambda|)$ so that if $u \in N$ is a unitary satisfying $|\tau(u) - \lambda| < \varepsilon$, then there is a unitary $\tilde{u} \in N$ with $\tau(\tilde{u}) = \lambda$ and $\|u - \tilde{u}\|_2 < K(|\lambda|)\sqrt{\varepsilon}$.

For $|\lambda| = 1$, take $\tilde{u} = \lambda 1$ and $K(1) = \sqrt{2}$. When $|\lambda| < 1$, follow the proof of Theorem 13.4.2 by making the following choices. Take t arcs (t replaces 8) where $\cos(2\pi/t) > |\lambda|$. Then choose an integer s to satisfy $2s(1 - \cos(\pi/t)) > 1$ ($sn\varepsilon$ replaces $10n\varepsilon$). Now choose $\delta > 0$ (δ replaces $1/150$) to satisfy the analogues of (13.4.4) and (13.4.5):

$$-(t-1)s\delta + (1 - (t-1)s\delta)\cos(\pi/t) > |\lambda| + 3\delta \tag{13.4.13}$$

and

$$-(t-2)s\delta + (1 - (t-2)s\delta)\cos(2\pi/t) > |\lambda| + 3\delta \tag{13.4.14}$$

The proof then proceeds as in Theorem 13.4.2 by considering the ranges $\varepsilon < \delta$ and $\varepsilon \geq \delta$. The constraints on these constants imposed above will force s, $t \to \infty$ and $\delta \to 0$ as $|\lambda| \to 1$, and the constant $K(|\lambda|)$ that would emerge in Corollary 13.4.3 would tend to ∞ as $|\lambda| \to 1$. Since $K(1) = \sqrt{2}$, it seems likely that there should be an absolute constant K to replace $K(|\lambda|)$, but our methods do not give one. $\square$

13.5 Irreducible hyperfinites in Γ factors

In this section we use the existence of an irreducible hyperfinite subfactor to construct such a subfactor with some additional properties in the case that N has property Γ.

We will use the notation $[x, y] = xy - yx$ for the commutator of two operators.

Lemma 13.5.1. *Let N be type II$_1$ factor with property Γ and let $N = \mathbb{M}_r \otimes M$ be a tensor product decomposition of N. Given $x_1, \ldots, x_n \in N$, $k \in \mathbb{N}$, and*

$\varepsilon > 0$, there exists a set of mutually orthogonal projections $\{p_s\}_{s=1}^k \in M$, each of trace k^{-1}, such that

$$\|[1 \otimes p_s, x_t]\|_2 < \varepsilon, \qquad 1 \le s \le k, \quad 1 \le t \le n. \qquad (13.5.1)$$

Proof. Write each x_t as an $r \times r$ matrix over M, and let $\{y_i\}_{i=1}^{nr^2}$ be a listing of all the resulting matrix entries in M. By Theorem 13.4.5, M has property Γ so choose a set $\{p_s\}_{s=1}^k \subseteq M$ of mutually orthogonal projections of trace k^{-1} satisfying

$$\|[p_s, y_t]\|_2 < r^{-1}\varepsilon, \qquad 1 \le s \le k, \quad 1 \le t \le nr^2, \qquad (13.5.2)$$

from Theorem 13.4.6. Inequality (13.5.1) follows directly from this.   □

Since the projections found above have equal trace, they may be viewed as the minimal projections in the diagonal of an $n \times n$ matrix subalgebra of M; we will use this subsequently.

The next two lemmas will handle the inductive step in the proof of Theorem 13.5.4.

Lemma 13.5.2. *Let N be a separable II_1 factor with property Γ and let $\mathbb{M}_r$ be an $r \times r$ matrix subfactor of N such that N has the form $\mathbb{M}_r \otimes M$. Given $x_1, \ldots, x_n \in N$, there exists a matrix subfactor $Q \subseteq M$ of dimension $n! \times n!$ with the following property. For each integer k satisfying $1 \le k \le n$, there are k orthogonal projections $p_1, \ldots, p_k$ in the diagonal of Q, each of trace k^{-1}, such that*

$$\|[x_j, I_r \otimes p_i]\|_2 < n^{-1}, \qquad 1 \le i \le k, \quad 1 \le j \le n. \qquad (13.5.3)$$

Proof. We will construct Q as $Q_1 \otimes Q_2 \otimes \ldots \otimes Q_n$ where each Q_m is an $m \times m$ matrix subfactor, and where the diagonal of Q is the tensor product of the diagonals of the Q_m's. We take $Q_1 = \mathbb{C}1$. Assume that $Q_1, \ldots, Q_{k-1}$ have been constructed so that the desired projections of trace m^{-1} for (13.5.3) are in the diagonal of Q_m for $1 \le m \le k-1$. We now apply Lemma 13.5.1 with $\varepsilon = n^{-1}$ and $\mathbb{M}_r$ replaced by $\mathbb{M}_r \otimes Q_1 \otimes \ldots \otimes Q_{k-1}$ to obtain a matrix subfactor Q_k in the commutant of this algebra whose minimal diagonal projections of trace k^{-1} satisfy (13.5.3). This inductive step allows us to arrive at $Q = Q_1 \otimes \ldots \otimes Q_n$ whose diagonal contains all the required projections to satisfy (13.5.3).   □

Lemma 13.5.3. *Let $\mathbb{M}_r$ be a matrix subalgebra of a separable II_1 factor N so that $N = \mathbb{M}_r \otimes M$. Given $\varepsilon > 0$ and $y_1, \ldots, y_k \in N$, there exists a matrix subalgebra $\mathbb{M}_q \subseteq M$ such that the diagonal $\mathbb{D}_{rq} = \mathbb{D}_r \otimes \mathbb{D}_q$ of $\mathbb{M}_r \otimes \mathbb{M}_q$ has the following property:*

$$\|\mathbb{E}_{\mathbb{D}_{rq}}(y_i) - \mathbb{E}_{\mathbb{D}_{rq}' \cap N}(y_i)\|_2 < \varepsilon \qquad (13.5.4)$$

for $1 \le i \le k$.

Proof. Choose a masa B in M, which is isomorphic to $L^\infty[0,1]$, by Theorem 3.5.2, and where the trace corresponds to Lebesgue measure. By dividing $[0,1]$ into 2^n intervals of equal length, we obtain an increasing sequence $B_1 \subseteq B_2 \subseteq \ldots \subseteq B$ of finite dimensional subalgebras, where the minimal projections in each B_n are equivalent in M. Thus there is an increasing sequence of matrix subfactors $M_1 \subseteq M_2 \subseteq \ldots \subseteq M$ with B_n the diagonal of M_n. The construction of the B_n's implies that their union is weakly dense in B. Then $\bigcup_{n \geq 1} \mathbb{D}_r \otimes B_n$ is weakly dense in the masa $\mathbb{D}_r \otimes B$ in N. By Lemma 3.6.8,

$$\lim_{n \to \infty} \|\mathbb{E}_{\mathbb{D}_r \otimes B_n}(y_i) - \mathbb{E}_{(\mathbb{D}_r \otimes B_n)' \cap N}(y_i)\|_2 = 0 \qquad (13.5.5)$$

for $1 \leq i \leq k$. If we take $\mathbb{M}_q$ to be M_n for a sufficiently large choice of n, then (13.5.4) follows from (13.5.5). □

The next theorem is a refinement of [30, Theorem 5.3], which gave the same conclusion except for the statement that A is a masa in N.

Theorem 13.5.4. *Let N be a separable II_1 factor with property Γ. There is a masa A in N and an irreducible hyperfinite subfactor R of N such that A is a Cartan masa in R, and if $m, n \in \mathbb{N}$, $x_1, \cdots, x_m \in N$ and $\varepsilon > 0$, then there exist mutually orthogonal projections $p_1, \ldots, p_n$ in A, each of trace n^{-1}, such that*

$$\|p_i x_j - x_j p_i\|_2 < \varepsilon, \quad for \quad 1 \leq i \leq n, \quad 1 \leq j \leq m. \qquad (13.5.6)$$

Proof. We will construct $A \subseteq R$ as the ultraweak closure of an ascending union of diagonals A_n of matrix subalgebras M_n of N, which are inductively defined. We choose a sequence $\{y_t\}_{t=1}^\infty$ from the unit ball of N with $y_1 = 1$, and which is $\| \cdot \|_2$-norm dense in the unit ball. For this choice, the induction hypotheses on A_n and M_n are

(i) $A_1 = M_1 = \mathbb{C}1$;

(ii) $M_{n-1} \subseteq M_n$ and $A_{n-1} \subseteq A_n$ with $M_n \cong M_{n-1} \otimes \mathbb{M}_{m_n}$ and $A_n \cong A_{n-1} \otimes \mathbb{D}_{m_n}$ for some $m_n \in \mathbb{N}$ for $n \geq 1$;

(iii) for each $k \leq n$ there exist mutually orthogonal projections $p_1, \cdots, p_k \in A_n$, $\tau(p_i) = k^{-1}$, satisfying

$$\|[p_i, y_t]\|_2 < n^{-1}, \quad 1 \leq i \leq k, \quad 1 \leq t \leq n; \qquad (13.5.7)$$

(iv) for $1 \leq t \leq n$,

$$\|\mathbb{E}_{A'_n \cap N}(y_t) - \mathbb{E}_{A_n}(y_t)\|_2 < 2^{-n}. \qquad (13.5.8)$$

Before we construct these algebras, we first see why their existence implies the conclusion of the theorem. Suppose that we have algebras $A_n \subseteq M_n$ for $n \geq 1$ as described in (i)–(iv). The weak closure of $\bigcup_{n \geq 1} M_n$ is a hyperfinite subfactor R containing the abelian subalgebra $A = \bigcup_{n \geq 1} \bar{A}_n$. Inequality (13.5.8) implies that A is a masa in N by Corollary 3.6.9. Then any $x \in R' \cap N$ commutes with A and so lies in $A \subseteq R$. Thus x is central in the factor R, so is a scalar, and it follows that $R' \cap N = \mathbb{C}1$. The permutation matrices in M_n generate this algebra and also normalise A_r for each $r \geq n$, from the nature of the tensor product construction of these algebras in (ii). Thus $M_n \subseteq \mathcal{N}(A, R)''$, for all $n \geq 1$, showing that A is Cartan in R. Finally, (iii) shows that A contains enough projections to meet Dixmier's criterion for property Γ in Theorem 13.4.6. Thus it only remains to construct inductively the algebras satisfying (i)–(iv).

Suppose that $A_k \subseteq M_k$ for $k \leq n-1$ have been constructed. We first tensor M_{n-1} with a matrix algebra R_1 so that the tensor product of the diagonals contains projections for which the inequalities of (13.5.7) are satisfied; this is possible by Lemma 13.5.2. Using Lemma 13.5.3 with $k = n$ and $\varepsilon = 2^{-n}$, we then tensor by a further matrix algebra R_2 so that (13.5.8) is satisfied when M_n is $M_{n-1} \otimes R_1 \otimes R_2$ and A_n is the tensor product of their diagonals. This finishes the inductive step and completes the proof. □

The following observation is immediate from Theorem 13.5.4. We refer to Appendix A for the definition of A^ω and N^ω.

Corollary 13.5.5. *If N is a separable II_1 factor with property Γ, then there is a masa A in N with $N' \cap A^\omega \neq \mathbb{C}1$ in N^ω.*

Proof. Let A be the masa constructed in Theorem 13.5.4 and let $\{y_i\}_{i=1}^\infty$ be the set of elements in the proof of that result. Then, for each $n \in \mathbb{N}$, there exist projections $p_1^n, \ldots, p_{2n}^n \in A$ of trace $(2n)^{-1}$ satisfying

$$\|[p_i^n y_j, y_j p_i^n]\|_2 < n^{-1}, \qquad 1 \leq i \leq 2n, \quad 1 \leq j \leq n. \tag{13.5.9}$$

We may then define a sequence of trace zero unitaries $u_n \in A$ by $u_n = 1 - 2\sum_{k=1}^n p_k^n$ for $n \geq 1$, and these asymptotically commute with all elements of N, but form a non-trivial element of N^ω since the tracial condition rules out the possibility that $\lim_{n \to \infty} \|u_n - \lambda_n 1\|_2 = 0$ for any sequence of scalars $\{\lambda_n\}_{n=1}^\infty$. □

Remark 13.5.6. The results of this chapter are based on the methods of Popa (see [139, 140, 141]) with the Γ factor section from Christensen, Pop, Sinclair and Smith [30, 31]. The results in this chapter have been used in cohomology calculations in von Neumann algebras (see [172] and [30]) and in Shen's proof that separable Cartan and Γ factors are singly generated (see [171]). □

Chapter 14

Maximal injective subalgebras

14.1 Introduction

The hyperfinite, or injective, von Neumann subalgebras of a type II_1 factor are closed under weak limits of unions of ascending chains so maximal injective von Neumann algebras exist in all type II_1 factors [36]. However explicit examples are hard to find. Popa [137] showed that the generator masa in a free group factor is a maximal injective von Neumann subalgebra in the factor. His methods show (Theorem 14.1.1) that the masas in the free group factors that arise from prime elements in the group are maximal injective von Neumann subalgebras of the free group factors. Note that in each type II_1 factor N there are injective subfactors R with $R' \cap N = \mathbb{C}1$ by Theorem 13.2.3, and if M is an injective von Neumann subalgebra of N with $R \subseteq M$, then $M' \cap N \subseteq R' \cap N = \mathbb{C}1$ so M is a maximal injective subfactor of N with trivial relative commutant. What is amazing is that there are maximal injective abelian von Neumann algebras [137, Corollary 3.3]. In the same paper other results about maximal injective von Neumann subalgebras of II_1 factors are also discussed.

Section 14.2 is devoted to a proof of the following:

Theorem 14.1.1. *Let $2 \leq n \leq \infty$. If a is a generator of $\mathbb{F}_n$, then the abelian von Neumann algebra A generated by a is a maximal injective von Neumann subalgebra of $L(\mathbb{F}_n)$.*

This is a direct consequence of Theorem 14.2.5 since each von Neumann subalgebra of $L(\mathbb{F}_n)$ properly containing A has a central summand that is a factor that does not have property Γ so is not injective. The proof of Theorem 14.2.5 depends on checking the asymptotic properties of commutators in the $\| \cdot \|_2$, which are handled in the ultrapower, so providing the link with property Γ via McDuff's results [114]. The combinatorial ideas are in Lemma 14.2.2 in the free group and are turned into orthogonality and $\| \cdot \|_2$-norm estimates in von

Neumann algebras in Lemma 14.2.3. The orthogonality and $\|\cdot\|_2$-norm estimates are lifted to properties of various subalgebras in $L(\mathbb{F}_n)^\omega$ in Lemma 14.2.4. The existence of non-zero atomic parts in several von Neumann subalgebras is an important tool in the proof.

In Section 14.3, we extend these results by showing how to obtain concrete examples of injective subfactors. The ones we present arise as crossed products by free groups, and their properties heavily depend on Theorem 14.1.1. We also discuss some basic results about crossed products, but the reader should consult [104, 105] for a more detailed survey of this subject.

14.2   Maximal injectivity and masas

Before proving the result that a generator masa in $L(\mathbb{F}_n)$ is a maximal injective von Neumann subalgebra (Theorem 14.1.1) we shall prove that, in group von Neumann algebras, masas arising from subgroups restrict the central structure of von Neumann subalgebras containing them. The condition on the pair $H \subseteq G$ in the next theorem is similar to conditions (C1)–(C4) from Section 3.3. Indeed, it is a simple algebraic exercise to see that (C1) is implied by the hypothesis of Theorem 14.2.1.

Theorem 14.2.1. *Let G be a countable ICC group and let H be an infinite abelian subgroup of G with the property that $gHg^{-1} \cap H = \{e\}$ for all g in $G \setminus H$. Let $A = L(H)$ be the von Neumann subalgebra of $L(G)$ generated by H. If N is a von Neumann subalgebra of $L(G)$ containing A, then there is a partition $\{e_k : k \geq 0\}$ of the identity 1 in the centre Z of N such that $Ne_0 = Ae_0$, and Ne_k is a II_1 factor with the von Neumann algebra $(N' \cap A^\omega)e_k$ having non-zero atomic part for all $k \geq 1$.*

Proof. As noted above, the hypotheses on G and H imply that condition (C1) of Section 3.3 is satisfied, and so A is a masa in $L(G)$ by Lemma 3.3.1. Thus A is a masa in N and A contains the centre Z of N. Let e_0 be the maximal projection p in Z such that $Np = Ap$ and let $e = 1 - e_0$. It is possible that $e_0 = 0$. We will show that eZ is an atomic (abelian) von Neumann algebra, and we will obtain a contradiction by supposing that this is not so. Then there is a non-zero projection q in eZ such that qZ is diffuse and $(e - q)Z$ is atomic. Let $B = (1 - q)A + qZ$ so $Z \subseteq B \subseteq A$ and $qB = qZ$ is diffuse. For all g in $G \setminus H$ the abelian von Neumann algebras gAg^{-1} and A are orthogonal by hypothesis. Hence gBg^{-1} and B are orthogonal for all g in $G \setminus H$. Let $\mathcal{N}(B)$ be the unitary normaliser of B in $L(G)$. By Lemma 6.3.2 with $N = N_0 = B$, the element $g \in G \setminus H$ is orthogonal to $\mathcal{N}(B)''$ so g is orthogonal to $B' \cap L(G)$, since this algebra is contained in $\mathcal{N}(B)''$. Thus $B' \cap L(G)$ is contained in $L^2(H) = L^2(A)$, which is the orthogonal complement of the linear span of $G \setminus H$, and so $B' \cap L(G)$ is contained in A. Since q is in Z with $qB = qZ$, we have

$$qA \supseteq q(B' \cap L(G)) = q(Z' \cap L(G)) \supseteq qN \qquad (14.2.1)$$

so that $qA = qN$. This contradicts the maximality of e_0 with $e_0A = e_0N$ so $(1 - e_0)Z = eZ$ is an atomic abelian von Neumann algebra.

Let $\{e_k\}_{k \geq 1}$ be the atoms of $(1 - e_0)Z$. Then $\sum_{k \geq 0} e_k = 1$. The von Neumann algebra N is diffuse since A is a masa in N and each of the factors e_kN is of type II$_1$ for $k \geq 1$.

Suppose that $(N' \cap A^\omega)e_k$ is diffuse for some $k \geq 1$ and let

$$B = (N' \cap A^\omega)e_k + A^\omega(1 - e_k) \qquad (14.2.2)$$

in $L(G)^\omega$. Then $B \subseteq A^\omega$ is diffuse. Let $g \in G \backslash H$. The algebras A^ω and $gA^\omega g^{-1}$ are orthogonal von Neumann algebras since A and gAg^{-1} are orthogonal. Hence B and gBg^{-1} are orthogonal von Neumann algebras. Using Lemma 6.3.2 again shows that g is orthogonal to $\mathcal{N}(B)''$ in $L(G)^\omega$ so that g is orthogonal to $B' \cap L(G)^\omega$ for all g in $G \backslash H$. Moreover

$$Ne_k \subseteq ((N' \cap A^\omega)' \cap L(G)^\omega)e_k = (B' \cap L(G)^\omega)e_k$$
$$\subseteq B' \cap L(G)^\omega \qquad (14.2.3)$$

since $((N' \cap A^\omega)'e_k = B'e_k$ by (14.2.2). Hence $G \backslash H$ is orthogonal in $L^2(L(G))$ to Ne_k so that Ne_k is contained in $L^2(H) = L^2(A)$ and Ne_k is contained in A. This contradicts the definition of e_k for $k \geq 1$. $\qquad \square$

Throughout the remainder of this section we shall use the following definitions and notation. Let $2 \leq n \leq \infty$, let the free group $\mathbb{F}_n$ have n generators $a, v_1, v_2, \ldots$, and let $L^2(\mathbb{F}_n)$ denote the Hilbert space arising from the trace τ on $N = L(\mathbb{F}_n)$. If W is a subset of $\mathbb{F}_n$, define the orthogonal projection p_W on $L^2(\mathbb{F}_n)$ by

$$p_W(x) = \sum_{g \in W} \alpha_g g \quad \text{if} \quad x = \sum_{g \in \mathbb{F}_n} \alpha_g g \in L^2(\mathbb{F}_n). \qquad (14.2.4)$$

If W is empty, then p_W is the zero projection. Observe that if V and W are subsets of $\mathbb{F}_n$, then $p_V p_W = p_{V \cap W}$. The following two elementary equalities of p_W will be used frequently in the subsequent calculations:

$$p_{gWg^{-1}}(x) = gp_W(g^{-1}xg)g^{-1} \quad \text{and} \quad p_W(x)^* = p_{W^{-1}}(x^*) \qquad (14.2.5)$$

for all $W \subseteq \mathbb{F}_n$, $g \in \mathbb{F}_n$ and $x \in L^2(\mathbb{F}_n)$.

Let $Gp(a)$ denote the subgroup generated by a. A word g in $\mathbb{F}_n$ is said to begin exactly (end exactly) with a^j for some $j \in \mathbb{Z}$ if $g = a^j h$ ($g = ha^j$) with no cancellation in $a^j h$ (ha^j) so $|g| = |j| + |h|$ and h cannot be written $h = ah_1$ ($h = h_1a$) with $|h| = 1 + |h_1|$. If h begins exactly with a^j, then $h = a^j v_{i_1} v_{i_2} \cdots v_{i_k}$, where v_{i_1} is not equal to a. In the next lemma, and for the rest of this section, U_m is the set of all words of length less than m that are not powers of a, W_m is the set of all words that are not powers of a that begin exactly with a^k for $0 \leq |k| < m$ and V_m is the set of all words that are not powers of a that begin exactly and end exactly with powers of a greater than or equal to m.

Lemma 14.2.2. *With the notation just described,*

(i) *if m is a positive integer, then $W_m \subseteq W_{m+1}$ and $U_m \subseteq U_{m+1}$;*

(ii) $\bigcup_{m=1}^{\infty} U_m = \bigcup_{m=1}^{\infty} W_m = \mathbb{F}_n \setminus Gp(a)$;

(iii) $U_m V_m U_m \cap V_m = \emptyset$ *and* $U_m V_m \cap V_m U_m = \emptyset$.

Proof. Parts (i) and (ii) are clear. Let g_1 begin exactly with a^j and g_2 end exactly with a^k with $|g_1| < m$ and $|g_2| < m$. If h is in V_m, then h begins exactly with a^s and ends exactly with a^t, where $|s|, |t| \geq m$, and h contains a generator not equal to a. In $g_1 h g_2$ the cancellation in $g_1 h$, and $h g_2$, stops at the generator in g_1, and g_2, not equal to a since $|s| \geq m$ and $|t| \geq m$. Hence $g_1 h g_2$ begins exactly with a^j and ends exactly with a^k, where $|j|, |k| < m$, so $g_1 h g_2$ is not in V_m. Since $U_m^{-1} = U_m$, the second inequality follows from the first. $\square$

Lemma 14.2.3. *For each positive integer m, let V_m and U_m be as defined before Lemma 14.2.2.*

(i) *Let m be a positive integer, let $\varepsilon > 0$ and let $x \in L^2(\mathbb{F}_n)$ with $\|x\|_2 \leq 1$. If*

$$\|a^j x - x a^j\|_2 \leq \varepsilon \quad \text{for} \quad 1 \leq j \leq 2m^2, \tag{14.2.6}$$

then

$$\|p_{W_m \cup W_m^{-1}}(x)\|_2^2 \leq 4(\varepsilon^2 + (2m+1)^{-1}). \tag{14.2.7}$$

(ii) *If $y \in L(\mathbb{F}_n)$ with $\mathbb{E}_A(y) = 0$, then*

$$\lim_{m \to \infty} \|y - p_{W_m}(y)\|_2 = \lim_{m \to \infty} \|y - p_{U_m}(y)\|_2 = 0. \tag{14.2.8}$$

(iii) *If $x_j, y_j \in L(\mathbb{F}_n)$ for $j = 1, 2$, then $p_{V_m}(x_1) p_{U_m}(y_1)$ and $p_{U_m}(y_2) p_{V_m}(x_2)$ are orthogonal.*

Proof. (i)   If $1 \leq j \leq m$, then using $(\alpha + \beta)^2 \leq 2(\alpha^2 + \beta^2)$ for $0 \leq \alpha, \beta$,

$$\|p_{W_m}(x)\|_2^2 \leq (\|p_{W_m}(x - a^{-2jm} x a^{2jm})\|_2 + \|p_{W_m}(a^{-2jm} x a^{2jm})\|_2)^2$$
$$\leq 2\|x - a^{-2jm} x a^{2jm}\|_2^2 + 2\|p_{W_m}(a^{-2jm} x a^{2jm})\|_2^2$$
$$\leq 2\varepsilon^2 + 2\|(a^{-2jm} p_{a^{2jm} W_m a^{-2jm}}(x) a^{2jm}\|_2^2$$
$$\leq 2\varepsilon^2 + 2\|p_{a^{2jm} W_m a^{-2jm}}(x)\|_2^2 \tag{14.2.9}$$

by (14.2.6). If $g \in a^{2jm} W_m a^{-2jm}$, then $g = a^{2jm} a^k h a^{-2jm}$, where $|k| < m$ and h does not begin with a, so g starts with a^{2jm+k} and $(2j-1)m < 2jm + k < (2j+1)m$. The open intervals $((2j-1)m, (2j+1)m)$ in $\mathbb{Z}$ are pairwise disjoint so that

$$a^{2im} W_m a^{-2im} \cap a^{2jm} W_m a^{-2jm} = \emptyset \tag{14.2.10}$$

and

$$p_{a^{2im} W_m a^{-2im}} p_{a^{2jm} W_m a^{-2jm}} = 0 \tag{14.2.11}$$

for $0 \leq i \neq j \leq m$. Using this orthogonality and summing (14.2.9) over $|j| \leq m$ yields

$$(2m+1)\|p_{W_m}(x)\|_2^2 \leq 2(2m+1)\varepsilon^2 + 2 \sum_{|j| \leq m} \|p_{a^{2jm}W_m a^{-2jm}}(x)\|_2^2$$

$$\leq 2(2m+1)\varepsilon^2 + 2 \left\| \sum_{|j| \leq m} p_{a^{2jm}W_m a^{-2jm}}(x) \right\|_2^2$$

$$\leq 2(2m+1)\varepsilon^2 + 2. \tag{14.2.12}$$

and hence

$$\|p_{W_m}(x)\|_2^2 \leq 2(\varepsilon^2 + (2m+1)^{-1}). \tag{14.2.13}$$

The element x^* satisfies the same hypothesis as x and $p_{W_m}(x^*) = p_{W_m^{-1}}(x)^*$ so that

$$\|p_{W_m \cup W_m^{-1}}(x)\|_2^2 = \|p_{W_m}(x)\|_2^2 + \|p_{W_m^{-1} \setminus W_m}(x)\|_2^2$$

$$\leq \|p_{W_m}(x)\|_2^2 + \|p_{W_m^{-1}}(x)\|_2^2$$

$$\leq 4(\varepsilon^2 + (2m+1)^{-1}) \tag{14.2.14}$$

by Lemma 14.2.2 and equation (14.2.12).

(ii)   If y is considered as an element of $L^2(\mathbb{F}_n)$, then $p_{Gp(a)}(y) = \mathbb{E}_A(y) = 0$, y is supported on $\mathbb{F}_n \setminus Gp(a) = \bigcup_{m=1}^{\infty} W_m = \bigcup_{m=1}^{\infty} U_m$ and hence $\lim_{m \to \infty} \|y - p_{W_m}(y)\|_2 = \lim_{m \to \infty} \|y - p_{U_m}(y)\|_2 = 0$.

(iii)   The support of the element $p_W(x)$ is contained in W. Hence the support of $p_{V_m}(x_1)p_{U_M}(y_1)$ is contained in $V_m U_m$ and the support of $p_{U_m}(y_2)p_{V_m}(x_2)$ is contained in $U_m V_m$. By Lemma 14.2.2 these supports are disjoint and hence the elements are orthogonal. $\qquad \square$

Lemma 14.2.4. *Let x be in $L(\mathbb{F}_n)^{\omega}$ with $\mathbb{E}_A(x) = 0$ and $ax = xa$. If y_j are in $L(\mathbb{F}_n)$ with $\mathbb{E}_A(y_j) = 0$ for $j = 1, 2$, then $y_1 x$ and xy_2 are orthogonal in $L(\mathbb{F}_n)^{\omega}$ and*

$$\|y_1 x - xy_2\|_{\omega,2}^2 = \|y_1 x\|_{\omega,2}^2 + \|xy_2\|_{\omega,2}^2. \tag{14.2.15}$$

Proof. Let $\varepsilon > 0$. Using the result from Appendix A, Lemma A.5.4, that $\mathbb{E}_{A^{\omega}}(z_r) = (\mathbb{E}_A(z_r))$ for all $(z_r) \in L(\mathbb{F}_n)^{\omega}$, we may replace x by $(x_r - \mathbb{E}_A(x_r))$ and so assume that $\mathbb{E}_A(x_r) = 0$ for all $r \in \mathbb{N}$. By scaling we assume that $x = (x_r)$ with $\|x_r\| \leq 1$ for all $r \in \mathbb{N}$ and $\|y_j\| \leq 1$ for $j = 1, 2$. Both y_1 and y_2 satisfy Lemma 14.2.3 so there is an m_0 in $\mathbb{N}$ such that

$$\|y_1 - p_{U_k} y_1\|_2 < \varepsilon \text{ and } \|y_2 - p_{U_k} y_2\|_2 < \varepsilon \tag{14.2.16}$$

for all $k \geq m_0$. Let m be a positive integer with $m \geq m_0$ and $(2m+1) \geq \varepsilon^{-2}$, and let

$$T = \{r \in \mathbb{N} : \|a^j x_r - x_r a^j\|_2 \leq \varepsilon \text{ for } 1 \leq j \leq 2m^2\}.$$

If $r \in T$, then

$$\|x_r - p_{V_m} x_r\|_2^2 = \|p_{W_m \cup W_m^{-1}}(x_r)\|_2^2 \leq 4(\varepsilon^2 + (2m+1)^{-1}) \leq 8\varepsilon^2 \tag{14.2.17}$$

by Lemma 14.2.3 since $x_r = p_{V_m}(x_r) + p_{W_m \cup W_m^{-1}}(x_r)$ using $p_{Gp(a)}(x_r) = 0$.
The scaling on x_r and y_j, and the Cauchy–Schwarz inequality imply that

$$
\begin{aligned}
|\tau(y_1 x_r (x_r y_2)^*)| \leq & \|y_1 - p_{U_m}(y_1)\|_2 + \|y_2^* - p_{U_m}(y_2)^*\|_2 + 2\|x_r - p_{V_m}(x_r)\|_2 \\
& + |\tau((p_{U_m}(y_1) p_{V_m}(x_r))[p_{V_m}(x_r) p_{U_m}(y_2)]^*)| \\
\leq & \, 2\varepsilon + 2^{5/2}\varepsilon \leq 8\varepsilon
\end{aligned}
\tag{14.2.18}
$$

for all $r \in T$ by (14.2.16) and Lemma 14.2.3.

Let χ_T be the characteristic function of T and assume that $\varepsilon < 1$. For each $j \in \mathbb{N}$,

$$
\omega(\|a^j x_r - x_r a^j\|_2) = \|a^j x - x a^j\|_{\omega,2} = 0
\tag{14.2.19}
$$

and so $\omega(\chi_T) = 1$. Hence

$$
\begin{aligned}
|\tau_\omega(y_1 x (x y_2)^*)| &= |\omega(\tau(y_1 x_r (x_r y_2)^*))| \\
&= |\omega(\chi_T(r)\tau(y_1 x_r (x_r y_2)^*))| \leq 8\varepsilon
\end{aligned}
\tag{14.2.20}
$$

by (14.2.18). □

Examining the proof of Lemma 14.2.4 indicates that minor changes to the notation show that if x_j are in $L(\mathbb{F}_n)^\omega$ with $\mathbb{E}_{A^\omega}(x_j) = 0$ and $ax_j = x_j a$ for $j = 1, 2$ with the remaining hypotheses as in Lemma 14.2.4, then $y_1 x_1$ and $x_2 y_2$ are orthogonal in $L(\mathbb{F}_n)^\omega$ and

$$
\|y_1 x_1 - x_2 y_2\|_{\omega,2}^2 = \|y_1 x_1\|_{\omega,2}^2 + \|x_2 y_2\|_{\omega,2}^2.
\tag{14.2.21}
$$

Theorem 14.2.5. *Let $2 \leq n \leq \infty$, let a be a generator of $\mathbb{F}_n$, and let A be the von Neumann algebra generated by a. If N is a von Neumann subalgebra of $L(\mathbb{F}_n)$ containing A, then N is a (central) direct sum of a subalgebra of A and a sequence of type II_1 factors that do not have property Γ.*

Proof. By Theorem 14.2.1 there are projections e_k $(k \geq 0)$ in the centre Z of N such that $\sum_{k \geq 0} e_k = 1$, $Ne_0 = Ae_0$ and Ne_k is a type II_1 factor for each $k \geq 1$. Suppose that Ne_k has property Γ for some e_k with $k \geq 1$. By McDuff's result, (see [114] or Lemma A.7.3), it follows that $(e_k Ne_k)' \cap (e_k Ne_k)^\omega$ has no atoms. By Theorem 14.2.1 the von Neumann algebra $(N' \cap A^\omega)e_k$ has non-zero atomic part. Since

$$
(N' \cap A^\omega)e_k \subseteq e_k(N' \cap N^\omega)e_k = e_k N' e_k \cap (e_k Ne_k)^\omega
\tag{14.2.22}
$$

there is an x in $(e_k Ne_k)' \cap (e_k Ne_k)^\omega$ that is not in A^ω, since the later is continuous. By subtracting $\mathbb{E}_{A^\omega}(x)$ from x, we may assume that $\mathbb{E}_{A^\omega}(x) = 0$. Then $xw = wx$ for all w in $e_k N$ and $xa = ax$, since $A \subseteq N$.

Choose a unitary w in the von Neumann algebra $e_k Ne_k = Ne_k$ that is orthogonal to A, or equivalently $\mathbb{E}_A(w) = 0$. This choice can be made as follows, for example. Let f_1 and f_2 be orthogonal projections in A with $f_1 + f_2 = e_k$ and $\tau(f_1) = \tau(f_2)$ so f_1 and f_2 are equivalent projections in Ne_k. Let v be a partial isometry in Ne_k with $vv^* = f_1$ and $v^*v = f_2$, and let $w = v + v^*$. Then

w is a self-adjoint unitary in the factor Ne_k with $\mathbb{E}_A(w) = 0$ since $\mathbb{E}_A(v) = 0$. By Lemma 14.2.4 with $w = y_1 = y_2$ orthogonality gives

$$2\|x\|_2^2 = \|wx\|_2^2 + \|xw\|_2^2 = \|wx - xw\|_2^2 = 0 \qquad (14.2.23)$$

because $xw = wx$. Thus $x = 0$ and this contradicts the assumption that $x \notin A^\omega$. $\qquad\square$

Theorem 14.2.5 may be proved with A replaced by the von Neumann algebra B generated by a prime element u in $\mathbb{F}_n$. Before starting the calculations one replaces u by an element v in its conjugacy class of minimal length and so cyclically reduced. It would be interesting to know for which groups G and prime elements u Theorem 14.2.5 is true.

14.3 Maximal injectivity of subfactors

In this section we will give some non-trivial examples of maximal injective subfactors. If N is a II_1 factor with an irreducible hyperfinite subfactor R, then any von Neumann algebra M lying between R and N is automatically a factor, since any central element would lie in $R' \cap N = \mathbb{C}1$. Thus Zorn's lemma guarantees the existence of maximal injective subfactors in any II_1 factor. Our objective is to give some concrete examples of this phenomenon. These will be drawn from crossed products by actions of groups by outer automorphisms. Blattner [6] was the first to observe that each countable discrete group acts by outer automorphisms on the hyperfinite II_1 factor R. One way to see this is by expressing R as $\bigotimes_{g \in G} R_g$, where each R_g is a copy of R; this isomorphism is possible because of the uniqueness of the hyperfinite II_1 factor R. The Bernoulli shift action of G is to let each $h \in G$ act as a permutation of the constituent factors, sending R_g to R_{hg}. Rather than give the argument that this is an outer action, which is notationally complex, we will instead look at the flip automorphism θ of $R \overline{\otimes} R$ which sends $x \otimes y$ to $y \otimes x$, since the case above works in the same way. If θ were implemented by a unitary $u \in R \overline{\otimes} R$, then conjugation by u would extend θ to an automorphism ϕ of $B(H) \overline{\otimes} B(H)$, where $H = L^2(R)$. Then ϕ would fix $R' \otimes 1 \subseteq (R \overline{\otimes} R)'$, and so would give an isomorphism between $(R \cup R')'' \otimes 1$ and $R' \overline{\otimes} R$. The former is $B(H) \otimes 1$ while the latter is a II_1 factor, showing that θ cannot be inner.

We will require a theorem of Choda [16] concerning von Neumann algebras which lie between a factor M and its crossed product $M \rtimes_\alpha G$ by an outer action of a discrete group G. We will prove this using the following result from [209] which is valid for general discrete groups but is notationally easier in the countable case. We consider a factor $M \subseteq B(H)$, and we list the group elements by

$$G = \{g_1, g_2, g_3, \ldots\}.$$

We regard the elements of $B(H \otimes_2 \ell^2(G))$ as infinite matrices with entries from $B(H)$. If $\alpha \colon G \to \mathrm{Aut}\,(M)$ is an outer action of G on M, then there are diagonal representations $\rho\colon M \to B(H \otimes_2 \ell^2(G))$ and $\sigma\colon M' \to B(H \otimes_2 \ell^2(G))$ given respectively by

$$\rho(x) = \begin{pmatrix} \alpha_{g_1}(x) & & \\ & \alpha_{g_2}(x) & \\ & & \ddots \end{pmatrix}, \quad x \in M, \qquad (14.3.1)$$

and

$$\sigma(y) = \begin{pmatrix} y & & \\ & y & \\ & & \ddots \end{pmatrix}, \quad y \in M'. \qquad (14.3.2)$$

With this notation, we have

Lemma 14.3.1. *Let α be an outer action of a countable discrete group G on a II_1 factor $M \subseteq B(H)$. Then the von Neumann algebra generated by $\rho(M)$ and $\sigma(M')$ is*

$$Q = \left\{ \begin{pmatrix} t_1 & & \\ & t_2 & \\ & & \ddots \end{pmatrix} : t_i \in B(H), \ \sup_{i \geq 1} \|t_i\| < \infty \right\},$$

a countable direct sum of copies of $B(H)$.

Proof. It is clear that $W^*(\rho(M), \sigma(M')) \subseteq Q$. Once we have shown the inclusion $W^*(\rho(M), \sigma(M'))' \subseteq Q'$, the result will follow from the double commutant theorem, Theorem 2.2.1.

Let $t \in B(H \otimes_2 \ell^2(G))$ lie in the commutant of $W^*(\rho(M), \sigma(M'))$, and view t as an infinite matrix (t_{ij}) with $t_{ij} \in B(H)$. Commutation with $\sigma(M')$ shows that each t_{ij} lies in M. Since each t_{ii} commutes with $\{\alpha_{g_i}(x)\colon x \in M\} = M$, we conclude that $t_{ii} \in \mathbb{C}1$ since M is a factor. If $i \neq j$, then

$$\alpha_{g_i}(x)t_{ij} = t_{ij}\alpha_{g_j}(x), \quad x \in M, \qquad (14.3.3)$$

which becomes

$$\alpha_{g_i g_j^{-1}}(x)t_{ij} = t_{ij}x, \quad x \in M, \qquad (14.3.4)$$

after replacing x by $\alpha_{g_j^{-1}}(x)$ in (14.3.3). We now have the situation of an outer automorphism $\theta = \alpha_{g_i g_j^{-1}}$, and an element $s = t_{ij} \in M$ such that

$$\theta(x)s = sx, \quad x \in M. \qquad (14.3.5)$$

It follows that $s^*sx = s^*\theta(x)s$ for $x \in M$, implying that $s^*s \in M$ commutes with M, so $s^*s = \lambda 1$ for some $\lambda \geq 0$. If $\lambda > 0$, then $s/\sqrt{\lambda} \in M$ is a unitary implementing θ, which is impossible. Thus $\lambda = 0$ and so $t_{ij} = 0$ for $i \neq j$. We have shown that t is a diagonal matrix with scalar entries, so $t \in Q'$. This completes the proof. □

Remark 14.3.2. (i) For the purposes of Theorem 14.3.3, we note that Lemma 14.3.1 is valid for any ordering of the group elements in the diagonal representation, in particular for the representation π of M in (14.3.11).

(ii) Suppose that we have two sequences of vectors $\eta_i, \zeta_i \in L^2(M)$ which are each square summable in norm. Then a relation

$$\sum_{i=1}^{\infty} \langle y\alpha_{g_i}(x)\eta_i, \zeta_i \rangle = 0, \quad x \in M, \ y \in M', \tag{14.3.6}$$

can be written in matrix form as

$$\left\langle \begin{pmatrix} y\alpha_{g_1}(x) & \\ & y\alpha_{g_2}(x) \\ & & \ddots \end{pmatrix} \begin{pmatrix} \eta_1 \\ \eta_2 \\ \vdots \end{pmatrix}, \begin{pmatrix} \zeta_1 \\ \zeta_2 \\ \vdots \end{pmatrix} \right\rangle = 0, \quad x \in M, \ y \in M'. \tag{14.3.7}$$

Lemma 14.3.1 then implies that

$$\left\langle \begin{pmatrix} t_1 & \\ & t_2 \\ & & \ddots \end{pmatrix} \begin{pmatrix} \eta_1 \\ \eta_2 \\ \vdots \end{pmatrix}, \begin{pmatrix} \zeta_1 \\ \zeta_2 \\ \vdots \end{pmatrix} \right\rangle = 0, \quad t_i \in B(H), \tag{14.3.8}$$

which is equivalent to

$$\sum_{i=1}^{\infty} \langle t_i\eta_i, \zeta_i \rangle = 0, \quad t_i \in B(H). \tag{14.3.9}$$

We will use this in (14.3.14). □

We come now to a theorem due to Choda [16]. If M is a II$_1$ factor represented on $L^2(M)$ and G is a countable discrete group acting by outer automorphisms on M by $\alpha \colon G \to \mathrm{Aut}\,(M)$, then the crossed product $M \rtimes_\alpha G$ is the von Neumann algebra in $B(L^2(M) \otimes_2 \ell^2(G))$ generated by $\pi(M)$ and the unitary operators

$$\lambda_g(\eta \otimes \delta_g) = \eta \otimes \delta_{gh}, \quad g, h \in G, \ \eta \in L^2(M), \tag{14.3.10}$$

where

$$\pi(x)(\eta \otimes \delta_g) = \alpha_{g^{-1}}(x)\eta \otimes \delta_g, \quad x \in M, \ g \in G, \ \eta \in L^2(M). \tag{14.3.11}$$

These definitions are made to ensure that $\lambda_g\pi(x)\lambda_{g^{-1}} = \pi(\alpha_g(x))$ for $x \in M$ and $g \in G$, as can easily be checked. In particular, we note that conjugation by λ_g leaves invariant M when identified with $\pi(M)$.

The vector $\xi \otimes \delta_e$ is both cyclic and separating for $M \rtimes_\alpha G$, and the associated vector state is a trace on $M \rtimes_\alpha G$. Each operator $t \in M \rtimes_\alpha G$ can then be identified with a vector $t(\xi \otimes \delta_e) \in L^2(M) \otimes_2 \ell^2(G)$ and has the form $\sum_{g \in G} \eta_g \otimes \delta_g$ for $\eta_g \in L^2(M)$ satisfying $\sum_{g \in G} \|\eta_g\|^2 < \infty$. Since η_g is $\mathbb{E}_M(t\lambda_{g^{-1}})$, we actually obtain $\eta_g \in M$. Thus each $t \in M \rtimes_\alpha G$ has a unique Fourier series $\sum_{g \in G} x_g \lambda_g$ where $x_g \in M$ and $\sum_{g \in G} \|x_g\|_2^2 < \infty$. As we noted in Section 3.2, this series converges in $L^2(M \rtimes_\alpha G)$ but does not converge strongly or weakly in general [115]. We refer to [104, 105] for a more detailed discussion of crossed products.

Theorem 14.3.3. *Let M be a II_1 factor and let G be a countable discrete group acting on M by outer automorphisms. Let H be a subgroup of G. If N is a von Neumann algebra such that $M \rtimes_\alpha H \subseteq N \subseteq M \rtimes_\alpha G$, then there exists a group K lying between H and G such that $N = M \rtimes_\alpha K$.*

Proof. Let K be the set of group elements $g \in G$ such that some $n \in N$ has a non-zero λ_g-coefficient in its Fourier series. Clearly $H \subseteq K$, but at the moment K is only a set and not a subgroup of G. We will first show that $\lambda_k \in L^2(N)$ for each $k \in K$. Fix an arbitrary vector $\sum_{g \in G} \mu_g \lambda_g \in L^2(N)^\perp$, where $\mu_g \in L^2(M)$ and $\sum_{g \in G} \|\mu_g\|_2^2 < \infty$.

Since $M \subseteq N$, $L^2(N)$ is invariant under left and right multiplication by M. If $n \in N$ has Fourier series $\sum_{g \in G} n_g \lambda_g$ where $n_g \in M$, and if $x, y \in M$, then ynx^* has Fourier series $\sum_{g \in G} y n_g \alpha_g(x^*) \lambda_g$, and so

$$\sum_{g \in G} \langle y J \alpha_g(x) J n_g \xi, \mu_g \rangle = 0. \qquad (14.3.12)$$

Applying J to (14.3.12) gives

$$\sum_{g \in G} \langle Jy J \alpha_g(x) J n_g \xi, J \mu_g \rangle = 0 \qquad (14.3.13)$$

for $x, y \in M$. Since $JMJ = M'$, we may apply Lemma 14.3.1 and Remark 14.3.2 (ii) to conclude that

$$\sum_{g \in G} \langle t_g J n_g \xi, J \mu_g \rangle = 0 \qquad (14.3.14)$$

whenever $t_g \in B(L^2(M))$ and $\sup_{g \in G} \|t_g\| < \infty$. For a fixed $k \in K$, we may choose $n \in N$ such that $n_k \neq 0$. Then let $t_g = 0$ for $g \neq k$, while selecting t_k to satisfy $t_k J n_k \xi = J \mu_k$. It follows from (14.3.14) that $\mu_k = 0$, and thus $\lambda_k \perp \sum_{g \in G} \mu_g \lambda_g$. We conclude that $\lambda_k \in L^2(N)$ for each $k \in K$. Now the expectation $\mathbb{E}_N$ of $M \rtimes_\alpha G$ onto N is also the Hilbert space projection of $L^2(M \rtimes_\alpha G)$ onto $L^2(N)$, and so $\mathbb{E}_N(\lambda_k) = \lambda_k$, showing that $\lambda_k \in N$. Consequently K is a subgroup of G, and we see that $M \rtimes_\alpha K \subseteq N$. The reverse inclusion follows trivially from the definition of K, proving the result. $\qquad\square$

Before we can state and prove Theorem 14.3.5, the main result of this section, we must first see when crossed products are injective.

Theorem 14.3.4. *Let G be a discrete group acting by automorphisms α_g on a II_1 factor M.*

(i) *If α_g is an outer automorphism for each $g \neq e$, then $M' \cap (M \rtimes_\alpha G) = \mathbb{C}1$ and $M \rtimes_\alpha G$ is a II_1 factor;*

(ii) *$M \rtimes_\alpha G$ is injective if and only if M is injective and G is amenable.*

Proof. (i)   Suppose that $z \in M' \cap (M \rtimes_\alpha G)$, and let $z = \sum_{g \in G} z_g \lambda_g$ be the $\| \cdot \|_2$-norm convergent Fourier series. For each $x \in M$, the relation $xz = zx$ becomes

$$\sum_{g \in G} (xz_g - z_g \alpha_g(x))\lambda_g = 0, \tag{14.3.15}$$

showing that $xz_g = z_g \alpha_g(x)$ for all $x \in M$. As in the proof of Lemma 14.3.1, we conclude that α_g is inner unless $z_g = 0$. It follows that $M' \cap (M \rtimes_\alpha G) = \mathbb{C}1$ since M is a factor. Since the centre of $M \rtimes_\alpha G$ is contained in the relative commutant of M, we see that $M \rtimes_\alpha G$ is also a factor and we have established above that it has a trace.

(ii)   Suppose that $M \rtimes_\alpha G$ is injective. This algebra contains both M and $L(G)$ so the conditional expectations onto these subalgebras show that they are injective. Amenability of G follows from Theorem 3.8.2.

Conversely, suppose that M is injective and that G is amenable. By Theorem 3.8.1 and the subsequent remarks, M' is injective. The operators λ_g conjugate M to itself and so have the same property for M'. If we let μ denote an invariant mean on $\ell^\infty(G)$, then we may define a contraction $\mathbb{E}: M' \to M'$ by the formula

$$\mathbb{E}(x)(\phi) = \mu(g \mapsto \phi(\lambda_g x \lambda_{g^{-1}})) \tag{14.3.16}$$

for $x \in M'$ and $\phi \in M'_*$. The invariance of μ shows that each $\mathbb{E}(x)$ commutes with $L(G)$, and thus $\mathbb{E}$ is the expectation of M' onto $(M \rtimes_\alpha G)'$ which proves injectivity of this last algebra. Injectivity of $M \rtimes_\alpha G$ now follows from Theorem 3.8.1. $\qquad \square$

Theorem 14.3.5. *For $2 \leq n \leq \infty$, let $\mathbb{F}_n$ act on the hyperfinite II_1 factor R by outer automorphisms, and let H be the abelian group generated by a single generator. Then $R \rtimes_\alpha H$ is a maximal injective subfactor in $R \rtimes_\alpha \mathbb{F}_n$ and has trivial relative commutant.*

Proof. By Theorem 14.3.4, $R \rtimes_\alpha H$ is injective and also a subfactor of $R \rtimes_\alpha \mathbb{F}_n$. Moreover, it has trivial relative commutant because R has this property. By Theorem 14.3.3, any von Neumann algebra between these crossed products has the form $R \rtimes_\alpha K$ for a group K lying between H and $\mathbb{F}_n$. If this is injective, then $L(K)$ is injective by Theorem 14.3.4, which forces $K = H$ since $L(H)$ is maximal injective in $L(\mathbb{F}_n)$ by Theorem 14.1.1. The result follows from this. $\qquad \square$

Chapter 15

Masas in non-separable factors

15.1 Introduction

Up to this point, our discussion of masas has focused on the separable case. However, all von Neumann algebras possess masas, and there are some interesting phenomena which appear only when we leave the separable setting. The ultrapower examples of Appendix A are important non-separable algebras which play a role even for separable algebras, so strict adherence to separability is not generally possible. In this chapter we investigate masas in non-separable factors, and the results that we present below are all due to Popa [136, 138]. Many of them concern the algebra N^ω which is discussed in Appendix A.

The main results for N^ω in this chapter are Theorems 15.2.3 and 15.2.8. They can be summarised as follows:

> If N is a separable II_1 factor, then all masas in N^ω are non-separable, N^ω has no Cartan subalgebras and N^ω is prime.

This is proved in Section 15.2. The third section is devoted to showing that, for an uncountable set S, masas in $L(\mathbb{F}_S)$ are separable and that this factor has no Cartan masas. These algebras were the first examples of the absence of Cartan masas. The algebras $L(\mathbb{F}_n)$, $n \geq 2$, also have this property [199], but in the separable case the proof is substantially more difficult. We draw attention to the difference between N^ω and $L(\mathbb{F}_S)$: the first has no separable masas and the second has no non-separable ones.

15.2 Masas in N^ω

Recall that Theorem 9.6.3 (i) in Section 9.6 states that

if A is a masa in a separable II_1 *factor and* u *is a unitary in* N *with* $\|(I - \mathbb{E}_A)\mathbb{E}_{uAu^*}\|_{\infty, 2} < \delta$, *then there is a unitary* $w \in \mathcal{N}(A)$ *such that* $\|u - w\|_2 < 5\delta$.

The following lemma and corollary are due to Popa [150].

Lemma 15.2.1. *Let* N *be a separable* II_1 *factor. If* A *is a masa in* N, *then* $\mathcal{N}(A)^\omega = \mathcal{N}(A^\omega)$.

Proof. If $u_n \in \mathcal{N}(A)$ for all n, then

$$\pi(u_n)\pi(a_n)\pi(u_n)^* = \pi(u_n a_n u_n^*) \in A^\omega \tag{15.2.1}$$

for all $(a_n) \in \ell^\infty(A)$, so that $\pi(u_n) \in \mathcal{N}(A^\omega)$.

Let $u \in \mathcal{N}(A^\omega)$ and let u_n be unitaries in N such that $u = \pi(u_n)$ by Theorem A.5.3 (ii) on representations in ultrapowers. Suppose that $\lim_\omega \|(I - \mathbb{E}_{u_n A u_n^*})\mathbb{E}_A\|_{\infty, 2} > 0$, and let $\varepsilon > 0$ be such that

$$\lim_\omega \|(I - \mathbb{E}_{u_n A u_n^*})\mathbb{E}_A)\|_{\infty, 2} > \varepsilon > 0. \tag{15.2.2}$$

For each n, choose an $a_n \in A$ with $\|a_n\| \leq 1$ and

$$\|(I - \mathbb{E}_{u_n A u_n^*})(a_n)\|_2 \geq \|(I - \mathbb{E}_{u_n A u_n^*})\mathbb{E}_A\|_{\infty, 2} - 2^{-n} \tag{15.2.3}$$

and let $a = \pi(a_n) \in A^\omega$. By Theorem A.5.3 and Lemma A.5.4,

$$\begin{aligned}
\|(I - \mathbb{E}_{uA^\omega u^*})(a)\|_{2, \omega} &= \|(I - \mathbb{E}_{(u_n A u_n^*)^\omega})(a)\|_{2, \omega} \\
&\geq \lim_\omega \left(\|(I - \mathbb{E}_{u_n A u_n^*})(a_n)\|_2 - 2^{-n} \right) \\
&= \lim_\omega \|(I - \mathbb{E}_{u_n A u_n^*})\mathbb{E}_A\|_{\infty, 2} > \varepsilon.
\end{aligned} \tag{15.2.4}$$

This inequality (15.2.4) contradicts $uA^\omega u^* = A^\omega$ and hence

$$\lim_\omega \|(I - \mathbb{E}_{u_n A u_n^*})\mathbb{E}_A\|_{\infty, 2} = 0. \tag{15.2.5}$$

Let $\delta_n = \|(I - \mathbb{E}_{u_n A u_n^*})\mathbb{E}_A\|_{\infty, 2}$ for all n. By Theorem 9.6.3 (i), for each n there is a unitary $w_n \in \mathcal{N}(A)$ such that $\|w_n - u_n\| \leq 5\delta_n$, so that

$$\|\pi(w_n) - u\|_{2, \omega} = \lim_\omega \|w_n - u_n\|_2 = 0. \tag{15.2.6}$$

Hence $u = \pi(w_n) \in \mathcal{N}(A)^\omega$ as required. $\square$

Corollary 15.2.2. *Let* A *be a masa in a separable* II_1 *factor* N. *Then* A *is singular in* N *if, and only if,* A^ω *is singular in* N^ω.

Proof. Suppose that A^ω is singular. If $u \in \mathcal{N}(A)$, then $uA^\omega u^* = (uAu^*)^\omega = A^\omega$, so that $u \in A^\omega$ and hence $u \in A$. Thus A is singular. The other direction is a direct application of Lemma 15.2.1. $\square$

Note that $\mathcal{N}(A^\omega)$ is relatively small in N^ω. If A is a Cartan subalgebra in N, then $\mathcal{N}(A)'' = N$ but $\mathcal{N}(A^\omega) \neq N^\omega$, since N^ω has no Cartan subalgebras by Theorem 15.2.8.

The algebras N^ω are rather extraordinary. Contrast the conclusion that masas in N^ω are non-separable with the situation in $L(\mathbb{F}_S)$, where masas are separable even when the algebra is non-separable, equivalent to S being uncountable.

Theorem 15.2.3. *If N is a separable II_1 factor, then the masas in N^ω are non-separable.*

Proof. Suppose that the masa B in N^ω is separable and let u be a unitary generator of B (see Theorem 3.5.2). By Theorem A.5.3 on lifting unitaries in the ultrapower, there are unitaries $u_n \in N$ such that $u = \pi(u_n)$. For each n, let B_n be a masa in N containing u_n. Then $B_0 = \bigoplus_n B_n$ is a masa in $\ell^\infty(N)$ so that $\pi(B_0)$ is an abelian von Neumann subalgebra of N^ω containing u, so containing B. Since B is a masa, $B = \pi(B_n)$.

Using the observation that $(B_n, \tau) \cong (L^\infty[0,1], \int_0^1 \cdot\, dt)$ (see Theorem 3.5.2), for each n choose an abelian subalgebra A_n of B_n with $\dim(A_n) = 2^n$ and each minimal projection in A_n of trace 2^{-n}. For each n, the finite tracial algebra (A_n, τ) is isomorphic to

$$C_n = \left(\bigotimes_1^n \mathbb{C}e_0 \oplus \mathbb{C}e_1, \tau \right), \tag{15.2.7}$$

where e_j, $j = 1, 2$, are projections with $e_0 + e_1 = 1$, $\tau(e_j) = 1/2$ and τ is tensorial on the tensor product. If $\alpha = (\alpha_1, \alpha_2, \ldots) \in \{0, 1\}^\infty$ is an infinite binary sequence, let $u(\alpha)_n$ be the self-adjoint unitary in A_n corresponding to the unitary

$$\otimes_{j=1}^n \left(e_0 + (-1)^{\alpha_j} e_1 \right) \in C_n. \tag{15.2.8}$$

Let $\tilde{u}(\alpha) = (u(\alpha)_1, u(\alpha)_2, \ldots) \in \ell^\infty(N_n)$ and let $u(\alpha) = \pi(\tilde{u}(\alpha))$. Then $u(\alpha)$ is a unitary in N_n^ω for all α.

If $\alpha, \beta \in \{0, 1\}^\infty$ with $\alpha_j \neq \beta_j$ for some $j \in \mathbb{N}$, then $\tau(u(\alpha)_n u(\beta)_n^*) = 0$ for all $n \geq j$, because this contains the factor $\tau(e_0 - e_1) = 0$ arising from the jth place. Hence

$$\tau_\omega(u(\alpha)u(\beta)^*) = \lim_\omega \tau(u(\alpha)_n u(\beta)_n^*) = 0. \tag{15.2.9}$$

Thus $\{u(\alpha) : \alpha \in \{0, 1\}^\infty\}$ is an uncountable orthonormal set in $L^2(N^\omega)$ that is contained in the masa B. $\qquad\square$

Lemma 15.2.4. *Let N be a separable II_1 factor. If $e_1, \ldots, e_k$ and $f_1, \ldots, f_k$ are projections in N with $\sum_j e_j = \sum_j f_j = 1$ and $\tau(e_j) = \tau(f_j) = k^{-1}$ for $1 \leq j \leq k$, then there is a unitary $u \in N$ such that $u e_j u^* = f_j$ for $1 \leq j \leq k$ and*

$$\|1 - u\|_2 \leq 2^{1/2} k^{1/2} \max\{\|e_j - f_j\|_2 : 1 \leq j \leq k\}. \tag{15.2.10}$$

Proof. Let $\delta = \max\{\|e_j - f_j\|_2 : 1 \le j \le k\}$. By the comparison Lemma 5.2.9 for projections, there is a partial isometry $v_j \in N$ with $v_j^* v_j = e_j$, $v_j v_j^* = f_j$ and $\|e_j - v_j\|_2 \le 2^{1/2} \|e_j - f_j\|_2$ for $1 \le j \le k$, since e_j is equivalent to f_j. Let $u = \sum_j v_j$. Then u is a unitary in N and

$$ue_j = v_j e_j = v_j = f_j v_j = f_j u \tag{15.2.11}$$

for all $1 \le j \le k$ by the orthogonality of the families $\{e_j : 1 \le j \le k\}$ and $\{f_j : 1 \le j \le k\}$. Moreover,

$$\|1 - u\|_2^2 = \Big\| \sum_j (e_j - v_j) \Big\|_2^2 = \sum_j \|e_j - v_j\|_2^2 \le 2k\delta^2, \tag{15.2.12}$$

since $e_i - v_i$ and $e_j - v_j$ are orthogonal for $i \ne j$. □

The estimate on $\|1 - u\|_2$ in (15.2.12) is not required subsequently. The following theorem indicates that separable diffuse abelian von Neumann subalgebras of N^ω behave like finite dimensional abelian subalgebras of a separable II_1 factor with all their minimal projections of equal trace.

Theorem 15.2.5. *Let N be a separable II_1 factor.*

(i) *If A and B are separable diffuse abelian von Neumann subalgebras of N^ω, then there is a unitary $u \in N^\omega$ such that $uAu^* = B$.*

(ii) *If A is a separable diffuse abelian von Neumann subalgebra of N^ω, then there is a hyperfinite subfactor R of N^ω with A a Cartan masa in R (arising from the diagonals of a $2^n \times 2^n$ matrix approximation to R).*

Proof. (i) By Theorem 3.5.2, the separable diffuse abelian von Neumann algebras (A, τ) and (B, τ) are both isomorphic to $(L^\infty[0, 1], \int_0^1 \cdot \, dt)$. Choose sets of mutually orthogonal projections $\{e_j^{(k)} : 1 \le j \le 2^k\}$ and $\{f_j^{(k)} : 1 \le j \le 2^k\}$ in A and B for all $k \ge 0$ such that

(1) $\sum_{j=1}^{2^k} e_j^{(k)} = 1 = \sum_{j=1}^{2^k} f_j^{(k)}$, $\tau_\omega(e_j^{(k)}) = 2^{-k} = \tau_\omega(f_j^{(k)})$,

(2) $e_j^{(k)} = e_{2j-1}^{(k+1)} + e_{2j}^{(k+1)}$ and $f_j^{(k)} = f_{2j-1}^{(k+1)} + f_{2j}^{(k+1)}$
for $1 \le j \le 2^k$ and $k \ge 1$,

(3) $A = \{e_j^{(k)} : 1 \le j \le 2^k, k \ge 0\}''$ and $B = \{f_j^{(k)} : 1 \le j \le 2^k, k \ge 0\}''$.

By the representation Theorem A.5.3 for lifting from the ultrapower, choose sequences of projections $e_{n,j}^{(k)}$ and $f_{n,j}^{(k)}$ in N such that

(4) $\sum_{j=1}^{2^k} e_{n,j}^{(k)} = 1 = \sum_{j=1}^{2^k} f_{n,j}^{(k)}$, $\tau(e_{n,j}^{(k)}) = 2^{-k} = \tau(f_{n,j}^{(k)})$,

(5) $e_{n,j}^{(k)} = e_{n,2j-1}^{(k+1)} + e_{n,2j}^{(k+1)}$ and $f_{n,j}^{(k)} = f_{n,2j-1}^{(k+1)} + f_{n,2j}^{(k+1)}$
for $1 \le j \le 2^k$, $k \ge 0$ and $n \ge 1$,

(6) $\pi(e_{n,j}^{(k)}) = e_j^{(k)}$ and $\pi(f_{n,j}^{(k)}) = f_j^{(k)}$ for $1 \leq j \leq 2^k$ and $k \geq 0$.

By the previous Lemma 15.2.4 for each $n \in \mathbb{N}$, there is a unitary $u_n \in N$ such that $u_n e_{n,j}^{(n)} u_n^* = f_{n,j}^{(n)}$ for $1 \leq j \leq 2^n$. The equations in (5) imply that $u_n e_{n,j}^{(k)} u_n^* = f_{n,j}^{(k)}$ for $1 \leq j \leq 2^k$ and $1 \leq k \leq n$, by a downward induction on k from n.

Let $u = \pi(u_n)$. Then

$$u e_j^{(k)} u^* = \pi(u_n)\pi(e_{n,j}^{(k)})\pi(u_n^*) = \pi(u_n e_{n,j}^{(k)} u_n^*) = \pi(f_{n,j}^{(n)}) = f_j^{(k)} \qquad (15.2.13)$$

for $1 \leq j \leq k$ and $k \geq 0$. Since $A = \{e_j^{(k)} : 1 \leq j \leq 2^k, \ k \geq 0\}''$ and $B = \{f_j^{(k)} : 1 \leq j \leq 2^k, \ k \geq 0\}''$, by (iii) above, $uAu^* = B$ as required.

(ii)   Construct a hyperfinite subfactor R in N^ω by inductively choosing matrix subalgebras $M_1, M_2, \ldots$, each isomorphic to the algebra $\mathbb{M}_2$ of 2×2 matrices, such that $M_n \subseteq \{M_1, \ldots, M_{n-1}\}' \cap N^\omega$ for all $n \geq 2$ using the observation that $\{M_1, \ldots, M_{n-1}\}' \cap N^\omega$ is a factor. Let $R = \left(\bigcup_n M_n\right)'' = \left(\bigotimes_n M_n\right)''$. There is a Cartan masa B in R, for example the masa arising from the diagonals in the tensor product. Thus B is a separable diffuse abelian von Neumann subalgebra of N^ω. By (i) there is a unitary $u \in N^\omega$ such that $A = uBu^*$. Hence uRu^* is a hyperfinite subfactor of N^ω containing A as a Cartan subalgebra.   □

We will require the following useful little lemma.

Lemma 15.2.6. *Let N be a separable* II_1 *factor and let M be a matrix subalgebra of N. If A and B are mutually orthogonal masas in M, then A is orthogonal to $B' \cap N$.*

Proof. If $e_1, \ldots, e_k$ are the minimal projections in B, which is the diagonal of M with respect to this basis then, by Remark 3.6.7,

$$\mathbb{E}_{B' \cap N}(x) = \sum_{j=1}^k e_j x e_j \quad \text{for all} \quad x \in N. \qquad (15.2.14)$$

From this it follows that $\mathbb{E}_{B' \cap N}(x) \in M$ for all $x \in M$. Thus the restriction of $\mathbb{E}_{B' \cap N}$ to M is the conditional expectation $\mathbb{E}_{B' \cap M} = \mathbb{E}_B$ from M onto $B' \cap M = B$ by the uniqueness of the conditional expectation. If $x \in A$, then

$$\mathbb{E}_{B' \cap N}(x) = \mathbb{E}_{B' \cap M}(x) = \mathbb{E}_B(x) = \tau(x)1, \qquad (15.2.15)$$

since A and B are orthogonal in M. Hence A and $B' \cap N$ are orthogonal in N by the definition of orthogonality (see also Lemma 6.3.2).   □

When we apply the next result subsequently, we will take M in its statement to be the ultrapower N^ω.

Lemma 15.2.7. *Let M be an arbitrary* II_1 *factor and let R be a hyperfinite subfactor of M. There exist two masas A and B in R such that A is orthogonal to $B' \cap M$ as von Neumann algebras, and A is the (diagonal) Cartan masa arising from $R \cong \left(\bigotimes_n N_n\right)''$, where $N_n \cong \mathbb{M}_2$ for all n.*

Proof. Let A_0 and B_0 be two orthogonal masas of the algebra of 2×2 matrices $\mathbb{M}_2$. For example, let A_0 be the diagonal masa $\mathbb{D}_2$ and let B_0 be wA_0w^*, where w is rotation through $\pi/4$. Let N_j, A_j and B_j be sequences of algebras with the triples $A_j \subseteq N_j$ and $B_j \subseteq N_j$ isomorphic with $A_0 \subset \mathbb{M}_2$ and $B_0 \subseteq \mathbb{M}_2$. With the tensorial trace, there is an isomorphism θ from $\overline{\bigotimes}_j N_j$ onto R. Let $A = \theta(\overline{\bigotimes}_j A_j)$ and $B = \theta(\overline{\bigotimes}_j B_j)$. Then A and B are (Cartan) masas in R, since $\overline{\bigotimes}_j A_j$ and $\overline{\bigotimes}_j B_j$ are (Cartan) masas in $\overline{\bigotimes}_j N_j$. In the algebra $M_n = \bigotimes_1^n N_j$, the algebras $\widetilde{A}_n = \bigotimes_1^n A_j$ and $\widetilde{B}_n = \bigotimes_1^n B_j$ are masas and $\widetilde{A}_n$ is orthogonal to $\widetilde{B}_n$, since each A_j is orthogonal to all the B_i. By the Lemma 15.2.6 above, applied to $\widetilde{A}_n$, $\widetilde{B}_n \subseteq M_n \cong \mathbb{M}_{2^n} \subseteq M$, the algebra $\widetilde{A}_n$ is orthogonal to $\widetilde{B}_n' \cap M$ for all n. Since $\widetilde{B}_n' \cap M \supseteq B' \cap M$, it follows that $\widetilde{A}_n$ is orthogonal to $B' \cap M$ for all n, so that $A = (\bigcup_n \widetilde{A}_n)''$ is orthogonal to $B' \cap M$. $\square$

A II_1 factor M is said to be *prime* if M cannot be written as a tensor product $M = M_1 \overline{\otimes} M_2$, where M_1 and M_2 are II_1 factors. Since each factor M can be expressed as $N \otimes \mathbb{M}_n$, where N is isomorphic to pMp for any projection $p \in M$ of trace $1/n$, this definition is equivalent to requiring that these trivial decompositions are the only ones possible.

The conclusion below that N^ω has no Cartan subalgebras is due to Popa [138] and that N^ω is prime is due to Fang, Ge and Li [67].

Theorem 15.2.8. *If N is a separable II_1 factor, then N^ω has no Cartan subalgebras and N^ω is prime.*

Proof. Let A be a masa in N^ω and let B be a separable diffuse (abelian) von Neumann subalgebra of A. By Theorem 15.2.5 (ii) there is a hyperfinite subfactor $R \subseteq N^\omega$ with B a Cartan subalgebra of R arising from a $2^n \times 2^n$ matrix approximation to R. By Lemma 15.2.7, there is a masa B_0 in R with B_0 orthogonal to $B' \cap N^\omega$. Since $B \subseteq A$, $B' \cap N^\omega \supseteq A' \cap N^\omega = A$, so that B_0 is orthogonal to A. Since B_0 and B are separable diffuse abelian von Neumann subalgebras of N^ω, by Theorem 15.2.5 (i), there is a unitary $u \in N^\omega$ such that $uBu^* = B_0$. Thus B is a diffuse abelian subalgebra of A and uBu^* is orthogonal to A so that u is orthogonal to $\mathcal{N}(A)''$ by Lemma 6.3.2 on orthogonality. Hence $\mathcal{N}(A)'' \neq N^\omega$ and A is not a Cartan masa.

Suppose that N^ω is not prime and let $N^\omega = M_1 \overline{\otimes} M_2$, where M_1 and M_2 are II_1 factors. Choose separable diffuse abelian von Neumann subalgebras $A_1 \subseteq M_1$ and $A_2 \subseteq M_2$ using standard techniques. By Theorem 15.2.5 (i) there is a unitary $u \in N^\omega$ such that $uA_1u^* = A_2 \subseteq M_2$, which is orthogonal to M_1. Here M_1 is identified with $M_1 \otimes 1$ and M_2 with $1 \otimes M_2$. Since A_1 is diffuse and contained in M_1, u is orthogonal to $\mathcal{N}(M_1)''$ in N^ω by Lemma 6.3.2 on orthogonality. However $\mathcal{N}(M_1)'' = M_1 \overline{\otimes} M_2 = N^\omega$, which is a contradiction. $\square$

If A is a masa in a separable II_1 factor N, then A^ω is a masa in N^ω. This and property Γ suggests asking how large is $A' \cap N^\omega$ in relation to A^ω. When

N has property Γ, then $A' \cap N^\omega \supseteq N' \cap N^\omega$ so the question is mainly of interest when N does not have property Γ.

The following result is surely known to experts but we do not know of a reference. The proof uses the idea of Theorem 15.2.3.

Lemma 15.2.9. *If N is a separable II$_1$ factor and A is a masa in N, then there is an uncountable set of mutually orthogonal unitaries in $A' \cap N^\omega$ all orthogonal to A^ω.*

Proof. By Theorem 3.5.2, each masa A in N is a diffuse abelian separable von Neumann subalgebra of N. By Theorem 12.3.1, there is a masa B in N and a hyperfinite subfactor R of N such that B is the Cartan subalgebra of R arising from the diagonals of the matrix algebras in $R = \bigotimes_k M_k$, where $M_k \cong \mathbb{M}_2$ for all $k \in \mathbb{N}$. Since B is a diffuse abelian separable von Neumann subalgebra of N, Theorem 15.2.5 (i) implies that there is a unitary $u \in N^\omega$ such that $A = uBu^*$. The hypotheses and conclusions of this Lemma 15.2.9 are invariant under unitary equivalence within N^ω, so it is sufficient to prove the result for B.

By the construction of B and R in Theorem 12.3.1, $R = \bigotimes_k M_k$ with M_k having matrix units $e_{ij}^{(k)}$ $(1 \leq i, j \leq 2)$ and $B = \{e_{ii}^{(k)} : 1 \leq i \leq 2^k, \ k \in \mathbb{N}\}''$. It seems desirable to have the superscript as (k) rather than k as powers appear subsequently in the proof and (k) specifies the tensor factor it is contained in. Let $B_k = \{e_{ii}^{(k)} : 1 \leq i \leq 2^k\}''$ and let

$$u_k = 1 \otimes \ldots 1 \otimes (e_{01}^{(k)} + e_{10}^{(k)}) \otimes 1 \otimes \ldots \in R \qquad (15.2.16)$$

for all $k \in \mathbb{N}$, where the superscript specifies where it is in the tensor product. Then

$$u_n b = b u_n \quad \text{for all } b \in B_k \text{ and all } 1 \leq k < n \qquad (15.2.17)$$

by the construction of u_n and B_k. If $f_0^{(k)} = 1 \otimes \ldots 1 \otimes e_{00}^{(k)} \otimes 1 \otimes \ldots \in B$ and $f_1^{(k)} = 1 \otimes \ldots 1 \otimes e_{11}^{(k)} \otimes 1 \otimes \ldots \in B$, then

$$
\begin{aligned}
\mathbb{E}_B(u_k) &= \mathbb{E}_B(f_0^{(k)} u_k + f_1^{(k)} u_k) \\
&= f_0^{(k)} \mathbb{E}_B(1 \otimes \ldots 1 \otimes e_{01}^{(k)} \otimes 1 \otimes \ldots) + f_1^{(k)} \mathbb{E}_B(1 \otimes \ldots 1 \otimes e_{10}^{(k)} \otimes 1 \otimes \ldots) \\
&= \mathbb{E}_B(1 \otimes \ldots 1 \otimes e_{01}^{(k)} \otimes 1 \otimes \ldots) f_0^{(k)} + \mathbb{E}_B(1 \otimes \ldots 1 \otimes e_{10}^{(k)} \otimes 1 \otimes \ldots) f_1^{(k)} \\
&= \mathbb{E}_B((1 \otimes \ldots 1 \otimes e_{01}^{(k)} \otimes 1 \otimes \ldots) f_0^{(k)} + (1 \otimes \ldots 1 \otimes e_{10}^{(k)} \otimes 1 \otimes \ldots) f_1^{(k)}) \\
&= 0, \qquad\qquad\qquad\qquad\qquad\qquad\qquad\qquad\qquad\qquad\qquad (15.2.18)
\end{aligned}
$$

since B is abelian.

If $\alpha = (\alpha_1, \alpha_2, \ldots) \in \{0, 1\}^\infty$ is an infinite binary sequence, let $w(\alpha)_n$ be the self-adjoint unitary in R given by

$$w(\alpha)_n = \bigotimes_{j=1}^{n} 1 \otimes \bigotimes_{k=1}^{\infty} u_k^{\alpha_k} \quad \text{for all } n \in \mathbb{N}. \qquad (15.2.19)$$

The values $\alpha_k = 0, 1$ produce respectively factors of 1 or u_k in the tensor product expansion of $w(\alpha)_n$, but in the $(n+k)^{\text{th}}$ position, and so we identify the matrix algebras M_k and M_{n+k} and their matrix units. Let

$$\widetilde{w}(\alpha) = \big(w(\alpha)_1, w(\alpha)_2, \ldots\big) \in \ell^\infty(N) \tag{15.2.20}$$

and let

$$w(\alpha) = \pi(\widetilde{w}(\alpha)) \quad \text{for all } \alpha \in \{0,1\}^\infty. \tag{15.2.21}$$

Then $w(\alpha)$ is a unitary in N^ω for all α.

If $\alpha, \beta \in \{0,1\}^\infty$ with $\alpha_j \neq \beta_j$ for some $j \in \mathbb{N}$, then $w(\alpha)_n w(\beta)_n^* = w(\gamma)_n$ for $\gamma = \alpha + \beta$, with the calculation done *mod* 2, so that $w(\alpha)_n w(\beta)_n^*$ contains the tensor $e_{01}^j + e_{10}^j$ in the $(n+j)^{\text{th}}$ place for all n. Hence $\mathbb{E}_B\big(w(\alpha)_n w(\beta)_n^*\big) = 0$ for all n by equation (15.2.18). By Theorem A.5.3,

$$\mathbb{E}_{B^\omega}(w_\alpha w_\beta^*) = \lim_\omega \mathbb{E}_B\big(w(\alpha)_n w(\beta)_n^*\big) = 0. \tag{15.2.22}$$

If $b \in B_j$ for some $j \in \mathbb{N}$, then

$$bw_\alpha = \pi\big(bw(\alpha)_n\big) = \pi\big(w(\alpha)_n b\big) = w_\alpha b \tag{15.2.23}$$

by equation (15.2.17). Hence w_α $(\alpha \in \{0,1\}^\infty \setminus \{(0,\ldots)\})$ is the required uncountable set of unitaries in N^ω. $\qquad\square$

Observe that the proof shows that there is an isomorphism

$$\chi \colon \mathbb{Z}_2^\infty \to \mathcal{U}(A' \cap N^\omega)$$

with $\mathbb{E}_{A^\omega}(\chi(u)) = 0$ for all $u \neq e$ in the group $\mathbb{Z}_2^\infty$. Modifying the approximation to R permits non-commutative discrete groups to be embedded in the same way.

15.3 Masas in $L(\mathbb{F}_S)$

If S is a non-empty set, let $\mathbb{F}_S$ denote the free group on the symbols in S. Note that $L(\mathbb{F}_S)$ is separable if, and only if, S is countable. This section is devoted to a proof of the following assertion:

> The von Neumann algebra $L(\mathbb{F}_S)$ for S uncountable is a non-separable II_1 factor in which all masas are separable and which has no Cartan masas.

This is in contrast to the situation of N^ω, discussed in Section 15.2, where all masas were non-separable.

Lemma 15.3.1. *Let S be a set containing at least two elements and let $\emptyset \neq S_0 \subseteq S$. If B a diffuse von Neumann subalgebra of $L(\mathbb{F}_{S_0})$, then the normaliser $\mathcal{N}(B)$ of B in $L(\mathbb{F}_S)$ is contained in $L(\mathbb{F}_{S_0})$.*

Proof. If $g \in \mathbb{F}_S \setminus \mathbb{F}_{S_0}$, then $g\mathbb{F}_{S_0}g^{-1} \cap \mathbb{F}_{S_0} = \{1\}$ since all non-trivial words in $g\mathbb{F}_{S_0}g^{-1}$ contain at least two symbols of $S \setminus S_0$. By Lemma 6.3.2, it follows that $\mathcal{N}(B)'' \subseteq L(\mathbb{F}_{S_0})$. □

Corollary 15.3.2. *Let S be a set containing at least two elements and let $\emptyset \neq S_0 \subseteq S$. Each (singular) masa in $L(\mathbb{F}_{S_0})$ is a (singular) masa in $L(\mathbb{F}_S)$.*

Proof. If B is a masa in $L(\mathbb{F}_{S_0})$, then B is diffuse. If u is a unitary in $L(\mathbb{F}_S)$ with $uBu^* = B$, then $u \in L(\mathbb{F}_{S_0})$ by Lemma 15.3.1. Each unitary $v \in B' \cap L(\mathbb{F}_S)$ normalises B and so lies in $L(\mathbb{F}_{S_0})$. Thus $v \in B' \cap L(\mathbb{F}_{S_0}) = B$, showing that B is also a masa in $L(\mathbb{F}_S)$. Moreover, the assumption of singularity in $L(\mathbb{F}_{S_0})$ implies the same in $L(\mathbb{F}_S)$ by the relation

$$B = \mathcal{N}(B, L(\mathbb{F}_{S_0}))'' = \mathcal{N}(B, L(\mathbb{F}_S))'' \tag{15.3.1}$$

□

Corollary 15.3.3. *Let S be an uncountable set.*

(i) *If B is a diffuse abelian von Neumann subalgebra of $L(\mathbb{F}_S)$, then there is a countable subset W of S such that $B' \cap L(\mathbb{F}_S) \subseteq L(\mathbb{F}_W)$, and $B' \cap L(\mathbb{F}_S)$ is separable.*

(ii) *If B is an abelian von Neumann subalgebra of $L(\mathbb{F}_S)$, then B is separable.*

Proof. (i)   Since B is a diffuse von Neumann algebra, there is a family of projections $\{e_j^n : 1 \leq j \leq 2^n, n \geq 0\}$ in B such that

(1) $e_1^0 = 1$,

(2) $e_{2j-1}^{n+1} + e_{2j}^{n+1} = e_j^n$,

(3) $\tau(e_j^n) = 2^{-n}$,

for $1 \leq j \leq 2^n$ and $n \geq 0$. Let

$$B_0 = \{e_j^n : 1 \leq j \leq 2^n, n \geq 0\}''$$

be the abelian von Neumann subalgebra of B generated by these elements, so $B_0 \cong L^\infty[0,1]$ (see Theorem 3.5.2). Since $e_j^n \in L(\mathbb{F}_S) \subseteq \ell^2(\mathbb{F}_S)$, it is supported on a countable subset W_j^n of S, when e_j^n is thought of as a function on S. Let $W = \bigcup_{j,n} W_j^n$. This is a countable subset of S, so that each element of B_0 is supported on W, and hence $B_0 \subseteq L(\mathbb{F}_W) \subseteq \ell^2(\mathbb{F}_W)$. Thus

$$B' \cap L(\mathbb{F}_S) \subseteq B_0' \cap L(\mathbb{F}_S) \subseteq \mathcal{N}(B_0, L(\mathbb{F}_S))'' \subseteq L(\mathbb{F}_W)$$

by Lemma 15.3.1. This completes the proof as $L(\mathbb{F}_W)$ is separable, so $B' \cap L(\mathbb{F}_S)$ is separable.

(ii)   Now suppose that $B \subseteq L(\mathbb{F}_S)$ is abelian. By Zorn's lemma, choose a masa B_0 containing B. Then B_0 is diffuse, so part (i) implies that $B_0' \cap L(\mathbb{F}_S) (= B_0)$ is separable. Separability of B follows. □

Note the difference in the sizes of masas in $L(\mathbb{F}_S)$ and in N^ω for each separable II$_1$ factor N.

Theorem 15.3.4. *Let S be an uncountable set. Each masa B in $L(\mathbb{F}_S)$ is contained in $L(\mathbb{F}_W)$ for some countable subset W of S and $L(\mathbb{F}_S)$ has no Cartan subalgebras.*

Proof. Let B be a masa in $L(\mathbb{F}_S)$. By Corollary 15.3.3 (i) there exists a countable subset W of S with $B \subseteq L(\mathbb{F}_S)$, so B is separable. By Lemma 15.3.1

$$\mathcal{N}(B)'' \subseteq L(\mathbb{F}_W) \subsetneqq L(\mathbb{F}_S) \qquad\qquad (15.3.2)$$

and so B is not Cartan. $\square$

Chapter 16

Singly generated II_1 factors

16.1 Introduction

Some of the technical results proved in these notes concerning masas with special properties have been used to study the generator problem in separable II_1 factors. Certain of these results are presented in this chapter using an invariant $\mathcal{G}$ which was introduced and studied by Shen [171].

The generator problem in von Neumann algebras asks whether each separably acting von Neumann algebra is generated by a single element [77]. The answer is positive for factors for type I, II_∞ and III, and so we discuss only the remaining case of II_1 factors in this chapter. For an overview of this problem, we refer to the book by Topping [196, Chapter 10], as well as some of the original papers of Douglas and Pearcy [53], Ge and Popa [71], Ge and Shen [81], Olsen and Zame [124], Pearcy [129], Popa [141], Shen [171] and Wogen [210]. The main progress on the case of II_1 factors can be found in [63, 71, 81, 141, 171] and much of this chapter is drawn from these sources.

The classes of separable II_1 factors that are known to be singly generated are as follows:

 (i) those with Cartan masas, [141], (see Corollary 16.8.4);

 (ii) tensor products of II_1 factors, [71], (see Theorem 16.8.1);

 (iii) those with property Γ, [71], (see Theorem 16.8.5).

The recent theory developed by Shen [171] has evolved from these earlier results and now gives a unified approach to the three cases above, as well as proving single generation in other contexts. Dykema and Haagerup [61] have obtained a specific quasinilpotent operator which generates $L(\mathbb{F}_2)$. The importance of this lies in a concrete description of the operator, since any factor generated by two unitaries is also generated by their two self-adjoint logarithms x_1 and x_2, and single generation is obtained from $x_1 + ix_2$. More generally, any factor with a single generator x is also generated by two hermitians $x + x^*$ and $i(x - x^*)$, and vice versa. The paper [98] by Jung contains many results related to those in the latter sections of this chapter. His approach is quite different, being based upon free probability theory.

The basic underlying idea for single generation is in the following lemma which is a $k \times k$ version of the 2×2 version due to Pearcy [129].

Lemma 16.1.1. *If N is a von Neumann algebra generated by k self-adjoint elements, then $\mathbb{M}_k \otimes N$ is generated by two self-adjoint elements.*

Proof. Let $x_1, \ldots, x_k$ be k self-adjoint generators of N and let $\{e_{ij} : 1 \leq i, j \leq k\}$ be matrix units of $\mathbb{M}_k$. By scaling and adding multiples of 1, we may assume that $j - 1/2 \leq x_j \leq j$ for $1 \leq j \leq k$. Let

$$y_1 = \sum_{j=1}^{k} e_{jj} \otimes x_j \quad \text{and} \quad y_2 = \sum_{i=1}^{k-1} (e_{i,i+1} + e_{i+1,i}) \otimes 1. \qquad (16.1.1)$$

By the choice of x_j, the von Neumann algebra generated by y_1 contains $e_{jj} \otimes x_j$ and $e_{jj} \otimes 1$ for $1 \leq j \leq k$, since $e_{jj} \otimes 1$ is the spectral projection of y_1 for the interval $[j - 1/2, j]$. From this it follows that

$$e_{j,j+1} \otimes 1 = (e_{jj} \otimes 1)(y_2 \otimes 1)(e_{j+1,j+1} \otimes 1) \in \{y_1, y_2\}'' \qquad (16.1.2)$$

for $1 \leq j \leq k$ so that $e_{ij} \otimes 1 \in \{y_1, y_2\}''$ for $1 \leq i, j \leq n$. Hence $\mathbb{M}_k \otimes N \subseteq \{y_1, y_2\}''$. $\qquad \square$

The use of matrix units in this lemma will be a recurring theme throughout the chapter, whose contents we now describe.

We introduce Shen's invariant $\mathcal{G}(\cdot)$ in Section 16.2. This is defined in terms of decompositions of 1 as finite sums of orthogonal projections, and we make the useful reduction to considering only those decompositions where the projections have equal trace. The relationship of this invariant to the problem of single generation is established in Theorem 16.6.2. There it is shown that if a factor N satisfies $\mathcal{G}(N) < 1/2$, then N is singly generated. Section 16.3 calculates $\mathcal{G}(\cdot)$ in various situations, and is a prelude to Section 16.4 where a formula is obtained for expressing $G(N_t)$ in terms of $G(N)$, where N_t is a compression of N (see Theorem 16.4.5). This is applied in Section 16.5 to the interpolated free group factors.

Section 16.6 is devoted to the problem of single generation. In general, the results are for factors, which are assumed to be finitely generated. However, Theorem 16.7.5 shows that it is possible, in certain circumstances, to obtain single generation without this hypothesis. The technical results of Sections 16.6 and 16.7 are applied in the last section to a range of examples, including tensor products and crossed products.

16.2 Notation and definitions

Throughout this chapter N will denote a separable finite von Neumann algebra with a faithful normal trace τ; subsequently it will frequently be a II$_1$ factor as well.

Definition 16.2.1. (1)   For each positive integer k, let $\mathcal{P}_k$ denote the set of all sets $\{p_1, \ldots, p_k\}$ of mutually orthogonal (non-zero) projections $p_i \in N$ with $\sum_i p_i = 1$ and let $\mathcal{P}_k^{eq}$ be the set of all $\{p_1, \ldots, p_k\} \in \mathcal{P}_k$ with $\tau(p_i) = 1/k$ for $1 \le i \le k$. Note that $\mathcal{P}_k^{eq}$ could be empty, as it is in $\mathbb{M}_n$ for $k > n$.

Let $\mathcal{P} = \bigcup_{k \in \mathbb{N}} \mathcal{P}_k$. Define $\preccurlyeq$ on $\mathcal{P}$ by $P \preccurlyeq Q$ if and only if

$$q = \sum_{p \in P, p \le q} p \quad \text{for all } q \in Q. \tag{16.2.1}$$

Note that $\preccurlyeq$ is a partial order on $\mathcal{P}$.

(2)   Let

$$\mathcal{I}(x; P) = \mathcal{I}(x; \{p_i\}) = \sum_{p_i x p_j \neq 0} \tau(p_i)\tau(p_j) \tag{16.2.2}$$

for all $x \in N$, all $k \in \mathbb{N}$ and all $P = \{p_i\} \in \mathcal{P}_k$.

If $\{p_i\} \in \mathcal{P}_k^{eq}$, then

$$\mathcal{I}(x; \{p_i\}) = k^{-2}|\{(i, j) : p_i x p_j \neq 0\}|, \tag{16.2.3}$$

where $|\cdot|$ denotes the cardinality of a set. This indicates the importance of the cardinality of certain sets, which play a role subsequently.

Let

$$\mathcal{I}(X; P) = \sum_{x \in X} \mathcal{I}(x, P) \tag{16.2.4}$$

for all finite subsets X of N and all $P \in \mathcal{P}$.

For each finite subset X of N and each $P \in \mathcal{P}$, it follows that $\mathcal{I}(X; P) \le |X|$. If X is a finite subset of N and $P \in \mathcal{P}_k^{eq}(N)$, then $\mathcal{I}(X; P) \ge k^{-2}|X|$.

Note that if $P = \{p_i\} \in \mathcal{P}_k(N)$, then $k^{-1} \le \sum_i \tau(p_i)^2$ with equality if, and only if, $P \in \mathcal{P}_k^{eq}(N)$.

(3)   Let

$$\mathcal{I}(X) = \inf\{\mathcal{I}(X; P) : P \in \mathcal{P}\}$$
$$= \inf\{\sum_{x \in X} \sum_{p_i x p_j \neq 0} \tau(p_i)\tau(p_j) : \{p_i\} \in \mathcal{P}_k, \ k \in \mathbb{N}\} \tag{16.2.5}$$

for all finite subsets X of N.

(4)   The support $\mathcal{S}(x; P)$ of an element $x \in N$ on $P \in \mathcal{P}$ is defined by

$$\mathcal{S}(x; P) = \vee\{p \in P : px \neq 0 \text{ or } xp \neq 0\}. \tag{16.2.6}$$

Note that for self-adjoint elements x the condition is $px \neq 0$.

The proof of the following lemma is a simple calculation about rectangles in a square.

Lemma 16.2.2. *Let N be a separable finite von Neumann algebra with a faithful normal trace τ. If $P, Q \in \mathcal{P}$ with $P \preceq Q$, then*

$$\mathcal{I}(X; P) \leq \mathcal{I}(X; Q) \quad \text{for all finite subsets } X \subseteq N. \tag{16.2.7}$$

Proof. This follows by induction using the following observation for a single element $x \in N$. The following pairs of projections are thought of as being in P and their sums as being in Q. If e_1, e_2 and f_1, f_2 are orthogonal pairs of projections with $(e_1 + e_2)x(f_1 + f_2) \neq 0$, then

$$\sum_{e_i x f_j \neq 0} \tau(e_i)\tau(f_j) \leq \tau(e_1 + e_2)\tau(f_1 + f_2). \tag{16.2.8}$$

$\square$

Remark 16.2.3. (1) If $\{p_i\} \in \mathcal{P}_k$ and z is in the centre $Z(N)$ of N, then $\mathcal{I}(z; \{p_i\}) = \sum_{zp_i \neq 0} \tau(p_i)^2$. If N is diffuse, then $\mathcal{P}_k^{eq} \neq \emptyset$ so that $\mathcal{I}(1; k) = k^{-1}$ by a standard minimisation argument. If N is a diffuse von Neumann algebra, then $\mathcal{I}(X) = 0$ for all finite subsets X of the $Z(N)$.
(2)   The definition of $\mathcal{S}(x; P)$ leads to two easy lower estimates on its trace for all $x \in N$. Let $x \in N$. If $P = \{p_i\} \in \mathcal{P}$, then

$$\mathcal{I}(x; P) = \sum_{p_i x p_j \neq 0} \tau(p_i)\tau(p_j) \leq \Big(\sum_{p_i x \neq 0} \tau(p_i) \Big)\Big(\sum_{x p_j \neq 0} \tau(p_j) \Big)$$
$$\leq \tau(\mathcal{S}(x; P))^2. \tag{16.2.9}$$

Similarly, by orthogonality,

$$\|x\|_2^2 = \Big\| \sum_{p_i x \neq 0} p_i x \Big\|_2^2 = \sum_{p_i x \neq 0} \|p_i x\|_2^2 \leq \|x\|^2 \sum_{p_i x \neq 0} \tau(p_i)$$
$$\leq \|x\|^2 \tau(\mathcal{S}(x; P)) \tag{16.2.10}$$

so that $\|x\|_2^2 / \|x\|^2 \leq \tau(\mathcal{S}(x; P))$.
(3)   Let X be a finite subset of N, let $P \in \mathcal{P}(N)$ and let $q = \vee_{x \in X} \mathcal{S}(x; P)$. Then $x = qx = xq$ for all $x \in X$ so that $y = qy = yq$ for all $y \in \text{Alg}(X \cup X^*)$. If X generates N as a weakly closed *-algebra, then $q = 1$.
(4)   Note that if X generates N, if $k \in \mathbb{N}$ and $P = \{p_i\} \in \mathcal{P}_k^{eq}(N)$, then $\mathcal{I}(X; P) \geq 1/k$. If $\mathcal{I}(X; P) < 1/k$, then $\mathcal{I}(X; P) = m/k^2$, where $m \in \mathbb{N}$ with $m < k$. Hence there is an i $(1 \leq i \leq k)$ such that $x p_i = p_i x = 0$ for all $x \in X$. This implies that p_i is orthogonal to $\text{Alg}(X)$ so $\text{Alg}(X)$ cannot be weakly dense in N. $\square$

The following lemma shows that for diffuse algebras, $\mathcal{I}$ is attained by families in $\mathcal{P}^{eq}$, those with minimal projections of equal trace, which is useful subsequently in II_1 factors. The case $n = 1$ is used frequently.

Lemma 16.2.4. *Let N be a separable diffuse finite von Neumann algebra with a faithful normal trace τ. If X is a finite subset of N and $n \in \mathbb{N}$, then*

$$\mathcal{I}(X) = \lim_k \inf_{P \in \mathcal{P}_{nk}^{eq}} \mathcal{I}(X; P). \tag{16.2.11}$$

Proof. Let $\varepsilon > 0$, let $k \in \mathbb{N}$ and let $\{p_i\} \in \mathcal{P}_k$ be such that

$$\mathcal{I}(X; \{p_i\}) \leq \mathcal{I}(X) + \varepsilon. \tag{16.2.12}$$

Assume that $\|x\| \leq 1$ for all $x \in X$. Let B be a masa in N containing $\{p_1, \ldots, p_k\}$ and identify (B, τ) with $(L^\infty[0,1], \int \cdot \, dt)$ using Theorem 3.5.2. Let

$$\delta = \varepsilon/2|X|k > 0. \tag{16.2.13}$$

By closely approximating $\tau(p_i)$, $1 \leq i \leq k$ by rational numbers with a common denominator of the form $n\ell$, there is an $\ell_0 \in \mathbb{N}$ such that if $\ell \geq \ell_0$, then there are $r_1, \ldots, r_k \in \mathbb{N}$ so that $0 < r_i/n\ell \leq \tau(p_i) \leq r_i/n\ell + \delta$ for $1 \leq i \leq k$. Choose mutually orthogonal projections $q_1, \ldots, q_k \in B$ such that $0 \leq q_i \leq p_i$ and $\tau(q_i) = r_i/n\ell$ for $1 \leq i \leq k$. Let $q_{k+1} = 1 - \sum_{i=1}^k q_i$. Then

$$\tau(q_{k+1}) \leq 1 - \sum_{i=1}^k r_i/n\ell \leq k\delta \tag{16.2.14}$$

and $\{q_i : 1 \leq i \leq k+1\} \in \mathcal{P}_{k+1}$. If $1 \leq i, j \leq k$, then $q_i x q_j = q_i p_i x p_j q_j \neq 0$ implies that $p_i x p_j \neq 0$. Hence

$$\begin{aligned}
\mathcal{I}(X; \{p_i\}) &= \sum_{x \in X} \sum_{p_i x p_j \neq 0} \tau(p_i)\tau(p_j) \\
&\geq \sum_{x \in X} \sum_{p_i x p_j \neq 0; 1 \leq i,j \leq k} \tau(q_i)\tau(q_j) \\
&\geq \sum_{x \in X} \sum_{q_i x q_j \neq 0; 1 \leq i,j \leq k} \tau(q_i)\tau(q_j) \\
&= \mathcal{I}(X; \{q_j\}) - \beta,
\end{aligned} \tag{16.2.15}$$

where

$$\begin{aligned}
\beta &= \sum_{x \in X} \Big(\sum_{q_i x q_{k+1} \neq 0} + \sum_{q_{k+1} x q_i \neq 0} \Big) \tau(q_i)\tau(q_{k+1}) \\
&\leq \sum_{x \in X} 2\tau(q_{k+1}) \\
&\leq 2|X|k\delta = \varepsilon
\end{aligned} \tag{16.2.16}$$

by inequality (16.2.14) and the choice of δ in (16.2.13). Hence

$$\mathcal{I}(X; \{q_j\}) \leq \mathcal{I}(X) + 2\varepsilon \tag{16.2.17}$$

by inequalities (16.2.15) and (16.2.16). Using the isomorphism $B \cong L^\infty[0,1]$ again, there is a refinement $\{e_i\} \in \mathcal{P}_{n\ell}^{eq}$ of $\{q_j\}$, so $\{e_i\} \preccurlyeq \{q_j\}$, such that

$$q_j = \sum_{e_i \leq q_j} e_i \quad \text{for } 1 \leq j \leq k+1 \tag{16.2.18}$$

and
$$\tau(e_i) = (n\ell)^{-1} \text{ for } 1 \le i \le n\ell. \tag{16.2.19}$$

Hence

$$\mathcal{I}(X) \le \mathcal{I}(X; \{e_i\}) \le \mathcal{I}(X; \{q_j\}) \le \mathcal{I}(X) + 2\varepsilon \tag{16.2.20}$$

by Remark 16.2.3 (3) and inequality (16.2.17). This proves the limits by Lemma 16.2.2. $\qquad\square$

Definition 16.2.5. If N is finitely generated, let

$$\mathcal{G}(N) = \inf\{\mathcal{I}(X) : X \text{ is finite and generates } N\}. \tag{16.2.21}$$

If N is not finitely generated, let $\mathcal{G}(N) = \infty$. Note that if N is finitely generated, then

$$\mathcal{G}(N) = \inf\{ \sum_{x \in X} \sum_{p_i x p_j \ne 0} \tau(p_i)\tau(p_j) :$$
$$x \in X, k \in \mathbb{N}, \{p_i\} \in \mathcal{P}_k, X \text{ a generating set of } N\}. \tag{16.2.22}$$

If N has n generators, then
$$0 \le \mathcal{G}(N) \le n \tag{16.2.23}$$

by (16.2.9), since the right-hand side of that inequality is bounded by 1. $\qquad\square$

The following corollary shows that for diffuse algebras, $\mathcal{G}$ is attained by families in $\mathcal{P}^{eq}$ as a limit. It is an immediate consequence of the definition 16.2.5 and Lemma 16.2.4.

Corollary 16.2.6. *Let N be a finitely generated separable diffuse finite von Neumann algebra with a faithful normal trace τ. If $n \in \mathbb{N}$, then*

$$\mathcal{G}(N) = \lim_k \inf_{\{X \cup X^*\}''=N} \inf_{P \in \mathcal{P}^{eq}_{nk}} \mathcal{I}(X; P). \tag{16.2.24}$$

16.3 Examples and basic lemmas

In this section we discuss various elementary results for $\mathcal{G}(N)$. We consider the behaviour of this invariant for tensor products and direct sums, and we also show that $\mathcal{G}(R) = 0$ for the hyperfinite II_1 factor. Such a result also holds in the abelian case.

Lemma 16.3.1. *If N is a diffuse separable abelian von Neumann algebra with a faithful normal trace, then $\mathcal{G}(N) = 0$.*

Proof. The algebra (N, τ) is isomorphic to $(L^\infty[0,1], \int \cdot \, dt)$ by Theorem 3.5.2. The latter is singly generated by $x = e^{2\pi it}$ or $x = t$ so that

$$0 \le \mathcal{G}(N) \le \mathcal{I}(x) = 0 \tag{16.3.1}$$

by Corollary 16.2.6. $\qquad\square$

It is well known that $\mathbb{M}_k$ is generated by the elements e_{11} and $v = \sum_{i=1}^{k-1}(e_{i,i+1}+e_{i+1,i})$. For notational convenience it is useful to regard the indices of e_{ij} in $\{1, \ldots, k\}$ as being handled by arithmetic (*mod k*). We shall often use this convention subsequently in this chapter. The matrices

$$w = \sum_{i=1}^{k-1} e_{i,i+1} \text{ and } v = w + w^* \qquad (16.3.2)$$

will be required in the next lemma and subsequently.

Lemma 16.3.2. *Let $\mathbb{M}_k$ be the algebra of $k \times k$ matrices over $\mathbb{C}$, let $1 \le t \le k$ and let $\{e_{ij} : 1 \le i, j \le k\}$ be a set of matrix units for $\mathbb{M}_k$.*

(i) *The set of two self-adjoint elements e_{tt} and v algebraically generate $\mathbb{M}_k$. Moreover,*

$$\mathcal{I}(\{e_{tt}, v\}; \{e_{ii}\}) = 2k^{-1} - k^{-2} \qquad (16.3.3)$$

and

$$\mathcal{S}(\{e_{tt}, v\}; \{e_{ii}\}) = 1. \qquad (16.3.4)$$

(ii) *The matrix w also generates $\mathbb{M}_k$ and gives a value of $(k-1)/k^2$ for $\mathcal{I}(\{w\}; \{e_{ii}\})$.*

Proof. (i) By permuting the integers $1, 2, \ldots, k$ cyclically, we assume $t = 1$ and we note that v is unchanged because the sum in (16.3.2) is over all pairs $(i, i+1)$. As the two elements are self-adjoint, the algebra and *-algebra they generate are equal. It is enough to show that e_{1j} is in this algebra for $1 \le j \le k$. This follows from the equations

$$e_{11}v = e_{1,2} \qquad (16.3.5)$$

and

$$e_{1j}v = e_{1,j+1} + e_{1,j-1} \qquad (16.3.6)$$

for $2 \le j \le k - 1$. Here $\mathcal{I}(e_{11}; \{e_{ii}\}) = k^{-2}$ and the definitions of v and $\mathcal{I}$ give (16.3.3). The definition of $\mathcal{S}$ leads to the final equality.

(ii) This is a series of simple matrix calculations which we will state but not carry out. The only self-adjoint matrix x which satisfies $xw, xw^*, wx, w^*x = 0$ is $x = 0$. If p is the identity of $C^*(w)$, then this shows that $1 - p = 0$ and so 1 lies in $C^*(w)$. Then verify that the only self-adjoint matrices commuting with w are scalars, and the result follows from the double commutant theorem, Theorem 2.2.1. The quantity $(k-1)/k^2$ is obtained by observing that w has $k - 1$ non-zero entries. $\square$

Lemma 16.3.2 (ii) provides an estimate on $\mathcal{G}(\mathbb{M}_k)$, since

$$\mathcal{G}(\mathbb{M}_k) \le \mathcal{I}(\mathbb{M}_k; \{e_{i,i}\}) \le k^{-1} - k^{-2}. \qquad (16.3.7)$$

Note that in the following lemma if N_i has identity 1_i and normal trace τ_i for $i = 1, 2$, then the trace on N is given by

$$\tau(x_1 \oplus x_2) = \tau(1_1)\tau_1(x_1) + \tau(1_2)\tau_2(x_2). \tag{16.3.8}$$

Lemma 16.3.3. *If N_i $(i = 1, 2)$ are separable finite von Neumann algebras with traces as above, then*

$$\mathcal{G}(N_1 \oplus N_2) = \tau(1_1)^2\mathcal{G}(N_1) + \tau(1_2)^2\mathcal{G}(N_2). \tag{16.3.9}$$

Proof. The case where either $\mathcal{G}(N_1)$ or $\mathcal{G}(N_2)$ is infinite is clear so assume both are finite and let $\varepsilon > 0$. By definition there are finite sets of generators X_1 and X_2 of N_1 and N_2, respectively, integers k, ℓ, and $\{p_i\} \in \mathcal{P}_k(N_1)$ and $\{q_j\} \in \mathcal{P}_\ell(N_2)$ such that

$$\mathcal{I}(X_1; \{p_i\}) < \mathcal{G}(N_1) + \varepsilon, \quad \mathcal{I}(X_2; \{q_j\}) < \mathcal{G}(N_2) + \varepsilon, \tag{16.3.10}$$

$$\sum_i p_i = 1_1 \quad \text{and} \quad \sum_j q_j = 1_2. \tag{16.3.11}$$

The set $Z = (X_1 \oplus 0) \cup (0 \oplus X_2)$ generates $N_1 \oplus N_2$ and

$$Q = \{p_i \oplus 0, \, 0 \oplus q_j : 1 \le i \le k, \, 1 \le j \le \ell\}$$

is in $\mathcal{P}_{k+\ell}(N_1 \oplus N_2)$. The projections $p_i \oplus 0$ do not interact with $0 \oplus X_2$ and the projections $0 \oplus q_j$ do not interact with $X_1 \oplus 0$. Hence using the trace τ as defined in equation (16.3.8)

$$\begin{aligned}
\mathcal{I}(Z; Q) &= \tau(1_1)^2\mathcal{I}(X_1, \{p_i\}) + \tau(1_2)^2\mathcal{I}(X_2; \{q_j\}) \\
&\le \tau(1_1)^2\mathcal{G}(N_1) + \tau(1_2)^2\mathcal{G}(N_2) + 2\varepsilon.
\end{aligned} \tag{16.3.12}$$

The definition of $\mathcal{G}(N_1 \oplus N_2)$ gives

$$\mathcal{G}(N_1 \oplus N_2) \le \tau(1_1)^2\mathcal{G}(N_1) + \tau(1_2)^2\mathcal{G}(N_2) + 2\varepsilon. \tag{16.3.13}$$

Let $\varepsilon > 0$. By definition of $\mathcal{G}$, there is a finite generating set X of $N_1 \oplus N_2$ and $P = \{p_i\} \in \mathcal{P}_k(N_1 \oplus N_2)$ such that

$$\mathcal{I}(X; P) < \mathcal{G}(N_1 \oplus N_2) + \varepsilon. \tag{16.3.14}$$

Let

$$p_i^{(1)} = (1_1 \oplus 0)p_i, \quad p_i^{(2)} = (0 \oplus 1_2)p_i, \tag{16.3.15}$$

and

$$x^{(1)} = (1_1 \oplus 0)x, \quad x^{(2)} = (0 \oplus 1_2)x, \tag{16.3.16}$$

for all $p_i \in P$ and $x \in X$. Then $x = x^{(1)} + x^{(2)}$ for all $x \in X$ and $p_i = p_i^{(1)} + p_i^{(2)}$ with

$$\tau(p_i) = \tau(1_1)\tau_{N_1}(p_i^{(1)}) + \tau(1_2)\tau_{N_2}(p_i^{(2)}) \tag{16.3.17}$$

for all $p_i \in P$. Let $P_1 = \{p_i^{(1)}\}$, $P_2 = \{p_i^{(2)}\}$, $X_1 = \{x^{(1)} \colon x \in X\}$ and $X_2 = \{x^{(2)} \colon x \in X\}$, where we are ignoring the zero projections that may occur for $p_i^{(1)}$ or $p_i^{(2)}$. Then X_1 generates N_1, X_2 generates N_2, $P \in \mathcal{P}(N_1)$ and $P_2 \in \mathcal{P}(N_2)$. If $x \in X$, then

$$p_i x p_j = p_i^{(1)} x^{(1)} p_j^{(1)} + p_i^{(2)} x^{(2)} p_j^{(2)} \tag{16.3.18}$$

so that $p_i x p_j \neq 0$ if, and only if, $p_i^{(1)} x^{(1)} p_j^{(1)} \neq 0$ or $p_i^{(2)} x^{(2)} p_j^{(2)} \neq 0$. Hence

$$
\begin{aligned}
\mathcal{I}(X;P) &= \sum_{x \in X} \sum \{\tau(p_i)\tau(p_j) \colon p_i^{(\ell)} x p_j^{(\ell)} \neq 0 \ \text{for} \ \ell = 1 \ \text{or} \ 2\} \\
&\geq \tau(1_1)^2 \sum_{x \in X} \sum_{p_i^{(1)} x^{(1)} p_j^{(1)} \neq 0} \tau_{N_1}(p_i^{(1)}) \tau_{N_1}(p_j^{(1)}) \\
&\quad + \tau(1_2)^2 \sum_{x \in X} \sum_{p_i^{(2)} x^{(2)} p_j^{(2)} \neq 0} \tau_{N_2}(p_i^{(2)}) \tau_{N_2}(p_j^{(2)}) \\
&= \tau(1_1)^2 \mathcal{I}(X_1; P_1) + \tau(1_2)^2 \mathcal{I}(X_2; P_2) \\
&\geq \tau(1_1)^2 \mathcal{G}(N_1) + \tau(1_2) \mathcal{G}(N_2), \tag{16.3.19}
\end{aligned}
$$

completing the proof. $\qquad\square$

Lemma 16.3.3 provides an easy way of showing that if N is a diffuse abelian separable (finite) von Neumann algebra, then $\mathcal{G}(N) = 0$, as was proved in Lemma 16.3.1. Let p be a projection of trace $1/2$ in N, let $N_1 = pN$ and $N_2 = (1 - p)N$. Then $N = N_1 \oplus N_2$ and $N \cong N_1 \cong N_2$. Since

$$0 \leq \mathcal{G}(N) = (1/2)^2 \mathcal{G}(N) + (1/2)^2 \mathcal{G}(N_2) = 1/2 \mathcal{G}(N), \tag{16.3.20}$$

we have $\mathcal{G}(N) = 0$.

Lemma 16.3.4. *If N_1 and N_2 are diffuse finitely generated finite von Neumann algebras with faithful normal traces τ_1 and τ_2, then $\mathcal{G}(N_1 \overline{\otimes} N_2) = 0$.*

Proof. Let X and Y be generating sets of N_1 and N_2, let $k, \ell \in \mathbb{N}$, and let $\{p_i\} \in \mathcal{P}_k(N_1)$ and $\{q_s\} \in \mathcal{P}_\ell(N_2)$. Then

$$Z = \{x \otimes 1, \ 1 \otimes y \colon x \in X, \ y \in Y\}$$

is a generating set for $N_1 \overline{\otimes} N_2$ and $\{p_i \otimes q_s\} \in \mathcal{P}_{k\ell}$. If $x \in X$ and $y \in Y$, then

$$(p_i \otimes q_s)(x \otimes 1)(p_j \otimes q_t) = p_i x p_j \otimes q_s q_t \neq 0 \tag{16.3.21}$$

if, and only if, $s = t$ and $p_i x p_j \neq 0$, and similarly

$$(p_i \otimes q_s)(1 \otimes y)(p_j \otimes q_t) = p_i p_j \otimes q_s y q_t \neq 0 \tag{16.3.22}$$

if, and only if, $i = j$ and $q_s y q p_t \neq 0$. Summing over the elements of X and Y yields

$$
\mathcal{I}(Z; \{p_i \otimes q_s\}) = \sum_{x \in X} \sum_{s=1}^{\ell} \tau(q_s)^2 \sum_{p_i x p_j \neq 0} \tau(p_i) \tau(p_j)
$$

$$
+ \sum_{y \in Y} \sum_{i=1}^{k} \tau(p_i)^2 \sum_{q_s y q_t \neq 0} \tau(q_s) \tau(q_t)
$$

$$
= \mathcal{I}(1; \{q_s\}) \mathcal{I}(X; \{p_i\}) + \mathcal{I}(1; \{p_i\}) \mathcal{I}(Y; \{q_s\}). \qquad (16.3.23)
$$

If N_1 and N_2 are diffuse, $\varepsilon > 0$ and $n \in \mathbb{N}$, then there exist integers $k, \ell \geq n$ and sets of projections $\{p_i\} \in \mathcal{P}_k^{eq}(N_1)$ and $\{q_s\} \in \mathcal{P}_\ell^{eq}(N_2)$ such that

$$
\mathcal{I}(X; \{p_i\}) < \mathcal{G}(N_1) + \varepsilon \quad \text{and} \quad \mathcal{I}(Y; \{q_s\}) < \mathcal{G}(N_2) + \varepsilon \qquad (16.3.24)
$$

by Lemma 16.2.4. It follows from equation (16.3.23) that

$$
\mathcal{G}(N_1 \overline{\otimes} N_2) \leq \mathcal{I}(Z; \{p_i \otimes q_s\}) \leq n^{-1} \big(\mathcal{G}(N_1) + \mathcal{G}(N_2) + 2\varepsilon \big), \qquad (16.3.25)
$$

since $\mathcal{I}(1_1; \{p_i\}) = k^{-1}$ and $\mathcal{I}(1_2; \{q_s\}) = \ell^{-1}$. This proves the result. $\qquad \square$

Minor alterations to the proof above yield $\mathcal{G}(\mathbb{M}_k \otimes N_2) \leq k^{-1} \mathcal{G}(N_2)$ for all $k \in \mathbb{N}$. This inequality is improved in the following lemma, and a more detailed discussion will be given in Section 16.4.

Lemma 16.3.5. *If N is a diffuse finitely generated von Neumann algebra with a faithful normal trace τ, then*

$$
\mathcal{G}(\mathbb{M}_n \otimes N) \leq n^{-2} \mathcal{G}(N) \qquad (16.3.26)
$$

for all $n \in \mathbb{N}$.

Proof. Let $\varepsilon > 0$. By Lemma 16.2.4 choose a finite generating set X for N, a $k \in \mathbb{N}$ with $k > \varepsilon^{-1}$ and a $P = \{p_1, \ldots, p_k\} \in \mathcal{P}_k^{eq}(N)$ such that

$$
\mathcal{I}(X; P) \leq \mathcal{G}(N) + \varepsilon. \qquad (16.3.27)
$$

Then the set

$$
Y = \big(e_{11} \otimes X \big) \cup \{ e_{i,i+1} \otimes 1 \colon 1 \leq i \leq n - 1 \}
$$

generates $\mathbb{M}_n \otimes N$, since $\{ e_{i,i+1} \colon 1 \leq i \leq n - 1 \}$ generates $\mathbb{M}_n$. Let

$$
Q = \{ e_{ii} \otimes p_j \colon 1 \leq i \leq n, 1 \leq j \leq k \}.
$$

Then $Q \in \mathcal{P}_{kn}^{eq}(\mathbb{M}_n \otimes N)$. By definition of $\mathcal{I}$,

$$
\mathcal{I}(e_{11} \otimes X; Q) = (nk)^{-2} \sum_{x \in X} |\{ (s,t) \colon p_s x p_t \neq 0 \text{ for some } x \in X \}|
$$

$$
= n^{-2} \mathcal{I}(X; P), \qquad (16.3.28)
$$

since only the projections $e_{11} \otimes p_j$ from Q contribute to $\mathcal{I}(e_{11} \otimes X; Q)$. In a similar way,

$$\mathcal{I}(e_{i,i+1} \otimes 1; Q) = (n)^{-2}(k)^{-1} \text{ for all } 1 \leq i \leq n-1, \tag{16.3.29}$$

since only the pairs of projections $(e_{ii} \otimes p_j, e_{i+1,i+1} \otimes p_j)$ are relevant. Hence, from (16.3.28) and (16.3.29),

$$\mathcal{I}(Y;Q) = n^{-2}\mathcal{I}(X;P) + ((n-1)k)(n)^{-2}(k)^{-1} < n^{-2}\mathcal{G}(N) + 2\varepsilon n^{-1}, \tag{16.3.30}$$

which gives the result. $\qquad\qquad\qquad\qquad\qquad\qquad\qquad\qquad\qquad\qquad\square$

There are several proofs of the following lemma, each of which emphasises some aspect of the hyperfinite II$_1$ factor R. The proof below is given to motivate the proofs of later results. Example 3.4.3 exhibits R as the factor generated by a certain pair of unitaries, and so is singly generated. Uniqueness of R also gives $R \cong \mathbb{M}_n \otimes R$ for any integer $n \geq 1$, so the result could be deduced from these facts using Lemma 16.3.5.

Lemma 16.3.6. *Let R be the hyperfinite* II$_1$ *factor. If $k \in \mathbb{N}$ and if $\{e_{ij} : 1 \leq i,j \leq k\}$ are matrix units in R, then there are self-adjoint generators x_1, x_2 in R such that $\mathcal{I}(\{x_1, x_2\}; \{e_{ii}\}) < 3k^{-1}$. Hence R is singly generated and $\mathcal{G}(R) = 0$.*

Proof. Regard $R \cong \left(\otimes_{n=1}^{\infty} \mathbb{M}_k^{(n)} \right)^{-w}$, where $M_k^{(1)} = \{e_{ij} : 1 \leq i,j \leq k\}''$ and $\mathbb{M}_k^{(n)}$ is the algebra of $k \times k$ matrices with normalised trace with (n) indicating the position in the tensor product. Let $e_{ij}^{(1)} = e_{ij}$ $(1 \leq i,j \leq k)$ and $\{e_{ij}^{(n)} : 1 \leq i,j \leq k\}$ be a set of matrix units for $\mathbb{M}_k^{(n)}$ $(n \geq 2)$. Identify $\mathbb{M}_k^{(n)}$ and $\otimes_{j=1}^n \mathbb{M}_k^{(j)}$ with the corresponding subalgebras in R. Let

$$y_1 = \sum_{j=1}^{k-1} \left(e_{j,j+1}^{(1)} + e_{j+1,j}^{(1)} \right), \tag{16.3.31}$$

$$y_{n+1} = e_{22}^{(1)} \otimes e_{22}^{(2)} \ldots \otimes e_{22}^{(n)} \otimes \left(\sum_{j=1}^{k-1} (e_{j,j+1}^{(n+1)} + e_{j+1,j}^{(n+1)}) \right) \text{ for all } n \in \mathbb{N}, \tag{16.3.32}$$

$$x_1 = e_{11}^{(1)} + \sum_{n=1}^{\infty} 2^{-n} e_{22}^{(1)} \otimes e_{22}^{(2)} \ldots \otimes e_{22}^{(n)} \otimes e_{11}^{(n+1)}, \tag{16.3.33}$$

and

$$x_2 = \sum_{n=1}^{\infty} y_n, \tag{16.3.34}$$

where the series defining x_2 converges strongly.

The spectral properties of the self-adjoint element x_1 imply that

$$e_{11}^{(1)}, \; e_{22}^{(1)} \otimes e_{22}^{(2)} \ldots \otimes e_{22}^{(n)} \otimes e_{11}^{(n+1)} \in \{x_1\}'' \text{ for all } n \in \mathbb{N}.$$

By equations (16.3.31) to (16.3.34) and by considering products of the form

$$\left(e_{22}^{(1)} \otimes \ldots \otimes e_{22}^{(n)} \otimes e_{11}^{(n+1)}\right)x_2,$$

it follows that

$$e_{22}^{(1)} \otimes \ldots \otimes e_{22}^{(n)} \otimes e_{12}^{(n+1)}, \; e_{22}^{(1)} \otimes \ldots \otimes e_{22}^{(n)} \otimes e_{13}^{(n+1)}, \ldots \in \{x_1, x_2\}''$$

for all $n \in \mathbb{N}$, since

$$\left(e_{22}^{(1)} \otimes \ldots \otimes e_{22}^{(n)} \otimes e_{1j}^{(n+1)}\right)x_2 = 2^{-n} e_{22}^{(1)} \otimes \ldots \otimes e_{22}^{(n)} \otimes \left(e_{1,j+1}^{(n+1)} + e_{1,j-1}^{(n+1)}\right)$$

for $2 \le j \le k-1$. The algebra $M_k^{(1)}$ is contained in $\{x_1, x_2\}''$ as in Lemma 16.3.2. A repetitive argument now yields $M_k^{(2)}, M_k^{(3)}, \ldots$ are contained in $\{x_1, x_2\}''$. Hence $\{x_1, x_2\}'' = R$.

Using $x_1 \in \{e_{ii}^{(1)} : 1 \le i \le k\}'$,

$$e_{ii}^{(1)} x_1 e_{jj}^{(1)} = e_{ii}^{(1)} e_{jj}^{(1)} x_1 \ne 0 \iff i = j$$

so that $\mathcal{I}(x_1; \{e_{ii}^{(1)}\}) = k^{-1}$.

Note that

$$e_{ii}^{(1)} x_2 e_{jj}^{(1)} \ne 0 \iff \text{ one of } e_{ii}^{(1)} y_n e_{jj}^{(1)} \ne 0 \; (n \in \mathbb{N}).$$

By Lemma 16.3.5, y_1 contributes the set $\{(i,j): |i - j| = 1, 1 \le i,j \le k\}$ of size $2(k-1)$ and all the other y_n $(n \ge 2)$ contribute only $\{(2,2)\}$. Thus $\mathcal{I}(x_2; \{e_{ii}^{(1)}\}) = k^{-2}(2k-1)$ and

$$\mathcal{I}(\{x_1, x_2\}; \{e_{ii}\}) = \mathcal{I}(\{x_1, x_2\}; \{e_{ii}^{(1)}\}) = 3k^{-1} - k^{-2}.$$

$\square$

16.4   The scaling formula for $\mathcal{G}$

This section deals with the problem of how $\mathcal{G}(\cdot)$ behaves with respect to scaling. We will make this precise after some further discussion, but the typical question concerns the relationship between $\mathcal{G}(N)$ and $\mathcal{G}(pNp)$ where $p \in N$ is a projection.

Two projections $p, q \in N$ of equal trace $t \in [0, 1]$ define isomorphic compressions since a partial isometry for the equivalence of p and q also implements a spatial isomorphism between pNp and qNq. Without ambiguity, we may then let N_t denote pNp where $\tau(p) = t$. For integers n, we let N_n be $\mathbb{M}_n \otimes N$, and we extend the definition of N_t to all non-integer values by first choosing an integer $n > t$ and letting N_t be $\mathbb{M}_n \otimes N_{t/n}$. There is no ambiguity in the choice of n, since if $m < n$ are admissible choices then we can regard $\mathbb{M}_m \otimes N$ as embedded in the upper $m \times m$ block of $\mathbb{M}_n \otimes N$. If p and q in N have respective traces t/m and t/n, then the projections $1_m \otimes p$ and $1_n \otimes q$ have equal traces in N, so their compressions are isomorphic. Thus the choice of m or n does not affect the definition of N_t. Direct calculation shows that

$$(N_s)_t \cong N_{st} \quad \text{for } s, t > 0. \tag{16.4.1}$$

Definition 16.4.1. With the above notation, we define, for $t > 0$, the *t-scaling* of N to be N_t. The set

$$\{t > 0 \colon N_t \cong N\}$$

is a subgroup of $(\mathbb{R}^+, \times)$ from (16.4.1), and is called the *fundamental group* $\mathcal{F}(N)$ of N.   □

The fundamental group was introduced by Murray and von Neumann [118], who showed that $\mathcal{F}(R) = \mathbb{R}^+$, where R is the hyperfinite II$_1$ factor. The factors arising from property T groups were shown by Connes to have countable fundamental groups [37], though no specific values other than 1 could be given. Very recently, Popa [147] has proved that if G is a countable multiplicative subgroup of $\mathbb{R}^+$, then there is a separable II$_1$ factor N with $\mathcal{F}(N) = G$ (see also [91, 146]). Moreover, if the separability requirement is dropped, then any subgroup of $\mathbb{R}^+$ is a possible fundamental group. The cardinality of such subgroups exceeds the cardinality of the set of separable factors, so non-separable factors are inevitable here.

The next lemma ensures that only the finitely generated case needs to be discussed in Theorem 16.4.5 and the proof contains a useful little calculation.

Lemma 16.4.2. *Let N be a II$_1$ factor and let $t > 0$. Then N_t is finitely generated if, and only if, N is finitely generated.*

Proof. (1)   If N is finitely generated by X and $n \in \mathbb{N}$, then $N_n = \mathbb{M}_n \otimes N$ is finitely generated by

$$Y = \{e_{11} \otimes X\} \cup \{e_{i,i+1} \otimes 1 \colon 1 \le i \le n - 1\},$$

where e_{ij} are the matrix units of $\mathbb{M}_n$.

(2)   Let N be finitely generated by $X = X^*$, let $0 < t < 1$ and let p be a projection in N such that $\tau(p) = t$. Then pNp is a representative of N_t.

Let $m = \lfloor t^{-1} \rfloor$ be the maximal integer such that $mt \le 1$. Choose orthogonal projections $p_1, \ldots, p_{m+1}$ in N such that

$$p_1 = p, \tag{16.4.2}$$

$$\tau(p_i) = \tau(p) \quad \text{for } 2 \le i \le m, \tag{16.4.3}$$

$$\sum_1^{m+1} p_i = 1. \tag{16.4.4}$$

Thus $0 \leq \tau(p_{m+1}) < \tau(p) = t$.

Choose partial isometries $v_2, \ldots, v_{m+1}$ in N such that

$$v_i v_i^* = p_i \quad \text{and} \quad v_i^* v_i = p_1 = p \quad \text{for} \quad 2 \leq i \leq m, \tag{16.4.5}$$

and

$$v_{m+1} v_{m+1}^* = p_{m+1} \quad \text{and} \quad v_{m+1}^* v_{m+1} = \widetilde{p_{m+1}} \leq p_1 = p. \tag{16.4.6}$$

Let

$$Y = \{v_i^* x v_j : 1 \leq i, j \leq m+1, \quad x \in X\}. \tag{16.4.7}$$

By the choice of v_i $(1 \leq i \leq m+1)$, we have $Y \subset pNp$. Inductive use of the equation

$$v_i^* x_1 x_2 v_j = \sum_{s=1}^{m+1} v_i^* x_1 v_s v_s^* x_2 v_j \quad \text{for all} \quad x_1, x_2 \in N \tag{16.4.8}$$

implies that $\text{Alg}(Y) = p \, \text{Alg}(X) p$. Hence pNp is generated by Y.

(3)   Let $0 < t < 1$ and let n be an integer with $nt > 1$. If N_t is finitely generated, then $(N_t)_n \cong N_{nt}$ is finitely generated by (1) so that $N \cong (N_{nt})_{1/nt}$ is finitely generated by (2).

Let $1 < t$ and let $n \in \mathbb{N}$ with $n > t$. If N is finitely generated, then N_n is finitely generated by (1) so that $N_t \cong (N_n)_{t/n}$ is finitely generated by (2). Conversely, if N_t is finitely generated, then $N \cong (N_t)_{1/t}$ is finitely generated by (2). □

The inequalities required in the proof of Theorem 16.4.5 are proved in the next two Lemmas 16.4.3 and 16.4.4.

Lemma 16.4.3. *Let N be a finitely generated* II_1 *factor. If $n \in \mathbb{N}$, then*

$$\mathcal{G}(N) \leq n^2 \mathcal{G}(\mathbb{M}_n \otimes N).$$

Proof. Let $\varepsilon > 0$. By Lemma 16.4.2 choose a finite generating set Y for $M = \mathbb{M}_n \otimes N$, an integer k and a set $P = \{p_t : 1 \leq t \leq nk\} \in \mathcal{P}_{nk}^{eq}(\mathbb{M}_n \otimes N)$ such that

$$\mathcal{I}(Y; P) < \mathcal{G}(\mathbb{M}_n \otimes N) + \varepsilon/n^2. \tag{16.4.9}$$

Since $\tau(p_s) = \tau(p_t)$ for $1 \leq s, t \leq nk$, we may find matrix units $\{p_{st} : 1 \leq s, t \leq nk\}$ with $p_{tt} = p_t$ for $1 \leq t \leq nk$. Let

$$f_{ij} = \sum_{s=1}^{k} p_{(i-1)k+s, (j-1)k+s} \quad \text{for} \quad 1 \leq i, j \leq n. \tag{16.4.10}$$

The family $\{f_{ij} : 1 \leq i, j \leq n\}$ is a set of matrix units for an $n \times n$ matrix subalgebra of $\mathbb{M}_n \otimes N$, because when calculating $f_{i,j} f_{\ell,m}$ as a double sum over s and t, say, the set $\{r \in \mathbb{Z} : -k+1 \leq r \leq k-1\} \cap k\mathbb{Z}$ equals $\{0\}$.

Let $\{e_{ij} \colon 1 \leq i, j \leq n\}$ be the standard matrix units of $\mathbb{M}_n$. Since

$$\tau_M(e_{11} \otimes 1) = \tau_M(f_{11}) = 1/n, \qquad (16.4.11)$$

there is a partial isometry $v \in M$ such that $vv^* = f_{11}$ and $v^*v = e_{11} \otimes 1$. Let

$$X = \{f_{1i}yf_{j1} \colon 1 \leq i, j \leq n, \quad y \in Y\}. \qquad (16.4.12)$$

Then v^*Xv is a generating set for $e_{11} \otimes N = (e_{11} \otimes 1)M(e_{11} \otimes 1)$ using (2) of the proof of the preceding Lemma 16.4.2. Let $q_j = v^*p_jv$ for $1 \leq j \leq k$. By definition of f_{11} in equation (16.4.10),

$$p_j \leq f_{11} = \sum_{i=1}^{k} p_{ii} \ (1 \leq j \leq k) \qquad (16.4.13)$$

so that $q_j \in e_{11} \otimes N$ for $1 \leq j \leq k$. Hence

$$P_0 = \{p_j \colon 1 \leq j \leq k\} \in \mathcal{P}_k^{eq}(f_{11}Mf_{11}) \qquad (16.4.14)$$

and

$$Q = \{q_j \colon 1 \leq j \leq k\} \in \mathcal{P}_k^{eq}(e_{11} \otimes N). \qquad (16.4.15)$$

Since the map $x \mapsto v^*xv \colon f_{11}Mf_{11} \to e_{11} \otimes N$ is an isomorphism that sends P_0 onto Q, we have

$$\mathcal{I}(v^*Xv; Q) = \mathcal{I}(X; P_0) \qquad (16.4.16)$$

with the first calculated in $e_{11} \otimes N$ and the second in $f_{11}Mf_{11}$. Since each projection p_j has trace k^{-1} in $f_{11}Mf_{11}$,

$$\mathcal{I}(f_{1i}yf_{j1}; P_0) = k^{-2}|\{(s, t) \colon p_s f_{1i}yf_{j1}p_t \neq 0; 1 \leq s, t \leq k\}|. \qquad (16.4.17)$$

The definition of f_{1i} and f_{j1} implies that

$$p_s f_{1i}yf_{j1}p_t = \sum_{\ell, m} p_s p_{\ell, (i-1)k+\ell} \, y p_{(j-1)k+m, m} p_t$$

$$= p_{s, (i-1)k+s} \, y p_{(j-1)k+t, t}, \qquad (16.4.18)$$

which is non-zero if, and only if,

$$p_{(i-1)k+s, (i-1)k+s} \, y p_{(j-1)k+t, (j-1)k+t} \neq 0. \qquad (16.4.19)$$

The mapping $(i, s) \mapsto (i-1)k+s$ from $\{1, \ldots, n\} \times \{1, \ldots, k\}$ to $\{1, \ldots, nk\}$ is a bijection with the sets $\{(i-1)k+s \colon 1 \leq s \leq k\}$ $(1 \leq i \leq n)$ forming a disjoint partition of $\{1, \ldots, nk\}$. Hence

$$\sum_{i, j} |\{(s, t) \colon p_s f_{1i}yf_{j1}p_t \neq 0; 1 \leq s, t \leq k\}|$$

$$= |\{(\ell, m) \colon p_\ell y p_m \neq 0; 1 \leq \ell, m \leq nk\}|. \qquad (16.4.20)$$

By this equation (16.4.20) and equations (16.2.3), (16.4.17)

$$\sum_{i,j=1}^{n} \mathcal{I}(f_{1i}yf_{j1}; P_0) = k^{-2}|\{(s,t): p_s y p_t \neq 0, \quad 1 \leq s,t \leq k\}|$$

$$= n^2 \mathcal{I}(y; P) \tag{16.4.21}$$

for each $y \in Y$ since $\tau(p_s) = (nk)^{-1}$. Summing this equation (16.4.21) over $y \in Y$ and using equation (16.4.16) gives

$$\mathcal{G}(N) \leq \mathcal{I}(v^* X v; Q) = \mathcal{I}(X; P_0) = \sum_{y \in Y} \sum_{i,j=1}^{n} \mathcal{I}(f_{1i}yf_{j1}; P_0)$$

$$= n^2 \sum_{y \in Y} \mathcal{I}(y; P) = n^2 \mathcal{I}(Y; P)$$

$$\leq n^2 \mathcal{G}(\mathbb{M}_n \otimes N) + \varepsilon. \tag{16.4.22}$$

This proves the lemma. □

Combining Lemmas 16.3.5 and 16.4.3 implies that $\mathcal{G}(N) = n^2 \mathcal{G}(\mathbb{M}_n \otimes N)$ for all $n \in \mathbb{N}$. This will be included in the conclusions of Theorem 16.4.5.

Lemma 16.4.4. *Let N be a finitely generated* II_1 *factor. If $0 < t < 1$, then* $\mathcal{G}(N) \leq t^2 \mathcal{G}(N_t)$.

Proof. Let p be a projection in N with $\tau(p) = t$ so that pNp is a representative of N_t. Let $\varepsilon > 0$. By Lemma 16.4.2 choose a finite generating set X of N_t, a positive integer k and a set $E = \{e_1, \ldots, e_k\} \in \mathcal{P}_k^{eq}(pNp)$ such that

$$\mathcal{I}(X; E) < \mathcal{G}(N_t) + t^2 \varepsilon. \tag{16.4.23}$$

Let $m = \lfloor t^{-1} \rfloor$ be the maximal integer such that $mt \leq 1$. Choose orthogonal projections $p_1, \ldots, p_{m+1}$ in N such that

$$p_1 = p, \tag{16.4.24}$$

$$\tau(p_i) = \tau(p) \quad \text{for} \quad 2 \leq i \leq m, \tag{16.4.25}$$

$$\sum_{1}^{m+1} p_i = 1. \tag{16.4.26}$$

Thus $0 \leq \tau(p_{m+1}) < \tau(p) = t$. Let $t^{-1}\tau(p_{m+1}) = r/k + \alpha$, where $0 \leq r \leq k-1$ and $0 \leq \alpha < 1/k$. Choose a projection $\widetilde{e_{r+1}}$ in the II_1 factor $e_{r+1}Ne_{r+1}$ such that $t^{-1}\tau(\widetilde{e_{r+1}}) = \alpha$. Then the projection $\widetilde{p_{m+1}} = \sum_1^r e_i + \widetilde{e_{r+1}}$ satisfies

$$\tau(\widetilde{p_{m+1}}) = t(r/k + \alpha) = \tau(p_{m+1}). \tag{16.4.27}$$

Choose partial isometries $v_2, \ldots, v_{m+1}$ in N such that

$$v_i v_i^* = p_i \quad \text{and} \quad v_i^* v_i = p_1 = p \quad \text{for} \quad 2 \leq i \leq m, \tag{16.4.28}$$

and

$$v_{m+1}v_{m+1}^* = p_{m+1} \quad \text{and} \quad v_{m+1}^*v_{m+1} = \widetilde{p_{m+1}} \leq p_1 = p. \qquad (16.4.29)$$

Let $v_1 = p_1 = p$. By equation (16.4.28), $v_i pNp v_j^* = p_i Np_j$ so the set $X \cup \{v_2, \ldots, v_m\}$ generates $(p_1 + \ldots + p_m)N(p_1 + \ldots + p_m)$, since X generates pNp. By equation (16.4.29),

$$v_{m+1}pNpv_{m+1}^* = p_{m+1}Np_{m+1} \subseteq \left(X \cup \{v_2, \ldots, v_{m+1}\} \right)'' \qquad (16.4.30)$$

and

$$v_{m+1}pN(p_2 + \ldots + p_m) = p_{m+1}N(p_2 + \ldots + p_m)$$
$$\subseteq \left(X \cup \{v_2, \ldots, v_{m+1}\} \right)''. \qquad (16.4.31)$$

Thus $Y = X \cup \{v_2, \ldots, v_{m+1}\}$ generates N, since $p_1 + \ldots + p_{m+1} = 1$.
    Let

$$F_0 = \{e_1, \ldots, e_r, \widetilde{e_{r+1}}, e_{r+1} - \widetilde{e_{r+1}}, e_{r+2}, \ldots, e_k\}. \qquad (16.4.32)$$

Choose an integer ℓ such that

$$\ell > k^{-1}\varepsilon^{-1}m \qquad (16.4.33)$$

and choose a refinement F of F_0 in $\mathcal{P}(pNp)$ such that each projection f_0 in F_0 is split into ℓ projections f in F of equal trace $\tau_{pNp}(f) = \ell^{-1}\tau_{pNp}(f_0)$ when $f \leq f_0$. Hence

$$\tau(f) \leq t\ell^{-1}k^{-1} \quad \text{for all} \quad f \in F. \qquad (16.4.34)$$

Note that, due to the role of $\widetilde{e_{r+1}}$, the projections in F may not all have equal trace.
    Let

$$Q = \{v_j f v_j^* : 1 \leq j \leq m+1, f \in F\}. \qquad (16.4.35)$$

Then $Q \in \mathcal{P}(N)$, since $\sum_j \sum_{f \in F} v_j f v_j^* = \sum_j p_j = 1$. By equation (16.4.34), $\tau(q) \leq tk^{-1}\ell^{-1}$ for all $q \in Q$. Since $pQ = F \cup \{0\}$ and X is contained in pNp, the trace scaling $\tau_{pNp}(x) = t^{-1}\tau(x)$ for all $x \in pNp$ gives

$$\mathcal{I}(X; Q) = t^{-2}\mathcal{I}(X; F) \qquad (16.4.36)$$

because $\mathcal{I}(X; Q)$ is calculated in N and $\mathcal{I}(X; F)$ is calculated in pNp. Hence

$$\mathcal{I}(X; Q) = t^{-2}\mathcal{I}(X; F) \leq t^{-2}\mathcal{I}(X; F_0)$$
$$\leq t^{-2}\mathcal{I}(X; E) < t^{-2}\mathcal{G}(N) + \varepsilon \qquad (16.4.37)$$

by equation (16.4.36) and Lemma 16.2.2, since $F \preccurlyeq F_0 \preccurlyeq E$.
    Let $2 \leq j \leq m$. If $q_s = v_s f_s v_s^*$ and $q_t = v_t f_t v_t^*$ for some f_s and f_t in F, then $q_s v_j q_t = v_s f_s v_s^* v_j v_t f_t v_t^* \neq 0$ if, and only if, $v_s^* v_j v_t = p_1 = p$ and $f_s = f_t$, since

f_s, $f_t \leq p$, $v_s^* v_s f_s = f_s$ and $f_t v_t^* v_t = f_t$. Here $v_s^* v_j v_t = p$ if, and only if, $s = j$ and $t = 1$. There are $(k+1)\ell$ values of $f_s = f_t$. Hence

$$\mathcal{I}(v_j; Q) = \sum_{q_s v_j q_t \neq 0} \tau(q_s)\tau(q_t)$$

$$\leq t^2(k\ell)^{-2}(k+1)\ell < 2t^2(k\ell)^{-1}, \qquad (16.4.38)$$

since $\tau(q_s) \leq (k\ell)^{-1}$. A similar final inequality holds for $\mathcal{I}(v_{m+1}; Q)$, using the equation (16.4.29) that defines v_{m+1} and $\widetilde{p_{m+1}} = e_1 + \ldots + e_r + \widetilde{e_{r+1}} \in F''$. Hence

$$\mathcal{I}(\{v_2, \ldots, v_{m+1}\}; Q) \leq 2t^2 mk^{-1}\ell^{-1} \qquad (16.4.39)$$

so that

$$\mathcal{G}(N) \leq \mathcal{I}(Y; Q) \leq t^{-2}\mathcal{G}(N_t) + \varepsilon + 2t^2 mk^{-1}\ell^{-1}$$

$$< t^{-2}\mathcal{G}(N_t) + 3\varepsilon. \qquad (16.4.40)$$

This proves the lemma. $\qquad \square$

Theorem 16.4.5. *If N is a separable II_1 factor, then*

$$\mathcal{G}(N_t) = t^{-2}\mathcal{G}(N) \quad \text{for all} \quad t > 0.$$

Proof. By Lemma 16.4.2, the algebras N and N_t may be assumed to be finitely generated. By Lemmas 16.3.5 and 16.4.3,

$$\mathcal{G}(N) \leq n^{-2}\mathcal{G}(N_{1/n}) \leq n^{-2}n^2\mathcal{G}((N_{1/n})_n) = \mathcal{G}(N) \qquad (16.4.41)$$

so that

$$\mathcal{G}(N) = n^{-2}\mathcal{G}(N_{1/n}) \quad \text{for all} \quad n \in \mathbb{N}. \qquad (16.4.42)$$

Replacing N by N_n in equation (16.4.42) gives

$$n^2\mathcal{G}(N_n) = \mathcal{G}((N_n)_{1/n}) = \mathcal{G}(N). \qquad (16.4.43)$$

If $m, n \in \mathbb{N}$, then

$$\mathcal{G}(N_{m/n}) = \mathcal{G}((N_m)_{1/n}) = n^2\mathcal{G}(N_m) = (m/n)^{-2}\mathcal{G}(N), \qquad (16.4.44)$$

using equation (16.4.43) for both integers.

Let $0 < t < 1$ and choose $0 < s < 1$ such that st is a rational number. By equation (16.4.44),

$$\mathcal{G}(N_{st}) = (st)^{-1}\mathcal{G}(N), \qquad (16.4.45)$$

so that

$$\mathcal{G}(N) \leq t^2\mathcal{G}(N_t) \leq s^2 t^2\mathcal{G}((N_t)_s) = (st)^2\mathcal{G}(N_{st}) = \mathcal{G}(N) \qquad (16.4.46)$$

by Lemma 16.4.4. Thus $\mathcal{G}(N) = t^2\mathcal{G}(N_t)$.

If $t > 1$, then $0 < t^{-1} < 1$. Applying the equation (16.4.46) to N_t gives $\mathcal{G}(N) = \mathcal{G}((N_t)_{1/t}) = t^{-2}\mathcal{G}(N_t)$ as required. $\qquad \square$

The following corollary is a direct consequence of Theorem 16.4.5 and Lemma 16.3.3, since one involves a factor $\tau(p_j)^{-2}$ and the other $\tau(p_j)^2$.

Corollary 16.4.6. *Let N be a separable II$_1$ factor, let $k \in \mathbb{N}$ and let $\{p_j\} \in \mathcal{P}_k(N)$. If $M = \sum_1^k p_j N p_j$ with the trace τ inherited from N, then*

$$\mathcal{G}(M) = k\mathcal{G}(N).$$

16.5 Interpolated free group factors and $\mathcal{G}$

In this section we investigate the relationships that hold for $\mathcal{G}(\cdot)$ on the free group factors and their generalisation to the interpolated free group factors. We begin by considering amalgamated free products.

Theorem 16.5.1. *Let M and N be separable II$_1$ factors, and let B be a diffuse von Neumann algebra with $B \subseteq M$ and $B \subseteq N$. Then*

$$\mathcal{G}(M \star_B N) \le \mathcal{G}(M) + \mathcal{G}(N).$$

Proof. Assume that M and N are finitely generated and let $\varepsilon > 0$. By Corollary 16.2.6, there is a $k \in \mathbb{N}$, finite generating sets X of M and Y of N, and $P \in \mathcal{P}_k^{eq}(M)$ and $Q \in \mathcal{P}_k^{eq}(N)$ such that

$$\mathcal{G}(M) \le \mathcal{I}(X; P) + \varepsilon \quad \text{and} \quad \mathcal{G}(N) \le \mathcal{I}(Y; Q) + \varepsilon. \qquad (16.5.1)$$

There is a set $E \in \mathcal{P}_k^{eq}(B)$, since B is diffuse. Since M and N are factors, there are unitaries $u \in M$ and $v \in N$ such that $uPu^* = E$ and $vQv^* = E$. The invariance of $\mathcal{G}$ under automorphisms implies that

$$\mathcal{G}(M) \le \mathcal{I}(uXu^*; E) + \varepsilon \quad \text{and} \quad \mathcal{G}(N) \le \mathcal{I}(vYv^*; E) + \varepsilon. \qquad (16.5.2)$$

The finite set $W = uXu^* \cup vYv^*$ generates $M \star_B N$ and

$$\begin{aligned}
\mathcal{G}(M \star_B N) \le I(W; E) &\le \mathcal{I}(uXu^*; E) + \mathcal{I}(vYv^*; E) \\
&\le \mathcal{G}(M) + \mathcal{G}(N) + 2\varepsilon. \qquad (16.5.3)
\end{aligned}$$

This proves the result.   □

If the free product is not amalgamated over a diffuse algebra, then the weaker formula $\mathcal{G}(M \star N) \le \mathcal{G}(M) + n$ holds, where n is the minimal number of generators in N. If the strong inequality $\mathcal{G}(M \star N) \le \mathcal{G}(M) + \mathcal{G}(N)$ did hold, then $\mathcal{G}(L(\mathbb{F}_2)) = 0$ and all the free group factors would be singly generated by subsequent results.

Dykema [58] and Rădulescu [157] defined and studied the interpolated free group factors $L\mathbb{F}_t$ for $t \ge 1$, which satisfy $(L\mathbb{F}_s) \star (L\mathbb{F}_t) = L\mathbb{F}_{s+t}$ for all $s, t \ge 1$ and $L\mathbb{F}_n$ is the group von Neumann algebra $L(\mathbb{F}_n)$ on the free group $\mathbb{F}_n$ for all $n \in \mathbb{N}$. After Voiculescu's preliminary research in [199], Dykema [58] and Rădulescu [157] proved the following scaling formula for this interpolated family of II$_1$ factors using free probability.

Theorem 16.5.2. *If* $L\mathbb{F}_t$ $(t \geq 1)$ *is the interpolated free group factor, then*

$$(L\mathbb{F}_t)_b = L\mathbb{F}_{1+b^{-2}(t-1)}$$

for all $t \geq 1$ *and all* $b > 0$.

Combining this scaling with the scaling of Theorem 16.4.5 leads to the following theorem because both involve a quadratic factor.

Theorem 16.5.3. *There is an* α *with* $0 \leq \alpha \leq 1/2$ *such that*

$$\mathcal{G}(L\mathbb{F}_{t+1}) = \alpha t \quad for\ all \quad t \geq 0. \tag{16.5.4}$$

Proof. Let

$$\alpha(t) = \mathcal{G}(L\mathbb{F}_{t+1}) \quad for\ all \quad t \geq 0. \tag{16.5.5}$$

If $b > 0$ and $r \geq 1$, then Theorem 16.4.5 applied to $N = L\mathbb{F}_r$ gives

$$b^{-2}\alpha(r-1) = b^{-2}\mathcal{G}(L\mathbb{F}_r) = \mathcal{G}((L\mathbb{F}_r)_b)$$
$$= \mathcal{G}(L\mathbb{F}_{1+b^{-2}(r-1)}) = \alpha(b^{-2}(r-1)). \tag{16.5.6}$$

Letting $r - 1 = t > 0$ and $b^{-2} = s > 0$ gives $s\alpha(t) = \alpha(st)$ for all $s, t > 0$. Hence $\alpha(t) = t\alpha$ for all $t > 0$, where $\alpha \geq 0$ is a constant. This is true for $t = 0$ also, because $\alpha(0) = \mathcal{G}(L(\mathbb{F}_1)) = \mathcal{G}(L^\infty[0,1]) = 0$ by Lemma 16.3.1.

If $n \in \mathbb{N}$, then $L(\mathbb{F}_{2n})$ is generated by the $2n$ generators $\{g_j\}$ $(1 \leq j \leq 2n)$ of the free group $\mathbb{F}_{2n}$. Since $\{g_j\}''$ is a masa in $L(\mathbb{F}_{2n})$ it is generated by a self-adjoint element h_j so $\{h_j\}'' = \{g_j\}''$ for each j. Thus $L(\mathbb{F}_{2n})$ is generated by the n elements $x_j = h_{2j-1} + ih_{2j}$ $(1 \leq j \leq n)$. Hence equation (16.2.23) implies that

$$(2n - 1)\alpha = \mathcal{G}(L(\mathbb{F}_{2n})) \leq n \quad for\ all \quad n \in \mathbb{N}. \tag{16.5.7}$$

From this it follows that $0 \leq \alpha \leq 1/2$. $\square$

Dykema and White [64] have proved that the DT-operator T, which is a single generator of $L(\mathbb{F}_2)$, (studied by Dykema and Haagerup [61]), has $\mathcal{I}(T) = 1/2$. This yields another proof of the above inequality.

As a corollary of the above results we obtain the dichotomy of Rădulescu [157] by noting that if $\alpha = 0$, then Theorem 16.6.2 implies that all the free group factors are singly generated.

Theorem 16.5.4. *Either the interpolated free group factors* $L\mathbb{F}_t$ $(t \geq 1)$ *are all singly generated or, if this is not the case, they are pairwise non-isomorphic for different* t.

Remark 16.5.5. Using the estimate $\alpha \leq 1/2$ and Theorems 16.5.3, 16.6.2 yields that
(1)   *if* $1 \leq t \leq 2$, *then* $L\mathbb{F}_t$ *is singly generated, and*
(2)   *if* $2 < t \leq 3$, *then* $L\mathbb{F}_t$ *is generated by three hermitian elements.* $\square$

If the interpolated free group factor LF_t for any $1 < t < \infty$ has a Cartan subalgebra or has non-trivial fundamental group, then $\mathcal{G}(LF_t) = 0$ so $\alpha = 0$. Hence $\mathcal{G}(LF_s) = 0$ for all $1 \leq s$, and all interpolated free products factors are singly generated by Theorem 16.6.2. This does not provide a proof of Voiculescu's Theorem [199] that $L(\mathbb{F}_n)$ does not have Cartan subalgebras unless one can prove that some $L(\mathbb{F}_n)$ is not singly generated.

16.6   Single generation

The following is the basic lemma on these invariants and generation by two hermitians or a single element. Other versions of this technique will be given later.

Lemma 16.6.1. *Let N be a finite diffuse von Neumann algebra with faithful normal trace τ and let B is a $k \times k$ matrix subalgebra of N with matrix units $\{e_{ij} : 1 \leq i, j \leq k\}$ and $\sum_i e_{ii} = 1$. Let $v = \sum_{i=1}^{k-1}(e_{i,i+1} + e_{i+1,i})$, let X be a finite subset of N with $\mathcal{I}(X; \{e_{ii}\}) = C$. If $C < 1/2 - 3/2k$, then there exists a self-adjoint element $y \in (X \cup B)''$ such that*

$$(\{y\} \cup B)'' = \{y, e_{11}, v\}'' = (X \cup B)''$$

and

$$S(y; \{e_{ii}\}) \leq \sum_1^{k-1} e_{ii}.$$

Proof. Let $X = \{x_1, \ldots, x_n\}$ and let

$$T = \{(i, j, \ell) : e_{ii}x_\ell e_{jj} \neq 0; 1 \leq i, j \leq k, 1 \leq \ell \leq n\}. \tag{16.6.1}$$

By definition of $\mathcal{I}(X; \{e_{ii}\})$,

$$|T| = k^2 C < (k^2 - 3k)/2. \tag{16.6.2}$$

The transpose map T is defined on the $(k-1) \times (k-1)$ lattice $\{(i, j) : 1 \leq i, j \leq k-1\}$ by $(i, j)^T = (j, i)$. Let

$$\mathcal{L} = \{(i, j) : i > j, 1 \leq i, j \leq k-1\} \tag{16.6.3}$$

be the strictly upper triangular lattice in the rectangular lattice. Since

$$|\mathcal{L}| = (k-1)(k-2)/2 > (k^2 - 3k)/2 > |T|, \tag{16.6.4}$$

there is an injective map

$$(i, j, \ell) \mapsto \big(s(i, j, \ell), t(i, j, \ell)\big)$$

from T into $\mathcal{L}$.

Let

$$y = \sum_{(i,j,\ell) \in \mathcal{T}} \left(e_{s(i,j,\ell),i} \, x_\ell \, e_{j,t(i,j,\ell)} + \left(e_{s(i,j,\ell),i} \, x_\ell \, e_{j,t(i,j,\ell)} \right)^* \right) \qquad (16.6.5)$$

and let $f = \sum_1^{k-1} e_{ii}$. By definition of y, f and $\mathcal{L}$,

$$fy = y = yf. \qquad (16.6.6)$$

By definition of y, it follows that $y \in C^*(X \cup B) \subseteq (X \cup B)''$. Let $1 \le \ell \le n$. Choose $1 \le i, j \le k$ such that $e_{ii} x_\ell e_{jj} \ne 0$ so that $(i, j, \ell) \in \mathcal{T}$. The definitions of $(s(\cdot), t(\cdot))$ and $\mathcal{L}$ give

$$e_{i,\,s(i,j,\ell)} y e_{t(i,j,\ell),\,j} = e_{ii} x_\ell e_{jj}, \qquad (16.6.7)$$

since $(s(i,j,\ell), t(i,j,\ell)) \in \mathcal{L}$ implies that $(s(i,j,\ell), t(i,j,\ell)) \notin \mathcal{L}^T$ and the map is one-to-one. Summing this equation (16.6.7) over all i, j with $(i, j, \ell) \in \mathcal{T}$ for fixed ℓ shows that $x_\ell \in \mathrm{Alg}\{e_{ij}, y\}$. Hence $(\{y\} \cup B)'' = \{y, e_{11}, v\}'' = (X \cup B)''$ as required, since $\{e_{11}, v\}'' = B$ by Lemma 16.1.1. $\qquad \square$

Note that we have proved that $C^*(\{y\} \cup B)) = C^*(X \cup B)$ in the above lemma.

Theorem 16.6.2. *Let N be a finitely generated II_1 factor. If $\mathcal{G}(N) < 1/2$, then N is singly generated.*

Proof. By Lemma 16.2.4 the infimum defining $\mathcal{G}(N)$ is attained in $\mathcal{P}^{eq}$ so there is a finite generating set X of N, an integer $k \in \mathbb{N}$ and $\{e_{ii}\} \in \mathcal{P}_k^{eq}(N)$ such that

$$\mathcal{G}(N) \le \mathcal{I}(X; \{e_{ii}\}) < 1/2 - 3/k < 1/2. \qquad (16.6.8)$$

Let $\{e_{ij} : 1 \le i, j \le k\}$ be a set of matrix units in N associated with $\{e_{ii}\}$ and let B be the $k \times k$ matrix subalgebra of N they generate. By Lemma 16.6.1 there is a self-adjoint element $y \in N = X'' = (B \cup X)''$ such that $N = (\{y\} \cup B)''$ and $\mathcal{S}(y; \{e_{ii}\}) \le \sum_1^{k-1} e_{ii}$. Thus e_{kk} is orthogonal to $\mathcal{S}(y; \{e_{ii}\})$ and so $e_{kk} y = y e_{kk} = 0$. By scaling, assume that $\|y\| = 1$.

Let

$$x_1 = 2e_{kk} + y \qquad (16.6.9)$$

and

$$x_2 = \sum_1^{k-1} (e_{i,\,i+1} + e_{i+1,\,i}). \qquad (16.6.10)$$

Then $e_{kk}, y \in \{x_1\}''$. Hence $\{x_1, x_2\}'' = (\{y\} \cup B)'' = N$ by Lemma 16.6.1 as required. $\qquad \square$

An easy calculation yields the estimate

$$\mathcal{I}(\{x_1, x_2\}; \{e_{ii}\}) < 1/2 + 2/k - 5/k^2. \tag{16.6.11}$$

Corollary 16.6.3. *Let N be a finitely generated II$_1$ factor.*

(1) *If the fundamental group $\mathcal{F}(N) \neq \{1\}$ is not trivial, then N is singly generated.*

(2) *If $k \in \mathbb{N}$ and if $\mathcal{G}(N) < k^2/2$, then $\mathcal{G}(\mathbb{M}_k \otimes N) \leq k^{-2}\mathcal{G}(N) < 1/2$ and $\mathbb{M}_k \otimes N$ is singly generated.*

(3) *If $k \in \mathbb{N}$ and N is generated by $k^2 - 1$ hermitians, then $\mathbb{M}_k \otimes N$ is singly generated.*

Proof. (1)   Let $t \in \mathcal{F}(N)$ with $t > 1$. By Theorem 16.4.5,

$$\mathcal{G}(N) = \mathcal{G}(N_t) = t^{-2}\mathcal{G}(N) \tag{16.6.12}$$

so that $\mathcal{G}(N) = 0$. Hence N is singly generated by Theorem 16.6.2.
(2)   Apply Theorem 16.4.5 to deduce that $\mathcal{G}(\mathbb{M}_k \otimes N) = k^{-2}\mathcal{G}(N) < 1/2$ from which the singly generated property follows by Theorem 16.6.2.
(3)   If k is odd with $k^2 - 1 = 2m$ even, then N is generated by the m elements $y_k = x_{2k-1} + ix_{2k}$, $1 \leq k \leq m$, so that $\mathcal{G}(N) \leq m = (k^2 - 1)/2 < k^2/2$.
    If k is even with $k^2 - 1 = 2m + 1$, then N is generated by the elements $y_k = x_{2k-1} + ix_{2k}$, $1 \leq k \leq m$, and $y_{m+1} = x_{2m+1}$. Let A be a masa in N containing x_{2m+1} and let $P \in \mathcal{P}_2^{eq}(A)$. Then $\mathcal{I}(y_j; P) \leq 1$ for $1 \leq j \leq m$ and $\mathcal{I}(y_{m+1}; P) \leq 1/2$ by direct calculation since y_{m+1} commutes with the projections of P. Thus

$$\begin{aligned} \mathcal{G}(N) &\leq \mathcal{I}(\{y_1, \ldots, y_{m+1}\}; P) \\ &\leq m + 1/2 = (k^2 - 1)/2 < k^2/2. \end{aligned} \tag{16.6.13}$$

Thus $\mathbb{M}_k \otimes N$ is singly generated in both cases.   □

Corollary 16.6.4. *If N is a finitely generated II$_1$ factor with $N \overline{\otimes} R \cong N$, then N is singly generated.*

Proof. Since $R \otimes \mathbb{M}_n \cong R$, we see that $\mathcal{F}(N) \neq \{1\}$, and the result follows from Corollary 16.6.3.   □

Recall that the ceiling function $\lceil \cdot \rceil \colon \mathbb{R} \to \mathbb{Z}$ is defined by

$$\lceil x \rceil = \min\{n \in \mathbb{Z} \colon x \leq n\}$$

so that $\lceil x \rceil - 1 < x \leq \lceil x \rceil$ for all $x \in \mathbb{R}$.
    The next lemma and theorem handle the case when $\mathcal{G}(N) < 1$ by an elementary variant of the same technique.

Lemma 16.6.5. *Let N be a finite diffuse von Neumann algebra with faithful normal trace τ and let B be a $k \times k$ matrix subalgebra of N with matrix units $\{e_{ij} : 1 \le i, j \le k\}$ and $\sum_{i=1}^{k} e_{ii} = 1$. Let $v = \sum_{i=1}^{k-1}(e_{i,i+1} + e_{i+1,i})$, let X be a finite subset of N with $\mathcal{I}(X; \{e_{ii}\}) = c^2$ and let $f = \sum_{i=1}^{\lceil ck \rceil} e_{ii}$. If $c \le 1 - 1/k$, then there exists an element $y \in (X \cup B)''$ such that*

$$(\{y\} \cup B)'' = \{y, e_{11}, v\}'' = (X \cup B)''$$

and

$$\mathcal{S}(y; \{e_{ii}\}) \le f.$$

Proof. Let $X = \{x_1, \ldots, x_n\}$ and let

$$\mathcal{T} = \{(i, j, \ell) : e_{ii} x_\ell e_{jj} \ne 0; \ 1 \le i, j \le k, 1 \le \ell \le n\}. \tag{16.6.14}$$

By definition of $\mathcal{I}(X; \{e_{ii}\})$ and c,

$$|\mathcal{T}| = k^2 \mathcal{I}(X; \{e_{ii}\}) = c^2 k^2 \le (k-1)^2. \tag{16.6.15}$$

Thus there is an injective map ϕ given by

$$(i, j, \ell) \mapsto \big(s(i, j, \ell), t(i, j, \ell)\big)$$

from $\mathcal{T}$ into

$$\mathcal{W} = \{(s, t) : 1 \le s, t \le k - 1\} \subseteq \{(s, t) : 1 \le s, t \le k\}.$$

Let

$$y = \sum_{(i,j,\ell) \in \mathcal{T}} e_{s(i,j,\ell), i} \, x_\ell \, e_{j, t(i,j,\ell)} \tag{16.6.16}$$

By definition of y, f and $\phi(\mathcal{T})$,

$$fy = y = yf. \tag{16.6.17}$$

By definition of y, it follows that $y \in C^*(X \cup B) \subseteq (X \cup B)''$. Let $1 \le \ell \le n$. Choose $1 \le i, j \le k$ such that $e_{ii} x_\ell e_{jj} \ne 0$ so that $(i, j, \ell) \in \mathcal{T}$. Since the map ϕ is injective, its definition implies that

$$e_{i, s(i,j,\ell)} y e_{t(i,j,\ell), j} = e_{ii} x_\ell e_{jj} \tag{16.6.18}$$

for all $(i, j, \ell) \in \mathcal{T}$. Summing this equation (16.6.18) over all i, j with $(i, j, \ell) \in \mathcal{T}$ for fixed ℓ shows that $x_\ell \in \mathrm{Alg}\{e_{ij}, y\}$. Hence $(\{y\} \cup B)'' = (X \cup B)''$ as required. The equation (16.6.18) implies that

$$\mathcal{S}(y; \{e_{ii}\}) \le f \tag{16.6.19}$$

from the definition of $\mathcal{S}$ in equation (16.2.6). $\qquad\square$

In Lemma 16.6.5 above, $\tau(\mathcal{S}(y; \{e_{ii}\}))$ and $\mathcal{I}(\{y, e_{11}, v\}; \{e_{ii}\})$ can be estimated as before.

Theorem 16.6.6. *Let N be a finitely generated II$_1$ factor. If $\mathcal{G}(N) < 1$, then N is generated by three hermitian elements.*

Proof. By Lemma 16.2.4 the infimum defining $\mathcal{G}(N)$ is attained in $\mathcal{P}^{eq}$ so there is a finite generating set X of N, an integer $k \in \mathbb{N}$ and $\{e_{ii}\} \in \mathcal{P}_k^{eq}(N)$ such that

$$\mathcal{G}(N) \leq \mathcal{I}(X; \{e_{ii}\}) \leq (1 - 1/k)^2 < 1. \qquad (16.6.20)$$

Let $\{e_{ij} : 1 \leq i, j \leq k\}$ be a set of matrix units in N associated with $\{e_{ii}\}$ and let B be the $k \times k$ matrix subalgebra of N they generate. By Lemma 16.6.1 there is an element $y \in N = X'' = (B \cup X)''$ such that $N = (\{y\} \cup B)''$ and $\mathcal{S}(y; \{e_{ii}\}) \leq f \leq \sum_{i=1}^{k-1} e_{ii}$, where f is the projection of Lemma 16.6.5 which satisfies inequality (16.6.19). Thus e_{kk} is orthogonal to $\mathcal{S}(y; \{e_{ii}\})$ and so $e_{kk}y = ye_{kk} = 0$. By scaling assume that $\|y\| = 1$.

Let

$$x_1 = 2e_{kk} + y \quad \text{and} \quad x_2 = \sum_{i=1}^{k-1}(e_{i,\,i+1} + e_{i+1,\,i}). \qquad (16.6.21)$$

Then $e_{kk}, y \in \{x_1\}''$, since $x_1^n = 2^n e_{kk} + y^n$ for all $n \in \mathbb{N}$. Hence

$$\{x_1, x_2\}'' = (\{y\} \cup B)'' = N \qquad (16.6.22)$$

by Lemma 16.6.5 as required. $\qquad\qquad\qquad\qquad\qquad\qquad\qquad\qquad\qquad\square$

The following generalisation of this is proved in [63].

Let N be a finitely generated II$_1$ factor. If $n \in \mathbb{N}$ and $\mathcal{G}(N) < n/2$, then N is generated by $n + 1$ hermitian elements.

16.7   Main technical lemmas

To this point, all results with the conclusion that N is singly generated have required the hypothesis that N is finitely generated. The main result of this section, Theorem 16.7.5, avoids this hypothesis by assuming that there is a link between suitable generating sets.

The main technical results of this chapter follow after a little elementary lemma on projections, which is just a version of the observation that the matrix

$$1/2 \begin{pmatrix} 1 + \cos\theta & \sin\theta \\ \sin\theta & 1 - \cos\theta \end{pmatrix}$$

is the projection from $\mathbb{R}^2$ onto the one dimensional space at angle $\theta/2$ to the x-axis. Below

$$p \sim \begin{pmatrix} 1 & 0 \\ 0 & 0 \end{pmatrix} \quad \text{and} \quad y \sim \begin{pmatrix} 0 & \sin\theta \\ \sin\theta & 0 \end{pmatrix}.$$

Lemma 16.7.1. *Let B be a unital C*-algebra. If $y = y^* \in B$ with $\|y\| \leq 1$ and p is a projection in B with*

$$py = y(1 - p), \tag{16.7.1}$$

then

$$q = 1/2\Big(p(1 + (1 - y^2)^{1/2})p + y + (1 - p)(1 - (1 - y^2)^{1/2}(1 - p)\Big) \tag{16.7.2}$$

is a projection in B.

Proof. Equation (16.7.1), and its conjugate, yield

$$py^2 = y(1 - p)y = y^2 p. \tag{16.7.3}$$

Let $c = (1 - y^2)^{1/2}$. Equation (16.7.3) shows that p commutes with c^2. Then $cp = pc$, since $c \in C^*(c^2)$ by the functional calculus, so that

$$
\begin{aligned}
q &= 1/2\big(p(1 + c) + y + (1 - p)(1 - c)\big) \\
&= 1/2(1 - c + 2pc + y). \tag{16.7.4}
\end{aligned}
$$

Using $c^2 + y^2 = 1$ and equation (16.7.1) to collect together the y terms gives

$$
\begin{aligned}
4q^2 &= 2 - 2c + 4pc^2 + 2c(py + yp) + 2y(1 - c) + 4pc(1 - c) \\
&= 2 - 2c + 4pc + 2cy + 2y(1 - c) \\
&= 2 - 2c + 4pc + 2y = 4q \tag{16.7.5}
\end{aligned}
$$

as required. $\qquad\square$

The following is similar to Lemma 16.6.5 on single generation but with the element y occurring there replaced by a projection q, which plays a role in subsequent calculations.

Lemma 16.7.2. *Let N be a finite diffuse von Neumann algebra with faithful normal trace τ and let B be a $k \times k$ matrix subalgebra of N with matrix units $\{e_{ij} : 1 \leq i, j \leq k\}$ and $\sum_{i=1}^{k} e_{ii} = 1$. Let $v = \sum_{i=1}^{k-1}(e_{i,i+1} + e_{i+1,i})$, let X be a finite subset of N with $\mathcal{I}(X; \{e_{ii}\}) = c^2$ and let $f = \sum_{i=1}^{2\lceil ck \rceil} e_{ii}$. If $c \leq 1/2 - 1/k$, then there exists a projection $q \in (X \cup B)''$ such that*

$$(\{q\} \cup B)'' = \{q, e_{11}, v\}'' = (X \cup B)'',$$

$$q \leq \mathcal{S}(q; \{e_{ii}\}) \leq f, \quad \tau(f) < 2(c + 1/k)$$

and

$$\mathcal{I}(\{q, e_{11}, v\}; \{e_{ii}\}) \leq 2c^2 + 2(c + 1)k^{-1} + k^{-2}.$$

Proof. Let $X = \{x_1, \ldots, x_n\}$ and let

$$\mathcal{T} = \{(i, j, \ell) : e_{ii} x_\ell e_{jj} \neq 0; \, 1 \leq i, j \leq k, 1 \leq \ell \leq n\}. \tag{16.7.6}$$

By definition of $\mathcal{I}(X; \{e_{ii}\})$,

$$|T| = k^2\mathcal{I}(X; \{e_{ii}\}) = c^2k^2. \tag{16.7.7}$$

Let

$$\mathcal{W} = \{(s,t): 1 \le s \le \lceil ck \rceil, \ \lceil ck \rceil + 1 \le t \le 2\lceil ck \rceil\}.$$

Since $k/2 - 1$ or $k/2 - 1/2$ is an integer and $ck \le k/2 - 1 \le k/2 - 1/2$ by hypothesis, $2\lceil ck \rceil + 1 \le k$ so that

$$\mathcal{W} \subseteq \{(s,t): 1 \le s, t \le k\}.$$

Because $|T| = c^2k^2 \le \lceil ck \rceil^2 = |\mathcal{W}|$, there is an injective map $\phi\colon\ T \to \mathcal{W}$ given by

$$(i, j, \ell) \mapsto \big(s(i, j, \ell), t(i, j, \ell)\big).$$

Let

$$y = \sum_{(i,j,\ell)\in T} \Big(e_{s(i,j,\ell),\, i}\, x_\ell\, e_{j,\, t(i,j,\ell)} + \big(e_{s(i,j,\ell),\, i}\, x_\ell\, e_{j,\, t(i,j,\ell)} \big)^* \Big) \tag{16.7.8}$$

$$q_1 = \sum_1^{\lceil ck \rceil} e_{ss} \quad\text{and}\quad q_2 = \sum_{\lceil ck \rceil + 1}^{2\lceil ck \rceil} e_{tt}. \tag{16.7.9}$$

By definition of y, q_1, q_2, f and $\phi(T)$, we obtain

$$q_1q_2 = 0, \quad (q_1 + q_2)y = fy = y = yf = y(q_1 + q_2) \tag{16.7.10}$$

and

$$q_1y = \sum_{(i,j,\ell)\in T} e_{s(i,j,\ell),\, i}\, x_\ell\, e_{j,\, t(i,j,\ell)} = yq_2. \tag{16.7.11}$$

Equation (16.7.11) and its conjugate imply that y^2 commutes with q_1 and q_2; more holds. Since the ranges of s and t are disjoint and the map ϕ is injective, the definition of y in (16.7.8) implies that

$$y^2 = \sum_{(i,j,\ell)\in T} \Big(e_{s(i,j,\ell),\, i}\, x_\ell\, e_{jj}\, x_\ell^*\, e_{i,\, s(i,j,\ell)} + e_{t(i,j,\ell),\, j}\, x_\ell^*\, e_{ii}\, x_\ell\, e_{j,\, t(i,j,\ell)} \Big)$$

$$\tag{16.7.12}$$

This equation (16.7.12) gives

$$y^2 \in \sum_i e_{ii} C^*(X \cup B) e_{ii}, \tag{16.7.13}$$

which will be used subsequently.

By scaling all the x_ℓ $(1 \le \ell \le n)$, if necessary, we assume that $\|y\| \le 1$ from now on. Let

$$q = 1/2\Big(q_1\big(1 + (1 - y^2)^{1/2}\big)q_1 + y + q_2\big(1 - (1 - y^2)^{1/2}\big)q_2 \Big). \tag{16.7.14}$$

Since q_1, q_2 and y lie in fNf from (16.7.10), equation (16.7.14) can be rewritten as

$$q = 1/2(q_1(f + (f - y^2)^{1/2})q_1 + y + (f - q_1)(f - (f - y^2)^{1/2})(f - q_1)). \quad (16.7.15)$$

If we then apply Lemma 16.7.1 in the algebra fNf, we see that q is a projection.

By definition of q_1, q_2 and y, it follows that $q \in C^*(X \cup B) \subseteq (X \cup B)''$. Let $1 \le \ell \le n$. Choose $1 \le i, j \le k$ such that $e_{ii}x_\ell e_{jj} \ne 0$ so that $(i, j, \ell) \in T$. Using $q_1 q_2 = 0$ eliminates the first and final terms in the definition of q, (16.7.14), leaving only the y term in

$$e_{i, s(i,j,\ell)}q e_{t(i,j,\ell), j} = 1/2 e_{i, s(i,j,\ell)}y e_{t(i,j,\ell), j} = 1/2 e_{ii}x_\ell e_{jj} \quad (16.7.16)$$

since the ranges of s and t are disjoint in $\{1, \ldots, k\}$ and the map ϕ is an injection. Summing this equation (16.7.16) over all i, j with $(i, j, \ell) \in T$ for fixed ℓ shows that $x_\ell \in \text{Alg}\{e_{ij}, q\}$. Hence

$$(\{q\} \cup B)'' = \{q, e_{11}, v\}'' = (X \cup B)'' \quad (16.7.17)$$

as required, since $\{e_{11}, v\}'' = B$ by Lemma 16.3.2.

The equation $(q_1 + q_2)q = q$ implies that

$$\mathcal{S}(q; \{e_{ii}\}) \le q_1 + q_2 \le f \quad (16.7.18)$$

from the definition of $\mathcal{S}$ in equation (16.2.6).

To estimate $\mathcal{I}(\{q, e_{11}, v\}; \{e_{ii}\})$ it is sufficient to estimate $\mathcal{I}(q, \{e_{ii}\})$ by counting various cases, because the other two terms are provided by Lemma 16.3.2. If (s, t) does not satisfy $1 \le s, t \le 2\lceil ck \rceil$, then $e_{ss}q e_{tt} = 0$ by definition of q_1, q_2 and y. Within this region, the estimate depends on which square generated by the two inequalities $1 \le \cdot \le \lceil ck \rceil$ and $\lceil ck \rceil + 1 \le \cdot \le 2\lceil ck \rceil$ contains the pair (s, t).

If $1 \le s, t \le \lceil ck \rceil$, then $e_{ss}q e_{tt} = 1/2 e_{ss}e_{tt}$ from the first term in the definition of q so is non-zero if, and only if, $s = t$. This contributes $\lceil ck \rceil$.

If $\lceil ck \rceil + 1 \le s, t \le 2\lceil ck \rceil$, then

$$e_{ss}q e_{tt} = e_{ss}(1 - y^2)^{1/2}e_{tt} = e_{ss}e_{tt}(1 - y^2)^{1/2} \quad (16.7.19)$$

by equation (16.7.14), which is zero if $s \ne t$, so contributes at most $\lceil ck \rceil$. Note that we have used $e_{ss}y e_{tt} = 0$ for s and t in the stated range, which follows from the definition of y in (16.7.8).

In the final two cases

$$e_{ss}q e_{tt} = 1/2 e_{ss}y e_{tt}$$
$$= 1/2 \sum_{(i,j,\ell) \in T} \left(e_{ss}e_{s(i,j,\ell), i} \, x_\ell \, e_{j, t(i,j,\ell)}e_{tt} + e_{ss}e_{t(i,j,\ell), i} \, x_\ell^* \, e_{s, t(i,j,\ell)}e_{tt} \right)$$
$$\ne 0 \quad (16.7.20)$$

if, and only if, either (s, t) or (t, s) is in the image of the injective function $(s(\cdot), t(\cdot))$ from $\mathcal{T}$ into $\mathcal{W}$. Since $\mathcal{W}$ is disjoint from its transpose $\{(t, s) \colon (s, t) \in \mathcal{W}\}$, only one of a pair (s, t) and (t, s) can be in $\mathcal{W}$ so this contributes $2|\mathcal{T}|$.

Hence

$$|\{(s, t) \colon e_{ss} q e_{tt} \neq 0\}| \leq 2|\mathcal{T}| + 2(\lceil ck \rceil) < 2(c^2 k^2 + ck + 1) \qquad (16.7.21)$$

so that

$$\mathcal{I}(q; \{e_{ii}\}) \leq 2(c^2 + ck^{-1} + k^{-2}) \qquad (16.7.22)$$

and

$$\mathcal{I}(\{q, e_{11}, v\}; \{e_{ii}\}) \leq 2c^2 + 2(c + 1)k^{-1} + k^{-2} \qquad (16.7.23)$$

by Lemma 16.3.2. $\qquad\square$

Note that we have proved that $C^*(\{q\} \cup B)) = C^*(X \cup B)$ in the above lemma. The above Lemma 16.7.2 can be used to show that if $\mathcal{G}(N) < 1/4$, then N is generated by a hermitian of the form $v = \sum_{i=1}^{k-1}(e_{i,i+1} + e_{i+1,i})$ and a self-adjoint of the form $q - e_{kk}$ or $2e_{kk} + q$, where $q e_{kk} = 0$.

The next lemma is an important step towards Theorem 16.7.5.

Lemma 16.7.3. *Let N be a separable II$_1$ factor and let $\varepsilon > 0$. If M is a diffuse finitely generated von Neumann subalgebra of N, then there exists a finitely generated irreducible subfactor M_ε of N with $M \subseteq M_\varepsilon$ and*

$$\mathcal{G}(M_\varepsilon) \leq \mathcal{G}(M) + \varepsilon.$$

Proof. By Lemma 16.2.4 the infimum defining $\mathcal{G}(M)$ is attained on $\mathcal{P}^{eq}$, since M is diffuse. Thus there is a finite set X of generators of M, an integer m and a set of projections $\{e_{ii} \colon 1 \leq i \leq m\} \in \mathcal{P}_m^{eq}(M)$ such that

$$\mathcal{I}(X; \{e_{ii}\}) \leq c + \varepsilon/2 \quad \text{and} \quad m \geq 8/\varepsilon. \qquad (16.7.24)$$

By Theorem 13.2.3, there is an irreducible hyperfinite subfactor R of N with $\{e_{ii}\} \subseteq R$. Let $\{e_{ij} \colon 1 \leq i, j \leq m\}$ be a set of matrix units in R with diagonal $\{e_{ii}\}$ and let $M_m = \{e_{ij} \colon 1 \leq i, j \leq m\}''$. If $R_1 = M_m' \cap R$, then $R = M_m \otimes R_1$. By the remarks preceding Lemma 16.3.6, R_1 is generated by two self-adjoint elements y_1, y_2 and these commute with M_m. By Lemma 16.3.2, M_m is generated as an algebra by e_{11} and $x = \sum_{i=1}^{m-1}(e_{i,i+1} + e_{i+1,i})$ with $\mathcal{I}(\{e_{11}, x\}; \{e_{ii}\}) < 2/m$. Let M_ε be the von Neumann subalgebra of N generated by M and R. This an irreducible subfactor of N, since R has this property, and M_ε is generated by $X \cup \{y_1, y_2, e_{11}, x\}$. Since y_1 and y_2 commute with $\{e_{ii}\}$,

$$\mathcal{I}(y_1; \{e_{ii}\}) = \mathcal{I}(y_2; \{e_{ii}\}) = m^{-1}. \qquad (16.7.25)$$

Hence

$$\begin{aligned}\mathcal{G}(M_\varepsilon) &\leq \mathcal{I}(X \cup \{y_1, y_2, e_{11}, x\}; \{e_{ii}\}) \\ &\leq c + \varepsilon/2 + 4/m < c + \varepsilon\end{aligned} \qquad (16.7.26)$$

as required. $\qquad\square$

The proof of the previous Lemma 16.7.3 showed that if X is a finite subset of N and $\{e_i : 1 \leq i \leq m\} \in \mathcal{P}_m^{eq}(N)$, then there is an irreducible hyperfinite subfactor R of N with hermitian generators y_1, y_2 such that $\{e_i : 1 \leq i \leq m\} \subseteq R$ and

$$\mathcal{G}((R \cup X)'') \leq \mathcal{I}(X \cup \{y_1, y_2\}; \{e_i\}) \leq \mathcal{I}(X; \{e_i\}) + 4m^{-1}. \qquad (16.7.27)$$

A set of matrix units $\{e_{ij} : 1 \leq i, j \leq n\}$ in a von Neumann algebra is assumed to have $\sum_i e_{ii} = 1$. When $\sum_i e_{ii} = p$ is just a projection in the algebra rather than the identity we say that it is a *subsystem of matrix units* and note that $\tau(e_{ii}) = n^{-1}\tau(p)$ for all i. The subsystems of matrix units $\{e_{ij}^{(k)}\}$ for $k \in \mathbb{N}$ constructed in the following lemma correspond to $e_{22}^{(1)} \otimes \ldots \otimes e_{22}^{(k-1)} \otimes e_{ij}^{(k)}$ in the proof of Lemma 16.3.6.

Lemma 16.7.4. *Let N be a separable II_1 factor and let $(N_{k,n})$ be a sequence of finitely generated von Neumann subalgebras of N such that*

(i) $N_{k,n} \cap N_{k+1,m}$ *is diffuse for all $k, m, n \in \mathbb{N}$ and*

(ii) $\lim_n \mathcal{G}(N_{k,n}) = 0$ *for all $k \in \mathbb{N}$.*

If $\varepsilon > 0$, there is a sequence n_k with $n_k \to \infty$ and a singly generated subfactor M_ε of N with

$$\bigcup_k N_{k,n_k} \subseteq M_\varepsilon \quad \text{and} \quad \mathcal{G}(M_\varepsilon) < \varepsilon.$$

Proof. Let $0 < \varepsilon < 1/8$. By induction on k, we construct sequences of

(0) positive integers n_k,

(1) positive integers m_k with $m_1 > 3/\varepsilon$,

(2) irreducible subfactors M_k of N with $N_{k,n_k} \subseteq M_k$,

(3) subsystems of matrix units $\{e_{ij}^{(k)} : 1 \leq i, j \leq m_k\}$ of $\{M_1, \ldots, M_k\}''$ and

(4) projections q_k in $\{M_1, \ldots, M_k\}''$

such that

(5) $e_{2,2}^{(k)} = \sum_{i=1}^{m_{k+1}} e_{ii}^{(k+1)}$,

(6) $q_{k+1} \leq e_{2,2}^{(k)}$ and $q_{k+1}e_{1,1}^{(k+1)} = q_{k+1}e_{2,2}^{(k+1)} = 0$, and

(7) $\{M_1, \ldots, M_k\}'' = \{q_1, \ldots q_k, e_{ij}^{(\ell)} : 1 \leq i, j \leq m_\ell, 1 \leq \ell \leq k\}''$ for all $k \in \mathbb{N}$.

By (ii), there is an n_1 such that $\mathcal{G}(N_{1,n_1}) < \varepsilon^2/2$. By Lemma 16.7.3 there is an irreducible finitely generated subfactor M_1 of N such that $N_{1,n_1} \subseteq M_1$ and $\mathcal{G}(M_1) < \varepsilon^2$. By Lemma 16.2.4, there is a finite generating set X_1 of M_1, an integer $m_1 > 3/\varepsilon$ and a set of projections $\{e_{ii}\}$ in $\mathcal{P}_{m_1}^{eq}(M_1)$ such that

$\mathcal{I}(X_1; \{e_{ii}\}) < \varepsilon^2$. Choose a set of matrix units $\{e_{ij}: 1 \leq i, j \leq m_1\}$ in M_1 for $\{e_{ii}: 1 \leq i \leq m_1\}$. Here $\varepsilon < 1/8 \leq 1/2 - 1/m_1$, and $2\lceil \varepsilon m_1 \rceil + 2 \leq m_1/2 + 2 < m_1 - 2$ so that Lemma 16.6.5 implies that there is a projection $q_1 \in M_1$ such that

$$\{q_1, e_{ij}: 1 \leq i, j \leq m_1\}'' = (X \cup \{e_{ij}\})'' = X'' = M_1 \tag{16.7.28}$$

and

$$q_1 \leq \sum_{i=1}^{m_1-2} e_{ii}. \tag{16.7.29}$$

By reordering the $\{e_{ij}: 1 \leq i, j \leq m_1\}$, assume that $q_1 e_{11} = 0$. This reordering does not affect the other parts of the construction. Let $e_{ij}^{(1)} = e_{ij}$ for $1 \leq i, j \leq m_1$. This is the construction for $k = 1$.

Suppose that the construction is completed to the kth stage. By hypothesis (ii) there is an integer n_{k+1} such that

$$\mathcal{G}(N_{k+1}, n_{k+1}) < 1/2 \left(\frac{1}{8m_1 \cdots m_k} \right)^2 \tag{16.7.30}$$

Hence by Lemma 16.7.3, there is an irreducible finitely generated subfactor M_{k+1} of N with $N_{k+1}, n_{k+1} \subseteq M_{k+1}$ and

$$\mathcal{G}(M_{k+1}) < \left(\frac{1}{8m_1 \cdots m_k} \right)^2. \tag{16.7.31}$$

Let

$$\mathbf{m}_k = m_1 \cdots m_k. \tag{16.7.32}$$

By Lemma 16.2.4 there is a finite generating set X_{k+1} of M_{k+1}, an integer m_{k+1} and a set of orthogonal equivalent projections $\{p_i\} = \{p_i: 1 \leq i \leq \mathbf{m}_k m_{k+1}\}$ in M_{k+1} such that

$$\mathcal{I}(X_{k+1}; \{p_i\}) \leq (1/(4\mathbf{m}_k))^2 \quad \text{and} \quad m_{k+1} > 10; \tag{16.7.33}$$

the final inequality is important for the inequality (16.7.43). Thus $\{p_i\} \in \mathcal{P}_{\mathbf{m}_k m_{k+1}}^{eq}(M_{k+1})$. Let

$$\mathbf{m}_{k+1} = \mathbf{m}_k m_{k+1} = m_1 \cdots m_{k+1}.$$

The factor M_k is an irreducible subfactor of N and hence $\{M_1, \ldots, M_k\}''$ is an (irreducible) subfactor of N. Let $\{e_{ij}^{(k+1)}: 1 \leq i, j \leq m_{k+1}\}$ be a set of matrix units in the factor $e_{22}^{(k)}\{M_1, \ldots M_k\}'' e_{22}^{(k)}$; so is a subsystem of matrix units in $\{M_1, \ldots, M_k\}''$. Thus (5) is satisfied.

Then

$$T_{k+1} = \{e_{i_1, 2}^{(1)} \cdots e_{i_k, 2}^{(k)} e_{st}^{(k+1)} e_{2, j_k}^{(k)} \cdots e_{2, j_1}^{(1)}:$$

$$1 \leq i_\ell, j_\ell \leq m_\ell; 1 \leq \ell \leq k; 1 \leq s, t \leq m_{k+1}\} \tag{16.7.34}$$

is a system of matrix units for an $\mathbf{m}_{k+1} \times \mathbf{m}_{k+1} = \mathbf{m}_k m_{k+1} \times \mathbf{m}_k m_{k+1}$ matrix subalgebra of $\{M_1, \ldots, M_k\}''$ with diagonal

$$P_{k+1} = \{e_{i_1,2}^{(1)} \cdots e_{i_k,2}^{(k)} e_{ss}^{(k+1)} e_{2,i_k}^{(k)} \cdots e_{2,i_1}^{(1)} :$$
$$1 \le i_\ell \le m_\ell; 1 \le \ell \le k; 1 \le s, t \le m_{k+1}\}; \tag{16.7.35}$$

thus $P_{k+1} \in \mathcal{P}^{eq}_{\mathbf{m}_{k+1}}(\{M_1, \ldots, M_k\}'')$. By the inductive hypothesis (5),

$$e_{st}^{(k+1)} \le e_{22}^{(k)} \le \ldots \le e_{22}^{(1)} \tag{16.7.36}$$

so that

$$e_{22}^{(1)} \cdots e_{22}^{(k)} e_{ss}^{(k+1)} e_{22}^{(k)} \cdots e_{22}^{(1)} = e_{ss}^{(k+1)} \tag{16.7.37}$$

for $1 \le s \le m_{k+1}$. Let

$$P_{k+1} = \{f_j : 1 \le j \le \mathbf{m}_{k+1}\} \tag{16.7.38}$$

with

$$f_j = e_{jj}^{(k+1)} \quad \text{for} \quad 1 \le j \le m_{k+1}. \tag{16.7.39}$$

The algebra $M_k \cap M_{k+1}$ is diffuse, since it contains the diffuse algebra $N_{k,n_k} \cap N_{k+1,n_{k+1}}$, so there is a set of projections $\{\widetilde{f_i}\} \in \mathcal{P}^{eq}_{\mathbf{m}_{k+1}}(M_k \cap M_{k+1})$ by Theorem 16.5.2, since a masa in $M_k \cap M_{k+1}$ is isomorphic to $(L^\infty[0,1], \int \cdot \, dt)$. Because $\{M_1, \ldots, M_k\}''$ and M_{k+1} are factors containing $\{\widetilde{f_i}\}$, there are unitaries $v_k \in \{M_1, \ldots, M_k\}''$ and $w_k \in M_{k+1}$ such that

$$v_k f_i v_k^* = \widetilde{f_i} \quad \text{and} \quad w_k \widetilde{f_i} w_k^* = p_i \tag{16.7.40}$$

so that

$$\{p_i\} = w_k v_k \{f_j\} v_k^* w_k^* = w_k v_k P_{k+1} v_k^* w_k^*. \tag{16.7.41}$$

From this, by the unitary invariance of $\mathcal{I}$,

$$\mathcal{I}(v_k^* w_k^* X_{k+1} w_k v_k; P_{k+1}) = \mathcal{I}(X_{k+1}; \{p_i\}) < (1/4\mathbf{m}_k)^2 < 1/8. \tag{16.7.42}$$

In the notation of Lemma 16.7.2, c and k are respectively $1/4\mathbf{m}_k$ and $\mathbf{m}_{k+1}$ so that

$$2\lceil ck \rceil + 2 = 2\lceil m_{k+1}/4 \rceil + 2 \le m_{k+1}/2 + 2 < m_{k+1} - 3. \tag{16.7.43}$$

The final inequality holds because $m_{k+1} > 10$, which was the reason for this restriction.

By Lemma 16.6.5, there is a projection $q_{k+1} \in \{v_k^* w_k^* X_{k+1} w_k v_k, T_{k+1}\}''$ such that

$$\{q_{k+1}, T_{k+1}\}'' = \{v_k^* w_k^* X_{k+1} w_k v_k, T_{k+1}\}'' \subseteq \{M_1, \ldots, M_{k+1}\}'' \tag{16.7.44}$$

and

$$q_{k+1} \leq \mathcal{S}(q_{k+1}; P_{k+1}) \leq \sum_{i=1}^{m_{k+1}-3} f_i. \tag{16.7.45}$$

By the choice of f_j for $1 \leq j \leq m_{k+1}$ in (16.7.39),

$$q_{k+1} \leq \sum_{i=1}^{m_{k+1}-3} f_i = \sum_{i=1}^{m_{k+1}-3} e_{ii}^{(k+1)} \leq e_{22}^{(k)} = \sum_{i=1}^{m_{k+1}} e_{ii}^{(k+1)} \tag{16.7.46}$$

and hence there are at least two different projections $f_s = e_{ss}^{(k+1)}$ and $f_t = e_{tt}^{(k+1)}$ with $q_{k+1}f_s = 0$ and $q_{k+1}f_t = 0$ and $1 \leq s < t \leq m_{k+1}$. By reordering the matrix units $\{e_{ij}^{(k+1)} : 1 \leq i,j \leq m_{k+1}\}$, if necessary, assume that $s = 1$ and $t = 2$ so that

$$q_{k+1}(e_{11}^{(k+1)} + e_{22}^{(k+1)}) = 0.$$

This reordering does not affect the construction so far. This proves (6).

By the inductive hypothesis (7) and equation (16.7.44)

$$\{q_1, \ldots, q_{k+1}, e_{ij}^{(\ell)} : 1 \leq i,j \leq m_\ell, 1 \leq \ell \leq k+1\}'' \subseteq \{M_1, \ldots, M_{k+1}\}''.$$

Because $T_{k+1} \subseteq \{M_1, \ldots, M_{k+1}\}''$, $X_{k+1}'' = M_{k+1}$ and $w_k v_k \in \{M_1, \ldots, M_{k+1}\}''$, we have

$$\begin{aligned}
\{M_1, \ldots, M_{k+1}\}'' &= \{M_1, \ldots, M_k, X_{k+1}, T_{k+1}\}'' \\
&= \{M_1, \ldots, M_k, v_k^* w_k^* X_{k+1} w_k v_k, T_{k+1}\}'' \\
&= \{M_1, \ldots, M_k, q_{k+1}, T_{k+1}\}'' \\
&= \{q_1, \ldots, q_{k+1}, e_{ij}^{(\ell)} : 1 \leq \ell \leq k+1\}''
\end{aligned} \tag{16.7.47}$$

by the inductive hypothesis (7), equation (16.7.44) and the definition of T_{k+1} in equation (16.7.34). This proves (7) and completes the induction.

Let

$$x_1 = \sum_{k=1}^{\infty} \left(2^{-k} e_{11}^{(k)} + 3^{-k} q_k\right) \tag{16.7.48}$$

and

$$x_2 = \sum_{k=1}^{\infty} \sum_{j=2}^{m_k} 2^{-k} \left(e_{j-1,j}^{(k)} + e_{j,j-1}^{(k)}\right). \tag{16.7.49}$$

By the inductive hypotheses (5) and (6), $\{e_{11}^{(k)}, q_k : k \in \mathbb{N}\}$ is a set of mutually orthogonal projections in N and hence is contained in the abelian von Neumann algebra $\{x_1\}''$ as the coefficients in (16.7.48) are all different.

From the orthogonality in (6), $e_{1,1}^{(k)} x_2 = 2^{-k} e_{1,2}^{(k)}$ from which if follows successively that $e_{2,2}^{(k)}, e_{2,3}^{(k)}, e_{3,3}^{(k)}, \ldots, e_{m_k-1, m_k}^{(k)}$ are in $\{x_1, x_2\}''$. Thus

$$\{M_1, \ldots, M_k\}'' = \{q_1, \ldots, q_k, e_{ij}^{(\ell)} : 1 \leq i, j \leq m_\ell, 1 \leq \ell \leq k\}''$$
$$\subseteq \{x_1, x_2\}'' \qquad (16.7.50)$$

for all $k \in \mathbb{N}$ so that $\{x_1, x_2\}'' = \left(\bigcup M_k \right)''$. Let $M_\varepsilon = \left(\bigcup M_k \right)''$.
From the definition of $\mathcal{I}$ it follows that

$$\mathcal{I}(x_1; \{e_{ii}^{(1)}\}) \leq m_1^{-1} \quad \text{and} \quad \mathcal{I}(x_2; \{e_{ii}^{(1)}\}) \leq 2m_1^{-1}. \qquad (16.7.51)$$

The first of these inequalities holds because x_1 and $e_{ii}^{(1)}$ commute and the second is valid as in the matrix algebra case (see Lemma 16.3.2) by the inductive conditions (5) and (6). Hence

$$\mathcal{I}(\{x_1, x_2\}; \{e_{ii}^{(1)}\}) \leq 3m_1^{-1} < \varepsilon \qquad (16.7.52)$$

so that $\mathcal{G}(M_\varepsilon) < \varepsilon$, proving the lemma. $\qquad \square$

Theorem 16.7.5. *Let N be a separable* II$_1$ *factor and let $\{N_k\}_{k=1}^\infty$ be a sequence of finitely generated von Neumann subalgebras of N such that $N = \left(\bigcup_{k=1}^\infty N_k \right)''$ and $N_k \cap N_{k+1}$ is diffuse for all $k \in \mathbb{N}$. If $\mathcal{G}(N_k) = 0$ for all $k \in \mathbb{N}$, then N is singly generated and $\mathcal{G}(N) = 0$.*

Proof. Let $N_{k,n} = N_k$ for all $k, n \in \mathbb{N}$. Let $\varepsilon > 0$. By Lemma 16.7.4, there exists a finitely generated irreducible subfactor M_ε of N with $\bigcup N_k = \bigcup N_{k, n_k} \subseteq M_\varepsilon$ and $\mathcal{G}(M_\varepsilon) < \varepsilon$. Hence $M_\varepsilon = N$ so that N is finitely generated with $\mathcal{G}(N) = 0$ since $\varepsilon > 0$ was arbitrary. $\qquad \square$

The following corollary of Lemma 16.7.4 is required to handle the case of II$_1$ factors with property Γ.

Corollary 16.7.6. *Let N be a separable* II$_1$ *factor, let A be a diffuse von Neumann subalgebra of N and let N_n be a sequence of von Neumann subalgebras of N all containing A with $\lim_n \mathcal{G}(N_n) = 0$. If $\varepsilon > 0$, there is a sequence n_k of integers and a finitely generated irreducible subfactor M_ε of N such that $\bigcup_k N_{n_k} \subseteq M_\varepsilon$ and $\mathcal{G}(M_\varepsilon) < \varepsilon$.*

Proof. Let $N_{k,n} = N_n$ for all $k, n \in \mathbb{N}$. Then $N_{k, m} \cap N_{k+1, n}$ is diffuse for all $k, m, n \in \mathbb{N}$ and $\lim_n \mathcal{G}(N_{k, n}) = 0$ for all $k \in \mathbb{N}$. The result now follows from Lemma 16.7.4. $\qquad \square$

16.8 Examples of singly generated II$_1$ factors

The following theorem was proved in Lemma 16.3.4 under the additional hypothesis that N_1 and N_2 are finitely generated.

Theorem 16.8.1. *Let N_1 and N_2 be a separable II$_1$ factors, then $N_1 \overline{\otimes} N_2$ is singly generated and $\mathcal{G}(N_1 \overline{\otimes} N_2) = 0$.*

Proof. Let $A_1 \subseteq N_1$ and $A_2 \subseteq N_2$ be masas and let $u_1 \in A_1$ and $u_2 \in A_2$ be Haar unitaries that generate these masas. The unitaries exist since $A_i \cong L^\infty[0,1]$ by Theorem 3.5.2. Let $\{x_j : j \in \mathbb{N}\}$ be a set of generators of N_1 and $\{y_j : j \in \mathbb{N}\}$ be a set of generators of N_2. Let

$$N_k = \{u_1, x_1, \ldots, x_k\}'' \overline{\otimes} \{u_2, y_1, \ldots, y_k\}''$$

for all $k \in \mathbb{N}$. Then N_k is a tensor product of two finitely generated diffuse von Neumann algebras so $\mathcal{G}(N_k) = 0$ by Lemma 16.3.4. The subalgebras N_k thus satisfy the conditions of Theorem 16.7.5 so that $N_1 \overline{\otimes} N_2$ is singly generated and $\mathcal{G}(N_1 \overline{\otimes} N_2) = 0$. $\qquad\square$

Lemma 16.8.2. *Let N be II$_1$ factor, let M be a finitely generated von Neumann subalgebra with $\mathcal{G}(M) = c$ and let u be a unitary in N. If there is a diffuse abelian von Neumann subalgebra $A \subseteq M$ such that $uAu^* \subseteq M$, then for each $\varepsilon > 0$, there is an irreducible finitely generated subfactor M_ε of N such that $(M \bigcup \{u\})'' \subseteq M_\varepsilon$ and $\mathcal{G}(M_\varepsilon) < c + \varepsilon$.*

Proof. By Lemma 16.7.3 there is a finitely generated irreducible subfactor M_0 of N with $M \subseteq M_0$ and $\mathcal{G}(M_0) < c + \varepsilon/2$. There is a finite set X of generators of M_0, an integer $k > 2/\varepsilon$ and a family of projections $\{p_j\} \in \mathcal{P}_k^{eq}(M_0)$ such that

$$\mathcal{I}(X; \{p_j\}) < c + \varepsilon/2. \tag{16.8.1}$$

Since A is diffuse and abelian, $(A, \tau) \cong (L^\infty[0,1], \int \cdot \, dt)$ so there is a family of mutually orthogonal projections $\{e_j\} \in \mathcal{P}_k^{eq}(A)$. Let $f_j = u^* e_j u \, (1 \le j \le k)$. Then $\{f_j\}$ is a set of mutually orthogonal projections in M by hypothesis, so is in $\mathcal{P}_k^{eq}(M)$.

Because M_0 is a factor containing $\{p_j\}$, $\{e_j\}$ and $\{f_j\}$, there are unitaries w_1 and w_2 in M_0 such that

$$e_j = w_1 p_j w_1^* \quad \text{and} \quad u^* e_j u = f_j = w_2 p_j w_2^* \quad (1 \le j \le k). \tag{16.8.2}$$

Hence

$$w_1^* u w_2 p_j = p_j w_1^* u w_2 \quad (1 \le j \le k) \tag{16.8.3}$$

and thus

$$\mathcal{I}(X \cup \{w_1^* u w_2\}; \{p_j\}) < c + \varepsilon/2 + 1/k < c + \varepsilon. \tag{16.8.4}$$

Let $M_\varepsilon = (X \cup \{w_1^* u w_2\})''$. By inequality (16.8.4) M_ε is finitely generated and $\mathcal{G}(M_\varepsilon) < c + \varepsilon$. From $w_1, w_2 \in M_0 \subseteq M_\varepsilon$, if follows that $u \in M_\varepsilon$ so that $(M \cup \{u\})'' \subseteq M_\varepsilon$, which is an irreducible subfactor of N. $\qquad\square$

Recall that $\mathcal{N}(M)$ is the unitary normaliser group

$$\{u \in \mathcal{U}(N) : uMu^* = M\}$$

of M in N.

Theorem 16.8.3. *Let M be a diffuse von Neumann subalgebra of a II_1 factor N with $\mathcal{N}(M)'' = N$. If $\mathcal{G}(M) = 0$, then $\mathcal{G}(N) = 0$ and N is singly generated.*

Proof. Let $\{u_k\colon k \in \mathbb{N}\}$ be a sequence of unitaries in $\mathcal{N}$ such that

$$\left(M \cup \{u_k\colon k \in \mathbb{N}\} \right)'' = N.$$

Let $M_k = \left(M \cup \{u_k\} \right)''$ for all $k \in \mathbb{N}$. Let A be a masa in M. By Theorem 3.5.2, A is a diffuse abelian subalgebra of M with $u_k A u_k^* \subseteq M$ for all $k \in \mathbb{N}$. By Lemma 16.7.3, for each $\varepsilon > 0$ there is an irreducible finitely generated subfactor $M_{k,\varepsilon}$ of N with $M_k \subseteq M_{k,\varepsilon}$ and $\mathcal{G}(M_{k,\varepsilon}) < \varepsilon$. By Theorem 16.7.5 $\left(\bigcup_k M_k \right)'' = N$ is finitely generated with $\mathcal{G}(N) = 0$ and so is singly generated. $\square$

From the previous Theorem 16.8.3 one obtains an immediate corollary because $\mathcal{G}(A) = 0$ for all masas in a separable II_1 factor.

Corollary 16.8.4. *Let N be a separable II_1 factor. If N has a Cartan subalgebra, then N is singly generated and $\mathcal{G}(N) = 0$.*

Theorem 16.8.5. *Let N be a separable II_1 factor. If N has property Γ, then N is singly generated and $\mathcal{G}(N) = 0$.*

Proof. By Theorem 13.5.4, there is a masa A in N such that if $m, n \in \mathbb{N}$, $x_1, \ldots x_m \in N$ and $\delta > 0$, then there exist mutually orthogonal projections $p_1, \ldots, p_n \in A$ each of trace $1/n$ such that

$$\|p_j x_i - x_i p_j\|_2 < \delta \quad \text{for all} \quad 1 \le j \le n, 1 \le i \le m. \tag{16.8.5}$$

Let $x \in N$ with $\|x\| \le 1$ and let $\varepsilon > 0$. If $r \in \mathbb{N}$, by inequality (16.8.5), choose $p_1^{(r)}, \ldots, p_r^{(r)}$ in $\mathcal{P}_r^{eq}(A)$ such that

$$\|p_j^{(r)} x - x p_j^{(r)}\|_2 < r^{-1} \tag{16.8.6}$$

for $1 \le j \le r$ so that

$$\left\| x - \sum_{j=1}^r p_j^{(r)} x p_j^{(r)} \right\|_2^2 = \left\| \sum_{j=1}^r p_j^{(r)} (p_j^{(r)} x - x p_j^{(r)}) \right\|_2^2$$

$$= \sum_{j=1}^r \| p_j^{(r)} (p_j^{(r)} x - x p_j^{(r)}) \|_2^2$$

$$\le \sum_{j=1}^r \| p_j^{(r)} x - x p_j^{(r)} \|_2^2 \le r^{-1}, \tag{16.8.7}$$

by orthogonality of $\{p_j^{(r)}\colon 1 \le j \le r\}$. If $x^{(r)} = \sum_j p_j^{(r)} x p_j^{(r)}$, then

$$\|x - x^{(r)}\|_2 \le r^{-1/2} \tag{16.8.8}$$

for all $r \in \mathbb{N}$. Let $N_r = \left(A \cup \{x^{(r)}\}\right)''$. If h is a hermitian generator of A, then $\mathcal{I}(h; \{p_j^{(r)}\}) = \mathcal{I}(x^{(r)}; \{p_j^{(r)}\}) = r^{-1}$, since $h, x^{(r)} \in \{p_j^{(r)}\}'$. Thus $\mathcal{G}(N_r) \leq 2r^{-1}$. By Lemma 16.7.3, there is a sequence r_n of integers and an irreducible finitely generated subfactor M_ε of N that contains $\bigcup_n N_{r_n}$ such that $\mathcal{G}(M_\varepsilon) < \varepsilon$. By the inequality (16.8.8), $\|x - x^{(r_n)}\|_2 < r_n^{-1/2}$ for all $n \in \mathbb{N}$ so that $x \in \left(\bigcup_n N_{r_n}\right)'' \subseteq M_\varepsilon$.

Let $\{x_k\}$ be a generating set for N and let $M_k = \left(A \cup \{x_k\}\right)''$ for all $k \in \mathbb{N}$. By the first part of the proof, for each $\varepsilon > 0$ and each $k \in \mathbb{N}$, there is a finitely generated irreducible subfactor $M_{k,\varepsilon}$ of N with $M_k \subseteq M_{k,\varepsilon}$ and $\mathcal{G}(M_{k,\varepsilon}) < \varepsilon$. By Theorem 16.7.5, N is singly generated and $\mathcal{G}(N) = 0$ as required.   □

Corollary 16.8.6. *Let M be a finitely generated* II$_1$ *factor, let G be a countable discrete group, and let $\alpha \colon G \to \mathrm{Aut}(M)$ be an action by outer automorphisms. Then $\mathcal{G}(M \rtimes_\alpha G) \leq \mathcal{G}(M)$. In particular, if $\mathcal{G}(M) = 0$ then $\mathcal{G}(M \rtimes_\alpha G) = 0$ and both algebras are singly generated.*

Proof. Given $\varepsilon > 0$, choose an integer k such that $k^2\varepsilon > 1$, and then choose orthogonal projections $P = \{p_1, \ldots, p_k\}$ of trace k^{-1} and a finite generating set X for M such that

$$\mathcal{I}(X; P) < \mathcal{G}(M) + \varepsilon. \tag{16.8.9}$$

Let $\{\lambda_g \colon g \in G\}$ be the representation of G in $M \rtimes_\alpha G$ such that $\lambda_g x \lambda_g^{-1} = \alpha_g(x)$ for $x \in M$. Each projection $\alpha_g(p_1)$ has trace k^{-1} and so is equivalent to p_1. If we enumerate the group elements as $\{g_n \colon n \geq 1\}$, then choose partial isometries $v_n \in M$, $n \geq 1$, so that

$$v_n v_n^* = p_1, \quad v_n^* v_n = \alpha_{g_n}(p_1) \tag{16.8.10}$$

for $n \geq 1$. Then $p_1 v_n \alpha_{g_n}(p_1) = v_n \neq 0$, so the element

$$y = \sum_{n=1}^{\infty} 2^{-n} v_n \lambda_{g_n} = \sum_{n=1}^{\infty} 2^{-n} p_1 v_n \lambda_g p_1 \tag{16.8.11}$$

has a non-zero λ_g-coefficient for all $g \in G$. Let $N = \{M \cup \{y\}\}''$ which lies between M and $M \rtimes_\alpha G$. By Theorem 14.3.3, $N = M \rtimes_\alpha K$ for a subgroup K of G. The form of y in (16.8.11) shows that K cannot be a proper subgroup of G, and thus $N = M \rtimes_\alpha G$. Let $Y = X \cup \{y\}$. Then Y generates $M \rtimes_\alpha G$ and y contributes only k^{-2} to $\mathcal{I}(Y; P)$ since $y = p_1 y p_1$ from (16.8.10). The first part follows from the estimate

$$\mathcal{G}(M \rtimes_\alpha G) \leq \mathcal{I}(Y; P) = \mathcal{I}(X; P) + k^{-2}$$
$$< \mathcal{G}(M) + \varepsilon + k^{-2} < \mathcal{G}(M) + 2\varepsilon. \tag{16.8.12}$$

When $\mathcal{G}(M) = 0$, (16.8.12) implies that $\mathcal{G}(M \rtimes_\alpha G) = 0$, and this algebra is then singly generated by Theorem 16.6.2.   □

Recall that a unitary u in a finite von Neumann algebra with trace τ is called a Haar unitary (with respect to τ) if $\tau(u^n) = 0$ for all $n \in \mathbb{N} \setminus \{0\}$. Typical of the theorems of Shen [171] that follow from the above results is the following, which is designed for application to certain I.C.C. groups (see [171, Theorem 5.2]).

Theorem 16.8.7. *Let N be a separable* II$_1$ *factor with a von Neumann subalgebra M and let u_k ($k \in \mathbb{N}$) be a sequence of Haar unitaries in N such that N is generated by $M \bigcup \{u_1, u_2, \ldots\}$ as a von Neumann algebra. If $\mathcal{G}(M) = 0$, $u_1 \in M$ and $u_{k+1} u_k u_{k+1}^* \in \left(M \bigcup \{u_1, \ldots, u_k\}\right)''$ for all $k \in \mathbb{N}$, then $\mathcal{G}(N) = 0$ and N is singly generated.*

Proof. By induction on k we construct sequences of irreducible finitely generated subfactors $\{N_{k,n} : n \in \mathbb{N}\}$ for each $k \in \mathbb{N}$ such that

(1) $\mathcal{G}(N_{k,n}) < n^{-1}$ and $\left(M \bigcup \{u_1, \ldots, u_k\}\right)'' \subseteq N_{k,n}$, and
(2) $N_{k,2n} \subseteq N_{k+1,n}$ for all $k, n \in \mathbb{N}$.

Regard $N_{0,k} = M$ for all $n \in \mathbb{N}$. Suppose that $N_{k,n}$ ($n \in \mathbb{N}$) have been constructed for some k. By Lemma 16.8.2, since $u_{k+1} \{u_k\}'' u_{k+1}^* \subseteq N_{k,2n}$ and $\mathcal{G}(N_{k,2n}) < (2n)^{-1}$, there is an irreducible finitely generated subfactor $N_{k+1,n}$ of N such that

$$\left(N_{k,2n} \cup \{u_{k+1}\}\right)'' \subseteq N_{k+1,n} \text{ and } \mathcal{G}(N_{k+1,n}) < n^{-1}.$$

The double sequence $N_{k,n}$ ($k, n \in \mathbb{N}$) satisfies the hypotheses of Lemma 16.7.4 so for each $\varepsilon > 0$ there is a sequence n_k of integers tending to infinity and an irreducible finitely generated subfactor M_ε of N such that

$$\bigcup_k N_{k,n_k} \subseteq M_\varepsilon \text{ and } \mathcal{G}(M_\varepsilon) < \varepsilon.$$

Since $M \bigcup \{u_1, u_2, \ldots\} \subseteq \bigcup_k N_{k,n_k} \subseteq M_\varepsilon$, the algebras M_ε and N are equal. $\quad\square$

Appendix A

The ultrapower and property Γ

A.1 Introduction

This appendix contains sections on: ultrafilters and characters of $\ell^\infty(\mathbb{N})$; a discussion of maximal ideals in finite von Neumann algebras; the construction of the ultrapowers N_n^ω and N^ω; property Γ and relative commutants in N^ω.

In these notes *ultrafilter* is used for *free ultrafilter in* $\mathbb{N}$ as these are the only ultrafilters discussed. See the article by Ge and Hadwin [78] for a detailed discussion of ultrafilters and ultraproducts directed at operator algebras, or the books [34, 90, 107, 204] for a discussion of filters and ultrafilters in set theory and general topology. It is convenient to think of ultrafilters as characters ω of $\ell^\infty(\mathbb{N})$ induced by points in $\beta\mathbb{N} \setminus \mathbb{N}$ so this relationship is discussed briefly in the second section.

Section A.3 on *maximal ideals* in a finite von Neumann algebra contains a theorem due to Wright [211] that the quotient of a finite von Neumann algebra by a maximal two-sided ideal is a finite factor with trace arising from the original algebra and maximal ideal; Wright actually proved this for AW^*-algebras. A theorem for AW^*-algebras that yields this was rediscovered by Feldman [68], though he does not state this exact result or examine the norm closed ideals as Wright does. This result for von Neumann algebras appears in Sakai's Yale notes [165] with no reference, and there is an account by Takesaki [187, p. 357]. The example of the Calkin algebra shows that *finite* is essential to obtain a von Neumann algebra as the quotient.

The ultrapower N_n^ω is defined in Section A.4 and the relative commutant lemma $\left(B_n^\omega\right)' \cap N_n^\omega = \left(B_n' \cap N_n\right)^\omega$ is proved. In Section A.5, we prove the useful technical lemma on lifting unitaries, projections, partial isometries and matrix units from N^ω to $\ell^\infty(N)$ due to Connes [36]. Property Γ is defined, R is shown to have property Γ and $L(\mathbb{F}_2)$ is shown not to have property Γ in Section A.7, and Dixmier's equivalent version of property Γ based on projections is proved in Theorem A.7.5 using a theorem of Fang, Ge and Li, Theorem A.6.5.

There are two natural sets of equivalent conditions on a II_1 factor that involve the ultrapower that will be briefly mentioned. One of these is used in these notes, a version sufficient for our use is proved in Section A.7, but we state the corresponding versions of both.

Theorem A.1.1. *Let N be a separable II_1 factor. The following conditions are equivalent:*

(i) $N \cong N\overline{\otimes}R$;

(ii) $N_\omega := N' \cap N^\omega$ *is non-commutative;*

(iii) N_ω *is a non-separable II_1 factor.*

Theorem A.1.1 is due to McDuff [114] and there are other important conditions that can be included due to Connes in the fundamental paper [36]; see also [189] and [67].

Property Γ was introduced by Murray and von Neumann to separate the hyperfinite factor R from the free group factor $L(\mathbb{F}_2)$ so giving two non-isomorphic separable II_1 factors [118]. A II_1 factor N is said to have property Γ if for each finite set $\{x_1, \dots, x_n\} \subseteq N$ and each $\varepsilon > 0$, there is a unitary u in N with $\tau(u) = 0$ and $\|ux_j - x_j u\|_2 < \varepsilon$ for all $1 \le j \le n$ (see Section A.7 for more on property Γ).

The following theorem is due to Dixmier [50] and Connes [36]. We use the notation $K(H)$ for the algebra of compact operators on a Hilbert space H.

Theorem A.1.2. *Let N be a separable II_1 factor. The following conditions are equivalent:*

(i) N *has property Γ;*

(ii) $N_\omega \ne \mathbb{C}1$;

(iii) N_ω *is a diffuse von Neumann algebra;*

(iv) *the C^*-algebra $C^*(N, N')$ generated by N and N' satisfies*

$$C^*(N, N') \cap K(L^2(N)) = \{0\}.$$

The reader is referred to the original papers, to the book of Takesaki [189] and to the introduction of [51] for more discussion and references. In Section A.6, we shall show that if $N_\omega \ne \mathbb{C}1$, then N_ω is diffuse.

A.2 Ultrafilters and characters

Let $\beta\mathbb{N}$ denote the Stone-Čech compactification of the set $\mathbb{N}$ of positive integers, identifying it with the carrier space of $\ell^\infty(\mathbb{N})$ (see [204]). The points ω in $\beta\mathbb{N} \setminus \mathbb{N}$ are the non-normal characters of $\ell^\infty(\mathbb{N})$ and these points may be identified with (free) ultrafilters (over $\mathbb{N}$), a notion that we now define.

A *(free) ultrafilter* is a non-empty set $\mathcal{F}$ of subsets of $\mathbb{N}$ satisfying

(1) if $X \in \mathcal{F}$ and $X \subseteq Y \subseteq \mathbb{N}$, then $Y \in \mathcal{F}$;

(2) if $X, Y \in \mathcal{F}$, then $X \cap Y \in \mathcal{F}$;

(3) $\emptyset \notin \mathcal{F}$;

(4) if $X \subseteq \mathbb{N}$, then $X \in \mathcal{F}$ or $\mathbb{N} \setminus X \in \mathcal{F}$; and

(5) no finite set is in $\mathcal{F}$.

A *filter* in $\mathbb{N}$ is a set $\mathcal{F}$ satisfying (1) to (3). An *ultrafilter* is a filter satisfying condition (4). Condition (5) ensures that $\mathcal{F}$ corresponds to a point in $\beta\mathbb{N} \setminus \mathbb{N}$ and excludes points of $\mathbb{N}$; the word *free* is used for this but all our ultrafilters are free and in $\mathbb{N}$ so the word *ultrafilter* has a restricted meaning here.

The bijection from $\beta\mathbb{N} \setminus \mathbb{N}$ onto the set of ultrafilters is given by $\omega \mapsto \mathcal{F}_\omega$, where

$$\mathcal{F}_\omega = \left\{ \{n \in \mathbb{N}: |x_n| < k^{-1}\}: k \in \mathbb{N}, x = (x_n) \in \ker \omega \right\}. \qquad (A.2.1)$$

The condition $|x_n| < k^{-1}$ can be replaced by $x_n < k^{-1}$ and $x_n \geq 0$, or $|x_n| < 1$ by removing the condition $x \geq 0$ and $k \in \mathbb{N}$. However, the version in equation (A.2.1) is convenient for showing that the map is onto. A direct calculation using the ideal property of $\ker \omega$ shows that if $\omega \in \beta\mathbb{N} \setminus \mathbb{N}$, then $\mathcal{F}_\omega$ is an ultrafilter. If $\omega_1 \neq \omega_2$ are in $\beta\mathbb{N} \setminus \mathbb{N}$, then there is an $x \in \ell^\infty(\mathbb{N})$ with $x \geq 0$, $\omega_1(x) = 0$ and $\omega_2(x) = 1$. Then $\{n \in \mathbb{N}: x_n < 1/2\}$ is in $\mathcal{F}_{\omega_1}$ but not in $\mathcal{F}_{\omega_2}$. This shows that the map is injective.

Let $\mathcal{F}$ be an ultrafilter and let

$$\mathcal{I} = \left\{ (x_n) \in \ell^\infty(\mathbb{N}): \{n \in \mathbb{N}: |x_n| < k^{-1}\} \in \mathcal{F} \text{ for all } k \in \mathbb{N} \right\}. \qquad (A.2.2)$$

Conditions (1) to (3) show that $\mathcal{I}$ is a proper ideal of $\ell^\infty(\mathbb{N})$, (4) that it is a maximal ideal and (5) that this ideal does not correspond to an element of $\mathbb{N}$. The character ω corresponding to this maximal ideal $\mathcal{I}$ has the property that $\mathcal{F}_\omega = \mathcal{F}$ as can be seen from the definitions in equations (A.2.1) and (A.2.2). Alternatively, ω can be constructed directly from $\mathcal{F}$ as follows. Let $x \in \ell^\infty(\mathbb{N})$. Then the collection

$$\left\{ \{x_n: n \in X\}^-: X \in \mathcal{F} \right\}$$

of closed subsets of $\{\lambda \in \mathbb{C}: |\lambda| \leq \|x\|\}$ has the finite intersection property so

$$Y_x = \bigcap_{X \in \mathcal{F}} \{x_n: n \in X\}^- \neq \emptyset. \qquad (A.2.3)$$

If Y_x is not a single point, then there are open subsets U and V of $\mathbb{C}$ such that $U \cap Y_x \neq \emptyset$, $V \cap Y_x \neq \emptyset$ and $\overline{U} \cap \overline{V} = \emptyset$. Let $W = \{n \in \mathbb{N}: x_n \in U\}$. From the definition of W, U and V, it follows that $W \neq \emptyset$ and $\mathbb{N} \setminus W \neq \emptyset$. Moreover $X \cap W \neq \emptyset$ and $X \cap (\mathbb{N} \setminus W) \neq \emptyset$ for all $X \in \mathcal{F}$ so that W and $\mathbb{N} \setminus W$ are in $\mathcal{F}$ contrary to (4). Hence Y_x is a single point denoted $\{\omega(x)\}$. From these definitions it follows that ω is a character of $\ell^\infty(\mathbb{N})$ with $\mathcal{F}_\omega = \mathcal{F}$.

Throughout this appendix, ω will denote a character in $\beta\mathbb{N} \setminus \mathbb{N}$ thought of as either a character or the ultrafilter $\mathcal{F}_\omega$. The quantity $\omega(x)$ is often denoted by $\omega(x_n)$, $\lim_\omega x_n$ or $\lim_{n\to\omega} x_n$. We shall frequently refer to ω as an ultrafilter as part of this identification.

The following elementary observation is useful:

> If $(x_n) \in \ell^\infty(\mathbb{N})$ with $\lim_{n\to\infty} x_n = 0$, then $\lim_\omega x_n = 0$ for all $\omega \in \beta\mathbb{N} \setminus \mathbb{N}$.

This is an easy consequence of (5) of the ultrafilter conditions as follows. Let $\|x_n\|_\infty \leq 1$, let $\varepsilon > 0$ and let $|x_n| < \varepsilon$ for all $n \geq k$. Let χ_k be the characteristic function of the set $\{1, \ldots, k\}$. Then

$$|x_n| \leq \chi_k(n)|x_n| + |(1 - \chi_k(n))x_n| \leq \chi_k(n) + \varepsilon \qquad \text{(A.2.4)}$$

so that $\lim_\omega |x_n| \leq \varepsilon$, since $\lim_{n\to\omega} \chi_k(n) = 0$.

A.3 Maximal quotients of finite algebras

This section is devoted to proving a result on which the definition of the ultra-power N^ω is based.

Theorem A.3.1. *If M is a finite von Neumann algebra and $\mathcal{I}$ is a maximal two-sided ideal in M, then $M/\mathcal{I}$ is a finite factor.*

This theorem is contained in a more general result, Theorem A.3.5. Note that the hypothesis *finite* is essential in deducing that the quotient is a von Neumann algebra; $K(H)$ is a maximal proper ideal in $B(H)$ for a separable Hilbert space H, but the Calkin algebra $C(H) := B(H)/K(H)$ is not a von Neumann algebra.

Before embarking on the proof of this theorem, we record one important fact about finite factors, although we will not make direct use of it. Recall that a C^*-algebra is said to be *simple* if it has no non-trivial norm closed ideals. Factors of types I_1 and II_∞ have non-trivial ideals: the algebra of compact operators in the first case and the norm closure of the operators with finite range projections in the second case [105, Section 6.8]. The situation is different for type III and finite factors, both of which are simple. We refer to [105, Section 6.8] for the type III case, but prove the finite case next.

Theorem A.3.2. *Let M be a finite factor with a faithful normalised normal trace τ. Then M is simple.*

Proof. Consider a non-zero norm closed ideal $\mathcal{I} \subseteq M$. Then $\mathcal{I}$ contains a non-zero element x which we may assume to be positive, otherwise replace it by x^*x. The trace τ takes the constant value $\tau(x) > 0$ on the set $\{uxu^* : u \in \mathcal{U}(M)\}$ and also on its norm closed convex hull $C(x)$, which lies in $\mathcal{I}$. By the Dixmier approximation theorem, Theorem 2.2.2, $C(x)$ contains an element of the centre of M and this must be $\tau(x)1$ since M is a factor. Thus $\mathcal{I}$ contains a non-zero multiple of 1 and so $\mathcal{I} = M$. This shows simplicity of M.   □

The following lemma handles a technical point which will arise in the proof of Theorem A.3.5.

Lemma A.3.3. *Let A be a unital C^*-algebra with a faithful normalised trace* τ.

(i) *If the closed unit ball of A is complete in the $\| \cdot \|_2$-norm, then A is a von Neumann algebra and τ is normal.*

(ii) *If A is a von Neumann algebra and τ is normal, then the closed unit ball of A is complete in the $\| \cdot \|_2$-norm.*

Proof. (i) The GNS representation with respect to τ allows us to assume that A is faithfully represented on $L^2(A)$. We let ξ denote the image of 1 in $L^2(A)$, and we have a conjugate linear isometry $J \colon L^2(A) \to L^2(A)$ defined initially on $A\xi$ by $J(a\xi) = a^*\xi$. Then $JAJ \subseteq A'$, so ξ is cyclic for A and for A', hence also separating for both algebras (see Lemma 2.3.1). It follows that ξ is also cyclic and separating for A''.

Consider $a \in A''$, $\|a\| \leq 1$. By the Kaplansky density theorem, Theorem 2.2.3, there is a sequence $\{a_n\}_{n=1}^{\infty}$ in the closed unit ball of A such that $\lim_{n \to \infty} a_n\xi = a\xi$ in $\| \cdot \|_2$-norm. Then $\{a_n\}_{n=1}^{\infty}$ is $\| \cdot \|_2$-Cauchy, so converges in $\| \cdot \|_2$-norm to $b\xi$ for some b in the closed unit ball of A, by hypothesis. Thus $(a - b)\xi = 0$, so $a = b$ since ξ is separating for A''. This shows that $A = A''$, and so A is a von Neumann algebra. We also have that τ is the vector functional $\langle \cdot \xi, \xi \rangle$, proving normality.

(ii) Suppose now that A is a von Neumann algebra with a faithful normal normalised trace τ. The mapping $a \mapsto a\xi$ takes the closed unit ball of A into the closed unit ball of $L^2(A)$, and the weak topology on the latter is defined by the functionals $\langle \cdot, y\xi \rangle$ for $y \in A$, since it is a $\| \cdot \|_2$-bounded set. If $a_\alpha \to a$ in the w^*-topology of the unit ball of A, then $\tau(a_\alpha y^*) \to \tau(ay^*)$ for each $y \in A$, showing that $a_\alpha\xi \to a\xi$ weakly in $L^2(A)$. Consequently the image of the closed unit ball of A is convex and weakly compact in $L^2(A)$, thus $\| \cdot \|_2$-complete.   □

Remark A.3.4. If A is a von Neumann algebra and τ is a non-normal faithful trace on A, them Lemma A.3.3 (i) shows that the closed unit ball is not $\| \cdot \|_2$-complete, otherwise normality of τ would follow. Thus the hypothesis of normality is essential in Lemma A.3.3 (ii). An example of this would be $A = L^\infty[0,1]$ with τ_1 the standard normal trace given by integration. Let τ_0 be any Hahn–Banach state extension of point evaluation at 0 on $C[0,1]$. Then $\tau = (\tau_0 + \tau_1)/2$ is a non-normal faithful trace on $L^\infty[0,1]$. For each $n \geq 1$, let f_n be the piecewise linear function whose graph joins the points $(0,1)$, $(1/n, 0)$ and $(1, 0)$. It is easy to check that $\{f_n\}_{n=1}^{\infty}$ is $\| \cdot \|_{2,\tau}$-Cauchy, and that the only candidate for a possible limit in $L^2(A, \tau)$ is 0. However, $\|f_n\|_{2,\tau}^2 \geq \tau_0(f_n^2)/2 = 1/2$, showing that the closed unit ball of A is not $\| \cdot \|_{2,\tau}$-complete.   □

Recall that an ideal in a von Neumann algebra is self-adjoint by the polar decomposition (see [105, p. 444]).

Throughout this section, let M denote a finite von Neumann algebra with centre Z and centre-valued trace $\mathbb{T}$. Let Ω be the carrier space of the C^*-algebra Z, that is the space of multiplicative linear functionals on Z, and write $\omega(z) = z(\omega)$ for all $z \in Z$ and $\omega \in \Omega$ by identifying Z and the algebra $C(\Omega)$ of all continuous functions on Ω. Let

$$|y|_{2,\mathbb{T}} = \left(\mathbb{T}(y^*y)\right)^{1/2} \quad \text{for all } y \in M. \tag{A.3.1}$$

This is a Z^+-valued modulus, or norm function, on M, that has the properties

$$|x + y|_{2,\mathbb{T}} \le |x|_{2,\mathbb{T}} + |y|_{2,\mathbb{T}} \tag{A.3.2}$$

and

$$|zx|_{2,\mathbb{T}} = |z||x|_{2,\mathbb{T}} \tag{A.3.3}$$

for all $z \in Z$, $x, y \in M$. Both of these follow directly from properties of $\mathbb{T}$ by acting with all homomorphisms $\omega \in \Omega$ on Z. For example, (A.3.2) is an inequality in an abelian C^*-algebra, so can be verified pointwise over the carrier space. It is then just a disguised version of the triangle inequality for the semi-norms arising from the semi-inner products $\langle x, y \rangle_\omega := \mathbb{T}(y^*x)(\omega)$ for $x, y \in M$ and $\omega \in \Omega$.

In subsequent applications M will usually be $\ell^\infty(\mathbb{N}) \overline{\otimes} N$ for a separable II_1 factor N.

In the next theorem we use the strong topology, so we assume that M is standardly represented on the Hilbert space $L^2(M)$. The result is not dependent on the particular representation chosen.

Theorem A.3.5. *Let M be a finite von Neumann algebra with centre Z, let Ω be the carrier space of Z and let $\mathcal{I}$ be a two-sided ideal in M.*

(i) *Then $\mathcal{I}$ is a maximal two-sided ideal in M if, and only if, there is an $\omega \in \Omega$ such that*

$$\mathcal{I} = \{x \in M : \mathbb{T}(x^*x)(\omega) = 0\}. \tag{A.3.4}$$

(ii) *If $\mathcal{I}$ is a maximal ideal in M with $\mathcal{I} = \{x \in M : \mathbb{T}(x^*x)(\omega) = 0\}$, then $M/\mathcal{I}$ is a finite factor with trace $\tau_\omega(x + \mathcal{I}) = \mathbb{T}(x)(\omega)$ for all $x \in M$, and τ_ω is normal.*

Proof. (i) Let $\mathcal{I}$ be a maximal two-sided ideal in M. Let $C(x)$ be the norm closed convex hull of the set $\{uxu^* : u \in \mathcal{U}(M)\}$ for all $x \in M$, where $\mathcal{U}(M)$ is the unitary group of M. Thus

$$C(x) = \overline{\text{conv}}^{\|\cdot\|}\{uxu^* : u \in \mathcal{U}(M)\}. \tag{A.3.5}$$

The norm continuity of the centre-valued trace implies that $\mathbb{T}(y) = \mathbb{T}(x)$ for all $y \in C(x)$. By the Dixmier Approximation Theorem, Theorem 2.2.2,

$$\mathcal{C}(x) \cap Z = \{\mathbb{T}(x)\} \quad \text{for all } x \in M. \tag{A.3.6}$$

Maximal ideals in unital Banach algebras are closed, so $C(x) \subseteq \mathcal{I}$ and

$$\mathbb{T}(x) \in \mathcal{I} \cap Z \text{ for all } x \in \mathcal{I}. \tag{A.3.7}$$

Since $\mathcal{I} \cap Z$ is an ideal in Z, there is an $\omega \in \Omega$ such that $\mathcal{I} \cap Z \subseteq \ker \omega$. Define

$$\mathcal{I}_\omega = \{x \in M \colon \mathbb{T}(x^*x)(\omega) = 0\}, \tag{A.3.8}$$

a two-sided ideal in M. If $x \in \mathcal{I}$, then $\mathbb{T}(x^*x) \in \mathcal{I} \cap Z \subseteq \mathcal{I}_\omega$ by equation (A.3.7) so that $\mathbb{T}(x^*x)(\omega) = 0$ and $x \in \mathcal{I}_\omega$. The maximality of $\mathcal{I}$ implies that $\mathcal{I} = \mathcal{I}_\omega$.

Conversely, suppose that $\mathcal{I}$ has the form given in equation (A.3.4) for a fixed $\omega \in \Omega$. Then choose a maximal proper ideal $\mathcal{J}$ which contains $\mathcal{I}$. The earlier part of the proof shows that $\mathcal{J}$ has the form

$$\mathcal{J} = \{x \in M \colon \mathbb{T}(x^*x)(\omega_0) = 0\}, \tag{A.3.9}$$

for some $\omega_0 \in \Omega$. Since

$$\ker \omega = \mathcal{I} \cap Z \subseteq \mathcal{J} \cap Z = \ker \omega_0, \tag{A.3.10}$$

we conclude that $\omega = \omega_0$, otherwise we would have a strict containment $\ker \omega \subsetneq \ker \omega_0$ of proper maximal ideals in Z. It follows that $\mathcal{I} = \mathcal{J}$, showing that $\mathcal{I}$ is maximal.

(ii)   The trace τ_ω is defined on the C^*-algebra $M/\mathcal{I}$ by

$$\tau_\omega(y + \mathcal{I}) = \mathbb{T}(y)(\omega) \text{ for all } y \in M. \tag{A.3.11}$$

It is clearly a faithful trace on the simple C^*-algebra $M/\mathcal{I}$.

To show that $M/\mathcal{I}$ is a von Neumann algebra it is sufficient, by Lemma A.3.3, to prove that the closed unit ball of $M/\mathcal{I}$ is complete in the $\|\cdot\|_2$-norm induced by τ_ω, and denoted by

$$\|x\|_{2,\,\omega} = \left(\tau_\omega(x^*x)\right)^{1/2} \ (x \in M/\mathcal{I}). \tag{A.3.12}$$

Let x_n be in the unit ball of $M/\mathcal{I}$ with $x_1 = 0$ and

$$\|x_n - x_{n-1}\|_{2,\,\omega} \le 2^{-n} \text{ for all } n \ge 1. \tag{A.3.13}$$

By a standard result in the theory of quotient C^*-algebras, if $x \in M/\mathcal{I}$, then there is a $y \in M$ with $x = y + \mathcal{I}$ and $\|y\| = \|x\|$ (see [187, p. 34]). Thus we may choose y_n in the closed unit ball of M such that $x_n = y_n + \mathcal{I}$ for all n, and $y_0 = 0$.

By induction on n we shall choose a sequence w_n in M with the properties that $w_0 = 0$, $w_1 = y_1$, $x_n = w_n + \mathcal{I}$, and

$$|w_n - w_{n-1}|_{2,\,\mathbb{T}} \le 2^{-n+1}1 \tag{A.3.14}$$

for all $n \in \mathbb{N}$.

Suppose that $w_0, \ldots, w_{n-1}$ have been chosen for some $n \geq 2$. Since x_{n-1} is represented by w_{n-1} in M and x_n by y_n,

$$
\begin{aligned}
|y_n - w_{n-1}|_{2,\mathrm{T}}(w) &= \Big(\mathrm{T}\big((y_n - w_{n-1})^*(y_n - w_{n-1})\big)\Big)^{1/2}(w) \\
&= \Big(\mathrm{T}\big((y_n - w_{n-1})^*(y_n - w_{n-1})\big)(w)\Big)^{1/2} \\
&= \|x_n - x_{n-1}\|_{2,w} < 2^{-n}
\end{aligned}
\tag{A.3.15}
$$

by the multiplicativity of w. Let

$$
U = \big\{\nu \in \Omega \colon |y_n - w_{n-1}|_{2,\mathrm{T}}(\nu) < 2^{-n+1}\big\}.
\tag{A.3.16}
$$

Then U is an open neighbourhood of w in the compact Hausdorff space Ω. By Urysohn's lemma, there is a $z \in Z \cong C(\Omega)$ such that

(i) $0 \leq z \leq 1$,

(ii) $z(\omega) = 1$, and

(iii) $z(\nu) = 0$ for all $\nu \in \Omega \setminus U$.

Let

$$
w_n = z y_n + (1 - z) w_{n-1},
\tag{A.3.17}
$$

which is a standard type of splitting in $\beta \mathbb{N}$ used in the theory of ultrapowers. Then

$$
y_n - w_n = (1 - z)(y_n - w_{n-1})
\tag{A.3.18}
$$

so that

$$
|y_n - w_n|_{2,\mathrm{T}}(\omega) = (1 - z)(\omega)|y_n - w_{n-1}|_{2,\mathrm{T}}(\omega) = 0,
\tag{A.3.19}
$$

by the Z-module property of $|\cdot|_{2,\mathrm{T}}$ and by the choice of z in (ii). Thus w_n represents x_n in M. The Z-module property of $|\cdot|_{2,\mathrm{T}}$ and $0 \leq z \leq 1$ imply that

$$
|w_n - w_{n-1}|_{2,\mathrm{T}} = |z(y_n - w_{n-1})|_{2,\mathrm{T}} = z|y_n - w_{n-1}|_{2,\mathrm{T}}.
\tag{A.3.20}
$$

If $\nu \in \Omega \setminus U$, then $z(\nu) = 0$ so that

$$
|w_n - w_{n-1}|_{2,\mathrm{T}}(\nu) = 0.
\tag{A.3.21}
$$

If $\nu \in U$, then $|y_n - w_{n-1}|_{2,\mathrm{T}}(\nu) < 2^{-n+1}$ so that

$$
|w_n - w_{n-1}|_{2,\mathrm{T}}(\nu) < 2^{-n+1}1.
\tag{A.3.22}
$$

This completes the inductive construction of w_n.

The sequence of w_n's is $\|\cdot\|_2$-Cauchy in the unit ball of M, and so converges to w in this ball by Lemma A.3.3. If $m \in M$, then

$$
\begin{aligned}
\|(w - w_n)m\xi\|_2^2 &= \tau((w - w_n)mm^*(w - w_n)^*) \leq \tau((w - w_n)\|m\|^2(w - w_n)^*) \\
&= \|m\|^2\|w - w_n\|_2^2,
\end{aligned}
\tag{A.3.23}
$$

and so the uniform bound $\|w_n\| \leq 1$ and the density of $M\xi$ in $L^2(M)$ show that $w_n \to w$ strongly.

Consider $z \in Z^+$, $\tau(z) \leq 1$. Then

$$
\begin{aligned}
\tau\Big(\mathbb{T}((w-w_n)^*(w-w_n))z\Big) &= \tau((w-w_n)^*(w-w_n)z) \\
&= \|(w-w_n)z^{1/2}\xi\|_2^2 \\
&= \lim_{j\to\infty} \|(w_j-w_n)z^{1/2}\xi\|_2^2 \\
&= \lim_{j\to\infty} \tau((w_j-w_n)^*(w_j-w_n)z) \\
&= \lim_{j\to\infty} \tau\Big(\mathbb{T}((w_j-w_n)^*(w_j-w_n))z\Big) \\
&= \lim_{j\to\infty} \| |w_j-w_n|_{2,\mathbb{T}} z^{1/2}\xi\|_2^2 \\
&\leq \limsup_{j\to\infty} \| |w_j-w_n|_{2,\mathbb{T}}\|_\infty^2,
\end{aligned}
\tag{A.3.24}
$$

where we have used several times that $\mathbb{T}$ is a trace preserving Z-module map, and also that $w_j \to w$ strongly in the third equality. If we take the supremum over all such elements $z \in Z^+$ in (A.3.24), then we obtain

$$
\| |w-w_n|_{2,\mathbb{T}}\|_\infty \leq \limsup_{j\to\infty} \| |w_j-w_n|_{2,\mathbb{T}}\|_\infty.
\tag{A.3.25}
$$

Since

$$
\begin{aligned}
|w_j - w_n|_{2,\mathbb{T}} &\leq \sum_{k=n}^{j-1} |w_{k+1}-w_k|_{2,\mathbb{T}} \\
&\leq \sum_{k=n}^{j-1} 2^{-k} 1 \leq 2^{-n+1},
\end{aligned}
\tag{A.3.26}
$$

by (A.3.22) and the triangle inequality (A.3.2), we obtain

$$
\| |w-w_n|_{2,\mathbb{T}}\|_\infty \leq 2^{-n+1} 1
\tag{A.3.27}
$$

from (A.3.25) and (A.3.26). Evaluation at ω in (A.3.27) gives

$$
\mathbb{T}((w-w_n)^*(w-w_n))(\omega) \leq 4^{-n+1},
\tag{A.3.28}
$$

and so $\|x-x_n\|_{2,\omega} \leq 2^{-n+1}$, where $x := w + \mathcal{I} \in M/\mathcal{I}$. Thus the unit ball of $M/\mathcal{I}$ is complete in the $\|\cdot\|_{2,\omega}$-norm, so $M/\mathcal{I}$ is a von Neumann algebra by Lemma A.3.3. Normality of τ_ω follows from Remark A.3.4. □

The method of splitting used in the construction of w_n in equation (A.3.17) is a standard technique in ultrapower constructions (see the next section or [114] for examples).

A.4 The algebra N_n^ω

The most important application of Theorem A.3.5 is when N is a separable II_1 factor, $M = N \overline{\otimes} \ell^\infty(\mathbb{N})$ and ω is in $\beta\mathbb{N} \setminus \mathbb{N}$. Here we introduce the more general algebras N_n^ω and define N^ω before studying it in the next section.

Definition A.4.1. Let N_n $(n \in \mathbb{N})$ be a sequence of finite factors with traces τ_n and let $\omega \in \beta\mathbb{N} \setminus \mathbb{N}$. If $M = \bigoplus_n N_n$, then M is a finite von Neumann algebra, since

$$(x_n) \mapsto \tau_j(x_j) \colon M \to \mathbb{C} \quad (j \in \mathbb{N}) \tag{A.4.1}$$

form a separating set of normal traces on M. A natural faithful normal representation of M is obtained by direct summing and representing it on $\bigoplus_n L^2(N_n, \tau_n)$ in the usual way. The centre $Z = \bigoplus_n \mathbb{C}1$ of M is isomorphic to $\ell^\infty(\mathbb{N})$. From the Z-module properties of $\mathbb{T}$ on M using the characteristic function of $\{j\}$ in Z, it follows that $\mathbb{T}(x)(j) = \tau_j(x_j)$ for all $x = (x_n) \in M$ and all $j \in \mathbb{N}$. Tautologically the topology of $\beta\mathbb{N}$ is the weak topology induced by $Z \cong C(\beta\mathbb{N})$, so

$$z(\omega) = \lim_\omega z_n \quad \text{for all} \quad z = (z_n) \in Z. \tag{A.4.2}$$

Hence

$$\mathbb{T}(x)(\omega) = \lim_\omega \mathbb{T}(x)(n) = \lim_\omega \tau_n(x_n) \tag{A.4.3}$$

for all $x = (x_n) \in M$ so that

$$\mathcal{I}_\omega = \big\{ (x_n) \in M \colon \lim_\omega \tau_n(x_n^* x_n) = 0 \big\}. \tag{A.4.4}$$

Define τ_ω on $M/\mathcal{I}_\omega$ by

$$\tau_\omega(x + \mathcal{I}_\omega) = \mathbb{T}(x)(\omega) = \lim_\omega \tau_n(x_n) \tag{A.4.5}$$

for all $x = (x_n) \in M$. Then $(M/\mathcal{I}_\omega, \tau_\omega)$ is a finite von Neumann algebra, which is denoted by $(N_n^\omega, \tau_\omega)$. If $N_n = N$ for all $n \in \mathbb{N}$, denote this algebra by N^ω.

Let π denote the quotient map $x \mapsto x + \mathcal{I}_\omega \colon M \to M/\mathcal{I}_\omega$. This *-homomorphism is not normal as $\ker \pi \cap Z = \ker \omega$ is not weakly closed in Z. The weakly closed maximal ideals in $\ell^\infty(\mathbb{N})$ are of the form $\{(x_n) \in \ell^\infty(\mathbb{N}) \colon x_j = 0\}$ for some $j \in \mathbb{N}$.

If B is a C^*-subalgebra and $\mathcal{I}$ is a closed two-sided ideal of a C^*-algebra M, then it is standard theory that $B + \mathcal{I}$ is a C^*-algebra (see [131, Corollary 1.5.8]) and so $B + \mathcal{I}/\mathcal{I} \cong B/B \cap \mathcal{I}$ is a C^*-subalgebra of $M/\mathcal{I}$. If (M_n, τ_n) is a von Neumann subalgebra of the II_1 factor (N_n, τ_n), for all $n \in \mathbb{N}$, and ω is an ultrafilter, then M_n^ω is naturally identified with a von Neumann subalgebra of N_n^ω by the isomorphism ψ

$$(y_n) + \Big(\bigoplus_n M_n\Big) \cap \mathcal{I}_\omega \mapsto (y_n) + \mathcal{I}_\omega \colon \bigoplus_n M_n / \Big(\bigoplus_n M_n\Big) \cap \mathcal{I}_\omega \to \Big(\bigoplus_n M_n\Big) + \mathcal{I}_\omega / \mathcal{I}_\omega.$$

This ψ is a *-isomorphism and the image is a C^*-algebra. If $\widetilde{M} = \bigoplus_n M_n$ and $\widetilde{N} = \bigoplus_n N_n$, then

$$\tau_{M_n^\omega}(\pi(x_n)) = \mathbb{T}_{Z(\widetilde{M})}(x_n)(\omega) = \lim_\omega \tau_n(x_n)$$
$$= \mathbb{T}_{Z(\widetilde{N})}(x_n)(\omega) = \tau_{N_n^\omega}(\pi(x_n)), \qquad (A.4.6)$$

for all $(x_n) \in \widetilde{M}$, where $\mathbb{T}_{Z(\widetilde{M})}$ and $\mathbb{T}_{Z(\widetilde{N})}$ are the centre-valued traces on $\widetilde{M}$ and $\widetilde{N}$, respectively. This shows that ψ is trace preserving so that $M_n^\omega \cong (\bigoplus_n M_n) + \mathcal{I}_\omega / \mathcal{I}_\omega$ is a von Neumann subalgebra of $N_n^\omega = \bigoplus_n N_n / \mathcal{I}_\omega$.

The following lemma finds the relative commutant of this embedding and shows that N_n^ω is a factor by showing that the operations of taking relative commutants and ultrapowers commute.

Lemma A.4.2. *If N_n is a finite factor with a von Neumann subalgebra B_n for each $n \in \mathbb{N}$, then N_n^ω is a finite factor and $(B_n^\omega)' \cap N_n^\omega = (B_n' \cap N_n)^\omega$.*

Proof. It is sufficient to prove that $((B_n)^\omega)' \cap (N_n)^\omega = (B_n' \cap N_n)^\omega$, since the factor part will follow from this on taking $B_n = N_n$ for all $n \in \mathbb{N}$ and using $\mathbb{C}^\omega = \ell^\infty(\mathbb{N}) / \ker \omega \cong \mathbb{C}$.

From $\bigoplus(B_n' \cap N_n) = (\bigoplus B_n') \cap \bigoplus N_n$ we have

$$(B_n' \cap N_n)^\omega \subseteq (B_n^\omega)' \cap N_n^\omega. \qquad (A.4.7)$$

If $x = (x_n) + \mathcal{I}_\omega \in (B_n^\omega)' \cap N_n^\omega \setminus (B_n' \cap N_n)^\omega$, then by equation (A.4.7)

$$(\mathbb{E}_{B_n' \cap N_n}(x_n)) + \mathcal{I}_\omega \in (B_n^\omega)'.$$

Let $\varepsilon > 0$. By Lemma 3.6.5 (ii), there is a unitary $u_n \in B_n$ such that

$$\|u_n x_n - x_n u_n\|_{2, \tau_n} + \varepsilon \geq 2^{1/2} \|x_n - \mathbb{E}_{B_n' \cap N_n}(x_n)\|_{2, \tau_n} \qquad (A.4.8)$$

for all $n \in \mathbb{N}$. Let $u = (u_n) + \mathcal{I}_\omega$; a unitary in B_n^ω. Taking $\lim_\omega$ in inequality (A.4.8) and using

$$\lim_\omega \|x_n\|_{2, \tau_n} = \left(\lim_\omega \tau_n(x_n^* x_n) \right)^{1/2} = \|x\|_{2, \omega} \qquad (A.4.9)$$

give

$$\|ux - xu\|_{2, \omega} + \varepsilon \geq 2^{1/2} \|x - ((\mathbb{E}_{B_n' \cap N}(x_n)) + \mathcal{I}_\omega)\|_{2, \omega}. \qquad (A.4.10)$$

Since $ux = xu$, this implies that $x = (\mathbb{E}_{B_n' \cap N}(x_n)) + \mathcal{I}_\omega$. $\qquad \square$

The next corollary follows directly from Lemma A.4.2; for an example related to it see Remark A.5.6.

Corollary A.4.3. *If N_n is a finite factor containing a masa B_n for each $n \in \mathbb{N}$, then B_n^ω is a masa in N_n^ω.*

Lemma A.4.4. *If N_n is a II_1 factor for all n, or a finite I_{k_n} factor with $k_n \in \mathbb{N}$ tending to infinity, then N_n^ω is non-separable and contains a masa with non-separable predual.*

Proof. The proof for N_n a I_{k_n} factor for each n needs minor modification from the proof here.

The proof is given for the case where each N_n contains a finite dimensional abelian *-subalgebra A_n with $\dim(A_n) = 2^n$ and each minimal projection in A_n has trace 2^{-n}. Clearly this is possible if N_n is a II_1 factor. For each n, the finite tracial algebra (A_n, τ) is isomorphic to

$$(B_n, \tau) = \Big(\bigotimes_1^n (\mathbb{C}e_0 \oplus \mathbb{C}e_1), \tau \Big), \qquad (A.4.11)$$

where the e_j's are projections with $\tau(e_j) = 1/2$ and τ is tensorial on the tensor product. If $\alpha = (\alpha_1, \alpha_2, \ldots) \in \{0,1\}^\infty$ is an infinite binary sequence, let $u(\alpha)_n$ be the self-adjoint unitary in A_n corresponding to the unitary

$$\otimes_{j=1}^n \big(e_0 + (-1)^{\alpha_j} e_1\big) \in B_n.$$

Let $\widetilde{u}(\alpha) = (u(\alpha)_1, u(\alpha)_2, \ldots) \in \ell^\infty(N_n)$ and let $u(\alpha) = \pi(\widetilde{u}(\alpha))$. Then $u(\alpha)$ is a unitary in N_n^ω for all α.

If $\alpha, \beta \in \{0,1\}^\infty$ with $\alpha_j \neq \beta_j$ for some $j \in \mathbb{N}$, then $\tau(u(\alpha)_n u(\beta)_n^*) = 0$ for all $n \geq j$, because this contains the factor $\tau(e_0 - e_1) = 0$ arising from the jth place. Hence

$$\tau_\omega(u(\alpha)u(\beta)^*) = \lim_\omega \tau(u(\alpha)_n u(\beta)_n^*) = 0. \qquad (A.4.12)$$

Thus $\{u(\alpha) : \alpha \in \{0,1\}^\infty\}$ is an uncountable orthonormal set in $L^2(N_n^\omega)$ that is contained in the abelian von Neumann algebra A_n^ω. Thus there is masa B in N_n^ω containing A_n^ω, so B and N_n^ω have non-separable preduals. $\square$

Remark A.4.5. (1) Note that Popa [138] has shown that if N_n is a separable II_1 factor N for all n, then all masas in N^ω have non-separable predual; this is proved in Theorem 15.2.3.

(2) If the N_n are permitted to vary, then for different $\omega \in \beta\mathbb{N} \setminus \mathbb{N}$ totally different N_n^ω may result.

For example, if N is a separable II_1 factor, we may define

$$N_n = \begin{cases} \mathbb{M}_2 & \text{if } n \text{ is odd}, \\ N & \text{if } n \text{ is even}. \end{cases} \qquad (A.4.13)$$

Then

$$N_n^\omega = \begin{cases} \mathbb{M}_2 & \text{if } \{\text{all odd integers}\} \in \omega, \\ N^\omega & \text{if } \{\text{all even integers}\} \in \omega. \end{cases} \qquad (A.4.14)$$

(3)   Let u_n be unitaries in N_n and let B_n be von Neumann subalgebras of N_n for all n. If $u = \pi(u_n)$, then $uB_n^{\omega}u^* = (u_nB_nu_n^*)^{\omega}$, since

$$uB_n^{\omega}u^* = \pi(u_n)\pi\left(\bigoplus B_n\right)\pi(u_n)^*$$
$$= \pi\left(\bigoplus u_nB_nu_n^*\right) = (u_nB_nu_n^*)^{\omega}. \qquad (A.4.15)$$

$\square$

A.5   The ultrapower N^{ω}

Throughout this section N will denote a separable II_1 factor with trace τ, ω will be in $\beta\mathbb{N}\setminus\mathbb{N}$ and n will be the running index in $\mathbb{N}$. The centre of $\ell^{\infty}(N) \cong N\overline{\otimes}\ell^{\infty}(\mathbb{C})$ is denoted by Z and the centre-valued trace by $\mathbb{T}$. Let

$$\mathcal{I}_{\omega} = \{x \in \ell^{\infty}(N): \mathbb{T}(xx^*)(\omega) = 0\},$$
$$N^{\omega} = \ell^{\infty}(N)/\mathcal{I}_{\omega},$$
$$\pi: \ell^{\infty}(N) \to N^{\omega}: x \to x + \mathcal{I}_{\omega},$$
$$\tau_{\omega}(x + \mathcal{I}_{\omega}) = \lim_{\omega} \tau(x_n) \quad \text{and}$$
$$\|x + \mathcal{I}_{\omega}\|_{2,\omega} = \left(\tau_{\omega}(xx^* + \mathcal{I}_{\omega})\right)^{1/2} \quad \text{for all} \quad x = (x_n) \in \ell^{\infty}(N).$$

In the previous section we showed that $(N^{\omega}, \tau_{\omega})$ is a non-separable II_1 factor, called the *ultrapower* of N.

When representing an $x \in N^{\omega}$ as $x = \pi(x_n)$, we shall *always* assume that $(x_n) \in \ell^{\infty}(N)$ satisfies $\|x_n\| \leq \|x\|$ for all $n \in \mathbb{N}$, by the standard result in the theory of quotient of C^*-algebras that there is an $(x_n) \in \ell^{\infty}(N)$ with $x = \pi(x_n)$ and $\|(x_n)\| = \|x\|$ (see [187, p. 34]). Expressions like $\pi((x_n))$ will usually be written $\pi(x_n)$ by omitting the second pair of brackets.

There is a natural trace preserving embedding of N into N^{ω} given by

$$N \hookrightarrow N^{\omega}: x \mapsto (x, x, \ldots) + \mathcal{I}_{\omega}.$$

Let $N_{\omega} = N' \cap N^{\omega}$, which is called the *central sequence algebra* of N.

Give that N_n^{ω} depends on ω if the finite von Neumann algebras N_n vary, a natural question is: *does N^{ω} depend on ω?* Using a result of Keisler from model theory [11], which assumes the continuum hypothesis, Ge and Hadwin [78, Theorem 3.2] proved the following theorem.

Theorem A.5.1. *Let N be a finite von Neumann algebra with a fixed faithful normal trace. If ω and α are in $\beta\mathbb{N}\setminus\mathbb{N}$, then there is an isomorphism from N^{ω} onto N^{α} that is the identity map on N and hence maps N_{ω} onto N_{α}.*

In studying N^{ω}, it is useful to lift operators in this quotient algebra to sequences of operators from N in such a way as to preserve structure. For example, we would like a projection to lift to a sequence of projections, rather than to a sequence of operators which are asymptotically projections. We show in Theorem A.5.3 that such liftings are always possible. To prove this for projections and partial isometries in N^ω, an approximation lemma is required that is similar to results in Section 5.3. More general versions of these results can be found in, for example, [189, p. 96].

Lemma A.5.2. *Let N be a II_1 factor. Let $y \in N$ satisfy $0 \le y \le 1$, and let $\delta = \|y - y^2\|_2^{1/2}$. If e is the spectral projection for y corresponding to the interval $[1 - \delta, 1]$, then $\|y - e\|_2 \le 3^{1/2}\|y - y^2\|_2^{1/2}$.*

Proof. The state τ on $\{y\}''$ induces a probability measure μ on $[0, 1]$ by the spectral theorem such that

$$\tau(f(y)) = \int_0^1 f(t) d\mu(t) \tag{A.5.1}$$

for all bounded Borel functions f on $[0, 1]$, since the spectrum of y is contained in $[0, 1]$. Then

$$\tau\big(y^2(1 - y)^2\big) = \int_0^1 t^2(1 - t)^2 \, d\mu(t) = \delta^4. \tag{A.5.2}$$

The function $t \mapsto t^2(1 - t)^2$ is symmetric about $t = 1/2$ and increases from 0 to $1/2$ so that $\delta^2(1 - \delta)^2 \le t^2(1 - t)^2$ for $\delta \le t \le 1 - \delta$. Hence

$$\delta^2(1 - \delta)^2\mu\big([\delta, \, 1 - \delta]\big) \le \delta^4 \tag{A.5.3}$$

and

$$(1 - \delta)^2\mu\big([\delta, \, 1 - \delta]\big) \le \delta^2. \tag{A.5.4}$$

Since the characteristic function for the interval $[1 - \delta, \, 1]$ is associated with e,

$$
\begin{aligned}
\|y - e\|_2^2 &= \tau(y^2 - 2ey + e) \\
&= \tau\big((1 - e)y^2\big) + \tau\big(e(y - 1)^2\big) \\
&= \int_0^{1-\delta} t^2 \, d\mu(t) + \int_{1-\delta}^1 (t - 1)^2 \, d\mu(t) \\
&\le \int_0^\delta t^2 \, d\mu(t) + \int_\delta^{1-\delta} t^2 \, d\mu(t) + \delta^2 \\
&\le \delta^2 + (1 - \delta)^2\mu\big([\delta, \, 1 - \delta]\big) + \delta^2 \\
&\le 3\delta^2
\end{aligned}
\tag{A.5.5}
$$

by inequality (A.5.4). $\qquad\square$

The following representation theorem lifts various elements from N^ω back to $\ell^\infty(N)$ and is useful as a technical tool. The second case could be easily deduced from the fourth but is sufficiently important to warrant its own statement.

Theorem A.5.3. *Let N be a separable II_1 factor.*

(i) *There is a natural monomorphism* $\theta \mapsto \theta_\omega$ *from* $\mathrm{Aut}(N)$ *into* $\mathrm{Aut}(N^\omega)$ *sending* $\mathrm{Inn}(N)$ *into* $\mathrm{Inn}(N^\omega)$ *defined by*

$$\theta_\omega(x) = \pi\big(\theta(x_n)\big) \quad \text{for all} \quad x = (x_n) + \mathcal{I}_\omega \in N^\omega. \tag{A.5.6}$$

(ii) *Each unitary* u *in* N^ω *has a representation* $u = \pi(u_n)$ *with* u_n *a unitary in* N *and* $\tau(u_n) = \tau_\omega(u)$ *for all* $n \in \mathbb{N}$. *In particular, if* $\tau_\omega(u) = 0$, *then* u_n *may be chosen to satisfy* $\tau(u_n) = 0$ *for all* $n \in \mathbb{N}$.

(iii) *Each projection* $p \in N^\omega$ *has a representation* $p = \pi(p_n)$ *with* p_n *a projection in* N *with* $\tau(p_n) = \tau_\omega(p)$ *for all* $n \in \mathbb{N}$.

(iv) *If* v *is a partial isometry in* N^ω *with* $vv^* = p$ *and* $v^*v = q$, *then there are partial isometries* v_n *in* N *for all* $n \in \mathbb{N}$ *such that* $\pi(v_n) = v$ *and* $\tau(v_n v_n^*) = \tau_\omega(p)$ *for all* $n \in \mathbb{N}$.

(v) *If* p *and* q *are orthogonal projections in* N^ω *with projections* $p_n \in N$ *such that* $\tau(p_n) = \tau_\omega(p)$ *for all* n *and* $\pi(p_n) = p$, *then there are projections* $q_n \in N$ *with* $q_n p_n = 0$ *and* $\tau(q_n) = \tau_\omega(q)$ *for all* n, *and* $\pi(q_n) = q$.

(vi) *If* e_{ij} $(1 \le i, j \le m)$ *are matrix units in* N^ω, *then there are matrix units* $e_{n;\,ij}$ $(1 \le i, j \le m)$ *in* N *for all* $n \in \mathbb{N}$ *such that* $\pi(e_{n;\,ij}) = e_{ij}$ $(1 \le i, j \le m)$.

Proof. (i)   The automorphism θ of N lifts to an automorphism θ^∞ of $\ell^\infty(N)$ by $\theta^\infty\big((x_n)\big) = \big(\theta(x_n)\big)$ for all $(x_n) \in \ell^\infty(N)$ and the trace preserving property of θ implies that $\theta^\infty(\mathcal{I}_\omega) = \mathcal{I}_\omega$ from which the other properties follow directly.
(ii)   If $u \in N^\omega$ is a unitary, then choose a self-adjoint element $h \in N^\omega$ such that $e^{ih} = u$. Lift h to a self-adjoint sequence $(h_n) \in N \overline{\otimes} \ell^\infty(\mathbb{N})$, and then let $u_n = e^{ih_n}$. The continuous functional calculus commutes with *-homomorphisms in the sense that $\rho(f(x)) = f(\rho(x))$ for x self-adjoint, ρ a *-homomorphism, and $f \in C_b(\mathbb{R})$. Thus $\pi(u_n) = u$.
If $\tau(u) = 0$, then for the second part apply Corollary 13.4.3 using $\tau_\omega(u) = \lim_\omega \tau(u_n) = 0$ to modify each initial unitary u_n to a unitary $\tilde{u}_n$ with $\tau(\tilde{u}_n) = 0$ and $\|u_n - \tilde{u}_n\|_2 < 26|\tau(u_n)|$ for all n. In general, apply Remark 13.4.9 to conclude that the unitaries u_n can be chosen to satisfy $\tau(u_n) = \tau_\omega(u)$. The constant 26 must be replaced by a constant K dependent on $|\tau_\omega(u)|$.
(iii)   Let $\pi(x_n) = p$ with $\|x_n\| \le 1$ for all n. Let $y_n = x_n^* x_n$ so that $0 \le y_n \le 1$ for all n and $\pi(y_n) = \pi(x_n)^*\pi(x_n) = p$. Let $\delta_n = \|y_n - y_n^2\|_2^{1/2}$ and let e_n be the spectral projection for y_n corresponding to the interval $[1 - \delta_n, 1]$ for all n. By Lemma A.5.2 (i),

$$\|y_n - e_n\|_2 \le \sqrt{3}\delta_n. \tag{A.5.7}$$

Since $\pi(y_n^2) = p^2 = p = \pi(y_n)$, we have $\lim_\omega \|y_n - y_n^2\|_2 = \lim_\omega \delta_n^2 = 0$ so that $\lim_\omega \delta_n = 0$ by the homomorphism property of $\lim_\omega$. Thus $\pi(e_n) = \pi(y_n) = p$ by inequality (A.5.7). Let $\varepsilon_n = \tau_\omega(p) - \tau(e_n)$ for all n. Then $\lim_\omega \varepsilon_n = 0$, since $p = \pi(e_n)$. If $\tau(e_n) < \tau_\omega(p)$, choose a projection $q_n \in N$ orthogonal to e_n with $\tau(q_n) = \tau_\omega(p) - \tau(e_n)$ and let $p_n = e_n + q_n$. If $\tau(e_n) = \tau_\omega(p)$, choose $p_n = e_n$.

If $\tau(e_n) > \tau_\omega(p)$, choose a projection $q_n \le e_n$ with $\tau(q_n) = \tau(e_n) - \tau_\omega(p)$ and let $p_n = e_n - q_n$. From $\tau(q_n) \le |\varepsilon_n|$ for all n, it follows that $\lim_\omega \tau(q_n) = 0$ so that $(q_n) \in \mathcal{I}_\omega$. Hence $\pi(p_n) = \pi(e_n) = p$ as required.

(iv)   By (iii), represent $p = \pi(p_n)$ and $q = \pi(q_n)$ with p_n and q_n projections in N and $\tau(p_n) = \tau_\omega(p) = \tau_\omega(q) = \tau(q_n)$ for all n. Represent $v = \pi(x_n)$ with $\|x_n\| \le 1$ for all n and let $y_n = p_n x_n q_n$ for all n. Then

$$\pi(y_n) = \pi(p_n)\pi(x_n)\pi(q_n) = pvq = v. \tag{A.5.8}$$

Let $y_n = r_n h_n$ be the polar decomposition of y_n with r_n a partial isometry and $h_n = (y_n^* y_n)^{1/2}$ with $0 \le h_n \le 1$ for all n. Then $h_n^2 = y_n^* y_n$ for all n so that

$$\pi(h_n^2) = \pi(y_n)^* \pi(y_n) = v^* v = q = q^2 = \pi(h_n^4) \tag{A.5.9}$$

and $\lim_\omega \|h_n^2 - h_n^4\|_2 = 0$. Let $\delta_n = \|h_n^2 - h_n^4\|_2^{1/2}$ so that $0 \le \delta_n \le 1$ for all n and $\lim_\omega \delta_n^2 = 0$. Hence $\lim_\omega \delta_n = 0$.

Let e_n be the spectral projection for h_n^2 corresponding to the interval $[1 - \delta_n, 1]$ so that $e_n \le p_n$ for all n, since p_n is the support projection for h_n^2. By Lemma A.5.2, $\|h_n^2 - e_n\|_2 \le 3^{1/2}\delta_n$ so that $\pi(e_n) = \pi(h_n^2) = p$. Let $w_n = v_n e_n$ for all n. Then

$$\pi(w_n) = \pi(v_n)\pi(e_n) = vp = v. \tag{A.5.10}$$

By their definitions,

$$w_n^* w_n = e_n \le p_n \quad \text{and} \quad w_n w_n^* = v_n e_n v_n^* \le q_n. \tag{A.5.11}$$

Since $\tau(p_n - e_n) = \tau(q_n - w_n w_n^*)$, the projections $p_n - e_n$ and $q_n - w_n w_n^*$ are equivalent in N and there is a partial isometry w_n^0 in N such that

$$w_n^{0*} w_n^0 = p_n - e_n \quad \text{and} \quad w_n^0 w_n^{0*} = q_n - w_n w_n^*. \tag{A.5.12}$$

Let $v_n = w_n + w_n^0$. Then $v_n^* v_n = p_n$ and $v_n v_n^* = q_n$ by equations (A.5.12). From equation (A.5.12) and $\pi(e_n) = p = \pi(p_n)$ it follows that

$$\lim_\omega \|w_n^0\|_2^2 = \lim_\omega \tau(p_n - e_n) = \lim_\omega \|p_n - e_n\|_2^2 = 0. \tag{A.5.13}$$

Hence $\pi(w_n^0) = 0$ and $\pi(v_n) = \pi(w_n) + \pi(w_n^0) = v$ as required.

(v)   The proof is a modified version of (iii). Let $\pi(x_n) = q$ with $\|x_n\| \le 1$ for all n. Let $y_n = (1-p_n)x_n^* x_n(1-p_n)$ for all n. Then $\pi(y_n) = (1-p)q^* q(1-p) = q$ so that $\pi(y_n^2) = q$. Let

$$\delta_n = \|y_n - y_n^2\|_2^{1/2} \tag{A.5.14}$$

and let e_n be the spectral projection for y_n corresponding to the interval $[1 - \delta_n, 1]$ for all n. Then $\|y_n - e_n\|_2 \le 3^{1/2}\delta_n$ for all n by Lemma A.5.2 and $\lim_\omega \delta_n^2 = \lim_\omega \delta_n = 0$ by the equation $\pi(y_n - y_n^2) = 0$. Thus $\pi(e_n) = \pi(y_n) = q$. Let $\varepsilon_n = \tau_\omega(q) - \tau(e_n)$ for all n. Then $\lim_\omega \varepsilon_n = 0 = \lim_\omega |\varepsilon_n|$, since $\pi(e_n) = q$. The projection $e_n \le (1-p_n)$ and $\tau(1-p_n) = 1 - \tau_\omega(p) \ge \tau_\omega(q)$. If $\tau(e_n) < \tau_\omega(q)$, choose a projection $q_n^0 \le 1 - p_n - e_n$ with $\tau(q_n^0) = \tau_\omega(q) - \tau(e_n)$ and let

$q_n = e_n + q_n^0$. If $\tau(e_n) = \tau_\omega(q)$, let $q_n = e_n$. If $\tau(e_n) > \tau_\omega(q)$, choose a projection $q_n^0 \leq e_n$ with $\tau(q_n^0) = \tau(e_n) - \tau_\omega(q)$ and let $q_n = e_n - q_n^0$. Then $\tau(q_n) = \tau_\omega(q)$ and $\tau(q_n^0) = |\tau_\omega(q) - \tau(e_n)| = |\varepsilon_n|$ for all n. Hence $\lim_\omega \tau(q_n^0) = 0$ and

$$\pi(q_n) = \pi(e_n \pm q_n^0) = \pi(e_n) = q. \qquad (A.5.15)$$

(vi)   Induction on i using (iii) and (v) shows that there are projections $e_{n;ii}$ $(1 \leq i \leq m, n \in \mathbb{N})$ in N such that $\pi(e_{n;ii}) = e_{ii}$ and $\tau(e_{n;ii}) = \tau_\omega(e_{ii})$ for $1 \leq i \leq m$. Using (iv), the elements $e_{n;i1}$ (or $e_{n;1i}$) are constructed for $1 \leq i \leq m$ and $n \in \mathbb{N}$, from which all $e_{n;ij}$ are obtained by $e_{n;ij} = e_{n;i1}e_{n;j1}^*$ $(1 \leq i,j \leq m)$.   □

One observation that comes from Theorem A.5.3 is that cutdowns and ultrapowers commute. If $p = \pi(p_n) \in N^\omega$ is a projection with p_n projections in N and $\tau(p_n) = \tau_\omega(p)$ for all n, then $pN^\omega p \cong (p_n N p_n)^\omega$ in natural way.

The method of Theorem A.5.3 (iii) yields the following: *If B is a diffuse von Neumann subalgebra of N and p is a projection in B^ω, then there are projections $p_n \in B$ with $\tau(p_n) = \tau_\omega(p)$ for all n such that $\pi(p_n) = p$.*

Conditional expectations are the basic tool for understanding subalgebras. The case of varying B_n in the lemma below is required subsequently in the study of normalisers. Before dealing with how they lift from N to N^ω, as $N^\omega \to B_n^\omega$ and $N^\omega \to B$, here is a little notation.

Let ψ be a bounded linear operator from N into N and let $(x_n) \in \ell^\infty(N)$. Then $\{\psi(x_n) : n \in \mathbb{N}\}$ is contained in the weakly compact subset

$$\{y \in N : \|y\| \leq \|\psi\| \, \|(x_n)\|_\infty\}$$

, so that $\lim_\omega \psi(x_n)$ exists as a weak limit in N with

$$\|\lim_\omega \psi(x_n)\| \leq \|\psi\| \, \|(x_n)\|$$

by the universal convergence property of $\lim_\omega$ with respect to compact sets. The map $\lim_\omega \psi \colon (x_n) \mapsto \lim_\omega \psi(x_n)$ is thus a continuous linear operator from $\ell^\infty(N)$ into N with $\|\lim_\omega \psi\| = \|\psi\|$.

If $x \in N$, regard x as the element $\pi(x)$, where $(x) \in \ell^\infty(N)$ is the constant sequence x.

Lemma A.5.4. *Let N be a II$_1$ factor and let $B, B_1, B_2, \ldots$ be von Neumann subalgebras of N. If $x = \pi(x_n) \in N^\omega$ with $(x_n) \in \ell^\infty(N)$, then*

$$\mathbb{E}_{B_n^\omega}(x) = \pi(\mathbb{E}_{B_n}(x_n)) \qquad (A.5.16)$$

and

$$\mathbb{E}_B(x) = \lim_\omega \mathbb{E}_B(x_n) = \pi(\lim_\omega \mathbb{E}_B(x_n)). \qquad (A.5.17)$$

Proof. Both parts will follow from Lemma 3.6.2.
(i)   If $x = \pi(x_n)$ with $(x_n) \in \ell^\infty(N)$, let $\phi(x) = \pi(\mathbb{E}_{B_n}(x_n))$. The module and trace preserving properties of $\mathbb{E}_B$ on N imply that

$$\phi(b_1 x b_2) = \pi\big(\mathbb{E}_{B_n}(b_{1,n} x_n b_{2,n})\big)$$
$$= \pi\big(b_{1,n}\mathbb{E}_{B_n}(x_n)b_{2,n}\big) = b_1 \phi(x) b_2 \qquad (A.5.18)$$

and

$$\tau_\omega(\phi(x)) = \lim_\omega \tau\left(\mathbb{E}_{B_n}(x_n)\right) = \lim_\omega \tau(x_n) = \tau_\omega(x) \tag{A.5.19}$$

for all $x = \pi(x_n) \in N^\omega$ and $b_1 = \pi(b_{1,n})$, $b_2 = \pi(b_{2,n}) \in B_n^\omega$. Thus $\phi = \mathbb{E}_{B_n^\omega}$ by Lemma 3.6.2, proving (A.5.16).

(ii)   If $x = \pi(x_n)$, let $\phi(x) = \lim_\omega \mathbb{E}_B(x_n)$. Since $\tau_\omega = \tau$ on N in N^ω using the normality of τ, we obtain $\phi(b_1 x b_2) = b_1 \phi(x) b_2$ and

$$\tau_\omega(\phi(x)) = \tau_\omega(\lim_\omega \mathbb{E}_B(x_n)) = \tau(\lim_\omega \mathbb{E}_B(x_n))$$
$$= \lim_\omega \tau(\mathbb{E}_B(x_n)) = \lim_\omega \tau(x_n) = \tau_\omega(x) \tag{A.5.20}$$

for all $x = \pi(x_n) \in N^\omega$ and $b_1, b_2 \in B$. Hence $\phi = \lim_\omega \mathbb{E}_B$ as required. $\square$

The following lemma is a possible approach to proving Theorem 13.4.5 (see [30]), although we chose a different route in Section 13.4. Below we identify $\mathbb{M}_k$ with a subalgebra of M so that $\mathbb{M}_k \subseteq M \subseteq M^\omega$.

Lemma A.5.5. *Let M be a* II_1 *factor, let* $\mathbb{M}_k \subseteq M$ *with* $k \in \mathbb{N}$, *let* $N = \mathbb{M}_k' \cap M$ *with* $M = \mathbb{M}_k \otimes N$. *Then* $M^\omega = \mathbb{M}_k \otimes N^\omega$.

Proof. If $x = \pi(x_n) \in M^\omega$, then

$$x_n = \sum_{i,j} e_{ij} \otimes x_{n,ij} = \sum_{i,j} (e_{ij} \otimes 1)(1 \otimes x_{n,ij}) \tag{A.5.21}$$

with e_{ij} $(1 \leq i, j \leq k)$ the matrix units of $\mathbb{M}_k$ and $x_{n,ij} \in N$ for all n, i, j. Hence

$$x = \pi(x_n) = \sum_{i,j} (e_{ij} \otimes 1)\pi(1 \otimes x_{n,ij}) \in \mathbb{M}_k \otimes N. \tag{A.5.22}$$

This proves the lemma. $\square$

From the lemma above and the tautological statement $(\mathbb{C}1)^\omega \cong \mathbb{C}1$, it follows that $(\mathbb{M}_k)^\omega = \mathbb{M}_k$ for each $k \in \mathbb{N}$. In general, tensor products and the ultrapower do not behave nicely; see [67].

Remark A.5.6. Relative commutants play an important role in the study of von Neumann subalgebras of II_1 factors. From Corollary A.4.3, it follows that if B is a masa in N, then B^ω is a masa in N^ω, or $B^\omega = (B^\omega)' \cap N^\omega$. The algebra $B' \cap N^\omega$, where B is a masa in N, is generally much larger than B^ω even when $N' \cap N^\omega = \mathbb{C}1$. Moreover if B is a masa in the hyperfinite II_1 factor R, then $B' \cap R^\omega \supseteq R' \cap R^\omega$, and the latter algebra is large.

Here is an example in a free group factor showing this explicitly. Let a and b be the generators of the free group $\mathbb{F}_2$ and let

$$y_{n,k} = \sum_{j=1}^{n} a^j b^k a^{-j} \quad \text{for all } n, k \in \mathbb{N}. \tag{A.5.23}$$

Since the elements $a^j b^k a^{-j}$ are pairwise orthogonal in N for different pairs j, k,

$$\|y_{n,k}\|_2 = n^{1/2} \tag{A.5.24}$$

and

$$y_{n,k_1} \perp y_{m,k_2} \text{ for } k_1 \neq k_2 \tag{A.5.25}$$

for all $m, n \in \mathbb{N}$. For each pair $k, n \in \mathbb{N}$ the elements $a^j b^k a^{-j}$ $(1 \leq j \leq n)$ form the generators of a free subgroup of $\mathbb{F}_2$ and hence

$$\|y_{n,k}\| = \|\sum_{j=1}^{n} a^j b^k a^{-j}\| = 2(n-1)^{1/2} \tag{A.5.26}$$

by a theorem of Akemann and Ostrand [1]. For each pair $k, n \in \mathbb{N}$, cancelling the common terms in each defining sum yields

$$a y_{n,k} a^{-1} - y_{n,k} = a^{n+1} b^k a^{-n-1} - a b^k a^{-1} \tag{A.5.27}$$

so that, by orthogonality,

$$\|a y_{n,k} a^{-1} - y_{n,k}\|_2 = \|a y_{n,k} a^{-1} - y_{n,k}\|_2 = 2^{1/2}. \tag{A.5.28}$$

Let $x_{n,k} = 2^{-1}(n-1)^{-1/2} y_{n,k}$ for all n, k. Then

$$\|x_{n,k}\| = 1$$
$$\|x_{n,k}\|_2 = 2^{-1} n^{1/2} (n-1)^{-1/2} \text{ and}$$
$$\|a x_{n,k} - x_{n,k} a\|_2 = 2^{-1/2} (n-1)^{-1/2} \quad \text{for all } n, k \in \mathbb{N}. \tag{A.5.29}$$

Let $x_k = \pi(x_{n,k})$ for all $k \in \mathbb{N}$. Then $\|x_k\| = 1$, $\|x_k\|_{2,\omega} = 2^{-1}$ and $a x_k = x_k a$ for all $k \in \mathbb{N}$ by inequalities (A.5.29).

If $A = \{a\}''$, then $\mathbb{E}_A(x_{n,k}) = 0$ for all $n, k \in \mathbb{N}$ so that $\mathbb{E}_{A^\omega}(x_k) = \pi(\mathbb{E}_A(x_{n,k})) = 0$ for all k by Lemma A.5.4. The orthogonality in (A.5.25) implies that x_k $(k \in \mathbb{N})$ are mutually orthogonal elements in N^ω. This shows that $A' \cap N^\omega$ is larger than A^ω. The elements x_k are easily seen to be orthogonal to N. $\qquad\square$

A.6   Relative commutants in N^ω

This section is devoted to the proof of a Theorem of Fang, Ge and Li [67], Theorem A.6.5, that is a generalisation of a result of Dixmier [50]. These results follow from some lemmas that depend on the averaging used in conditional expectations (see Lemma 3.6.5).

Lemma A.6.1. *Let N be a* II_1 *factor and let M be an irreducible subfactor. If p is a projection in N with $\tau(p) = r$ and if $\varepsilon > 0$, then there is a unitary u in M such that*

$$\|pu - up\|_2 \geq (r - r^2)^{1/2} - \varepsilon. \tag{A.6.1}$$

Proof. This follows directly from the inequality

$$2^{1/2}\|p - \mathbb{E}_{M'\cap N}(p)\|_2 \le \|up - pu\|_2 + \varepsilon \qquad \text{(A.6.2)}$$

which has already appeared as (3.6.35) in Lemma 3.6.5 (ii). Just substitute the relations

$$\mathbb{E}_{M'\cap N}(p) = \mathbb{E}_{\mathbb{C}1}(p) = \tau(p)1 \qquad \text{(A.6.3)}$$

and

$$\|p - \mathbb{E}_{M'\cap N}(p)\|_2^2 = \tau(p) - \tau(p)^2 = r - r^2 \qquad \text{(A.6.4)}$$

into (A.6.2). $\qquad\qquad\qquad\qquad\qquad\qquad\qquad\qquad\qquad\qquad\qquad\square$

Lemma A.6.2. *Let N be a II_1 factor and let M be an irreducible subfactor. If p is a projection in N^ω with $\tau_\omega(p) = r$, then there is a unitary u in M^ω such that*

$$\|pu - up\|_{2,\,\omega} \ge (r - r^2)^{1/2}. \qquad \text{(A.6.5)}$$

Proof. By Theorem A.5.3 (iii), represent $p = \pi(p_n)$ with p_n projections in N satisfying $\tau(p_n) = \tau_\omega(p) = r$ for all n. For each n choose a unitary u_n in M, by Lemma A.6.1, such that

$$\|u_n p_n - p_n u_n\|_2 \ge (r - r^2)^{1/2} - 2^{-n}. \qquad \text{(A.6.6)}$$

Then

$$\|up - pu\|_{2,\,\omega} = \lim_\omega \|u_n p_n - p_n u_n\|_2 \ge (r - r^2)^{1/2}, \qquad \text{(A.6.7)}$$

as required. $\qquad\qquad\qquad\qquad\qquad\qquad\qquad\qquad\qquad\qquad\qquad\qquad\square$

Lemma A.6.3. *Let N be a II_1 factor and let M be an irreducible subfactor. Let p be a projection in $M' \cap N^\omega$ with projections p_n in N satisfying $p = \pi(p_n)$ and $\tau(p_n) = \tau_\omega(p)$ for all n. Let $v_1, \ldots, v_m$ be in M and let $\varepsilon > 0$. If q is a projection in N, there is a $k \in \mathbb{N}$ such that*

(i) $|\tau(p_k q) - \tau_\omega(p)\tau(q)| < \varepsilon$ *and*

(ii) $\|p_k v_i - v_i p_k\|_2 < \varepsilon$ *for $1 \le i \le m$.*

Proof. By Lemma 3.6.5 (i), there are unitaries $u_1, \ldots, u_\ell$ in M and positive real numbers $\alpha_1, \ldots, \alpha_\ell$ such that $\sum_{j=1}^\ell \alpha_j = 1$ and

$$\Big\| \sum_{j=1}^\ell \alpha_j u_j q u_j^* - \tau(q)1 \Big\|_2 < \varepsilon/2. \qquad \text{(A.6.8)}$$

From the hypotheses $pu_j = u_j p$ for $1 \le j \le \ell$, $pv_i = v_i p$ for $1 \le i \le m$ and the definition of $\| \cdot \|_{2,\,\omega}$, there is an $k \in \mathbb{N}$ such that

$$\|u_j p_k - p_k u_j\|_2 < \varepsilon/2 \quad \text{for } 1 \le j \le \ell \qquad \text{(A.6.9)}$$

and
$$\|v_i p_k - p_k v_i\|_2 < \varepsilon/2 \quad \text{for } 1 \leq i \leq m, \tag{A.6.10}$$
since $\pi(p_n) = p$. Hence

$$\left\| \sum_{j=1}^{\ell} \alpha_j u_j p_k q u_j^* - \tau(q) p_k \right\|_2 \leq \left\| \sum_{j=1}^{\ell} \alpha_j u_j p_k q u_j^* - \sum \alpha_j p_k u_j q u_j^* \right\|_2$$

$$+ \left\| p_k \left(\sum_{j=1}^{\ell} \alpha_j u_j q u_j^* - \tau(q)1 \right) \right\|_2$$

$$< \sum_{j=1}^{\ell} \alpha_j \|u_j p_k - p_k u_j\|_2 \|q u_j\|_2 + \varepsilon/2$$

$$< \sum_{j=1}^{\ell} \alpha_j \varepsilon/2 + \varepsilon/2 = \varepsilon. \tag{A.6.11}$$

Since $\tau(u_j p_k q u_j^*) = \tau(p_k q)$ and $\tau_\omega(p) = \tau(p_k)$ for all $k \in \mathbb{N}$, taking the trace in inequality (A.6.11) gives

$$|\tau(p_k q) - \tau_\omega(p)\tau(q)| = |\tau(p_k q) - \tau(p_k)\tau(q)| < \varepsilon \tag{A.6.12}$$

as required. □

Lemma A.6.4. *If p and q are projections in finite von Neumann algebra with trace τ, then*

$$\|p \vee q - q\|_2 \geq \|p(1-q)\|_2. \tag{A.6.13}$$

Proof. The positivity of the functional $x \mapsto \tau\big((p \vee q - q)x(p \vee q - q)\big)$ applied to $p \leq 1$ gives

$$\|p(1-q)\|_2^2 = \tau\big((1-q)p(1-q)\big) = \tau\big((p \vee q - q)p(p \vee q - q)\big)$$
$$\leq \tau\big((p \vee q - q)1(p \vee q - q)\big) = \tau(p \vee q - q)$$
$$= \|p \vee q - q\|_2^2, \tag{A.6.14}$$

completing the proof. □

Theorem A.6.5. *Let N be a II_1 factor and let M be an irreducible subfactor. If $M' \cap N^\omega \neq \mathbb{C}1$, then $M' \cap N^\omega$ is diffuse.*

Proof. Suppose that there is a minimal projection p in the finite von Neumann algebra $M' \cap N^\omega$ with $p \neq 0, 1$. Let q be the central support of p in $M' \cap N^\omega$. Then $q(M' \cap N^\omega)q$ is a finite type I factor with minimal projection p, so there is an integer k such that $q(M' \cap N^\omega)q \cong \mathbb{M}_k p \cong \mathbb{M}_k$. Let $\tau_\omega(p) = r > 0$. Then $\tau_\omega(q) = k\tau(p) = kr$. Let $\mathcal{P}$ be the manifold of all minimal projections in $q(M' \cap N^\omega)q$ in the metric given by the $\|\cdot\|_{2,\omega}$-norm. Since $\mathcal{P}$ is separable, there is a countable dense sequence $e^{(1)}, e^{(2)}, \ldots$ in $\mathcal{P}$. By the representation

Theorem A.5.3 (iii) for projections in N^ω, choose sequences of projections (p_n), (q_n) and $(e_n^{(j)})$ for each j such that

(1) $p = \pi(p_n)$, $q = \pi(q_n)$ and $e^{(j)} = \pi(e_n^{(j)})$ for all j with

(2) $\tau(p_n) = \tau_\omega(p)$, $\tau(q_n) = \tau_\omega(q)$ and $\tau(e_n^{(j)}) = \tau_\omega(e^{(j)})$ for all j and n.

Note that $\tau_\omega(q) = kr$ and $\tau_\omega(e^{(j)}) = \tau_\omega(p) = r$ for all j. By Lemma A.6.2, there is a unitary $u^{(j)} \in M^\omega$ such that

$$\|u^{(j)}e^{(j)} - e^{(j)}u^{(j)}\|_{2,\omega} \geq (r - r^2)^{1/2}. \qquad (A.6.15)$$

Represent $u^{(j)} = \pi(u_n^{(j)})$ with $u_n^{(j)}$ unitaries in M for all j and n. For each $n \in \mathbb{N}$, apply Lemma A.6.3 to $q_n \in N$ and $u_k^{(j)} \in M$ for $1 \leq j,\, k \leq n$ to obtain an integer m_n (denoted k in that lemma) such that

$$|\tau(p_{m_n}q_n) - \tau(p_{m_n})\tau(q_n)| < 1/n \qquad (A.6.16)$$

and

$$\|p_{m_n}u_k^{(j)} - u_k^{(j)}p_{m_n}\|_2 < 1/n \quad \text{for } 1 \leq j,\, k \leq n. \qquad (A.6.17)$$

It follows from these relations that

$$\tau(p_{m_n}q_n) > \tau(p_{m_n})\tau(q_n) - 1/n = kr^2 - 1/n. \qquad (A.6.18)$$

Now define $f = \pi(p_{m_n})$ in N^ω. Clearly $f \in M' \cap N^\omega$, since $p_{m_n} \in M' \cap N$, and $\tau_\omega(f) = \lim_\omega \tau(p_{m_n}) = r$. From equation (A.6.18),

$$\tau_\omega(fq) = \lim_\omega \tau(p_{m_n}q_n) \geq kr^2. \qquad (A.6.19)$$

Hence $fq \neq 0$. Since q is a central projection in $M' \cap N^\omega$, fq is a non-zero projection in $q(M' \cap N^\omega)q \cong \mathbb{M}_k$ with

$$\tau_\omega(fq) \leq \tau_\omega(f) = r = \tau_\omega(p) \qquad (A.6.20)$$

so that fq is a minimal projection in $q(M' \cap N^\omega)q$ with $\tau_\omega(fq) = r = \tau_\omega(f) = \tau_\omega(p)$, and hence $fq = f$. Thus $f \in \mathcal{P}$. From equation (A.6.17) $fu^{(j)} - u^{(j)}f = 0$. By the choice of $u^{(j)}$,

$$(r - r^2)^{1/2} \leq \|e^{(j)}u^{(j)} - u^{(j)}e^{(j)}\|_{2,\omega}$$
$$\leq \|(e^{(j)} - f)u^{(j)} - u^{(j)}(e^{(j)} - f)\|_{2,\omega} \leq 2\|e^{(j)} - f\|_{2,\omega} \qquad (A.6.21)$$

for all j. This contradicts the $\|\cdot\|_{2,\omega}$-norm density of $\{e^{(j)}: j \in \mathbb{N}\}$. Hence there is no minimal projection in $M' \cap N^\omega$. $\square$

A.7   Property Γ revisited

Let us first recall from Section 13.4 that a II_1 factor N has property Γ if, for each finite subset $\{x_1, \ldots, x_n\} \subseteq N$ and each $\varepsilon > 0$, there is a unitary u in N with $\tau(u) = 0$ and

$$\|ux_j - x_ju\|_2 < \varepsilon \text{ for all } 1 \leq j \leq n. \qquad (A.7.1)$$

We showed in Remark 13.4.7 that the hyperfinite factor R has property Γ using the construction by an increasing sequence of matrix subalgebras. A different way is to take a countable I.C.C. group G which is the union of an increasing sequence $G_1 \subseteq G_2 \subseteq \ldots$ of finite subgroups. An example would be the group S_∞ of finite permutations of $\mathbb{N}$ with S_n the subgroup of permutations which fix each $k \geq n+1$. Any finite subset $\{x_1, \ldots x_m\}$ of $\mathbb{C}G$ will lie in $\mathbb{C}G_n$ for a sufficiently large choice of n. Any unitary $h_n \in G \setminus G_n$ will have the property that $\tau(h_n) = 0$ and $\|x_i h_n - h_n x_i\|_2 = 0$ for $1 \leq i \leq m$, and a simple approximation argument using the $\|\cdot\|_2$-density of $\mathbb{C}G$ in $\ell^2(G)$ then shows that $L(G)$ has property Γ. Of course, $L(G)$ is the hyperfinite II_1 factor since the algebras $L(G_n)$ are finite dimensional.

In their initial papers on the subject [116, 117, 118, 202], Murray and von Neumann showed the existence of II_1 factors through various constructions, including the hyperfinite or 'approximately finite' factor R as the weak closure of an increasing sequence of matrix algebras. They knew that all separable hyperfinite factors were isomorphic, but they were faced with the problem of deciding whether there was more than one isomorphism class of general II_1 factors. In a letter to Halperin dated 22nd February 1940, von Neumann reports that he and Murray had

> ... finally succeeded in showing that a certain operator (ring) of class (II_1) is not "approximately finite" ...

(see [158, p. 134]). The example given in the letter was $L(\mathbb{Z}_3 \star \mathbb{Z}_2)$, and property Γ was introduced to distinguish this factor from R. We will show that $L(\mathbb{F}_k)$ does not have property Γ for $2 \leq k \leq \infty$ by using the following lemma, which is motivated by the original approach of Murray and von Neumann [118].

Lemma A.7.1. *Let G be a countable I.C.C. group. If S is a subset of G and g_1, g_2, g_3 in G satisfy*

(1) $S \cup g_1 S g_1^{-1} = G \setminus \{1\}$

and

(2) $S, g_2 S g_2^{-1}, g_3 S g_3^{-1}$ *are pairwise disjoint,*

then, for all $x \in L(G)$,

$$\|x - \tau(x)1\|_2 \leq 14 \max\left\{\|g_j x - x g_j\|_2 : j = 1, 2, 3\right\}. \qquad (\text{A.7.2})$$

Proof. Let $x = \sum_g x_g g \in L(G)$ and let

$$\nu(W) = \sum_{g \in W} |x_g|^2 \quad \text{for all} \quad W \subseteq G. \qquad (\text{A.7.3})$$

Let

$$k = \nu(G \setminus \{1\}) = \|x - \tau(x)1\|_2^2 \qquad (\text{A.7.4})$$

and

$$\delta_j = \|g_j x - x g_j\|_2 \quad \text{for} \quad j = 1, 2, 3. \tag{A.7.5}$$

Then

$$\delta_j^2 = \|\sum_g (x_g g_j g - x_g g g_j)\|_2^2 = \|\sum_g (x_g - x_{g_j g g_j^{-1}}) g_j g\|_2^2$$

$$= \sum_g |x_g - x_{g_j g g_j^{-1}}|^2. \tag{A.7.6}$$

for each j. From $g_j S g_j^{-1} \subseteq G \setminus \{1\}$ and (A.7.6), it follows that

$$|\nu(S) - \nu(g_j S g_j^{-1})| = |\nu(S)^{1/2} + \nu(g_j S g_j^{-1})^{1/2}| \cdot |\nu(S)^{1/2} - \nu(g_j S g_j^{-1})^{1/2}|$$

$$\leq 2\nu(G \setminus \{1\})^{1/2} \Big| \Big(\sum_{g \in S} |x_g|^2\Big)^{1/2} - \Big(\sum_{g \in S} |x_{g_j g g_j^{-1}}|^2\Big)^{1/2} \Big|$$

$$\leq 2k^{1/2} \Big(\sum_{g \in S} |x_g - x_{g_j g g_j^{-1}}|^2\Big)^{1/2} \leq 2k^{1/2} \delta_j. \tag{A.7.7}$$

From (1), (2) and inequality (A.7.7),

$$k = \nu(G \setminus \{1\}) \leq \nu(S) + \nu(g_1 S g_1^{-1}) \leq 2\nu(S) + 2k^{1/2} \delta_1 \tag{A.7.8}$$

and

$$k \geq \nu(S) + \nu(g_2 S g_2^{-1}) + \nu(g_3 S g_3^{-1}) \geq 3\nu(S) - 2k^{1/2} \delta_2 - 2k^{1/2} \delta_3. \tag{A.7.9}$$

From these inequalities (A.7.8) and (A.7.9),

$$3(k - 2k^{1/2} \delta_1) \leq 6\nu(S) \leq 2k + 4k^{1/2}(\delta_2 + \delta_3) \tag{A.7.10}$$

and this simplifies to

$$k^{1/2} \leq 6\delta_1 + 4(\delta_2 + \delta_3). \tag{A.7.11}$$

Thus

$$\|x - \tau(x)1\|_2 = k^{1/2} \leq 6\delta_1 + 4(\delta_2 + \delta_3) \leq 14 \max\{\delta_j : j = 1, 2, 3\} \tag{A.7.12}$$

as required. $\square$

Theorem A.7.2. *The free group von Neumann algebra* $L(\mathbb{F}_k)$ *does not have property* Γ *for all* $2 \leq k \leq \infty$.

Proof. Let $\mathbb{F}_k$ have generators $g_1, g_2, \ldots$ and if $k = 2$ let g_3 be g_2^{-1}. Let u be a unitary in $L(\mathbb{F}_k)$ with

$$\|u g_j - g_j u\|_2 \leq (28)^{-1} \quad \text{for} \quad j = 1, 2, 3. \tag{A.7.13}$$

In Lemma A.7.1, let $G = \mathbb{F}_k$ and let S be the set of all words ending in a non-zero power of g_1. Then $g_1 S g_1^{-1}$ contains all words in $\mathbb{F}_k$ that do not end exactly in g_1^{-1} so that $S \cup g_1 S g_1^{-1} = \mathbb{F}_k \setminus \{e\}$. Clearly S, $g_2 S g_2^{-1}$ and $g_3 S g_3^{-1}$ are pairwise disjoint. By Lemma A.7.1,

$$1 - |\tau(u)| \le \|u - \tau(u)1\|_2 \le 1/2 \qquad (A.7.14)$$

so that $|\tau(u)| \ge 1/2$. The condition $\tau(u) = 0$ required in property Γ is impossible so $L(\mathbb{F}_k)$ does not have property Γ. □

Lemma A.7.3. *A separable* II_1 *factor* N *has property* Γ *if, and only if,* $N_\omega = N' \cap N^\omega \ne \mathbb{C}1$.

Proof. Let N have property Γ and let $\{x_j : j \in \mathbb{N}\}$ be a countable subset of the unit ball $\mathcal{B}$ of N that is dense in $\mathcal{B}$ in the $\|\cdot\|_2$-norm. By property Γ, choose a sequence of unitaries u_n in N such that $\|u_n x_j - x_j u_n\|_2 < 2^{-n}$ for $1 \le j \le n$ and $\tau(u_n) = 0$ for all n. From this inequality and the $\|\cdot\|_2$-norm density of $\{x_j : j \in \mathbb{N}\}$ in $\mathcal{B}$, it follows that $\|u_n x - x u_n\|_2 \to 0$ as $n \to \infty$, and hence that $\lim_\omega \|u_n x - x u_n\|_2 = 0$ for all $x \in \mathcal{B}$. Let $u = \pi(u_n) \in N^\omega$. Then u is a unitary in N^ω and $\tau_\omega(u) = \lim_\omega \tau(u_n) = 0$. From above, $\|ux - xu\|_{2, \omega} = 0$ so that $u \in N' \cap N^\omega \setminus \mathbb{C}1$.

Conversely, suppose that $N' \cap N^\omega \ne \mathbb{C}1$. This algebra is diffuse, by Theorem A.6.5, so there is a unitary u in $N' \cap N^\omega$ with $\tau_\omega(u) = 0$. By Theorem A.5.3 (ii), represent $u = \pi(u_n)$ with u_n unitary in N and $\tau(u_n) = 0$ for all n. Let $\varepsilon > 0$ and let $x_1, \dots, x_k \in N$. If

$$\sum_{j=1}^k \|u_n x_j - x_j u_n\|_2 \ge \varepsilon \quad \text{for all} \quad n \in \mathbb{N}, \qquad (A.7.15)$$

then

$$\varepsilon \le \lim_\omega \sum_{j=1}^k \|u_n x_j - x_j u_n\|_2$$
$$= \sum_{j=1}^k \|ux_j - x_j u\|_{2, \omega} = 0. \qquad (A.7.16)$$

This is a contradiction, so there exists an n such that $\|u_n x_j - x_j u_n\|_2 < \varepsilon$ for all $1 \le j \le k$. Hence N has property Γ. □

Before proving a useful equivalent version of property Γ in Theorem A.7.5, a standard little technical result is required.

Lemma A.7.4. *Let* M *be a finite von Neumann algebra with faithful normal trace* τ *and let* B *be a diffuse von Neumann subalgebra of* M. *If* A *is a masa in* B *and* $k \in \mathbb{N}$, *then there are mutually orthogonal projections* p_j $(1 \le j \le k)$ *in* A *with* $\sum_j p_j = 1$ *and* $\tau(p_j) = k^{-1}$ *for* $1 \le j \le k$.

Proof. If p were a minimal non-zero projection in A, then $Ap = \mathbb{C}p$. Since B is diffuse, $pBp \neq \mathbb{C}p$, so we may choose a self-adjoint non-scalar element $h \in pBp$, whereupon $(A \cup \{h\})''$ is an abelian subalgebra of B which strictly contains the masa A. This contradiction implies that A is diffuse. As in the proof of Corollary 15.3.3, A contains a diffuse separable subalgebra A_0, and the result follows from the isomorphism $A_0 \cong L^\infty[0,1]$ of Theorem 3.5.2. □

We conclude this appendix with the following equivalent version of property Γ due to Dixmier [50]. This has proved to be very useful in the study of Γ factors (see, for example, [30]).

Theorem A.7.5. *A separable* II_1 *factor N has property Γ if, and only if, for each m, $k \in \mathbb{N}$, $x_1, \ldots, x_m \in N$ and $\varepsilon > 0$, there exist mutually orthogonal projections $p_1, \ldots, p_k$ in N, each with trace k^{-1}, such that*

$$\|p_i x_j - x_j p_i\|_2 < \varepsilon \quad \text{for} \quad 1 \leq j \leq m, \ 1 \leq i \leq k. \tag{A.7.17}$$

Proof. Suppose that N has property Γ. Then $N_\omega = N' \cap N^\omega \neq \mathbb{C}1$ by Lemma A.7.3, so that $N' \cap N^\omega$ is diffuse by Theorem A.6.5 with $M = N$. Since $B = N' \cap N^\omega$ is a diffuse von Neumann subalgebra of N^ω, there are projections $q_1, \ldots, q_k$ in $N' \cap N^\omega$ with $\sum_i q_i = 1$ and $\tau_\omega(q_i) = k^{-1}$ for $1 \leq i \leq k$ by Lemma A.7.4. By Theorem A.5.3 (v), there are projections $q_{i,n} \in N$ with $\sum_{i=1}^n q_{i,n} = 1$ and $\tau(q_{i,n}) = k^{-1}$ for all i and n satisfying $q_i = \pi(q_{i,n})$ for $1 \leq i \leq k$.

Regarding $N \subseteq N^\omega$,

$$\|\pi(q_{i,n} x_j - x_j q_{i,n})\|_{2,\tau_\omega} = \|q_i x_j - x_j q_i\|_{2,\tau_\omega} = 0 \tag{A.7.18}$$

for $1 \leq i \leq k$ and $1 \leq j \leq m$, since $q_i \in N' \cap N^\omega$. By the properties of $\lim_\omega$, there is an n such that $\|q_{i,n} x_j - x_j q_{i,n}\|_2 \leq \varepsilon$ for $1 \leq i \leq k$ and $1 \leq j \leq m$. Let $p_i = q_{i,n}$.

Conversely choose a sequence $\{x_j : j \in \mathbb{N}\}$ in the unit ball of N that is dense in this unit ball in the $\|\cdot\|_2$ norm. By the hypothesis for $k = 2$, for each n choose two orthogonal projections p_n and q_n in N with $p_n + q_n = 1$ and $\tau(p_n) = \tau(q_n) = 1/2$ such that

$$\|p_n x_j - x_j p_n\|_2 < 2^{-n} \quad \text{for} \quad 1 \leq j \leq n, \tag{A.7.19}$$

and similarly for q_n. Let $p = \pi(p_n)$ and $q = \pi(q_n)$ be projections in N^ω. Then $\tau_\omega(p) = \tau_\omega(q) = 1/2$ and $p, q \in N' \cap N^\omega$, since $\|px_j - x_j p\|_{2,\omega} = \lim_\omega \|p_n x_j - x_j p_n\|_2 = 0$ for all j. Thus $N' \cap N^\omega \neq \mathbb{C}1$, and property Γ for N now follows from Lemma A.7.3. □

Appendix B

Unbounded operators

B.1 Introduction

For the most part, these notes are concerned with the bounded operators which constitute the von Neumann algebras under consideration. However, results from the theory of unbounded operators have played a role in Chapter 9, and knowledge of this topic is essential for reading the literature in this area. Since we feel that this theory is less well known than its counterpart for bounded operators, we include here a brief exposition of the main theorems required in these notes. Most of what is needed may be found in [104, Section 5.6], and we follow their development to a considerable extent. However, we have specific goals for the theory and we do not offer a comprehensive treatment. The main objective is to understand the operators that arise as unbounded left multiplication operators on II_1 factors.

Section B.2 contains the basic theory of closed and closable operators. In Section B.3, we develop as much of the functional calculus as we will need. We carry this out for positive operators, relating matters to the well known functional calculus for bounded positive operators. Along the way we establish the polar decomposition of a closed operator which appeared in Lemma 9.4.2. The important unbounded operators of II_1 factor theory are those that arise from vectors in $L^2(N)$ and $L^1(N)$, and we lay out their theory in Sections B.4 and B.5 respectively. We have developed this theory for factors, but we note that the results of these sections remain valid for a general finite von Neumann algebra N. The results in this more general situation can be easily deduced from the factor case by viewing N as a subalgebra of any containing finite factor, for example the free product factor $N \star N$.

B.2 Basic results

The setting is a Hilbert space H, a dense subspace $\mathcal{D}(T)$, and a linear operator $T\colon \mathcal{D}(T) \to H$. The space $\mathcal{D}(T)$ is called the *domain* of T. A fundamental object is the graph $\mathcal{G}(T)$ of such an operator. This is the subspace of $H \times H$ defined by

$$\mathcal{G}(T) = \{(\xi, T\xi) \colon \xi \in \mathcal{D}(T)\}.$$

If $\mathcal{G}(T)$ is a closed subspace then T is called closed. The class of closed operators is the one most amenable to study along the lines of the bounded theory, and is the one that will concern us here. Even if $\mathcal{G}(T)$ is not closed, its closure may be the graph of a linear operator, denoted $\overline{T}$, in which case T is called closable and $\overline{T}$ is its closure. Then the domain $\mathcal{D}(\overline{T})$ of $\overline{T}$ contains that of T, and $\overline{T}$ extends the definition of T to $\mathcal{D}(\overline{T})$. In general, when two operators S and T satisfy $\mathcal{D}(S) \subseteq \mathcal{D}(T)$ and S is the restriction of T to $\mathcal{D}(S)$, we describe this relationship by $S \subseteq T$. The equation $S = T$ carries the implicit requirement that the domains of the operators be equal, and it is certainly possible to have $S \subseteq T$ without equality. Another important object associated to a closed operator is a core. This is a subspace K of $\mathcal{D}(T)$ such that $\{(\xi, T\xi) \colon \xi \in K\}$ is dense in $\mathcal{G}(T)$. Of course $\mathcal{D}(T)$ is a core for T, but in specific examples it can be very useful to work with some naturally arising core rather than $\mathcal{D}(T)$.

In parallel to the bounded theory, the adjoint T^* is defined by the equation $\langle T\xi, \eta \rangle = \langle \xi, T^*\eta \rangle$, although some care is required to make sense of this. If $\mathcal{D}(T)$ is dense in H then the domain $\mathcal{D}(T^*)$ of T^* is defined to be those vectors η for which $\xi \mapsto \langle T\xi, \eta \rangle$ defines a bounded linear functional on $\mathcal{D}(T)$, which then extends to H. There is then a unique vector, denoted $T^*\eta$, such that

$$\langle T\xi, \eta \rangle = \langle \xi, T^*\eta \rangle, \qquad \xi \in \mathcal{D}(T), \quad \eta \in \mathcal{D}(T^*), \tag{B.2.1}$$

and it is clear that $T^* \colon \mathcal{D}(T^*) \to H$ is a linear operator. In general, we can say no more than $0 \in \mathcal{D}(T^*)$, but when T is densely defined and closed, we will establish below that T^* is densely defined. We note one relationship that is immediate from (B.2.1): if S and T are densely defined operators and $S \subseteq T$, then $T^* \subseteq S^*$.

Before embarking on the general theory we should indicate how these operators appear in relation to type II_1 factors. If N is such a factor with a faithful normal tracial state τ, then we assume that N is faithfully represented on $L^2(N, \tau)$ by left multiplication. If ξ denotes the vector 1 in $L^2(N, \tau)$, then N embeds as a dense subspace $N\xi = \{x\xi \colon x \in N\}$. Recall the isometric conjugate linear isometry J defined on $N\xi$ by $x\xi \mapsto x^*\xi$. Then each vector $\eta \in L^2(N)$ induces a linear operator ℓ_η with domain $N\xi$ defined by

$$\ell_\eta(x\xi) = Jx^*J\eta, \qquad x \in N. \tag{B.2.2}$$

If η happens to be $y\xi$ for some $y \in N$, then ℓ_η is left multiplication by y, but in general ℓ_η is unbounded. However, each ℓ_η will be closable, and so the theory of closed operators will be useful in analysing such operators.

The following example gives a good illustration of what can be expected from the theory of unbounded operators.

Example B.2.1. Each function $f \in L^\infty(\mathbb{R})$ defines a bounded multiplication operator M_f on $L^2(\mathbb{R})$ by $M_f(g) = fg$ for $g \in L^2(\mathbb{R})$. The same formula defines a linear operator M_f when f lies in the space $L^\infty_{\text{loc}}(\mathbb{R})$ of measurable functions bounded on compact sets, where the domain $\mathcal{D}(M_f)$ is $\{g \in L^2(\mathbb{R}) : fg \in L^2(\mathbb{R})\}$. This is a densely defined operator because $\mathcal{D}(M_f)$ contains $\chi_{[-n,n]}L^2(\mathbb{R})$ for all $n \geq 1$, and the union of these subspaces is dense in $L^2(\mathbb{R})$. If $h \in \mathcal{D}(M_f^*)$ then $g \mapsto \int f(x)g(x)\overline{h(x)}\, dx$ defines a bounded linear functional on $\mathcal{D}(M_f)$, and so it follows that $M_f^*(h) = \bar{f}h = M_{\bar{f}}(h)$. On the other hand, if $h \in \mathcal{D}(M_{\bar{f}})$ then $\bar{f}h \in L^2(\mathbb{R})$, and $g \mapsto \int f(x)g(x)\overline{h(x)}\, dx$ defines a bounded linear functional on $\mathcal{D}(M_f)$, placing h in $\mathcal{D}(M_f^*)$. It follows that $M_f^* = M_{\bar{f}}$, and M_f^* has dense domain.

Now suppose that $(h, k) \in \mathcal{G}(M_f)^\perp$ for some pair $h, k \in L^2(\mathbb{R})$. Then, for $g \in \mathcal{D}(M_f)$, we have

$$\int g(x)\overline{h(x)}\, dx + \int f(x)g(x)\overline{k(x)}\, dx = 0, \qquad \text{(B.2.3)}$$

and it follows that $\bar{f}k$ defines a bounded linear functional on $\mathcal{D}(M_f)$, so that $k \in \mathcal{D}(M_f^*) = \mathcal{D}(M_{\bar{f}})$. From (B.2.3) we also obtain that $h = -\bar{f}k$, and (h, k) has the form $(-M_{\bar{f}}k, k)$. It is clear from (B.2.3) that any element of the form $(-M_{\bar{f}}k, k)$ for $k \in \mathcal{D}(M_{\bar{f}})$ lies in $\mathcal{G}(M_f)^\perp$, and so $\mathcal{G}(M_f)^\perp = V\mathcal{G}(M_{\bar{f}})$, where $V : L^2(\mathbb{R}) \times L^2(\mathbb{R}) \to L^2(\mathbb{R}) \times L^2(\mathbb{R})$ is the unitary operator $(h, k) \mapsto (-k, h)$. If we interchange f and $\bar{f}$ then we see that $\mathcal{G}(M_f) = V^*\mathcal{G}(M_{\bar{f}})^\perp$ and is thus closed, so that each M_f, $f \in L^\infty_{\text{loc}}(\mathbb{R})$, is a closed operator. The relation $M_f^* = M_{\bar{f}}$ reduces to $M_f^* = M_f$ when f is real valued, so that M_f is self-adjoint in this case.

Now let K be the union of the subspaces $\chi_{[-n,n]}L^2(\mathbb{R})$, $n \geq 1$. We show that K is a core for each M_f, $f \in L^\infty_{\text{loc}}(\mathbb{R})$. If this is not so, then there is a non-zero element $(g, fg) \in \mathcal{G}(M_f)$ orthogonal to each element $(\chi_{[-n,n]}h, \chi_{[-n,n]}fh)$ where $h \in L^2(\mathbb{R})$ and $n \geq 1$. Thus

$$\int \chi_{[-n,n]}(x)g(x)(1 + |f(x)|^2)\overline{h(x)}\, dx = 0, \qquad h \in L^2(\mathbb{R}), \quad n \geq 1, \quad \text{(B.2.4)}$$

from which the contradiction $g(x) = 0$ a.e. follows easily. Thus K is indeed a core for M_f.

If $f \in L^\infty_{\text{loc}}(\mathbb{R})^+$, then M_f is self-adjoint and $\langle M_f g, g \rangle \geq 0$ for all $g \in \mathcal{D}(M_f)$, so that $M_f \geq 0$. Consequently $\langle (I + M_f)g, g \rangle \geq \|g\|_2^2$ for $g \in \mathcal{D}(M_f)$, and $\ker(I + M_f) = 0$. Moreover, for each $h \in L^2(\mathbb{R})$, $(1 + f)^{-1}h \in \mathcal{D}(I + M_f)$, and $I + M_f$ maps this vector to h. For each $g \in \mathcal{D}(I + M_f)$, the Cauchy–Schwarz inequality gives

$$\|(I + M_f)g\|_2\|g\|_2 \geq |\langle (I + M_f)g, g \rangle| \geq \|g\|_2^2, \qquad \text{(B.2.5)}$$

showing that there is a bounded inverse for $I + M_f$, easily seen to be $M_{(1+f)^{-1}}$. Since the study of the function f is equivalent to that of $(1 + f)^{-1}$, we expect to understand M_f by working with the bounded operator $(I + M_f)^{-1}$.

If $f \in L^\infty_{\text{loc}}(\mathbb{R})$ is real valued, then M_f is self-adjoint, and the functions $(f \pm i)^{-1}$ are bounded. Then, for each $h \in L^2(\mathbb{R})$, $(f \pm i)^{-1}h$ are in the domains of $M_f \pm iI$, and are mapped to h, showing that the ranges of $M_f \pm iI$ are $L^2(\mathbb{R})$. Since

$$|\langle (M_f \pm iI)h, h\rangle| = |\langle M_f h, h\rangle \pm i\|h\|_2^2| \geq \|h\|_2^2, \qquad h \in \mathcal{D}(M_f), \qquad \text{(B.2.6)}$$

we see that the operators $M_f \pm iI$ have trivial kernel.

The equation $M_f^* M_f = M_{|f|^2}$ holds for each $f \in L^\infty_{\text{loc}}(\mathbb{R})$, as we now show. The domain of the left-hand side is $\{g \in L^2(\mathbb{R}): fg \in L^2(\mathbb{R}) \text{ and } |f|^2 g \in L^2(\mathbb{R})\}$. By examining separately the sets where $|f(x)| \leq 1$ and $|f(x)| > 1$, this coincides with $\mathcal{D}(M_{|f|^2}) = \{g \in L^2(\mathbb{R}): |f|^2 g \in L^2(\mathbb{R})\}$, showing that $M_f^* M_f = M_{|f|^2}$. Thus $M_f^* M_f$ is a positive self-adjoint operator.

Now consider the functions $f(x) = e^x$ and $g(x) = e^{-x}$ in $L^\infty_{\text{loc}}(\mathbb{R})$. We should expect that $M_f M_g = M_g M_f = I$, and this is certainly true on their common domains. However, these three operators have the three distinct domains $\{h \in L^2(\mathbb{R}): e^{-x}h \in L^2(\mathbb{R})\}$, $\{h \in L^2(\mathbb{R}): e^x h \in L^2(\mathbb{R})\}$ and $L^2(\mathbb{R})$, so that no pair is equal. Thus care must be taken in making algebraic manipulations that are routine for bounded operators. These particular functions also point out the fact that $M_f M_g$ may not be closed, even though M_f and M_g are themselves closed.

Finally, we note that there is a polar decomposition for these multiplication operators. If $f \in L^\infty_{\text{loc}}(\mathbb{R})$, then define a bounded measurable function ϕ by

$$\phi(x) = \begin{cases} f(x)|f(x)|^{-1}, & \text{when } f(x) \neq 0, \\ 0, & \text{when } f(x) = 0. \end{cases}$$

Then $f = \phi|f|$, M_ϕ is a partial isometry, and $M_f = M_\phi M_{|f|}$ since both sides have domain $\mathcal{D}(M_f)$. $\qquad \square$

Several of the properties that we have highlighted in this example carry over to general closed operators, and we now investigate this. The unitary V on $L^2(\mathbb{R}) \times L^2(\mathbb{R})$ may be defined for a general Hilbert space by

$$V((\xi, \eta)) = (-\eta, \xi), \qquad (\xi, \eta) \in H \times H, \qquad \text{(B.2.7)}$$

and this is helpful below.

Lemma B.2.2. *Let T be a densely defined operator on a Hilbert space H. Then*

(i) *T^* is a closed operator;*

(ii) *T is closable if and only if T^* is densely defined, in which case $\overline{T} = T^{**}$.*

Proof. (i) Since T is densely defined, the operator T^* exists. If $\xi \in \mathcal{D}(T)$ and $\eta \in \mathcal{D}(T^*)$ then

$$\langle T\xi, \eta\rangle - \langle \xi, T^*\eta\rangle = 0, \qquad \text{(B.2.8)}$$

which can be rewritten as

$$\langle(-T\xi, \xi), (\eta, T^*\eta)\rangle = 0. \tag{B.2.9}$$

Thus $(\eta, T^*\eta) \perp V\mathcal{G}(T)$, and so $\mathcal{G}(T^*) \subseteq (V\mathcal{G}(T))^\perp$. On the other hand, suppose that $(\eta_1, \eta_2) \in (V\mathcal{G}(T))^\perp$. Then

$$\langle -T\xi, \eta_1\rangle + \langle \xi, \eta_2\rangle = 0, \qquad \xi \in \mathcal{D}(T), \tag{B.2.10}$$

from which it follows that $\xi \mapsto \langle T\xi, \eta_1\rangle$ is a bounded linear functional on $\mathcal{D}(T)$. Thus $\eta_1 \in \mathcal{D}(T^*)$, and the equation $T^*\eta_1 = \eta_2$ follows from (B.2.10). This shows that $(\eta_1, \eta_2) \in \mathcal{G}(T^*)$ and so $\mathcal{G}(T^*) = (V\mathcal{G}(T))^\perp$, which is always closed.
(ii) Suppose now that T^* is densely defined. From (i), T^{**} is closed and

$$\begin{aligned}
\mathcal{G}(T^{**}) &= (V\mathcal{G}(T^*))^\perp = (V(V(\mathcal{G}(T)))^\perp)^\perp \\
&= \mathcal{G}(T)^{\perp\perp} = \overline{\mathcal{G}(T)}, \tag{B.2.11}
\end{aligned}$$

using $V^2 = -I$, and $(V(\mathcal{G}(T)))^\perp = V(\mathcal{G}(T)^\perp)$. Thus T^{**} is the closure $\overline{T}$ of T.
Conversely suppose that T is closable with closure $\overline{T}$. Since $\overline{T}^* \subseteq T^*$, it suffices to show that $\overline{T}^*$ has dense domain, and thus we may assume that T is closed. Let $P\colon \mathcal{G}(T) \to H$ be the bounded operator

$$P((\xi, T\xi)) = \xi, \qquad \xi \in \mathcal{D}(T). \tag{B.2.12}$$

Since $\mathcal{G}(T)$ is a Hilbert space, there is a bounded adjoint $P^*\colon H \to \mathcal{G}(T)$. Only one element in $\mathcal{G}(T)$ has first coordinate 0, and so P is injective, showing that P^* has dense range. Let K be the subspace of H consisting of the second coordinates of elements in the range of P^*. Then K is dense in $\text{Ran}(T)$. If $\eta_2 \in K$, then there exist $\eta_0, \eta_1 \in H$ such that $P^*\eta_0 = (\eta_1, \eta_2)$. If $\xi \in \mathcal{D}(T)$ then

$$\begin{aligned}
\langle T\xi, \eta_2\rangle &= \langle(\xi, T\xi), (\eta_1, \eta_2)\rangle - \langle\xi, \eta_1\rangle \\
&= \langle(\xi, T\xi), P^*\eta_0\rangle - \langle\xi, \eta_1\rangle \\
&= \langle P(\xi, T\xi), \eta_0\rangle - \langle\xi, \eta_1\rangle \\
&= \langle\xi, \eta_0\rangle - \langle\xi, \eta_1\rangle, \tag{B.2.13}
\end{aligned}$$

showing that $\xi \mapsto \langle T\xi, \eta_2\rangle$ is a bounded linear functional on $\mathcal{D}(T)$. Thus $\eta_2 \in \mathcal{D}(T^*)$ and $K \subseteq \mathcal{D}(T^*)$. Each $\eta \in (\text{Ran }T)^\perp$ lies in $\mathcal{D}(T^*)$ since $\langle T\xi, \eta\rangle = 0$ for $\xi \in \mathcal{D}(T)$, so $\mathcal{D}(T^*)$ contains the algebraic sum $K + (\text{Ran }T)^\perp$, and T^* is thus densely defined, completing the proof. $\qquad\square$

Lemma B.2.3. *Let $S \in B(H)$ be positive and assume that S has trivial kernel. Then there is a densely defined positive operator $T\colon \mathcal{D}(T) \to H$ such that $ST = I$ on $\mathcal{D}(T)$ and $TS = I$ on H.*

Proof. We have already defined the unitary $V\colon H \times H \to H \times H$ by $(\xi, \eta) \mapsto (-\eta, \xi)$, and we now introduce a second unitary $W\colon H \times H \to H \times H$ given by $(\xi, \eta) \mapsto (\eta, \xi)$. Note that $V^2 = -I$, $W^2 = I$.

Since $(\text{Ran } S)^{\perp} = \ker S = \{0\}$, we may let $\mathcal{D}(T)$ be Ran S and define the linear operator T: Ran $S \to H$ to be the set-theoretic inverse of S. Then

$$\begin{aligned} \mathcal{G}(T) &= \{(\xi, T\xi)\colon \ \xi \in \text{Ran } S\} = \{(S\eta, \eta)\colon \ \eta \in H\} \\ &= W(\mathcal{G}(S)), \end{aligned} \tag{B.2.14}$$

showing that T is closed since $\mathcal{G}(S)$ is closed. From Lemma B.2.2, T^* is closed so (B.2.11) and (B.2.14) give

$$\begin{aligned} V(\mathcal{G}(T^*)) &= \mathcal{G}(T)^{\perp} = (W(\mathcal{G}(S)))^{\perp} \\ &= W(\mathcal{G}(S)^{\perp}) = WV(\mathcal{G}(S)) \\ &= WVW(\mathcal{G}(T)). \end{aligned} \tag{B.2.15}$$

Thus $\mathcal{G}(T^*) = VWVW(\mathcal{G}(T))$. An easy computation shows that $(VW)^2 = I$, and we conclude that $T = T^*$. If $\xi \in \mathcal{D}(T)$ then $\xi = S\eta$ for some $\eta \in H$. Thus

$$\langle T\xi, \xi \rangle = \langle TS\eta, S\eta \rangle = \langle \eta, S\eta \rangle \geq 0, \tag{B.2.16}$$

showing that $T \geq 0$. $\qquad\square$

Lemma B.2.4. *Let T be a closed densely defined operator on a Hilbert space H. Then T^*T is a densely defined positive operator and $\text{Ran}(I + T^*T) = H$.*

Proof. From the proof of Lemma B.2.2 there is a bounded operator P^*: $H \to \mathcal{G}(T)$ with dense range such that the second coordinates of $\{P^*\xi\colon \ \xi \in H\}$ lie in $\mathcal{D}(T^*)$. Thus the space consisting of first coordinates of elements in Ran P^* is dense in $\mathcal{D}(T)$, hence in H, and also lies in $\mathcal{D}(T^*T)$, showing that T^*T is densely defined. If $\xi \in H$ then $P^*\xi$ has the form $(\xi_1, T\xi_1)$ for some $\xi_1 \in \mathcal{D}(T)$. For $\eta \in \mathcal{D}(T)$, we have

$$\begin{aligned} \langle \xi, \eta \rangle &= \langle \xi, P(\eta, T\eta) \rangle = \langle P^*\xi, (\eta, T\eta) \rangle \\ &= \langle \xi_1, \eta \rangle + \langle T\xi_1, T\eta \rangle \\ &= \langle (I + T^*T)\xi_1, \eta \rangle. \end{aligned} \tag{B.2.17}$$

Since $\eta \in \mathcal{D}(T)$ was arbitrary, we conclude that $(I + T^*T)\xi_1 = \xi$, showing that the range of $I + T^*T$ is H. If $\xi \in \mathcal{D}(T^*T)$ then $\xi \in \mathcal{D}(T)$ and $T\xi \in \mathcal{D}(T^*)$. Thus

$$0 \leq \langle T\xi, T\xi \rangle = \langle \xi, T^*T\xi \rangle, \tag{B.2.18}$$

showing that $\|(I + T^*T)\xi\| \geq \|\xi\|$. Thus there is a bounded operator S: $H \to \mathcal{D}(T^*T)$ which is the inverse of $I + T^*T$. Moreover, if $\eta \in H$ then $\eta = (I + T^*T)\xi$ for some $\xi \in \mathcal{D}(T^*T)$, so that

$$\langle S\eta, \eta \rangle = \langle \xi, (I + T^*T)\xi \rangle = \|\xi\|^2 + \|T\xi\|^2 \geq 0, \tag{B.2.19}$$

showing that S is positive. Since S has trivial kernel, Lemma B.2.3 applies to this operator, and we conclude that $I + T^*T$ is both positive and self-adjoint. Then T^*T is self-adjoint (and, in particular, closed) and (B.2.18) shows that $T^*T \geq 0$. $\qquad\square$

B.3   The functional calculus

We now develop a functional calculus for densely defined positive operators T and we begin with the following observations. If $T \geq 0$ then so is $I + T$. Any $\xi \in (\mathrm{Ran}(I + T))^{\perp}$ lies in $\mathcal{D}((I + T)^*) = \mathcal{D}(I + T)$, and

$$0 = \langle (I + T)\xi, \xi \rangle = \langle \xi, \xi \rangle + \langle T\xi, \xi \rangle \geq \|\xi\|^2, \tag{B.3.1}$$

where the first equality reflects the fact that $(I + T)\xi \in \mathrm{Ran}(I + T)$. Thus $\xi = 0$ and $\mathrm{Ran}(I + T)$ is dense in H. The inequality

$$\langle (I + T)\xi, \xi \rangle \geq \|\xi\|^2 \tag{B.3.2}$$

from (B.3.1) is valid for any $\xi \in \mathcal{D}(T)$, and shows that $\ker(I + T) = 0$. It follows from (B.3.2) that $1 + T$ has a bounded inverse mapping $\mathrm{Ran}(I + T)$ to $\mathcal{D}(T)$, and this extends by continuity to an operator $S \in B(H)$. If $\xi = (1 + T)\eta \in \mathrm{Ran}(1 + T)$ then

$$\langle S\xi, \xi \rangle = \langle \eta, (I + T)\eta \rangle \geq 0, \tag{B.3.3}$$

and so $S \geq 0$. Then the range of S contains $\mathcal{D}(T)$ so is dense in H, and thus $\ker S = 0$. By Lemma B.2.3, S^{-1} is a densely defined positive operator and clearly $I + T \subseteq S^{-1}$. Taking the adjoint proves equality, and this shows that $\mathrm{Ran}(1 + T)$ is not only dense in H but equals H. In principle, all information about T is encoded in the bounded positive injective operator S, and we will use the customary functional calculus for S to obtain one for T. We will assume below that S is a bounded injective positive operator with $\|S\| \leq 1$, so that its spectrum $\sigma(S)$ lies in $[0,1]$. Recall that for each pair of vectors $\xi, \eta \in H$, there is a bounded linear functional on $C[0,1]$, a measure $\mu_{\xi,\eta}$, defined initially on polynomials p by $p \mapsto \langle p(S)\xi, \eta \rangle$, and extended to $C[0,1]$ by continuity. If $B[0,1]$ denotes the algebra of bounded Borel measurable functions on $[0,1]$, then these measures induce a contractive $*$-homomorphism of $B[0,1]$ into $B(H)$, where an operator $f(S)$ is associated to each $f \in B[0,1]$ by the defining equation

$$\langle f(S)\xi, \eta \rangle = \int f \, d\mu_{\xi,\eta}, \qquad \xi, \eta \in H. \tag{B.3.4}$$

For each Borel set $E \subseteq [0,1]$, the spectral projection of S for this set is the operator corresponding to the function χ_E. The positivity of S and the definition of these measures easily imply the following relations, where $\lambda \in \mathbb{C}$:

$$\mu_{\xi,\xi} \geq 0, \tag{B.3.5}$$

$$\mu_{\xi,\eta} = \bar{\mu}_{\eta,\xi}, \tag{B.3.6}$$

$$\mu_{\lambda\xi,\eta} = \lambda\mu_{\xi,\eta}, \tag{B.3.7}$$

$$\mu_{\xi_1+\xi_2,\eta} = \mu_{\xi_1,\eta} + \mu_{\xi_2,\eta}, \tag{B.3.8}$$

for all $\xi, \xi_1, \xi_2, \eta \in H$. Since S is injective, and $\chi_{\{0\}}$ gives the projection onto $\ker S$, we see that $f(S) = g(S)$ when f and g agree on $(0,1]$. Then the value of $f \in B[0,1]$ at 0 is irrelevant to the functional calculus, and so we could work equally well with the algebra $B(0,1]$ of bounded Borel functions on $(0,1]$. We now introduce the algebra $B_{\text{loc}}(0,1]$ of Borel measurable functions on $(0,1]$ which are bounded on compact subsets of $(0,1]$. Any continuous function on $(0,1]$ (such as t^{-1}) lies in this space and we will use (B.3.4) to extend the functional calculus to $B_{\text{loc}}(0,1]$, expecting unbounded operators to result from this. It will be convenient to have to hand some of the spectral projections of S so, for each integer $n \geq 1$, let P_n be the projection corresponding to the interval $[n^{-1},1]$. Since the sequence $\{\chi_{[n^{-1},1]}\}_{n=1}^{\infty}$ increases pointwise to $\chi_{(0,1]}$, (B.3.4) gives that $\{P_n\}_{n=1}^{\infty}$ is increasing with strong limit I. The equation (B.3.4) implies that each operator $f(S)$, for $f \in B(0,1]$, lies in the abelian von Neumann algebra A generated by S, which is equivalent to $U^*f(S)U = f(S)$ for all unitaries $U \in A'$. A closed operator R is *affiliated* to A if $U^*RU = R$ (with agreement of domains) for all unitaries $U \in A'$, and for bounded operators this is precisely the requirement of being in A. The operators $f(S)$, $f \in B_{\text{loc}}(0,1]$, which we now define, will all be affiliated to A.

Consider a fixed $f \in B_{\text{loc}}(0,1]$. We define the domain $\mathcal{D}(f(S))$ by

$$\mathcal{D}(f(S)) = \{\xi \in H: \quad \text{for all } \eta \in H, f \in L^1(|\mu_{\xi,\eta}|), \text{ and}$$

$$\text{there exists } k > 0 \text{ such that } \int |f| \, d|\mu_{\xi,\eta}| \leq k\|\eta\|\}. \tag{B.3.9}$$

The relation

$$|\mu_{\xi_1+\xi_2,\eta}| = |\mu_{\xi_1,\eta} + \mu_{\xi_2,\eta}| \leq |\mu_{\xi_1,\eta}| + |\mu_{\xi_2,\eta}| \tag{B.3.10}$$

implies that $\mathcal{D}(f(S))$ is a subspace of H, and we define a linear operator $f(S)$ on $\mathcal{D}(f(S))$ by

$$\langle f(S)\xi, \eta \rangle = \int f \, d\mu_{\xi,\eta}, \qquad \xi \in \mathcal{D}(f(S)), \quad \eta \in H. \tag{B.3.11}$$

The integral inequality in (B.3.9) guarantees that $\eta \mapsto \int f \, d\mu_{\xi,\eta}$ is a bounded conjugate linear functional on H, and defines uniquely the vector $f(S)\xi$.

Lemma B.3.1. *Let* $g \in B(0,1]$, $f \in B_{\text{loc}}(0,1]$ *and let* U *be a unitary commuting with* A, *the von Neumann algebra generated by* S.

(i) *For* $\xi, \eta \in H$,

$$\mu_{g(S)\xi,\eta} = \mu_{\xi,\bar{g}(S)\eta} = g\mu_{\xi,\eta}. \tag{B.3.12}$$

(ii) *If* $\xi \in \mathcal{D}(f(S))$ *then* $g(S)\xi \in \mathcal{D}(f(S))$.

(iii) *For each* $\xi \in H$, $U\xi \in \mathcal{D}(f(S))$ *if and only if* $\xi \in \mathcal{D}(f(S))$, *and for such vectors* $U^*f(S)U\xi = f(S)\xi$.

Proof. (i) If $h \in B(0,1]$ and $\xi, \eta \in H$, then

$$\int hg \, d\mu_{\xi,\eta} = \langle (hg)(S)\xi, \eta \rangle = \langle h(S)g(S)\xi, \eta \rangle$$

$$= \int h \, d\mu_{g(S)\xi,\eta}, \tag{B.3.13}$$

showing that $g\mu_{\xi,\eta} = \mu_{g(S)\xi,\eta}$. Since

$$\langle h(S)g(S)\xi, \eta \rangle = \langle h(S)\xi, \bar{g}(S)\eta \rangle = \int h \, d\mu_{\xi,\bar{g}(S)\eta}, \qquad (B.3.14)$$

the last two equations prove (i).

(ii) If $\xi \in \mathcal{D}(f(S))$ then $f \in L^1(|\mu_{\xi,\eta}|)$ for all $\eta \in H$, and there exists a constant $k > 0$ such that $\int |f| \, d|\mu_{\xi,\eta}| \leq k\|\eta\|$. From (i), $\mu_{g(S)\xi,\eta} = \mu_{\xi,\bar{g}(S)\eta}$, and so if $\xi \in \mathcal{D}(f(S))$ then $f \in L^1(|\mu_{g(S)\xi,\eta}|) = L^1(|\mu_{\xi,\bar{g}(S)\eta}|)$, and

$$\int |f| \, d|\mu_{g(S)\xi,\eta}| = \int |f| \, d|\mu_{\xi,\bar{g}(S)\eta}|$$
$$\leq k\|\bar{g}(S)\eta\| \leq k\|g\|_\infty \|\eta\|, \qquad (B.3.15)$$

proving that $g(S)\xi \in \mathcal{D}(f(S))$.

(iii) If $U \in A'$, then for each $h \in B(0,1]$,

$$\int h \, d\mu_{U\xi,\eta} = \langle h(S)U\xi, \eta \rangle = \langle Uh(S)\xi, \eta \rangle$$
$$= \langle h(S)\xi, U^*\eta \rangle = \int h \, d\mu_{\xi,U^*\eta}, \qquad \xi, \eta \in H, \qquad (B.3.16)$$

and thus $\mu_{U\xi,\eta} = \mu_{\xi,U^*\eta}$. If $\xi \in \mathcal{D}(f(S))$ then

$$\int |f| \, d|\mu_{U\xi,\eta}| = \int |f| \, d|\mu_{\xi,U^*\eta}| \leq k\|U^*\eta\| = k\|\eta\|, \qquad (B.3.17)$$

and so $U\xi \in \mathcal{D}(f(S))$. This applies equally to U^*, so if $U\xi \in \mathcal{D}(f(S))$ then so also is $U^*(U\xi) = \xi$. If $\xi \in \mathcal{D}(f(S))$ and $\eta \in H$ then

$$\langle U^*f(S)U\xi, \eta \rangle = \langle f(S)U\xi, U\eta \rangle$$
$$= \int f \, d\mu_{U\xi,U\eta} = \int f \, d\mu_{\xi,U^*U\eta}$$
$$= \langle f(S)\xi, \eta \rangle, \qquad (B.3.18)$$

and so $U^*f(S)U\xi = f(S)\xi$. $\square$

Lemma B.3.2. *If $f \in B_{\mathrm{loc}}(0,1]$, then*

(i) *$f(S)$ is a closed densely defined operator;*

(ii) *$f(S)^* = \bar{f}(S)$ and $f(S)$ is self-adjoint if f is real valued;*

(iii) *$f(S) \geq 0$ when $f \geq 0$;*

(iv) *$\bar{f}(S)f(S) = |f|^2(S)$.*

Proof. For a fixed but arbitrary integer $n \geq 1$, consider a vector ξ in the range of P_n. For each $\eta \in H$ and $g \in B(0,1]$,

$$\langle g(S)\xi, \eta \rangle = \langle g(S)\chi_{[n^{-1},1]}(S)\xi, \eta \rangle \tag{B.3.19}$$

and so $\mu_{\xi,\eta} = \chi_{[n^{-1},1]}\mu_{\xi,\eta}$. For each $\eta \in H$, there is a bounded measurable function h, $\|h\|_\infty \leq 1$, such that $|\mu_{\xi,\eta}| = h\mu_{\xi,\eta}$, and so

$$\int |f| \, d|\mu_{\xi,\eta}| = \int |f| h\chi_{[n^{-1},1]} \, d\mu_{\xi,\eta}$$
$$= |\langle (|f| h\chi_{[n^{-1},1]})(S)\xi, \eta \rangle|$$
$$\leq \||f|\chi_{[n^{-1},1]}\|_\infty \|\xi\| \|\eta\|, \tag{B.3.20}$$

since $f\chi_{[n^{-1},1]}$ is bounded. This shows that $\mathcal{D}(f(S))$ contains $P_n H$ for every $n \geq 1$, and so $\mathcal{D}(f(S))$ is dense in H.

It is clear from (B.3.9) that $\mathcal{D}(\bar{f}(S)) = \mathcal{D}(f(S))$. Consider a fixed vector $\eta \in \mathcal{D}(\bar{f}(S))$. From (B.3.9), we obtain a constant k such that

$$|\langle f(S)\xi, \eta \rangle| = \left| \int f \, d\mu_{\xi,\eta} \right| \leq \int |f| \, d|\mu_{\xi,\eta}|$$
$$= \int |f| \, d|\mu_{\eta,\xi}| \leq k\|\xi\| \tag{B.3.21}$$

for $\xi \in \mathcal{D}(f(S))$. Thus $\eta \in \mathcal{D}(f(S)^*)$ and $f(S)^*$ is densely defined, with $\mathcal{D}(\bar{f}(S)) \subseteq \mathcal{D}(f(S)^*)$. Moreover, if $\xi \in \mathcal{D}(f(S))$ and $\eta \in \mathcal{D}(\bar{f}(S))$, then

$$\langle f(S)\xi, \eta \rangle = \int f \, d\mu_{\xi,\eta} = \overline{\int \bar{f} \, d\mu_{\eta,\xi}}$$
$$= \overline{\langle \bar{f}(S)\eta, \xi \rangle} = \langle \xi, \bar{f}(S)\eta \rangle, \tag{B.3.22}$$

showing that $\bar{f}(S) \subseteq f(S)^*$. We now show equality.

Consider $\eta \in \mathcal{D}(f(S)^*)$. By definition of the adjoint, there exists $k > 0$ such that

$$|\langle f(S)\xi, \eta \rangle| = \left| \int f \, d\mu_{\xi,\eta} \right| \leq k\|\xi\|, \qquad \xi \in \mathcal{D}(f(S)). \tag{B.3.23}$$

There are measurable functions g, h of modulus 1 such that $|f| = fg$ and $h\mu_{\xi,\eta} = |\mu_{\xi,\eta}|$. Then

$$\int |f| \, d|\mu_{\eta,\xi}| = \int |f| \, d|\mu_{\xi,\eta}|$$
$$= \left| \int fgh \, d\mu_{\xi,\eta} \right|$$
$$= \left| \int f \, d\mu_{(gh)(S)\xi,\eta} \right|$$
$$\leq k\|\xi\| \tag{B.3.24}$$

from (B.3.23), since $(gh)(S)\xi \in \mathcal{D}(f(S))$, by Lemma B.3.1, and $\|(gh)(S)\| \leq 1$. Now if ξ is a unit vector in H and is the limit of a sequence of unit vectors $\xi_n \in \mathcal{D}(f(S))$, then $\lim\limits_{n\to\infty} |\mu_{\eta,\xi_n}| = |\mu_{\eta,\xi}|$ in norm and so, for each $r \geq 1$,

$$\int |f| \wedge r1 \, d|\mu_{\eta,\xi}| = \lim_{n\to\infty} \int |f| \wedge r1 \, d|\mu_{\eta,\xi_n}|$$
$$\leq k, \tag{B.3.25}$$

applying (B.3.24) to the vectors ξ_n. Letting $r \to \infty$ in (B.3.25) gives, by the monotone convergence theorem,

$$\int |f| \, d|\mu_{\eta,\xi}| \leq k \tag{B.3.26}$$

for all unit vectors $\xi \in H$. This shows that $\eta \in \mathcal{D}(f(S)) = \mathcal{D}(\bar{f}(S))$, and we conclude that $f(S)^* = \bar{f}(S)$. Replacing f by $\bar{f}$ gives $f(S) = \bar{f}(S)^*$ and thus $f(S)$, being the adjoint of a densely defined operator, is closed by Lemma B.2.2 (i). This proves (i), and the relation $f(S)^* = f(S)$ when f is real valued is now clear, proving (ii).

If $f \geq 0$ then $f(S)$ is self-adjoint, and for $\xi \in \mathcal{D}(f(S))$,

$$\langle f(S)\xi, \xi \rangle = \int f \, d\mu_{\xi,\xi} \geq 0, \tag{B.3.27}$$

proving (iii), since $\mu_{\xi,\xi} \geq 0$ from (B.3.5).

We now turn to the last statement of the lemma. If $\xi \in \mathcal{D}(|f|^2(S))$ then, for all $\eta \in H$, $|f|^2 \in L^1(|\mu_{\xi,\eta}|)$ for all $\eta \in H$. Let k be the constant associated to $|f|^2$ from (B.3.9). Since $|\mu_{\xi,\eta}|$ is a finite measure, it follows that $|f| \in L^1(|\mu_{\xi,\eta}|)$ by considering separately $E = \{t\colon |f(t)| \leq 1\}$ and its complement. Moreover, since $|\mu_{\xi,\eta}| = h\mu_{\xi,\eta}$ for some measurable h of modulus 1,

$$\int |f| \, d|\mu_{\xi,\eta}| \leq \int \chi_E \, d|\mu_{\xi,\eta}| + \int \chi_{E^c} |f|^2 \, d|\mu_{\xi,\eta}|$$
$$\leq |\langle h(S)\chi_E(S)\xi, \eta \rangle| + k\|\eta\|$$
$$\leq (k + \|\xi\|)\|\eta\|, \qquad \eta \in H. \tag{B.3.28}$$

The shows that $\xi \in \mathcal{D}(f(S))$.

For each $g \in B(0,1]$, $\eta \in H$,

$$\int g \, d\mu_{f(S)\xi,\eta} = \langle g(S)f(S)\xi, \eta \rangle = \langle f(S)\xi, \bar{g}(S)\eta \rangle$$
$$= \int f \, d\mu_{\xi,\bar{g}(S)\eta} = \int fg \, d\mu_{\xi,\eta}, \tag{B.3.29}$$

by Lemma B.3.1, showing that $\mu_{f(S)\xi,\eta} = f\mu_{\xi,\eta}$ by letting g vary. Then, for some measurable h of modulus 1,

$$\int |f| \, d|\mu_{f(S)\xi,\eta}| = \int |f|h \, d\mu_{f(S)\xi,\eta} = \int |f|hf \, d\mu_{\xi,\eta}$$
$$\leq \int |f|^2 \, d|\mu_{\xi,\eta}| \leq k\|\eta\|, \qquad \eta \in H, \tag{B.3.30}$$

showing that $f(S)\xi \in \mathcal{D}(\bar{f}(S))$. Moreover, if $\eta \in \mathcal{D}(f(S))$, then

$$\langle f(S)^* f(S)\xi, \eta \rangle = \langle f(S)\xi, f(S)\eta \rangle = \int f \, d\mu_{\xi, f(S)\eta}$$

$$= \overline{\int \bar{f} \, d\mu_{f(S)\eta, \xi}} = \int |f|^2 \, d\mu_{\xi, \eta}, \qquad \text{(B.3.31)}$$

showing that $f(S)^* f(S)$ and $|f|^2(S)$ agree on $\mathcal{D}(|f|^2(S))$. Thus $|f|^2(S) \subseteq f(S)^* f(S)$, and taking the adjoint proves equality, establishing (iv). $\qquad \square$

Lemma B.3.3. *The operators $f(S), f \in B_{\mathrm{loc}}(0,1]$, have a common core $C = \bigcup_{n \geq 1} P_n H$.*

Proof. Fix $f \in B_{\mathrm{loc}}(0,1]$, $\xi \in \mathcal{D}(f(S))$, and let $\eta \in H$ be arbitrary. From Lemma B.3.1 (ii), we have that $P_n \xi \in \mathcal{D}(f(S))$ for all $n \geq 1$. Since $f \in L^1(|\mu_{\xi,\eta}|)$, the dominated convergence theorem gives

$$\lim_{n \to \infty} \langle f(S) P_n \xi, \eta \rangle = \lim_{n \to \infty} \int f \chi_{[n^{-1}, 1]} \, d\mu_{\xi, \eta}$$

$$= \int f \, d\mu_{\xi, \eta}$$

$$= \langle f(S)\xi, \eta \rangle, \qquad \text{(B.3.32)}$$

showing that $f(S) P_n \xi \to f(S)\xi$ weakly. It follows that $(P_n \xi, f(S) P_n \xi) \to (\xi, f(S)\xi)$ weakly in $H \times H$, showing that $\mathcal{G}(f(S)|_C)$ is weakly dense in $\mathcal{G}(f(S))$. Moreover, $(\mathcal{G}(f(S)|_C)$ is convex and so it is also norm dense in $\mathcal{G}(f(S))$. $\qquad \square$

Remark B.3.4. Examining the proof of this lemma, it is clear that we could replace the P_n's by any sequence of operators $g_n(S)$ where each $g_n \in B(0,1]^+$ is compactly supported and the sequence $\{g_n\}_{n=1}^\infty$ increases pointwise to $\chi_{(0,1]}$. $\square$

Lemma B.3.5. *For each $f \in B_{\mathrm{loc}}(0,1]$, $f(S)$ is affiliated to A, the von Neumann algebra generated by S.*

Proof. By Lemma B.3.2 (i), $f(S)$ is closed, and Lemma B.3.1 (iii) then proves the result. $\qquad \square$

The next result considers two operators S and $g(S)$ where g is a bounded function, and links the functional calculus for S to that for $g(S)$.

Lemma B.3.6. *Let $g \colon [0,1] \to [0,1]$ be a continuous function such that $g^{-1}(\{0\}) = \{0\}$ and let S be an injective positive contraction. Then $g(S)$ is an injective positive contraction and, for each $f \in B_{\mathrm{loc}}(0,1]$, $f(g(S)) = (f \circ g)(S)$.*

Proof. Following previous notation, there exist measures $\mu_{\xi,\eta}$, $\nu_{\xi,\eta}$ for $\xi, \eta \in H$ such that

$$\langle f(S)\xi, \eta \rangle = \int f \, d\mu_{\xi, \eta}, \qquad \langle f(g(S))\xi, \eta \rangle = \int f \, d\nu_{\xi, \eta}, \qquad \text{(B.3.33)}$$

for all $f \in B(0, 1]$, since the hypotheses on g imply that $g(S)$ is an injective positive contraction. The equations (B.3.33) give the following relation between the measures, since $f(g(S)) = f \circ g(S)$ for bounded Borel functions by the bounded functional calculus:

$$\nu_{\xi, \eta}(E) = \mu_{\xi, \eta}(g^{-1}(E)), \qquad \xi, \eta \in H, \tag{B.3.34}$$

for all Borel sets $E \subseteq (0, 1]$. Simple measure theory then implies that $f \in L^1(|\nu_{\xi, \eta}|)$ if and only if $f \circ g \in L^1(|\mu_{\xi, \eta}|)$, and

$$\int |f \circ g| \, d|\mu_{\xi, \eta}| = \int |f| \, d|\nu_{\xi, \eta}| \tag{B.3.35}$$

is established by first considering linear combinations of characteristic functions and applying the dominated convergence theorem. Any bound of $k\|\eta\|$ for one side of (B.3.35) will also hold for the other, so that $\mathcal{D}((f \circ g)(S)) = \mathcal{D}(f(g(S)))$, and it follows from (B.3.33) and (B.3.35) that $(f \circ g)(S) = f(g(S))$. $\square$

We now apply these results to general closed operators. The following theorem contains the polar decomposition for closed operators.

Theorem B.3.7. (i) *If T is a positive operator then T has a unique positive square root.*
(ii) *If T is a closed operator and $|T|$ denotes the positive square root of T^*T, then there is a partial isometry v satisfying $T = v|T|$, and v has initial space $\overline{\mathrm{Ran}|T|}$ and final space $\overline{\mathrm{Ran}\, T}$. If $T = v_1 T_1$ with $T_1 \geq 0$ and v_1 a partial isometry, then $v_1 = v$ and $T_1 = |T|$.*
(iii) *Under the hypothesis of (ii), $T^* = |T|v^*$.*
(iv) *If T is a closed operator with polar decomposition $T = v|T|$, and T is affiliated to a von Neumann algebra B, then $v \in B$ and $|T|$ is affiliated to B.*

Proof. (i) From the discussion following Lemma B.2.4, the range of $I + T$ is H and there is a bounded positive contraction S with trivial kernel such that $S = (I + T)^{-1}$. Let $f(t) = t^{-1} - 1 \in B_{\mathrm{loc}}(0, 1]$. Since $f \geq 0$, we have $f(S) \geq 0$, and it follows from Lemma B.3.2 (iii), (iv) that $f^{1/2}(S) \geq 0$ and that $(f^{1/2}(S))^2 = f(S)$, so that $f(S)$ has a positive square root. If $\xi \in \mathcal{D}(f(S))$ then, for each $\eta \in H$,

$$\langle Sf(S)\xi, \eta \rangle = \langle f(S)\xi, S\eta \rangle = \int f \, d\mu_{\xi, S\eta}$$

$$= \int f \, d\mu_{S\xi, \eta} = \int tf(t) \, d\mu_{\xi, \eta}$$

$$= \langle (I - S)\xi, \eta \rangle, \tag{B.3.36}$$

so that $(Sf(S) + S)\xi = \xi$. The operator S maps into the domain of T so $\xi \in \mathcal{D}(T)$ and we may apply $I + T$ on the left to obtain

$$f(S)\xi + \xi = \xi + T\xi, \qquad \xi \in \mathcal{D}(f(S)). \tag{B.3.37}$$

This shows that $f(S) \subseteq T$, and we obtain equality by taking adjoints. Thus T has a positive square root.

We now show uniqueness. Suppose that R is another positive operator such that $R^2 = T$. Let $S_1 = (1 + R)^{-1}$, an injective positive contraction. We will apply the functional calculus already developed to S_1 and S. Then $R = (t^{-1} - 1)(S_1)$ so $T = R^2 = (t^{-1} - 1)^2(S_1)$. Thus

$$S = (I + T)^{-1} = ((t^{-1} - 1)^2 + 1)^{-1}(S_1)$$
$$= (t^2/(2t^2 - 2t + 1))(S_1). \tag{B.3.38}$$

Since the square root of T obtained above is $f^{1/2}(S)$, we have

$$f^{1/2}(S) = f^{1/2}(t^2/(2t^2 - 2t + 1)(S_1))$$
$$= \left(\frac{2t^2 - 2t + 1}{t^2} - 1\right)^{1/2}(S_1)$$
$$= (t^{-1} - 1)(S_1) = R, \tag{B.3.39}$$

where the second equality is Lemma B.3.6 applied to S_1 and the continuous function $g(t) = t^2/(2t^2 - 2t + 1)$. Uniqueness of the square root follows.

(ii) Recall, from Lemma B.2.2, the bounded operator $P \colon \mathcal{G}(T) \to H$ defined by $P(\xi, T\xi) = \xi$ for $\xi \in \mathcal{D}(T)$. Let $M \subseteq H$ be the linear subspace consisting of the first coordinates of the elements in the range of P^*. Since the range of P^* is dense in $\mathcal{G}(T)$, it follows that M is a core for T. If $\xi \in M$ then $(\xi, T\xi) = P^*(\zeta)$ for some $\zeta \in H$. For $\eta \in \mathcal{D}(T)$,

$$\langle \eta, \xi \rangle + \langle T\eta, T\xi \rangle = \langle (\eta, T\eta), P^*(\zeta) \rangle$$
$$= \langle P(\eta, T\eta), \zeta \rangle = \langle \eta, \zeta \rangle, \tag{B.3.40}$$

and this equation shows that $T\xi \in \mathcal{D}(T^*)$. Thus $\xi \in \mathcal{D}(T^*T)$, and $\mathcal{D}(T^*T)$ is a core for both T^*T and T, since we just proved that $M \subseteq \mathcal{D}(T^*T)$. By Lemma B.2.4, T^*T is a closed positive operator, and has a positive square root, which we denote $|T|$, by part (i). The relation $|T|^2 = T^*T$ contains implicitly the relation $\mathcal{D}(T^*T) \subseteq \mathcal{D}(|T|)$. Since both T^*T and $|T|$ were obtained from the functional calculus for the bounded operator $S = (1 + T^*T)^{-1}$, both $|T|$ and T^*T have a comma core, by Lemma B.3.3. We conclude from this discussion that $\mathcal{D}(T^*T)$ is a core for T, $|T|$ and T^*T.

For $\xi \in \mathcal{D}(T^*T)$, $T\xi$ lies in $\mathcal{D}(T^*)$ so

$$\langle T\xi, T\xi \rangle = \langle \xi, T^*T\xi \rangle = \langle \xi, |T|^2\xi \rangle$$
$$= \langle |T|\xi, |T|\xi \rangle. \tag{B.3.41}$$

Thus $\|T\xi\| = \||T|\xi\|$ for all $\xi \in \mathcal{D}(T^*T)$. Since $\{T\xi \colon \xi \in \mathcal{D}(T^*T)\}$ and $\{|T|\xi \colon \xi \in \mathcal{D}(T^*T)\}$ are respectively dense in Ran T and Ran$|T|$, it follows that there is a partial isometry v, with initial and final spaces respectively the closures of Ran$|T|$ and of Ran T, such that $T\xi = v|T|\xi$ for $\xi \in \mathcal{D}(T^*T)$. We now show that $T = v|T|$.

If $\xi \in \mathcal{D}(|T|)$ then, using the fact that $\mathcal{D}(T^*T)$ is a core for $|T|$, there is a sequence $\{\xi_n\}_{n=1}^\infty \in \mathcal{D}(T^*T)$ such that $\xi_n \to \xi$ and $|T|\xi_n \to |T|\xi$. Then $(\xi_n, T\xi_n) \in \mathcal{G}(T)$ and has limit $(\xi, v|T|\xi)$. Since $\mathcal{G}(T)$ is closed, $\xi \in \mathcal{D}(T)$ and $T\xi = v|T|\xi$. This shows that $v|T| \subseteq T$. Conversely suppose that $\xi \in \mathcal{D}(T)$. As before, we can find a sequence $\{\xi_n\}_{n=1}^\infty \in \mathcal{D}(T^*T)$ such that $\xi_n \to \xi$, $T\xi_n \to T\xi$. Then $v^*T\xi_n \to v^*T\xi$. Since v^*v is a projection whose range contains $\mathrm{Ran}|T|$, and $v^*T\xi_n = |T|\xi_n$ for $n \geq 1$, we conclude that $(\xi_n, v^*T\xi_n)$ must converge to the point $(\xi, |T|\xi) \in \mathcal{G}(|T|)$ which is closed. Then $\xi \in \mathcal{D}(|T|)$ and $T\xi = v|T|\xi$, proving that $T \subseteq v|T|$, and establishing equality.

Now suppose that $T = v_1 T_1$, where $T_1 \geq 0$ and v_1 is a partial isometry with initial space $\overline{\mathrm{Ran}\, T_1}$ and final space $\overline{\mathrm{Ran}\, T}$. Then $v_1^*T = T_1$. Consider $\xi \in \mathcal{D}(T)$ and $\eta \in \mathcal{D}(T^*T)$. Then $T\eta \in \mathcal{D}(T^*)$, so

$$\langle v_1^*T\xi, v_1^*T\eta \rangle = \langle T\xi, T\eta \rangle = \langle \xi, T^*T\eta \rangle. \qquad (B.3.42)$$

This shows that $v_1^*T\eta \in \mathcal{D}((v_1^*T)^*)$, and that $T_1\eta \in \mathcal{D}(T_1)$, so that $\eta \in \mathcal{D}(T_1^2)$. Moreover, from (B.3.42), $T_1^2\eta = T^*T\eta$, by letting ξ vary over $\mathcal{D}(T)$. Thus $T_1^2 = T^*T$ on $\mathcal{D}(T^*T)$, so $T^*T \subseteq T_1^2$. Both are self-adjoint, so taking the adjoint gives equality. Thus $|T|$ and T_1 are positive square roots of T^*T, and it follows from (i) that $|T| = T_1$. It is then easy to see that $v = v_1$, and the polar decomposition is thus unique.

(iii)   If $\xi \in \mathcal{D}(T)$ and $\eta \in \mathcal{D}(T^*)$, then

$$\langle \xi, T^*\eta \rangle = \langle T\xi, \eta \rangle = \langle v|T|\xi, \eta \rangle$$
$$= \langle |T|\xi, v^*\eta \rangle. \qquad (B.3.43)$$

Since $\mathcal{D}(T) = \mathcal{D}(|T|)$, it follows that $v^*\eta \in \mathcal{D}(|T|^*) = \mathcal{D}(|T|)$, and that

$$\langle \xi, T^*\eta \rangle = \langle \xi, |T|v^*\eta \rangle. \qquad (B.3.44)$$

Varying ξ over $\mathcal{D}(T)$ gives $T^*\eta = |T|v^*\eta$ for $\eta \in \mathcal{D}(T^*)$, and so $T^* \subseteq |T|v^*$. Conversely, suppose that $\eta \in \mathcal{D}(|T|v^*)$. Then $v^*\eta \in \mathcal{D}(|T|)$. If $\xi \in \mathcal{D}(T)$ then

$$\langle T\xi, \eta \rangle = \langle v|T|\xi, \eta \rangle = \langle |T|\xi, v^*\eta \rangle$$
$$= \langle \xi, |T|v^*\eta \rangle. \qquad (B.3.45)$$

From this, we conclude that $\eta \in \mathcal{D}(T^*)$ and that $T^*\eta = |T|v^*\eta$, proving the reverse containment $|T|v^* \subseteq T^*$. Thus, $T^* = |T|v^*$.

(iv)   Now suppose that T is closed and affiliated to a von Neumann algebra B. If $U \in B'$ is a unitary, and $T = v|T|$ is the polar decomposition of T, then

$$T = UTU^* = (UvU^*)(U|T|U^*) \qquad (B.3.46)$$

is also a polar decomposition for T. The uniqueness of (ii) then gives $UvU^* = v$ and $U|T|U^* = |T|$, proving the result. $\square$

We now describe the functional calculus for a positive densely defined operator T. We will use the functional calculus already developed for an injective positive operator S, which will be the inverse of $I + T$. In analogy to

$B(0,1]$ and $B_{\text{loc}}(0,1]$, we let $B[0,\infty)$ denote the bounded Borel measurable functions on $[0,\infty)$, and $B_{\text{loc}}[0,\infty)$ the Borel measurable functions which are bounded on compact sets. To relate these quantities we will use the homeomorphism $\phi\colon (0,1] \to [0,\infty)$ defined by $\phi(t) = t^{-1} - 1$, $t \in (0,1]$, with inverse $\psi(t) = (1+t)^{-1}$, $t \in [0,\infty)$. Then there if a $*$-isomorphism $f \mapsto \tilde{f}\colon B_{\text{loc}}[0,\infty) \to B_{\text{loc}}(0,1]$ defined by $\tilde{f} = f \circ \phi$ for $f \in B_{\text{loc}}[0,\infty)$.

Theorem B.3.8. *Let T be a densely defined positive operator. Then, to each $f \in B_{\text{loc}}[0,\infty)$, there is assigned a closed densely defined operator $f(T)$ with the following properties:*

(i) *if $f(t) = 1$ then $f(T) = I$;*

(ii) *if $f(t) = t$ then $f(T) = T$;*

(iii) *if $f \in B[0,\infty)$ then $f(T) \in B(H)$ and $f \mapsto f(T)$ is a $*$-homomorphism from $B[0,\infty)$ into $B(H)$;*

(iv) *if $E_n = \chi_{[0,n]}(T)$, $n \geq 1$, then each E_n is a projection, the sequence $\{E_n\}_{n=1}^\infty$ converges monotonically and strongly to I, and $C = \bigcup_{n=1}^\infty E_n H$ is a core for each $f(T)$, $f \in B_{\text{loc}}[0,\infty)$;*

(v) *if $f, g \in B_{\text{loc}}[0,\infty)$ then $(f+g)(T)$ and $(fg)(T)$ are respectively the closures of $f(T) + g(T)$ and $f(T)g(T)$;*

(vi) *if $f \in B_{\text{loc}}[0,\infty)$, then $f(T)^* = \bar{f}(T)$, and $f(T)^*f(T) = |f|^2(T)$;*

(vii) *if T is affiliated to a von Neumann algebra B, then so is each $f(T)$ for $f \in B_{\text{loc}}[0,\infty)$.*

Proof. Let $S = (I + T)^{-1}$, an injective positive contraction. Since we have already developed the functional calculus for S, we define $f(T)$ to be $\tilde{f}(S)$ for $f \in B_{\text{loc}}[0,\infty)$. Parts (i), (ii), (iii) and (vi) are then immediate from Lemma B.3.2 and the bounded functional calculus for S. Lemma B.3.3 shows that each $f(T)$ $(= \tilde{f}(S))$, $f \in B_{\text{loc}}[0,\infty)$, has a core $C = \bigcup_{n=1}^\infty P_n H$ where P_n is the spectral projection of S for the interval $[n^{-1}, 1]$. Since ϕ maps this interval to $[0, n-1]$, we have $P_n = E_{n-1}$ for $n \geq 2$, and $C = \bigcup_{n=1}^\infty E_n H$. Then $\{E_n\}_{n=1}^\infty$ converges monotonically and strongly to I, since $\{P_n\}_{n=1}^\infty$ has this property. This proves (iv).

Now suppose that $f, g \in B_{\text{loc}}[0,\infty)$. Since $C \subseteq \mathcal{D}(f(T)) \cap \mathcal{D}(g(T)) = \mathcal{D}(f(T) + g(T))$, we see that the sum is densely defined. If $\xi \in \mathcal{D}(f(T) + g(T))$, then, from the proof of Lemma B.3.3, $E_n\xi \in \mathcal{D}(f(T)) \cap \mathcal{D}(g(T))$ and

$$\lim_{n\to\infty} (P_n\xi, (f(T) + g(T))P_n\xi) = (\xi, (f(T) + g(T))\xi) \tag{B.3.47}$$

weakly in $H \times H$. From the definition of these operators, (B.3.9) and (B.3.11), it is clear that $f(T) + g(T)$ and $(f + g)(T)$ agree on each $E_n H$ and thus on C. Then (B.3.47) shows that $\{(\xi, f(T)\xi + g(T)\xi)\colon \xi \in C\}$ is dense in both

$\overline{\mathcal{G}(f(T) + g(T))}$ and $\mathcal{G}((f + g)(T))$, proving that $(f + g)(T)$ is the closure of $f(T) + g(T)$.

We now consider the product $f(T)g(T)$. By Lemma B.3.1 (i), with g replaced by $\chi_{[(n+1)^{-1},1]}$ and then by $1 - \chi_{[(n+1)^{-1},1]}$, we see that $\mu_{E_n\xi,(I-E_n)\eta} = 0$ for all $\xi, \eta \in H$. It then follows from (B.3.11) then $E_n H$ is an invariant subspace for each $f(T)$, $f \in B_{\mathrm{loc}}[0,\infty)$, and each $n \geq 1$. Thus the product $f(T)g(T)$ is defined on C, and (B.3.11) shows that $f(T)g(T)$ and $(fg)(T)$ agree on C. Since $\overline{\mathcal{G}((fg)(T)|_C)} = \mathcal{G}((fg)(T))$, it follows that $\overline{\mathcal{G}(f(T)g(T))} \supseteq \mathcal{G}((fg)(T))$. Suppose that $\xi \in \mathcal{D}(f(T)g(T))$. Then $\xi \in \mathcal{D}(g(T))$ and $g(T)\xi \in \mathcal{D}(f(T))$. By (B.3.32), $f(T)E_n g(T)\xi \to f(T)g(T)\xi$ weakly. For each $\eta \in H$,

$$\langle E_n g(T)\xi, \eta \rangle = \langle \tilde{g}(S)\xi, E_n \eta \rangle = \int \tilde{g}\, d\mu_{\xi, E_n \eta}$$

$$= \int \tilde{g}\, d\mu_{E_n \xi, \eta} = \langle g(T)E_n \xi, \eta \rangle, \qquad (B.3.48)$$

using Lemma B.3.1 (i) for the function $\chi_{[(n+1)^{-1},1]}$. Thus

$$f(T)E_n g(T)\xi = f(T)g(T)E_n \xi = (fg)(T)E_n \xi, \qquad n \geq 1, \qquad (B.3.49)$$

since $f(T)g(T)$ and $(fg)(T)$ agree on C. Thus

$$\lim_{n\to\infty} (E_n \xi, (fg)(T)E_n \xi) = (\xi, f(T)g(T)\xi) \qquad (B.3.50)$$

weakly in $H \times H$, showing that $\mathcal{G}(f(T)g(T)) \subseteq \mathcal{G}((fg)(T))$, and proving equality of the closures.

Now suppose that T is affiliated to a von Neumann algebra B and let $U \in B'$ be an arbitrary unitary. Since $(I + T)S = I$, it follows that

$$I = U(I + T)U^* U S U^* = (I + T)U S U^*. \qquad (B.3.51)$$

Multiply on the left by S in (B.3.51) to obtain $S = U S U^*$, showing that $S \in B$. Since $f(T) = \tilde{f}(S)$ for each $f \in B_{\mathrm{loc}}[0,\infty)$, it follows from Lemma B.3.5 that $f(T)$ is affiliated to the von Neumann algebra generated by S, and hence to B. $\qquad \square$

A densely defined operator T is *symmetric* if $T \subseteq T^*$. Any self-adjoint operator is symmetric, but symmetric operators need not be self-adjoint. The following two results establish necessary and sufficient conditions for a symmetric operator to be self-adjoint.

Lemma B.3.9. *Let T be closed and symmetric. Then the ranges of $T \pm iI$ are closed.*

Proof. We consider $T + iI$, the other case being similar. If $\xi \in \mathcal{D}(T)$, then $\xi \in \mathcal{D}(T^*)$ and so $\langle T\xi, \xi \rangle = \langle \xi, T\xi \rangle$, and $\langle T\xi, \xi \rangle \in \mathbb{R}$. Then

$$\|\xi\|^2 \leq |\langle T\xi, \xi \rangle + i\langle \xi, \xi \rangle| = |\langle (T + iI)\xi, \xi \rangle| \leq \|(T + iI)\xi\| \|\xi\|, \qquad (B.3.52)$$

showing that $\|\xi\| \leq \|(T + iI)\xi\|$ for $\xi \in \mathcal{D}(T)$. The isomorphism $(\xi, \eta) \mapsto (\xi, \eta + i\xi)$ of $H \times H$ to itself has bounded inverse $(\xi, \eta) \mapsto (\xi, \eta - i\xi)$, and maps $\mathcal{G}(T)$ onto $\mathcal{G}(T + iI)$. The latter space is then closed in $H \times H$. If η is the limit of a sequence $\{(T + iI)\xi_n\}_{n=1}^{\infty}$ with $\xi_n \in \mathcal{D}(T)$, then the last inequality shows that $\{\xi_n\}_{n=1}^{\infty}$ is also a convergent sequence. Then $\{(\xi_n, (T + iI)\xi_n)\}_{n=1}^{\infty}$ converges to a point in $\mathcal{G}(T + iI)$ whose second coordinate is η. This shows that the range of $T + iI$ is closed. $\square$

Proposition B.3.10. *Let T be a closed symmetric operator. Then this operator is self-adjoint if and only if the ranges of $T \pm iI$ are both equal to H.*

Proof. Suppose first that T is self-adjoint. We will show that $T + iI$ has dense range, which is sufficient by Lemma B.3.9. The case of $T - iI$ is similar. Let $\eta \in H$ be a vector orthogonal to the range of $T + iI$. Then

$$0 = \langle T\xi + i\xi, \eta \rangle = \langle T\xi, \eta \rangle + i\langle \xi, \eta \rangle, \qquad \xi \in \mathcal{D}(T). \tag{B.3.53}$$

Thus $\xi \mapsto \langle T\xi, \eta \rangle$ is bounded on $\mathcal{D}(T)$, placing $\eta \in \mathcal{D}(T^*) = \mathcal{D}(T)$. Then (B.3.53) gives $T\eta = -i\eta$, since $\mathcal{D}(T)$ is dense in H. Now $\langle T\eta, \eta \rangle \in \mathbb{R}$ since $\langle T\eta, \eta \rangle = \langle \eta, T\eta \rangle$, but it also lies in $i\mathbb{R}$ since $\langle T\eta, \eta \rangle = -\langle i\eta, \eta \rangle$. This yields $\langle -i\eta, \eta \rangle = 0$, and $\eta = 0$. It follows that $\text{Ran}(T + iI)$ is dense in H, and hence equal to H.

Now suppose that the ranges of $T \pm iI$ both equal H. In order to show that $\mathcal{G}(T) = \mathcal{G}(T^*)$, we consider an element $(\eta, T^*\eta) \in \mathcal{G}(T^*)$ which is orthogonal to $\mathcal{G}(T)$, and we prove that $\eta = 0$. Then

$$\langle \xi, \eta \rangle + \langle T\xi, T^*\eta \rangle = 0, \qquad \xi \in \mathcal{D}(T). \tag{B.3.54}$$

The hypothesis on the ranges of $T \pm iI$ implies that there is a vector $\xi \in \mathcal{D}(T)$ such that $(T - iI)\xi \in \mathcal{D}(T)$ and $(T + iI)(T - iI)\xi = \eta$. For this ξ, (B.3.54) yields

$$\begin{aligned}
\langle \eta, \eta \rangle &= \langle (T^2 + I)\xi, \eta \rangle = \langle T^2\xi, \eta \rangle + \langle \xi, \eta \rangle \\
&= \langle T\xi, T^*\eta \rangle + \langle \xi, \eta \rangle = 0,
\end{aligned} \tag{B.3.55}$$

and $\eta = 0$ as required. Thus $T = T^*$. $\square$

This result concludes the general theory of unbounded operators that we will need subsequently. More comprehensive treatments are to be found in the standard texts on operator algebras, see for example [104, Chapter 5] or [184, Chapter 9].

B.4 Operators from $L^2(N)$

We are now ready to discuss the operators ℓ_η defined in (B.2.2) by

$$\ell_\eta(x\xi) = Jx^*J\eta, \qquad x \in N, \quad \eta \in L^2(N), \tag{B.4.1}$$

where N is a type II_1 factor with trace τ and ξ is the vector $1 \in L^2(N)$. If $y \in N$ then

$$
\begin{aligned}
|\langle \ell_\eta(x\xi), y\xi \rangle| &= |\langle y^* J x^* J \eta, \xi \rangle| \\
&= |\langle y^* \eta, x^* \xi \rangle| \leq \|y\| \|\eta\|_2 \|x\|_2,
\end{aligned} \tag{B.4.2}
$$

and so $y\xi \in \mathcal{D}(\ell_\eta^*)$. Thus ℓ_η^* is densely defined, so ℓ_η is closable, by Lemma B.2.2, and has closure ℓ_η^{**} which we denote by L_η. Each unitary in N' has the form JuJ for some unitary $u \in N$. Then

$$
\begin{aligned}
Ju^* J \ell_\eta JuJx\xi &= Ju^* J \ell_\eta(xu^*\xi) \\
&= Ju^* J Jux^* J \eta \\
&= Jx^* J \eta = \ell_\eta(x\xi), \qquad x \in N, \tag{B.4.3}
\end{aligned}
$$

proving that $Ju^* J \ell_\eta JuJ = \ell_\eta$. The two sides of this equation have closures respectively $Ju^* J L_\eta JuJ$ and L_η, showing that each L_η, $\eta \in L^2(N)$, is affiliated to N.

There is a natural positive cone $L^2(N)^+$ in $L^2(N)$, defined as the $\|\cdot\|_2$-norm closure of $N^+\xi$. If $x, y \in N^+$ then

$$
\begin{aligned}
\|x + y\|_2^2 &= \|x\|_2^2 + \|y\|_2^2 + 2\tau(xy) \\
&= \|x\|_2^2 + \|y\|_2^2 + 2\tau(y^{1/2}xy^{1/2}) \\
&\geq \|x\|_2^2 + \|y\|_2^2. \tag{B.4.4}
\end{aligned}
$$

If $\eta, -\eta \in L^2(N)^+$ then choose sequences $\{x_n\}_{n=1}^\infty$, $\{y_n\}_{n=1}^\infty \in N^+$ such that $\lim_{n\to\infty} x_n\xi = \eta$, $\lim_{n\to\infty} y_n\xi = -\eta$. Then (B.4.4) shows that $\lim_{n\to\infty} \|x_n\|_2 = \lim_{n\to\infty} \|y_n\|_2 = 0$, and $\eta = 0$. Thus $L^2(N)^+$ is a proper cone. If $x \in N$ is self-adjoint then $x = x^+ - x^-$ with $x^\pm \in N^+$ and $x^+x^- = 0$, by spectral theory. Then $\|x\|_2^2 = \|x^+\|_2^2 + \|x^-\|_2^2$, so that $\{x_n^\pm\xi\}_{n=1}^\infty$ are Cauchy sequences in $L^2(N)^+$ whenever $\{x_n\xi\}_{n=1}^\infty$ is Cauchy for $x_n \in N_{\text{s.a.}}$. It follows that any Cauchy sequence $\{x_n\xi\}_{n=1}^\infty$ can be written as the sum of four Cauchy sequences given by first expressing x_n as $y_n + iz_n$, $y_n, z_n \in N_{\text{s.a.}}$, and then as $x_n = y_n^+ - y_n^- + iz_n^+ - iz_n^-$. It is then clear, by taking limits, that $L^2(N)$ is the algebraic span of $L^2(N)^+$. Since $Jx\xi = x\xi$ for any self-adjoint $x \in N$, any vector $\eta \in L^2(N)^+$ satisfies $J\eta = \eta$.

The following theorem gives some of the main properties of the L_η operators. Clearly (vi) is a stronger statement than (i), but (vi) can only be proved once (i) has been established.

Theorem B.4.1. *The following statements hold for the operators L_η associated to vectors η:*

  (i) *for each $\eta \in L^2(N)$, $L_{J\eta} \subseteq L_\eta^*$;*

  (ii) *if $\eta \in L^2(N)$, then $\eta = J\eta$ if and only if L_η is self-adjoint;*

  (iii) *$\eta \in L^2(N)^+$ if and only if $L_\eta \geq 0$;*

(iv) *if a self-adjoint T is affiliated to N and $\xi \in \mathcal{D}(T)$, then $T = L_\eta$ for some $\eta \in L^2(N)$ satisfying $\eta = J\eta$. Moreover, if $T \geq 0$, then $\eta \in L^2(N)^+$;*

(v) *if $\eta \in L^2(N)$ and $L_\eta = v|L_\eta|$ is the polar decomposition, then $v \in N$ and $|L_\eta| = L_\zeta$ for some $\zeta \in L^2(N)^+$;*

(vi) *if $\eta \in L^2(N)$ then $L_\eta^* = L_{J\eta}$.*

Proof. (i) If $x, y \in N$, then from (B.4.2), $y\xi \in \mathcal{D}(\ell_\eta^*)$ and

$$\begin{aligned}
\langle x\xi, \ell_\eta^*(y\xi) \rangle &= \langle \ell_\eta(x\xi), y\xi \rangle = \langle Jx^*J\eta, y\xi \rangle \\
&= \langle \eta, JxJy\xi \rangle = \langle \eta, yx^*\xi \rangle \\
&= \langle xy^*\xi, J\eta \rangle = \langle JyJx\xi, J\eta \rangle \\
&= \langle x\xi, Jy^*JJ\eta \rangle = \langle x\xi, \ell_{J\eta}(y\xi) \rangle.
\end{aligned} \tag{B.4.5}$$

Letting x vary, we conclude that $\ell_\eta^*(y\xi) = \ell_{J\eta}(y\xi)$ for $y \in N$. Thus $\ell_{J\eta} \subseteq \ell_\eta^*$ so, taking adjoints, $L_\eta = \ell_\eta^{**} \subseteq \ell_{J\eta}^*$. Take adjoints again to obtain $L_{J\eta} \subseteq L_\eta^*$.
(ii) If L_η is self-adjoint, then from (i), $L_{J\eta} = L_\eta$. Then $J\eta = \eta$ by applying these operators to ξ.

Now suppose that $J\eta = \eta$. From (i), $L_\eta \subseteq L_\eta^*$, and so L_η is symmetric. From Lemma B.3.9 and Proposition B.3.10, it suffices to show that $L_\eta \pm iI$ have dense ranges. Fix a sequence $\{z_n\}_{n=1}^\infty \in N$ such that $\lim_{n \to \infty} \|z_n\xi - \eta\| = 0$. Applying J, we obtain $\lim_{n \to \infty} z_n^*\xi = \eta$ so we may replace z_n by $(z_n + z_n^*)/2$ and thus we may assume that $z_n \in N_{\mathrm{s.a.}}$. By spectral theory, each $z_n + i1$ has an inverse in N bounded in norm by 1. Let w_n denote $(z_n + i1)^{-1}$. If $x \in N$ is arbitrary, then

$$\begin{aligned}
(L_\eta + iI)(w_n x\xi) &= Jx^*JJw_n^*J\eta + iw_n x\xi \\
&= Jx^*JJw_n^*Jz_n\xi + iw_n x\xi + Jx^*JJw_n^*J(\eta - z_n\xi). \tag{B.4.6}
\end{aligned}$$

Since $\|w_n\| \leq 1$, the last term in (B.4.6) tends to 0 as $n \to \infty$, while

$$\begin{aligned}
Jx^*JJw_n^*Jz_n\xi + iw_n x\xi &= z_n w_n x\xi + iw_n x\xi \\
&= (z_n + i1)w_n x\xi = x\xi. \tag{B.4.7}
\end{aligned}$$

Thus $\lim_{n \to \infty} (L_\eta + iI)(w_n x\xi) = x\xi$, showing that $L_\eta + iI$ has dense range. A similar calculation shows that the same is true for $L_\eta - iI$, and thus L_η is self-adjoint.
(iii) If $L_\eta \geq 0$ then $L_\eta = L_\eta^*$, so $L_{J\eta} \subseteq L_\eta$ by (i). Apply these operators to ξ to obtain $J\eta = \eta$. As in (ii), we may choose a sequence $\{z_n\}_{n=1}^\infty \in N_{\mathrm{s.a.}}$ such that $\lim_{n \to \infty} \|\eta - z_n\xi\| = 0$. If we write $z_n = z_n^+ - z_n^-$ with $z_n^+ z_n^- = 0$, as in (ii), then the sequences $\{z_n^\pm \xi\}_{n=1}^\infty$ are Cauchy and converge to vectors $\eta^\pm \in L^2(N)^+$; clearly $\eta = \eta^+ - \eta^-$. Moreover, if $x \in N$, then

$$\begin{aligned}
\langle x\eta^+, \eta^- \rangle &= \lim_{n \to \infty} \langle xz_n^+\xi, z_n^-\xi \rangle = \lim_{n \to \infty} \tau(z_n^- x z_n^+) \\
&= \lim_{n \to \infty} \tau(xz_n^+ z_n^-) = 0, \tag{B.4.8}
\end{aligned}$$

so $\eta^- \perp \overline{N\eta^+}$. This space is invariant for N, so is the range of a projection $JpJ \in N'$ where $p \in N$ is a projection. We note that $J(1-p)J\eta^- = \eta^-$ while $\overline{N\eta^+}$ is the kernel of $J(1-p)J$. Then, for $x \in N$,

$$
\begin{aligned}
0 \le \langle L_\eta(1-p)x\xi), (1-p)x\xi \rangle &= \langle Jx^*JJ(1-p)J\eta, (1-p)x\xi \rangle \\
&= -\langle Jx^*J\eta^-, (1-p)x\xi \rangle = -\lim_{n\to\infty} \langle Jx^*Jz_n^-\xi, (1-p)x\xi \rangle \\
&= -\lim_{n\to\infty} \tau(x^*(1-p)z_n^-x) = -\lim_{n\to\infty} \tau(x^*z_n^-(1-p)x) \\
&= -\lim_{n\to\infty} \langle J(1-p)Jz_n^-\xi, xx^*\xi \rangle = -\langle \eta^-, xx^*\xi \rangle \\
&= -\langle \eta^-, JxJx\xi \rangle = -\langle L_{\eta^-}(x\xi), x\xi \rangle.
\end{aligned}
\tag{B.4.9}
$$

Since $\eta^- \in L^2(N)^+$, $L_{\eta^-} \ge 0$, and so (B.4.9) shows that $\eta^- \perp xx^*\xi$ for all $x \in N$, and hence $\eta^- \perp N\xi$, giving $\eta^- = 0$. Then $\eta = \eta^+ \in L^2(N)^+$.

(iv) Suppose that T is self-adjoint and is affiliated to N with $\xi \in \mathcal{D}(T)$. For all $u \in N'$, $u\xi \in \mathcal{D}(T)$ and so $\mathcal{D}(T)$ contains $N'\xi = N\xi$. Let $\eta = T\xi$. Then for each unitary $u \in N$,

$$
\begin{aligned}
Tu\xi = TJu^*J\xi &= Ju^*JT\xi \\
&= Ju^*JL_\eta\xi = L_\eta(u\xi),
\end{aligned}
\tag{B.4.10}
$$

and by linearity, this holds with u replaced by $x \in N$. Thus T and L_η agree on a core for L_η and so $L_\eta \subseteq T$. Thus $T \subseteq L_\eta^*$ and so $\eta = T\xi = L_\eta^*\xi = J\eta$, since $L_{J\eta} \subseteq L_\eta^*$ by (i). Thus L_η is self-adjoint by (ii). Taking the adjoint in $L_\eta \subseteq T$, we obtain $T \subseteq L_\eta$, proving equality. Now if $T \ge 0$ then $L_\eta \ge 0$ and $\eta \in L^2(N)^+$ by (iii).

(v) Let $L_\eta = v|L_\eta|$ be the polar decomposition. Since L_η is affiliated to N, so too is $|L_\eta|$, by Theorem B.3.7 (iv), and $v \in N$. By (iv), $|L_\eta| = L_\zeta$ for some $\zeta \in L^2(N)^+$.

(vi) Consider $\eta \in L^2(N)$. By (v) there exist $\zeta \in L^2(N)^+$ and $v \in N$ such that $L_\eta = vL_\zeta$. Since N is a type II_1 factor $1 - vv^*$ and $1 - v^*v$ are equivalent projections in N and there exists a partial isometry $w \in N$ so that these projections are respectively ww^* and w^*w. Then $u = v + w$ is a unitary such that $L_\eta = uL_\zeta$. Now $\eta_2 \in \mathcal{D}((uL_\zeta)^*)$ if and only if $\eta_1 \mapsto \langle uL_\zeta\eta_1, \eta_2 \rangle$ is continuous for $\eta_1 \in \mathcal{D}(L_\zeta)$, and this is equivalent to $u^*\eta_2$ being in $\mathcal{D}(L_\zeta^*)$. Thus $\mathcal{D}((uL_\zeta)^*) = u\mathcal{D}((L_\zeta)^*)$, and the relation $(uL_\zeta)^* = L_\zeta u^*$ follows since L_ζ is self-adjoint. Now $\mathcal{G}(L_\zeta u^*) = \{(u\eta_1, L_\zeta\eta_1): \eta \in \mathcal{D}(L_\zeta)\}$, and this is the image under a unitary transformation of $\mathcal{G}(L_\zeta)$. Since $N\xi$ is a core for L_ζ, the same is then true for $L_\zeta u^*$. Let $\zeta_1 = L_\zeta u^*\xi$. Then L_{ζ_1} and $L_\zeta u^*$ agree on their common core $N\xi$, and are thus equal. This means that $L_\eta^* = L_\zeta u^* = L_{\zeta_1}$, so by (i), $L_{J\eta} \subseteq L_\eta^* = L_{\zeta_1}$. Thus $\zeta_1 = J\eta$, and we have $L_{J\eta} = L_\eta^*$. $\qquad\square$

B.5   Operators from $L^1(N)$

We now turn to the space $L^1(N)$, for which we recall the following definition. If $x \in N$, then $\phi_x \in N_*$ denotes the bounded linear functional $y \mapsto \tau(xy)$, $y \in N$.

Let $x = v|x|$ be the polar decomposition of x. If $y \in N$, $\|y\| \leq 1$, then by the Cauchy–Schwarz inequality

$$
\begin{aligned}
|\phi_x(y)|^2 = |\tau(v|x|y)|^2 &= |\tau(v|x|^{1/2}|x|^{1/2}y)|^2 \\
&\leq \tau(|x|^{1/2}v^*v|x|^{1/2})\tau(|x|^{1/2}yy^*|x|^{1/2}) \\
&\leq \tau(|x|)^2,
\end{aligned}
\tag{B.5.1}
$$

and so $\|\phi_x\| \leq \tau(|x|)$. On the other hand,

$$
\phi_x(v^*) = \tau(xv^*) = \tau(v^*x) = \tau(|x|),
\tag{B.5.2}
$$

showing that $\|\phi_x\| = \tau(|x|)$. Thus $\|x\|_1 = \tau(|x|)$ defines a norm on N whose completion is $L^1(N)$. The identification of $x \in N$ with $\phi_x \in N_*$ gives an isometric isomorphism between $L^1(N)$ and N_*, since $\{\phi_x: \ x \in N\}$ is a norm dense subspace of N_*. The convex cone of elements in $L^1(N)$ corresponding to N_*^+ is denoted $L^1(N)^+$. Since the linear functional ϕ_x, defined above, is positive if and only if $x \in N^+$, this extends the notion of positivity from N to $L^1(N)$ in an unambiguous manner.

By analogy to $L^2(N)$ discussed in Section B.4, we wish to regard the elements of $L^1(N)$ as unbounded operators on $L^2(N)$, extending the identification of N with left multiplication operators on $L^2(N)$. To motivate the definition, consider a fixed $x \in N$ and its associated $\phi_x \in N_*$. Then, for $y, z \in N$, with y fixed,

$$
\begin{aligned}
|\phi_x(yz)| = |\tau(xyz)| &= |\langle xy\xi, z^*\xi \rangle| \\
&\leq \|xy\xi\|\|z\|_2.
\end{aligned}
\tag{B.5.3}
$$

Thus there is a constant $k = \|xy\xi\|$ such that $|\phi_x(yz)| \leq k\|z\|_2$ for all $z \in N$. For each $\phi \in N_*$ we will define an operator m_ϕ by first specifying the domain $\mathcal{D}(m_\phi)$ to be

$$
\{y\xi \in N\xi: \ |\phi(yz)| \leq k_y\|z\|_2 \text{ for a constant } k_y \text{ and for all } z \in N\}. \tag{B.5.4}
$$

This is a subspace of $L^2(N)$, although it must be proved to be dense. By definition of $\mathcal{D}(m_\phi)$, $z\xi \mapsto \phi(yz^*)$ is a bounded conjugate linear functional on $N\xi$ for $y \in \mathcal{D}(m_\phi)$, and so $m_\phi(y\xi)$ is defined by the relation

$$
\langle m_\phi(y\xi), z\xi \rangle = \phi(yz^*), \qquad z \in N. \tag{B.5.5}
$$

If $\phi = \phi_x$ for some $x \in N$, then $\mathcal{D}(m_{\phi_x}) = N\xi$ and

$$
\phi_x(yz^*) = \tau(xyz^*) = \tau(z^*xy) = \langle \ell_x(y\xi), z\xi \rangle \tag{B.5.6}
$$

for $y, z \in N$, and so $m_{\phi_x} = \ell_x$.

For $\phi \in N_*$, let $\phi^*(x) = \overline{\phi(x^*)}$, $x \in N$. We say that ϕ is self-adjoint if $\phi = \phi^*$.

In Lemma B.5.3, we will require the fact that each $\phi \in N_*$ is a vector functional. We prove this in Lemma B.5.2, using Lemma B.5.1 which is of independent interest. The first of these lemmas is a special case of a theorem in [109].

Lemma B.5.1. *Let N be a finite von Neumann algebra with normalised normal faithful trace τ. If $a, b \in N^+$, then*

$$\|a^{1/2} - b^{1/2}\|_2 \leq \sqrt{2}\|a - b\|_1^{1/2}. \tag{B.5.7}$$

Proof. First suppose that $a \geq b$. Then, by [131, Prop. 1.3.8], $a^{1/2} \geq b^{1/2}$, so write $h = a^{1/2} - b^{1/2} \geq 0$. Then

$$\begin{aligned}
\tau(a - b) &= \tau((h + b^{1/2})^2 - b) = \tau(h^2 + hb^{1/2} + b^{1/2}h) \\
&= \tau(h^2) + 2\tau(h^{1/2}b^{1/2}h^{1/2}) \\
&\geq \|h\|_2^2,
\end{aligned} \tag{B.5.8}$$

so $\|a - b\|_1 \geq \|a - b\|_2^2$.

In general, let $k = a - b$ and decompose k into its positive and negative parts $k = k^+ - k^-$. Then $a \leq b + k^+$, so

$$\|a^{\frac{1}{2}} - (b + k^+)^{1/2}\|_2 \leq \|a - b - k^+\|_1^{1/2} = \|k^-\|_1^{1/2}. \tag{B.5.9}$$

Also $b + k^+ \geq b$ so $\|(b + k^+)^{1/2} - b^{1/2}\|_2 \leq \|k^+\|_1^{1/2}$. It follows that

$$\begin{aligned}
\|a^{1/2} - b^{1/2}\|_2 &\leq \|a^{1/2} - (b + k^+)^{1/2}\|_2 + \|(b + k^+)^{1/2} - b^{1/2}\|_2 \\
&\leq \|k^-\|_1^{1/2} + \|k^+\|_1^{1/2} \leq \sqrt{2}(\|k^+\|_1 + \|k^-\|_1)^{1/2} \\
&= \sqrt{2}\|k\|_1^{1/2} = \sqrt{2}\|a - b\|_1^{1/2}
\end{aligned} \tag{B.5.10}$$

as desired. □

Lemma B.5.2. *If $\phi \in N_*$ then there exist vectors $\eta, \zeta \in L^2(N)$ such that $\phi(x) = \langle x\eta, \zeta \rangle$ for $x \in N$.*

Proof. First suppose that $\phi \geq 0$. Since the set of normal functionals of the form $\phi_x(\cdot) = \tau(\cdot x)$, $x \in N$, is norm dense in N_*, ϕ is the $\|\cdot\|_1$-norm limit of a sequence of such functionals. A Hahn–Banach separation argument for ϕ and the $\|\cdot\|_1$-closure of the cone $\{\phi_x : x \in N^+\}$ shows that the sequence can be chosen from this cone. Thus there is a sequence $a_n \in N^+$, $n \geq 1$, such that $\|\phi - \phi_{a_n}\|_1 \to 0$. By Lemma B.5.1, $\|a_n^{1/2} - a_m^{1/2}\|_2 \leq \sqrt{2}\|a_n - a_m\|_1^{1/2}$, and so $\{a_n^{1/2}\}_{n=1}^\infty$ is a Cauchy sequence in $L^2(N)$ with limit η. For $x \in N$,

$$\begin{aligned}
\phi(x) &= \lim_{n \to \infty} \tau(xa_n^{1/2}a_n^{1/2}) = \lim_{n \to \infty} \langle xa_n^{1/2}\xi, a_n^{1/2}\xi \rangle \\
&= \langle x\eta, \eta \rangle,
\end{aligned} \tag{B.5.11}$$

proving the result in this case.

Now consider a general $\phi \in N_*$. By w^*-compactness of the unit ball of N, there exists $y \in N$, $\|y\| \leq 1$, such that $\|\phi\| = \phi(y)$. Let $y = w|y|$ be the polar decomposition of y, where we may take w to be a unitary since N is finite. Now define $\psi \in N_*$ by $\psi(\cdot) = \phi(w\cdot)$. Then ψ attains its norm at the positive element

$|y|$, and so is a positive linear functional. From the first part, $\psi(\cdot)$ has the form $\langle \cdot \eta, \eta \rangle$ for some $\eta \in L^2(N)$, and

$$\phi(x) = \psi(w^*x) = \langle w^*x\eta, \eta \rangle = \langle x\eta, w\eta \rangle, \qquad x \in N. \tag{B.5.12}$$

The proof is completed by setting $\zeta = w\eta$. $\qquad\qquad\qquad\qquad\qquad\square$

Lemma B.5.3. *For each $\phi \in N_*$, the operator m_ϕ is densely defined and has a densely defined adjoint. Its closure, M_ϕ, satisfies $M_\phi \subseteq (m_{\phi^*})^* = (M_{\phi^*})^*$.*

Proof. Fix $\phi \in N_*$. Then there exist two vectors $\eta, \zeta \in L^2(N)$ such that $\phi(y) = \langle y\eta, \zeta \rangle$, $y \in N$, by Lemma B.5.2. Thus

$$\phi(yz) = \langle z\eta, y^*L_\zeta\xi \rangle, \qquad y, z \in N. \tag{B.5.13}$$

Theorem B.4.1 (v), applied to $J\zeta$, gives the polar decomposition $L_{J\zeta} = v|L_{J\zeta}|$ with $v \in N$, and $|L_{J\zeta}| = L_\mu$ for some $\mu \in L^2(N)^+$. By Theorems B.4.1 (vi) and B.3.7 (iii), we may take adjoints to obtain

$$L_\zeta = (L_{J\zeta})^* = (vL_\mu)^* = L_\mu v^*. \tag{B.5.14}$$

Thus (B.5.13) becomes

$$\phi(yz) = \langle z\eta, y^*L_\mu v^*\xi \rangle, \qquad y, z \in N, \tag{B.5.15}$$

where $L_\mu \geq 0$. Let $\{e_n\}_{n=1}^\infty$ be the spectral projections of L_μ corresponding to the intervals $[0, n]$, $n \geq 1$. Then each $e_n L_\mu$ is a bounded operator, denoted b_n, in N. For $y, z \in N$ with y fixed and z varying, we have

$$\begin{aligned} |\phi(e_n yz)| = |\langle z\eta, y^*e_n L_\mu v^*\xi \rangle| &= |\langle z\eta, y^*b_n v^*\xi \rangle| \\ &= |\langle \eta, z^*y^*b_n v^*\xi \rangle| \leq \|\eta\| \|z^*y^*b_n v^*\|_2 \\ &\leq \|\eta\| \|y^*b_n v^*\| \|z\|_2. \end{aligned} \tag{B.5.16}$$

This shows that $e_n y\xi \in \mathcal{D}(m_\phi)$ for all $n \geq 1$ and for all $y \in N$. Since $e_n \to 1$ strongly as $n \to \infty$, we conclude that m_ϕ has dense domain. Replace ϕ by ϕ^* to obtain the same result for m_{ϕ^*}. If $y\xi \in \mathcal{D}(m_\phi)$ and $z\xi \in \mathcal{D}(m_{\phi^*})$, then

$$|\langle m_\phi(y\xi), z\xi \rangle| = |\phi(yz^*)| = |\phi^*(zy^*)| \leq k\|y\|_2, \tag{B.5.17}$$

for some constant k by the definition of $\mathcal{D}(m_{\phi^*})$. Thus $z\xi \in \mathcal{D}((m_\phi)^*)$. In particular, $(m_\phi)^*$ is densely defined and $m_{\phi^*} \subseteq (m_\phi)^*$. Taking adjoints gives

$$M_\phi = (m_\phi)^{**} \subseteq (m_{\phi^*})^* = (M_{\phi^*})^*, \tag{B.5.18}$$

completing the proof. $\qquad\qquad\qquad\qquad\qquad\qquad\qquad\qquad\qquad\square$

Proposition B.5.4. *Let $\phi \in N_*$. Then*

(i) *ϕ is self-adjoint if and only if M_ϕ is self-adjoint;*

(ii) $\phi \geq 0$ if and only if $M_\phi \geq 0$.

Proof. (i) Suppose that M_ϕ is self-adjoint, and consider $y\xi, z\xi \in \mathcal{D}(m_\phi) \subseteq \mathcal{D}(M_\phi) = \mathcal{D}((M_\phi)^*)$. Then

$$\begin{aligned}
\phi(yz^*) = \langle m_\phi(y\xi), z\xi \rangle &= \langle M_\phi(y\xi), z\xi \rangle \\
&= \langle y\xi, M_\phi(z\xi) \rangle = \langle y\xi, m_\phi(z\xi) \rangle \\
&= \overline{\phi(zy^*)} = \phi^*(yz^*).
\end{aligned} \tag{B.5.19}$$

Since $\mathcal{D}(m_\phi)$ is dense in $L^2(N)$, the relation $\phi = \phi^*$ follows from (B.5.19) and the normality of ϕ.

Now suppose that $\phi = \phi^*$. By Lemma B.5.3, M_ϕ is closed and satisfies $M_\phi \subseteq (M_\phi)^*$, and thus M_ϕ is symmetric. To prove that M_ϕ is self-adjoint it suffices, by Lemma B.3.9 and Proposition B.3.10, to show that $M_\phi \pm iI$ have dense ranges. We consider only $M_\phi - iI$, as the other case is similar.

Suppose that $a \in N$, $x\xi \in \mathcal{D}(m_\phi)$ and $z \in N$. Then there is a constant $k \geq 0$ such that

$$|\phi(xaz^*)| \leq k\|az^*\|_2 \leq k\|a\|\|z\|_2, \tag{B.5.20}$$

showing that $xa\xi \in \mathcal{D}(m_\phi)$. Moreover,

$$\begin{aligned}
\langle m_\phi(xa\xi), z\xi \rangle = \phi(xaz^*) &= \langle m_\phi(x\xi), za^*\xi \rangle \\
&= \langle m_\phi(x\xi), JaJz\xi \rangle \\
&= \langle Ja^*Jm_\phi(x\xi), z\xi \rangle,
\end{aligned} \tag{B.5.21}$$

from which it follows that $Ja^*Jm_\phi(x\xi) = m_\phi(xa\xi)$. This shows that the range of m_ϕ is an invariant subspace for $JNJ = N'$, from which we conclude that the projection p onto the closure of the range of $M_\phi - iI$ lies in N. We wish to show that $p = 1$, so suppose not. Let $q = 1 - p \neq 0$. Then, for all $x\xi \in \mathcal{D}(m_\phi)$, $q(m_\phi(x\xi) - ix\xi) = 0$. For any $z \in N$, this gives

$$\begin{aligned}
\phi(x(zq)^*) = \langle m_\phi(x\xi), qz\xi \rangle &= \langle qm_\phi(x\xi), z\xi \rangle \\
&= \langle iqx\xi, z\xi \rangle = i\tau(qxz^*).
\end{aligned} \tag{B.5.22}$$

By density of $\mathcal{D}(m_\phi)$ and normality of ϕ and τ, (B.5.22) then holds for all $x, z \in N$. If we put $x = z = q$, then $\phi(q) = i\tau(q)$, while $\phi(q) = \phi^*(q) = \overline{\phi(q)} = -i\tau(q)$. This yields $\tau(q) = 0$ and $q = 0$, a contradiction. Thus M_ϕ is self-adjoint.

(ii) Suppose now that $\phi \geq 0$, so that M_ϕ is self-adjoint by (i). If $x\xi \in \mathcal{D}(m_\phi)$ then $x\xi \in \mathcal{D}(M_\phi)$, so

$$\langle M_\phi(x\xi), x\xi \rangle = \phi(xx^*) \geq 0. \tag{B.5.23}$$

The graph of m_ϕ is dense in that of M_ϕ, so if $\eta \in \mathcal{D}(M_\phi)$ then we may find a sequence $\{x_n\xi\}_{n=1}^\infty$ in $\mathcal{D}(m_\phi)$ such that $(x_n\xi, m_\phi(x_n\xi)) \to (\eta, M_\phi(\eta))$. Clearly

$$\langle M_\phi(\eta), \eta \rangle = \lim_{n \to \infty} \langle m_\phi(x_n\xi), x_n\xi \rangle \geq 0, \tag{B.5.24}$$

by (B.5.23), and so $M_\phi \geq 0$.

Now suppose that $M_\phi \geq 0$. From the proof of Lemma B.5.3, there is a sequence $\{e_n\}_{n=1}^\infty$ of projections in N, increasing strongly to 1, so that $e_n N\xi \subseteq \mathcal{D}(m_\phi)$ for all $n \geq 1$. Then

$$\phi(e_n xx^* e_n) = \langle m_\phi(e_n x\xi), e_n x\xi \rangle \geq 0 \qquad (\text{B.5.25})$$

for $x \in N$ and $n \geq 1$. Since ϕ may be expressed as a vector functional, the strong convergence of $\{e_n\}_{n=1}^\infty$ to 1 shows that $\phi(xx^*) = \lim_{n \to \infty} \phi(e_n xx^* e_n)$, and positivity of ϕ follows from (B.5.25), letting x vary over N. $\qquad \square$

Lemma B.5.5. *Let $\phi, \psi \in N_*$, and suppose that $M_\phi = M_\psi$. Then $\phi = \psi$.*

Proof. From the proof of Lemma B.5.3, there exist two increasing sequences $\{e_n\}_{n=1}^\infty$, $\{f_n\}_{n=1}^\infty$ of projections in N, both converging strongly to 1, such that $e_n N\xi \subseteq \mathcal{D}(m_\phi)$, $f_n N\xi \subseteq \mathcal{D}(m_\psi)$ for $n \geq 1$. Let $g_n = e_n \wedge f_n$, $n \geq 1$. Then $g_n N\xi \subseteq \mathcal{D}(m_\phi) \cap \mathcal{D}(m_\psi)$. For each $z \in N$, $n \geq 1$,

$$\phi(g_n z) = \langle m_\phi(g_n \xi), z^* \xi \rangle = \langle m_\psi(g_n \xi), z^* \xi \rangle$$
$$= \psi(g_n z). \qquad (\text{B.5.26})$$

The result follows from (B.5.26) by letting $n \to \infty$, since $g_n \to 1$ strongly. $\quad \square$

The inequality $\|ab\|_1 \leq \|a\|_2 \|b\|_2$, for $a, b \in N$, is a simple consequence of the definition

$$\|ab\|_1 = \sup\{|\tau(xab)|: \ \|x\| \leq 1\}$$
$$= \sup\{|\langle xa\xi, b^*\xi \rangle|: \ \|x\| \leq 1\}. \qquad (\text{B.5.27})$$

This gives a bilinear map $(\eta_1, \eta_2) \mapsto \eta_1 \eta_2$ of $L^2(N) \times L^2(N)$ into $L^1(N)$, the unique continuous extension of the map $(a, b) \mapsto ab$ on $N \times N$. The notation $\eta_1 \eta_2$ is used to denote the result of applying this map to a pair (η_1, η_2) for this reason. The spaces $L^2(N)$, $L^1(N)$ and N_* are all N-bimodules. Left and right multiplications on N by elements of N are bounded in the $\|\cdot\|_1$- and $\|\cdot\|_2$-norms, and so extend to bounded actions of N on these spaces satisfying

$$\|a\eta_1\|_1, \|\eta_1 a\|_1 \leq \|\eta_1\|_1 \|a\|, \qquad \eta_1 \in L^1(N), \qquad a \in N,$$
$$\|a\eta_2\|_2, \|\eta_2 a\|_2 \leq \|\eta_2\|_2 \|a\|, \qquad \eta_2 \in L^2(N), \qquad a \in N. \qquad (\text{B.5.28})$$

For $\phi \in N_*$ and $a \in N$, $a\phi$ and ϕa in N_* are defined respectively, for $x \in N$, by

$$(a\phi)(x) = \phi(xa), \qquad (\phi a)(x) = \phi(ax). \qquad (\text{B.5.29})$$

We observe that if $\phi \in N_*$, then ϕ is the vector functional $\langle \cdot \eta, \zeta \rangle$ for a pair of vectors $\eta, \zeta \in L^2(N)$. This suggests that the operators M_ϕ, L_η and L_ζ should be related, which we establish below. We first discuss some further properties of these operators.

Proposition B.5.6. *Let $\phi \in N_*$, $\eta \in L^2(N)$ be given, and let u and w be arbitrary unitaries in N and N' respectively.*

(i) $wM_\phi w^* = M_\phi$, and M_ϕ is affiliated to N.

(ii) $uL_\eta = L_{u\eta}$ and $L_\eta u = L_{Ju^* J\eta}$.

(iii) $uM_\phi = M_{u\phi}$ and $M_\phi u = M_{\phi u}$.

(iv) $|M_\phi| = M_\psi$ for some $\psi \in N_*^+$, and the polar decomposition is $M_\phi = vM_\psi$ for a partial isometry $v \in N$.

(v) $M_\phi^* = M_{\phi^*}$.

Proof. We may write $w = JvJ$ for some unitary $v \in N$. It is clear that the map $z\xi \mapsto \phi(xz^*)$, $z \in N$, is bounded in the $\| \cdot \|_2$-norm if and only if the same is true for the map $z\xi \mapsto \phi(xvz^*)$. Thus $wm_\phi w^*$ and m_ϕ have the same domain. If $x\xi \in \mathcal{D}(m_\phi)$ and $z\xi \in N\xi$, then

$$\langle wm_\phi w^*(x\xi), z\xi \rangle = \langle m_\phi(xv\xi), zv\xi \rangle$$
$$= \phi(xvv^* z^*) = \phi(xz^*)$$
$$= \langle m_\phi(x\xi), z\xi \rangle, \tag{B.5.30}$$

showing that $wm_\phi w^* = m_\phi$. Since $\mathcal{D}(m_\phi)$ is a core for both $wM_\phi w^*$ and M_ϕ, (i) follows.

(ii) Let $x \in N$ be arbitrary. Then

$$uL_\eta(x\xi) = uJx^* J\eta = Jx^* Ju\eta = L_{u\eta}(x\xi). \tag{B.5.31}$$

Thus uL_η and $L_{u\eta}$ are equal on their common core $N\xi$, and so $uL_\eta = L_{u\eta}$. Now, for $x \in N$,

$$L_\eta u(x\xi) = Jx^* JJu^* J\eta = L_{Ju^* J\eta}(x\xi), \tag{B.5.32}$$

from which it follows that $L_\eta u = L_{Ju^* J\eta}$.

(iii) Suppose now that $\phi \in N_*$ and $u \in N$ is a unitary. The graph of uM_ϕ is obtained from that of M_ϕ by applying the unitary operator $1 \times u$, and so $uM_\phi = M_{u\phi}$ will follow from $um_\phi = m_{u\phi}$. By definition, $x\xi \in \mathcal{D}(m_\phi)$ if and only if $z\xi \mapsto \phi(xz^*)$ is $\| \cdot \|_2$-norm bounded. This is equivalent to the $\| \cdot \|_2$-norm boundedness of $z\xi \mapsto \phi(xz^* u) = (u\phi)(xz^*)$, showing that m_ϕ and $m_{u\phi}$ have the same domain. If $x\xi \in \mathcal{D}(m_\phi)$ and $z \in N$, then

$$\langle m_{u\phi}(x\xi), z\xi \rangle = (u\phi)(xz^*) = \phi(xz^* u) = \langle m_\phi(x\xi), u^* z \rangle, \tag{B.5.33}$$

showing that $um_\phi = m_{u\phi}$, as desired.

The graph of $M_\phi u$ is obtained by applying the unitary $u^* \times 1$ to that of M_ϕ. Thus the relation $M_\phi u = M_{\phi u}$ will follow from $m_\phi u = m_{\phi u}$. Now $x\xi \in \mathcal{D}(m_\phi u)$ if and only if $ux\xi \in \mathcal{D}(m_\phi)$. This is equivalent to the $\| \cdot \|_2$-norm boundedness of $z \mapsto \phi(uxz^*)$ for $z \in N$, and this is the same as the boundedness of $z \mapsto (\phi u)(xz^*)$. Thus the domains of $m_\phi u$ and $m_{\phi u}$ are equal. Let $x\xi \in \mathcal{D}(m_\phi u)$ and let $z \in N$ be arbitrary. Then

$$\langle m_\phi u(x\xi), z\xi \rangle = \phi(uxz^*) = \langle m_{\phi u}(x\xi), z\xi \rangle, \tag{B.5.34}$$

proving that $m_\phi u = m_{\phi u}$.

(iv) Let $M_\phi = v|M_\phi|$ be the polar decomposition of M_ϕ. Then $v \in N$, by Theorem B.3.7 (iv). As in the proof of Theorem B.4.1 (vi), we may replace v by a unitary $u \in N$. Then $u^* M_\phi = |M_\phi|$, and thus $|M_\phi| = M_\psi$ where $\psi = u^*\phi$, by (iii). Proposition B.5.4 (ii) then shows that $\psi \geq 0$.

(v) Consider $\phi \in N_*$. From (iv), there exist $\psi \in N_*^+$ and a unitary $u \in N$ such that $M_\phi = uM_\psi$. If $\eta \in \mathcal{D}(uM_\psi)$ and $\zeta \in \mathcal{D}((uM_\psi)^*)$, then $\eta \mapsto \langle uM_\psi(\eta), \zeta \rangle$ is continuous on $\mathcal{D}(uM_\psi) = \mathcal{D}(M_\psi)$, and continuity of $\eta \mapsto \langle M_\psi(\eta), u^*\zeta \rangle$ follows. Thus $u^*\zeta \in \mathcal{D}(M_\psi^*)$. The argument is reversible, showing that $u\mathcal{D}(M_\psi^*) = \mathcal{D}((uM_\psi)^*)$. This gives $(uM_\psi)^* = M_\psi^* u^*$. The operator M_ψ is self-adjoint, by Proposition B.5.4 (i), so for $\eta \in \mathcal{D}(M_\phi) = \mathcal{D}(M_\psi)$ and $\zeta \in \mathcal{D}(M_\phi^*) = u\mathcal{D}(M_\psi)$, we have

$$\langle \eta, M_\phi^* \zeta \rangle = \langle \eta, M_\psi u^* \zeta \rangle = \langle \eta, M_{\psi u^*} \zeta \rangle, \qquad (B.5.35)$$

by (iii). It follows that $M_\phi^* = M_{\psi u^*}$. The relation $M_\phi = uM_\psi = M_{u\psi}$ gives $\phi = u\psi$, by Lemma B.5.5. Then

$$\phi^*(x) = \overline{\phi(x^*)} = \overline{\psi(x^* u)} = \psi(u^* x) = (\psi u^*)(x), \qquad x \in N, \qquad (B.5.36)$$

where we have used the self-adjointness of ψ. From this we conclude that $M_\phi^* = M_{\phi^*}$. $\qquad \square$

Two elements $a, b \in N$ define a vector functional $\phi \in N_*$ by

$$\phi(x) = \langle xa\xi, b\xi \rangle = \tau(xab^*), \qquad x \in N. \qquad (B.5.37)$$

In this case M_ϕ is a bounded operator, and

$$\begin{aligned} \langle M_\phi(x\xi), z\xi \rangle &= \phi(xz^*) = \tau(xz^* ab^*) = \tau(ab^* xz^*) \\ &= \langle L_a L_b^*(x\xi), z\xi \rangle, \qquad x, z \in N. \end{aligned} \qquad (B.5.38)$$

Thus $M_\phi = L_a L_b^*$. The next result is analogous, where we replace $a\xi$ and $b\xi$ by general vectors from $L^2(N)$.

Theorem B.5.7. *Let $\eta, \zeta \in L^2(N)$, and let ϕ be the associated vector functional $\langle \cdot \eta, \zeta \rangle$ in N_*. Then M_ϕ is the closure of $L_\eta L_\zeta^*$.*

Proof. We begin by showing that $L_\eta L_\zeta^*$ has a densely defined adjoint and is thus closable. Initially, suppose that $\eta, \zeta \in L^2(N)^+$, so that L_η and L_ζ are positive self-adjoint operators. Let $\{e_n\}_{n=1}^\infty$ and $\{f_n\}_{n=1}^\infty$ be the spectral projections of L_η and L_ζ respectively, corresponding to the intervals $[0, n]$. By Theorem B.3.8 (iv), $\bigcup_{n=1}^\infty e_n L^2(N)$ and $\bigcup_{n=1}^\infty f_n L^2(N)$ are dense in $L^2(N)$, are cores for L_η and L_ζ respectively, and the operators $L_\eta e_n$ and $L_\zeta f_n$ have bounded closures. In particular, each of these operators maps $N\xi$ to $N\xi$, and so $\bigcup_{n=1}^\infty f_n L^2(N)$ is contained in the domain of $L_\eta L_\zeta$. The same argument shows that $L_\zeta L_\eta$ is densely defined, since its domain contains $\bigcup_{n=1}^\infty e_n L^2(N)$. If $x, y \in N$ then

$$\langle L_\eta L_\zeta(f_n x\xi), e_m y\xi \rangle = \langle f_n x\xi, L_\zeta L_\eta e_m y\xi \rangle \qquad (B.5.39)$$

for all $m, n \geq 1$. This shows that $e_m y \in \mathcal{D}((L_\eta L_\zeta)^*)$ for $m \geq 1$ and $y \in N$, and thus $L_\eta L_\zeta$ has a densely defined adjoint, proving that $L_\eta L_\zeta$ is closable.

Now let $\eta, \zeta \in L^2(N)$ be general vectors. By Theorem B.4.1 and Proposition B.5.6, we have

$$L_\eta L_\zeta^* = L_\eta L_{J\zeta} = L_\eta u L_\mu = L_{Ju^* J\eta} L_\mu = v L_\nu L_\mu, \qquad (B.5.40)$$

where u and v are suitably chosen unitaries arising from the polar decomposition, and $\nu, \mu \in L^2(N)^+$. Then $L_\eta L_\zeta^*$ is closable, since we have already shown this for $L_\nu L_\mu$.

Now consider $x\xi \in \mathcal{D}(m_\phi)$ for some $x \in N$. By definition, there exists a constant k such that $|\phi(xz)| \leq k\|z\|_2$ for $z \in N$. Then

$$\begin{aligned}
\langle m_\phi(x\xi), z\xi \rangle &= \phi(xz^*) = \langle xz^*\eta, \zeta \rangle \\
&= \langle z^*\eta, x^*\zeta \rangle = \langle Jx^*\zeta, Jz^*\eta \rangle \\
&= \langle L_{J\zeta}(x\xi), \ell_{J\eta}(z\xi) \rangle, \qquad z \in N.
\end{aligned} \qquad (B.5.41)$$

Thus $L_{J\zeta}(x\xi) \in \mathcal{D}(\ell_{J\eta}^*)$, and so

$$\langle m_\phi(x\xi), z\xi \rangle = \langle \ell_{J\eta}^* L_{J\zeta}(x\xi), z\xi \rangle = \langle L_\eta L_\zeta^*(x\xi), z\xi \rangle, \qquad z \in N. \qquad (B.5.42)$$

Thus $m_\phi \subseteq L_\eta L_\zeta^* \subseteq \overline{L_\eta L_\zeta^*}$, and so $M_\phi \subseteq \overline{L_\eta L_\zeta^*}$. Let us now suppose that equality fails. From the polar decomposition $M_\phi = u M_\psi$ for a unitary $u \in N$ and some $\psi \in N_*^+$, we may multiply by u on the left to obtain

$$M_\psi \subsetneq u\overline{L_\eta L_\zeta^*} = \overline{u L_\eta L_\zeta^*} = \overline{L_{u\eta} L_\zeta^*}, \qquad (B.5.43)$$

using Proposition B.5.6 (ii), and the simple fact that $\overline{uT} = u\overline{T}$ for any closable operator T. This allows us to make the further assumption that $\phi \geq 0$, so that M_ϕ is self-adjoint. Positivity of ϕ implies that ϕ is also the vector functional $\langle \cdot\zeta, \eta \rangle$, so the first part of the proof also shows that $m_\phi \subseteq L_\zeta L_\eta^*$.

Now consider $(\eta_1, \eta_2) \in \mathcal{G}(M_\phi)^\perp$. Then

$$\langle \nu, \eta_1 \rangle + \langle M_\phi \nu, \eta_2 \rangle = 0, \qquad \nu \in \mathcal{D}(M_\phi), \qquad (B.5.44)$$

so $\nu \mapsto \langle M_\phi \nu, \eta_2 \rangle$ is continuous on $\mathcal{D}(M_\phi)$. Thus $\eta_2 \in \mathcal{D}(M_\phi^*) = \mathcal{D}(M_\phi)$, and (B.5.44) becomes

$$\langle \nu, \eta_1 \rangle + \langle \nu, M_\phi \eta_2 \rangle = 0, \qquad \nu \in \mathcal{D}(M_\phi), \qquad (B.5.45)$$

and this also holds for all $\nu \in L^2(N)$. From (B.5.45), $(\eta_2, -\eta_1) \in \mathcal{G}(M_\phi)$, so we may choose a sequence $\{y_n\}_{n=1}^\infty$ in N such that $y_n \xi \in \mathcal{D}(m_\phi)$ and $\lim_{n\to\infty} y_n \xi = \eta_2$, $\lim_{n\to\infty} m_\phi(y_n\xi) = -\eta_1$. For $\nu \in \mathcal{D}(L_\eta L_\zeta^*)$,

$$\begin{aligned}
\langle L_\eta L_\zeta^*(\nu), \eta_2 \rangle &= \lim_{n\to\infty} \langle L_\eta L_\zeta^*(\nu), y_n\xi \rangle = \lim_{n\to\infty} \langle L_\zeta^*(\nu), L_\eta^* y_n\xi \rangle \\
&= \lim_{n\to\infty} \langle \nu, L_\zeta L_\eta^* y_n\xi \rangle = \lim_{n\to\infty} \langle \nu, m_\phi(y_n\xi) \rangle \\
&= -\langle \nu, \eta_1 \rangle, \qquad (B.5.46)
\end{aligned}$$

where we have used $m_\phi \subseteq L_\zeta L_\eta^*$ to see that $y_n \xi \in \mathcal{D}(L_\eta^*)$ and $L_\eta^*(y_n \xi) \in \mathcal{D}(L_\zeta)$. Then (B.5.46) implies that (η_1, η_2) is orthogonal to $\mathcal{G}(L_\eta L_\zeta^*)$ and thus to its closure. This establishes that $\overline{\mathcal{G}(L_\eta L_\zeta^*)} \subseteq \mathcal{G}(M_\phi)$, and this contradiction completes the proof. $\qquad\qquad\square$

Remark B.5.8. In the previous result, if ϕ is the positive vector functional $\langle \cdot\eta, \eta \rangle$ then $M_\phi = L_\eta L_\eta^*$ since this product is already closed, by Lemma B.2.4. In general, for the vector functional $\langle \cdot\eta, \zeta \rangle$, it may not be true that $M_\phi = L_\eta L_\zeta^*$. To see this choose a projection $p \in N'$, $p \neq 0, 1$, and then construct vectors η, ζ so that $\eta, \zeta \notin N\xi$ while $p\eta = \eta$, $(1 - p)\zeta = \zeta$. Then $\phi = 0$ so M_ϕ is everywhere defined while L_ζ^* is not. Thus we cannot dispense with the closure in the relation $M_\phi = \overline{L_\eta L_\zeta^*}$. $\qquad\qquad\square$

The trace on N satisfies $|\tau(x)| \leq \|x\|_1$ for $x \in N$, and so has a unique continuous extension, also denoted by τ, to $L^1(N)$. The isometric identification of $L^1(N)$ with N_* pairs $\eta \in L^1(N)$ with the linear functional $\tau(\cdot\eta)$. If this is denoted ϕ_η, then the associated densely defined operator M_{ϕ_η} will be written M_η to avoid cumbersome notation. The product map $(\eta, \zeta) \to \eta\zeta$ of $L^2(N) \times L^2(N)$ into $L^1(N)$ sends the pair (η, ξ) to η, and so $L^2(N)$ can be regarded as a subspace of $L^1(N)$, albeit with a different norm.

Theorem B.5.9.   (i) $\eta \in L^1(N)^+$ *if and only if there is a sequence* $\{x_n\}_{n=1}^\infty$ *in* N^+ *such that* $\lim_{n\to\infty} \|\eta - x_n\|_1 = 0$.

   (ii) *If* $\eta \in L^2(N)$ *then* $M_\eta = L_\eta$.

   (iii) *The product map is a surjection.*

   (iv) $\eta \in L^1(N)^+$ *if and only if there exists* $\zeta \in L^2(N)^+$ *such that* $\eta = \zeta^2$.

   (v) *If* $\eta \in L^1(N)$ *then there exists* $\zeta \in L^1(N)^+$ *and a partial isometry* $v \in N$ *such that* $\eta = v\zeta$ *(polar decomposition).*

Proof. (i) If $\eta \in L^1(N)^+$, then the associated linear functional ϕ lies in N_*^+. A Hahn–Banach separation argument shows that $\{\tau(\cdot a): a \in N^+\}$ is norm dense in the convex cone N_*^+, so any sequence from this set converging in norm to ϕ will yield a sequence $\{x_n\}_{n=1}^\infty$ from N^+ so that $\|\eta - x_n\|_1 \to 0$. Conversely, if $\eta \in L^1(N)$ is the limit of a sequence from N^+, then the associated sequence in N_* converges in norm to the linear functional $\tau(\cdot\eta)$ which is then positive. Thus $\eta \in L^1(N)^+$.

(ii) Let $\eta \in L^2(N)$, and let ϕ be the linear functional associated to η regarded as an element of $L^1(N)$. Let $\{a_n\}_{n=1}^\infty$ be a sequence from N such that $\lim_{n\to\infty} \|\eta - a_n\xi\|_2 = 0$, and then this limit is also 0 for the $\|\cdot\|_1$-norm. If $x \in N$, then

$$\phi(x) = \lim_{n\to\infty} \tau(xa_n) = \lim_{n\to\infty} \langle xa_n\xi, \xi \rangle$$
$$= \langle x\eta, 1\xi \rangle, \qquad\qquad (B.5.47)$$

so by Theorem B.5.7, $M_\eta = \overline{L_\eta L_\xi} = \overline{L_\eta} = L_\eta$, since L_η is closed.

(iii) If $\eta \in L^1(N)$, let ϕ be the associated linear functional $\tau(\cdot\eta)$. Then there exist vectors $\mu, \nu \in L^2(N)$ such that $\tau(x\eta) = \langle x\mu, \nu \rangle$ for $x \in N$. Choose sequences $\{a_n\}_{n=1}^{\infty}, \{b_n\}_{n=1}^{\infty}$ from N which converge in the $\| \cdot \|_2$-norm to μ and ν respectively. Then let ϕ_n denote the linear functional $\phi_n(x) = \langle xa_n\xi, b_n\xi \rangle$, $x \in N$. Clearly $\lim_{n \to \infty} \|\phi - \phi_n\| = 0$. Since $(\phi - \phi_n)(x) = \tau(x(\xi - a_nb_n^*))$, continuity of the product map shows that $\xi = \mu(J\nu)$.

(iv) Let $\eta \in L^1(N)$ and suppose that $\eta = \zeta^2$ for some $\zeta \in L^2(N)^+$. Then ζ is the limit in the $\| \cdot \|_2$-norm of a sequence from $N^+\xi$, the square of which converges in the $\| \cdot \|_1$-norm to η. Thus $\eta \in L^1(N)^+$ by (i).

Conversely, suppose that $\eta \in L^1(N)^+$. Then $\tau(\cdot\eta)$ is a positive element in N_*, so is a multiple of a vector state, the vector being μ, say. By Theorem B.5.7, $M_\eta = L_\mu L_\mu^*$. By Theorem B.4.1, there exist $\zeta \in L^2(N)^+$ and a unitary $u \in N$ such that $L_\mu = L_\zeta u$. Then $M_\eta = L_\zeta^2$, and thus L_ζ is the positive square root of M_η. If $\psi \in N_*^+$ is the vector functional $\langle \cdot\zeta, \zeta \rangle$, then ψ is $\tau(\cdot\zeta^2)$ by first approximating ζ by elements of $N\xi$. Thus $M_{\zeta^2} = L_\zeta^2 = M_\eta$, and $\eta = \zeta^2$ from Lemma B.5.5.

(v) From Theorem B.5.7, if $\eta \in L^1(N)$, then there exist a unitary $u \in N$ and $\zeta \in L^1(N)^+$ such that $M_\eta = uM_\zeta = M_{u\zeta}$. Then $\eta = u\zeta$, again using Lemma B.5.5. $\square$

Appendix C

The trace revisited

C.1 Introduction

In Chapter 4, we gave a construction of the unique faithful semifinite normal trace Tr on $\langle N, e_B \rangle$ satisfying

$$Tr(xe_By) = \tau(xy), \qquad x, y \in N, \tag{C.1.1}$$

where B is a von Neumann subalgebra of a finite von Neumann algebra N with faithful normal normalised trace τ. There we began with equation (C.1.1), and used it to define the trace on $\langle N, e_B \rangle^+$. In this appendix, we show a different route, by starting with a faithful normal semifinite trace on $\langle N, e_B \rangle$, possible because this is a semifinite algebra, and modifying it until (C.1.1) is satisfied.

Section C.2 contains some preliminary lemmas on abelian von Neumann algebras, and we also include a basic version of the Radon–Nikodým theorem for positive linear functionals. These are used in Section C.3 to construct the required trace on $\langle N, e_B \rangle$. Uniqueness will require the increasing approximate identity $\{k_\lambda\}_{\lambda \in \Lambda}$ for Ne_BN from Lemma 4.3.5. The technicalities of Section C.2 are designed to circumvent the difficulty that the net $\{k_\lambda^2\}_{\lambda \in \Lambda}$ need not be increasing.

C.2 Preliminary lemmas

The first lemma concerns continuous functions on an interval, and will be used in applying the functional calculus to elements of Ne_BN.

Lemma C.2.1. *Let $d > 0$ and let $f \in C[0, d]^+$ be such that $\|f\|_\infty < 1$ and f is identically 0 on a neighbourhood of 0. Then there exists a polynomial $p(x)$, non-negative on $[0, d]$, and satisfying $p(0) = 0$, $\|p\|_\infty < 1$, and $p(x)^2 \geq f(x)$.*

Proof. Choose α so that $\|f\|_\infty < \alpha < 1$. Then $f(x) \leq \alpha 1$. Since f vanishes on a neighbourhood of 0, $f(x)^{1/2}/x$ defines a continuous function on $[0, d]$ which is

373

strictly below $\alpha^{1/2}/x$ on $(0, d]$. The latter function tends to ∞ at 0, so there exists a uniform positive gap between these functions. Thus there is a continuous function h such that $h(x) > f(x)^{1/2}/x$ on $[0, d]$ and $h(x) < \alpha^{1/2}/x$ on $(0, d]$. By the Stone–Weierstrass theorem, we may choose a polynomial $q(x)$ so close to $h(x)$ in $\| \cdot \|_\infty$-norm that

$$f(x)^{1/2}/x \le q(x) \le \alpha^{1/2}/x, \qquad x \in (0, d]. \tag{C.2.1}$$

Let $p(x) = xq(x) \ge 0$. Then (C.2.1) implies that

$$f(x) \le p(x)^2 \le \alpha 1, \qquad x \in [0, d], \tag{C.2.2}$$

and the proof is completed by observing that $p(0) = 0$. $\qquad\qquad\square$

In Lemma 4.3.5, it was shown that the set $\{x \in N e_B N : 0 \le x < 1\}$ formed an increasing approximate identity, denoted $\{k_\lambda\}_{\lambda \in \Lambda}$, for $N e_B N$ and that $k_\lambda \to 1$ strongly in $\langle N, e_B \rangle$. The following corollary will be useful in showing uniqueness of the trace on $\langle N, e_B \rangle$.

Corollary C.2.2. *For a fixed, but arbitrary $\lambda \in \Lambda$, let $f \in C[0, \|k_\lambda\|]^+$ satisfy $\|f\|_\infty < 1$ and suppose that f is identically 0 on a neighbourhood of 0. Then there exists $\mu \in \Lambda$ such that $k_\mu^2 \ge f(k_\lambda)$.*

Proof. Since the spectrum of k_λ lies in $[0, \|k_\lambda\|]$, we may apply the functional calculus on this interval. By Lemma C.2.1, choose a polynomial $p \ge 0$ such that $p(0) = 0$, $\|p\|_\infty < 1$, and $p^2 \ge f$. Then $p(k_\lambda) \ge 0$ and $\|p(k_\lambda)\| < 1$. Thus $p(k_\lambda)$ is k_μ for some $\mu \in \Lambda$, and $k_\mu^2 \ge f(k_\lambda)$ by the choice of p. $\qquad\square$

We now turn to abelian von Neumann algebras. The next two results are essentially the Radon–Nikodým theorem for positive linear functionals, and we follow the technique developed by von Neumann (see [164, p. 280]).

Our convention of using ξ to denote the image of 1 in $L^2(N)$ will be used below.

Lemma C.2.3. *Let A be an abelian von Neumann algebra and let σ, ρ be faithful normal positive linear functionals satisfying $\sigma \le \rho$. Then there exists $a \in A$, $0 \le a \le 1$, such that $\sigma(x) = \rho(ax)$ for $x \in A$. Moreover, the spectral projection of a for the one point set $\{0\}$ is 0.*

Proof. Since A is abelian, σ and ρ are both faithful normal traces on A. Each induces a $\| \cdot \|_2$-norm, denoted by a suitable subscript.

If $x \in A$ then there is a unitary $u \in A$ such that $x = u|x|$. Then, using the Cauchy–Schwarz inequality,

$$|\sigma(x)| = |\sigma(u|x|)| \le \sigma(1)^{1/2}\sigma(|x|^2)^{1/2} \le \sigma(1)^{1/2}\|x\|_{2,\rho}, \tag{C.2.3}$$

and so $x \mapsto \sigma(x)$ extends to a bounded linear functional on $L^2(A, \rho)$. Thus there exists a vector $\eta \in L^2(A, \rho)$ so that

$$\sigma(x) = \langle x\xi, \eta \rangle_\rho, \qquad x \in A. \tag{C.2.4}$$

Choose a sequence $a_n \in A$ so that $\lim_{n\to\infty} \|a_n \xi - \eta\|_{2,\rho} = 0$, and then let B be the separable abelian von Neumann subalgebra generated by $\{a_n\}_{n=1}^\infty$. Then $\eta \in L^2(B, \rho)$. Separability of B allows us to identify B with $L^\infty(\Omega, \mu)$ for some finite measure space (Ω, μ). Restricting σ and ρ to B, these now lie in $B_* = L^1(\Omega, \mu)$, and we may apply the Radon–Nikodým theorem for measures to conclude that $\eta \in B$ and satisfies $0 \le \eta \le 1$. Thus we may rename η as $a \in A$ with $0 \le a \le 1$. It follows from (C.2.4) that $\sigma(x) = \rho(ax)$ for $x \in A$.

Let p be the spectral projection of a for the one point set $\{0\}$. Then $pa = 0$, so $\sigma(p) = 0$ and the faithfulness of σ implies that $p = 0$. $\square$

Corollary C.2.4. *Let σ and ρ be faithful normal positive linear functionals on an abelian von Neumann algebra A. Then there exists $a \in A$, $0 \le a \le 1$, such that $\sigma((1-a)x) = \rho(ax)$ for $x \in A$. Moreover, the spectral projections of a for the single point sets $\{0\}$ and $\{1\}$ are both 0.*

Proof. We apply the previous lemma to $\sigma \le \sigma + \rho$ to obtain $a \in A$, $0 \le a \le 1$, so that

$$\sigma(x) = (\sigma + \rho)(ax), \qquad x \in A, \tag{C.2.5}$$

which implies that $\sigma((1 - a)x) = \rho(ax)$. If p_0 and p_1 are respectively the spectral projections of a for the one point sets $\{0\}$ and $\{1\}$, then $ap_1 = p_1$ and $(1 - a)p_1 = 0$, giving $\rho(p_1) = 0$. Similarly, $\sigma(p_0) = 0$, so faithfulness of σ and ρ implies that $p_0 = p_1 = 0$. $\square$

C.3 Construction of the trace

Throughout, N is a finite von Neumann algebra with a faithful normal normalised trace τ, and B is a von Neumann subalgebra of N. Since $\langle N, e_B \rangle' = JBJ$, a finite von Neumann algebra, we see that $\langle N, e_B \rangle$ is a semifinite von Neumann algebra. Moreover, e_B is finite projection because $e_B \langle N, e_B \rangle e_B = e_B B \cong B$ which is a finite algebra. Thus there is a semifinite normal faithful trace Tr on $\langle N, e_B \rangle$ such that $Tr(e_B) = 1$. However, this may not be the trace that we require and so some modifications are necessary. The first step is to obtain the correct trace on the subalgebra $e_B Z$ where Z is the centre of B. Note that the next lemma does not address the issues of semifiniteness and faithfulness.

Lemma C.3.1. *There exists a normal trace Tr on $\langle N, e_B \rangle$ such that*

$$Tr(e_B z) = \tau(z), \qquad z \in Z(B). \tag{C.3.1}$$

Proof. We first choose a normal faithful semifinite trace ϕ on $\langle N, e_B \rangle$ so that $\phi(e_B) = 1$. On Z, define a positive normal linear functional σ by

$$\sigma(z) = \phi(e_B z), \qquad z \in Z. \tag{C.3.2}$$

Thus σ and τ are faithful on Z so, by Corollary C.2.4, there exists $a \in Z$, $0 \le a \le 1$, such that

$$\sigma(z(1 - a)) = \tau(za), \qquad z \in Z, \tag{C.3.3}$$

and the spectral projections at the end points of $[0,1]$ are both 0. For each $n \geq 2$, let p_n be the spectral projection of a for the interval $[1/n, 1]$. Then p_n increases strongly to 1 as $n \to \infty$ and $p_n a$ is invertible in $Z p_n$ with inverse $b_n \in Z p_n$. Moreover, $\{b_n\}_{n=2}^{\infty}$ is increasing in Z, although not necessarily uniformly bounded.

The centre of $\langle N, e_B \rangle$ is also the centre of its commutant JBJ which is JZJ. Thus $J(1-a)b_n J$ is central in $\langle N, e_B \rangle$ so we may define, for $x \in \langle N, e_B \rangle^+$ and $n \geq 2$,

$$\phi_n(x) = \phi(J(1-a)b_n J x). \qquad (C.3.4)$$

Each ϕ_n is a normal semifinite trace, although not faithful. Moreover, $\phi_{n+1} \geq \phi_n$ for $n \geq 2$. If we define

$$Tr(x) = \lim_{n \to \infty} \phi_n(x), \qquad x \in \langle N, e_B \rangle^+, \qquad (C.3.5)$$

then this is the desired trace as we now show.

Observe that, for any $z \in Z$ and $x \in N$,

$$Jz^* J e_B x \xi = Jz^* J \mathbb{E}_B(x) \xi = \mathbb{E}_B(x) z \xi$$
$$= z \mathbb{E}_B(x) \xi = z e_B x \xi. \qquad (C.3.6)$$

Thus $Jz^* J e_B = z e_B$. It follows that, for $z \in Z$,

$$\phi_n(e_B z) = \phi(J(1-a)b_n J e_B z) = \phi(e_B(1-a)b_n z)$$
$$= \sigma((1-a)b_n z) = \tau(b_n a z)$$
$$= \tau(p_n z), \qquad (C.3.7)$$

and so $Tr(e_B z) = \tau(z)$, by letting $n \to \infty$. It is easy to check that the increasing limit of normal maps is normal. $\square$

Note that any trace ϕ on $\langle N, e_B \rangle^+$ for which $\phi(e_B) = 1$ satisfies $\phi(x^* e_B x) = \phi(e_B x x^* e_B) \leq \|x\|^2$, so can be extended by linearity to $N e_B N$.

Lemma C.3.2. *Let ϕ be a trace on $N e_B N$ such that $\phi(e_B z) = \tau(z)$ for $z \in Z$. Then*

$$\phi(x e_B y) = \tau(xy), \qquad x, y \in N. \qquad (C.3.8)$$

Proof. Consider first the subspace $e_B B$. This is a C^*-algebra isomorphic to B with identity e_B. The restriction of ϕ to $e_B B$ is then a state since $\phi(e_B) = 1$. In particular this restriction is bounded on $e_B B$. Since

$$\phi(e_B b) = \phi(u e_B b u^*) = \phi(e_B u b u^*) \qquad (C.3.9)$$

for $b \in B$ and $u \in \mathcal{U}(B)$, continuity implies that

$$\phi(e_B b) = \phi(e_B x) \qquad (C.3.10)$$

for any $x \in \overline{\mathrm{conv}}^{\|\cdot\|}\{ubu^* : u \in \mathcal{U}(B)\}$. There is an element $z \in Z$ which lies in this norm closed convex hull, by the Dixmier approximation theorem (Theorem 2.2.2), and $\tau(z) = \tau(b)$ by continuity of τ. Thus

$$\phi(e_B b) = \phi(e_B z) = \tau(z) = \tau(b), \qquad b \in B. \tag{C.3.11}$$

For $x, y \in N$,

$$\begin{aligned}
\phi(x e_B y) &= \phi(e_B y x e_B) = \phi(e_B \mathbb{E}_B(yx)) \\
&= \tau(\mathbb{E}_B(yx)) = \tau(yx) = \tau(xy),
\end{aligned} \tag{C.3.12}$$

proving the result. □

Since we have already shown the existence of a normal trace on $\langle N, e_B \rangle^+$ whose extension to $N e_B N$ satisfies (C.3.12), the following theorem gives the second proof of the existence of the canonical trace on $\langle N, e_B \rangle^+$.

We will require the results on the approximate identity $\{k_\lambda\}_{\lambda \in \Lambda}$ presented in Section C.2.

Theorem C.3.3. *There exists a normal semifinite faithful trace Tr on $\langle N, e_B \rangle^+$ such that $Tr(e_B) = 1$ and such that its linear extension to $N e_B N$ satisfies*

$$Tr(x e_B y) = \tau(xy), \qquad x, y \in N. \tag{C.3.13}$$

Moreover, Tr is unique with these properties.

Proof. By Lemmas C.3.1 and C.3.2, there is a normal trace Tr on $\langle N, e_B \rangle^+$ satisfying (C.3.13). We will show that it is semifinite and faithful, and then that it is unique. This will be accomplished by showing that

$$Tr(x) = \sup_{\mu \in \Lambda} Tr(x^{1/2} k_\mu^2 x^{1/2}), \qquad x \in \langle N, e_B \rangle^+. \tag{C.3.14}$$

Since $\{k_\lambda\}_{\lambda \in \Lambda}$ increases strongly to 1 by Lemma 4.3.5, normality of Tr shows that

$$Tr(x) = \lim_\lambda Tr(x^{1/2} k_\lambda x^{1/2}), \qquad x \in \langle N, e_B \rangle^+. \tag{C.3.15}$$

Fix an $x \in \langle N, e_B \rangle^+$, choose a number $\alpha < Tr(x)$, and then choose $\lambda \in \Lambda$ from (C.3.15) such that

$$\alpha < Tr(x^{1/2} k_\lambda x^{1/2}) \leq Tr(x), \tag{C.3.16}$$

the second inequality coming from $k_\lambda \leq 1$. Regard λ as fixed from now on. On the interval $[0, \|k_\lambda\|]$, choose an increasing sequence of continuous functions $f_n(t) \geq 0$ tending uniformly to t and such that each $f_n(t)$ vanishes on some neighbourhood of 0. Then $f_n(k_\lambda)$ tends upward to k_λ so, by normality of Tr, we may choose n so large that

$$\alpha < Tr(x^{1/2} f_n(k_\lambda) x^{1/2}) \leq Tr(x^{1/2} k_\lambda x^{1/2}) \leq Tr(x). \tag{C.3.17}$$

For this choice of n, Corollary C.2.2 allows us to choose $\mu \in \Lambda$ so that $k_\mu^2 \geq f_n(k_\lambda)$, and so

$$\alpha < Tr(x^{1/2}k_\mu^2 x^{1/2}) \leq Tr(x). \qquad \text{(C.3.18)}$$

This shows that (C.3.14) holds since $\alpha < Tr(x)$ was arbitrary.

Consider now $x \in \langle N, e_B \rangle^+$ with $x \neq 0$. If $Tr(x) = 0$ then $Tr(x^{1/2}k_\mu^2 x^{1/2}) = 0$ for all $\mu \in \Lambda$, by (C.3.14). Thus $Tr(k_\mu x k_\mu) = 0$ and $k_\mu x k_\mu \in Ne_B N$ (since $e_B \langle N, e_B \rangle e_B = e_B B$), so faithfulness of Tr on $Ne_B N$, proved in Lemma 4.3.2, shows that $k_\mu x k_\mu = 0$ for all $\mu \in \Lambda$. Thus $x^{1/2}k_\mu = 0$ for all μ from which it follows that $x = 0$ since $k_\mu \to 1$ strongly. This contradiction shows that $Tr(x) > 0$ and Tr is faithful.

We now show that Tr is semifinite. Fix $x \in \langle N, e_B \rangle^+$ and choose a maximal set of non-zero elements of the form $yk_\mu^2 y$, $y \in \langle N, e_B \rangle^+$, whose sum is dominated by x, and let x_0 be its sum. Each $yk_\mu^2 y$ has finite trace because $Tr(yk_\mu^2 y) = Tr(k_\mu y^2 k_\mu)$ and the latter element is in $Ne_B N$. If $x_0 \neq x$, then $Tr(x - x_0) > 0$, and so we may choose a k_ν so that $Tr((x-x_0)^{1/2}k_\nu^2(x-x_0)^{1/2}) > 0$. In particular, this is a non-zero element which can be added to the existing collection, contradicting maximality. Thus $x_0 = x$. By taking finite sums from this collection, we obtain an increasing net of finite trace operators whose upper bound is x, proving semifiniteness.

Returning to (C.3.14), we see that

$$Tr(x) = \sup_{\mu \in \Lambda} Tr(k_\mu x k_\mu), \qquad x \in \langle Ne_B \rangle^+, \qquad \text{(C.3.19)}$$

and so the values of Tr on $\langle N, e_B \rangle^+$ are determined by its values on $Ne_B N$. Thus there can only be one trace which satisfies the hypotheses and conclusions of the theorem. $\qquad \square$

Bibliography

[1] C. A. Akemann and P. A. Ostrand. Computing norms in group C^*-algebras. *Amer. J. Math.*, 98(4):1015–1047, 1976.

[2] S. Anastasio. Maximal abelian subalgebras in hyperfinite factors. *Amer. J. Math.*, 87:955–971, 1965.

[3] W. B. Arveson. Subalgebras of C^*-algebras. *Acta Math.*, 123:141–224, 1969.

[4] K. Beauchamp and R. Nicoara. Orthogonal maximal abelian *-subalgebras of the 6×6 matrices. Mathematics ArXiv math.OA/0609076.

[5] B. Blackadar. *Operator algebras*, volume 122 of *Encyclopaedia of Mathematical Sciences*. Springer-Verlag, Berlin, 2006. Theory of C^*-algebras and von Neumann algebras, Operator Algebras and Non-commutative Geometry, III.

[6] R. J. Blattner. Automorphic group representations. *Pacific J. Math.*, 8:665–677, 1958.

[7] D. P. Blecher and C. Le Merdy. *Operator algebras and their modules—an operator space approach*, volume 30 of *London Mathematical Society Monographs. New Series*. The Clarendon Press, Oxford University Press, Oxford, 2004.

[8] F. Boca and F. Rădulescu. Singularity of radial subalgebras in II_1 factors associated with free products of groups. *J. Funct. Anal.*, 103(1):138–159, 1992.

[9] N. Brown and N. Ozawa. C^*-algebras and finite dimensional approximations. Graduate Studies in Mathematics Series. American Mathematical Society, Providence, RI, to appear.

[10] D. Bures. *Abelian subalgebras of von Neumann algebras*. American Mathematical Society, Providence, RI, 1971. Memoirs of the American Mathematical Society, No. 110.

[11] C. C. Chang and H. J. Keisler. *Model theory*, volume 73 of *Studies in Logic and the Foundations of Mathematics*. North-Holland Publishing Co., Amsterdam, third edition, 1990.

[12] I. Chifan. On the normalizing algebra of a MASA in a II_1 factor. Mathematics ArXiv math.OA/0606225.

[13] W.-M. Ching. Non-isomorphic non-hyperfinite factors. *Canad. J. Math.*, 21:1293–1308, 1969.

[14] W.-M. Ching. A continuum of non-isomorphic non-hyperfinite factors. *Comm. Pure Appl. Math.*, 23:921–937, 1970.

[15] W.-M. Ching. Free products of von Neumann algebras. *Trans. Amer. Math. Soc.*, 178:147–163, 1973.

[16] H. Choda. A Galois correspondence in a von Neumann algebra. *Tôhoku Math. J. (2)*, 30(4):491–504, 1978.

[17] M.-D. Choi and E. Christensen. Completely order isomorphic and close C^*-algebras need not be $*$-isomorphic. *Bull. London Math. Soc.*, 15(6):604–610, 1983.

[18] M.-D. Choi and E. G. Effros. Separable nuclear C^*-algebras and injectivity. *Duke Math. J.*, 43(2):309–322, 1976.

[19] M.-D. Choi and E. G. Effros. Nuclear C^*-algebras and injectivity: the general case. *Indiana Univ. Math. J.*, 26(3):443–446, 1977.

[20] E. Christensen. Perturbations of type I von Neumann algebras. *J. London Math. Soc. (2)*, 9:395–405, 1974/75.

[21] E. Christensen. Perturbations of operator algebras. *Invent. Math.*, 43(1):1–13, 1977.

[22] E. Christensen. Perturbations of operator algebras. II. *Indiana Univ. Math. J.*, 26(5):891–904, 1977.

[23] E. Christensen. Extension of derivations. *J. Funct. Anal.*, 27(2):234–247, 1978.

[24] E. Christensen. Subalgebras of a finite algebra. *Math. Ann.*, 243(1):17–29, 1979.

[25] E. Christensen. Near inclusions of C^*-algebras. *Acta Math.*, 144(3-4):249–265, 1980.

[26] E. Christensen. Derivations and their relation to perturbations of operator algebras. In *Operator algebras and applications, Part 2 (Kingston, Ont., 1980)*, volume 38 of *Proc. Sympos. Pure Math.*, pages 261–273. American Mathematical Society, Providence, RI, 1982.

[27] E. Christensen. Extensions of derivations. II. *Math. Scand.*, 50(1):111–122, 1982.

[28] E. Christensen. Close operator algebras. In *Deformation theory of algebras and structures and applications (Il Ciocco, 1986)*, volume 247 of *NATO Adv. Sci. Inst. Ser. C Math. Phys. Sci.*, pages 537–556. Kluwer Academic Publishers, Dordrecht, 1988.

[29] E. Christensen. Finite von Neumann algebra factors with property Γ. *J. Funct. Anal.*, 186(2):366–380, 2001.

[30] E. Christensen, F. Pop, A. M. Sinclair, and R. R. Smith. Hochschild cohomology of factors with property Γ. *Ann. Math. (2)*, 158(2):635–659, 2003.

[31] E. Christensen, F. Pop, A. M. Sinclair, and R. R. Smith. Property Γ factors and the Hochschild cohomology problem. *Proc. Natl. Acad. Sci. USA*, 100(7):3865–3869 (electronic), 2003.

[32] J. M. Cohen. Radial functions on free products. *J. Funct. Anal.*, 59(2):167–174, 1984.

[33] J. M. Cohen and A. R. Trenholme. Orthogonal polynomials with a constant recursion formula and an application to harmonic analysis. *J. Funct. Anal.*, 59(2):175–184, 1984.

[34] W. W. Comfort and S. Negrepontis. *The theory of ultrafilters.* Springer-Verlag, New York, 1974. Die Grundlehren der mathematischen Wissenschaften, Band 211.

[35] A. Connes. Sur la classification des facteurs de type II. *C. R. Acad. Sci. Paris Sér. A-B*, 281(1):Aii, A13–A15, 1975.

[36] A. Connes. Classification of injective factors. Cases II_1, II_∞, III_λ, $\lambda \neq 1$. *Ann. Math. (2)*, 104(1):73–115, 1976.

[37] A. Connes. A factor of type II_1 with countable fundamental group. *J. Operator Theory*, 4(1):151–153, 1980.

[38] A. Connes. Classification des facteurs. In *Operator algebras and applications, Part 2 (Kingston, Ont., 1980)*, volume 38 of *Proc. Sympos. Pure Math.*, pages 43–109. American Mathematical Society, Providence, RI, 1982.

[39] A. Connes. Nombres de Betti L^2 et facteurs de type II_1 (d'après D. Gaboriau et S. Popa). *Astérisque*, (294):ix, 321–333, 2004.

[40] A. Connes, J. Feldman, and B. Weiss. An amenable equivalence relation is generated by a single transformation. *Ergodic Theory Dynamical Syst.*, 1(4):431–450, 1981.

[41] A. Connes and V. Jones. A II$_1$ factor with two nonconjugate Cartan subalgebras. *Bull. Amer. Math. Soc. (N.S.)*, 6(2):211–212, 1982.

[42] A. Connes and V. Jones. Property T for von Neumann algebras. *Bull. London Math. Soc.*, 17(1):57–62, 1985.

[43] M. Cowling and U. Haagerup. Completely bounded multipliers of the Fourier algebra of a simple Lie group of real rank one. *Invent. Math.*, 96(3):507–549, 1989.

[44] J. De Cannière and U. Haagerup. Multipliers of the Fourier algebras of some simple Lie groups and their discrete subgroups. *Amer. J. Math.*, 107(2):455–500, 1985.

[45] P. de la Harpe. Operator algebras, free groups and other groups. *Astérisque*, (232):121–153, 1995. Recent advances in operator algebras (Orléans, 1992).

[46] J. Dixmier. Les anneaux d'opérateurs de classe finie. *Ann. Sci. École Norm. Sup. (3)*, 66:209–261, 1949.

[47] J. Dixmier. Sous-anneaux abéliens maximaux dans les facteurs de type fini. *Ann. Math. (2)*, 59:279–286, 1954.

[48] J. Dixmier. *Les algèbres d'opérateurs dans l'espace hilbertien (Algèbres de von Neumann)*. Cahiers scientifiques, Fascicule XXV. Gauthier-Villars, Paris, 1957.

[49] J. Dixmier. *Les C^*-algèbres et leurs représentations*. Cahiers Scientifiques, Fasc. XXIX. Gauthier-Villars & Cie, Éditeur-Imprimeur, Paris, 1964.

[50] J. Dixmier. Quelques propriétés des suites centrales dans les facteurs de type II$_1$. *Invent. Math.*, 7:215–225, 1969.

[51] J. Dixmier. *von Neumann algebras*, volume 27 of *North-Holland Mathematical Library*. North-Holland Publishing Co., Amsterdam, 1981. With a preface by E. C. Lance, Translated from the second French edition by F. Jellett.

[52] J. Dixmier and E. C. Lance. Deux nouveaux facteurs de type II$_1$. *Invent. Math.*, 7:226–234, 1969.

[53] R. G. Douglas and C. Pearcy. Von Neumann algebras with a single generator. *Michigan Math. J.*, 16:21–26, 1969.

[54] H. A. Dye. On groups of measure preserving transformations. I. *Amer. J. Math.*, 81:119–159, 1959.

[55] H. A. Dye. On groups of measure preserving transformations. II. *Amer. J. Math.*, 85:551–576, 1963.

[56] K. Dykema. Free products of hyperfinite von Neumann algebras and free dimension. *Duke Math. J.*, 69(1):97–119, 1993.

[57] K. Dykema. On certain free product factors via an extended matrix model. *J. Funct. Anal.*, 112(1):31–60, 1993.

[58] K. Dykema. Interpolated free group factors. *Pacific J. Math.*, 163(1):123–135, 1994.

[59] K. Dykema. Two applications of free entropy. *Math. Ann.*, 308(3):547–558, 1997.

[60] K. Dykema and U. Haagerup. DT-operators and decomposability of Voiculescu's circular operator. *Amer. J. Math.*, 126(1):121–189, 2004.

[61] K. Dykema and U. Haagerup. Invariant subspaces of the quasinilpotent DT-operator. *J. Funct. Anal.*, 209(2):332–366, 2004.

[62] K. Dykema, A. M. Sinclair, and R. R. Smith. Values of the Pukánszky invariant in free group factors and the hyperfinite factor. *J. Funct. Anal.*, 240:373–398, 2006.

[63] K. Dykema, A. M. Sinclair, R. R. Smith, and S. A. White. Generators of II$_1$ factors. Mathematics ArXiv, arXiv:0706.1953.

[64] K. Dykema and S. A. White. Private communication.

[65] E. G. Effros and E. C. Lance. Tensor products of operator algebras. *Adv. Math.*, 25(1):1–34, 1977.

[66] E. G. Effros and Z.-J. Ruan. *Operator spaces*, volume 23 of *London Mathematical Society Monographs. New Series*. The Clarendon Press, Oxford University Press, New York, 2000.

[67] J. Fang, L. Ge, and W. Li. Central sequence algebras of von Neumann algebras. *Taiwanese J. Math.*, 10(1):187–200, 2006.

[68] J. Feldman. Embedding of AW^* algebras. *Duke Math. J.*, 23:303–307, 1956.

[69] A. Figà-Talamanca and M. A. Picardello. *Harmonic analysis on free groups*, volume 87 of *Lecture Notes in Pure and Applied Mathematics*. Marcel Dekker Inc., New York, 1983.

[70] D. Gaboriau. Coût des relations d'équivalence et des groupes. *Invent. Math.*, 139(1):41–98, 2000.

[71] D. Gaboriau and S. Popa. An uncountable family of nonorbit equivalent actions of $\mathbb{F}_n$. *J. Amer. Math. Soc.*, 18(3):547–559 (electronic), 2005.

[72] L. Ge. A splitting property for subalgebras of tensor products. *Bull. Amer. Math. Soc. (N.S.)*, 32(1):57–60, 1995.

[73] L. Ge. On maximal injective subalgebras of factors. *Adv. Math.*, 118(1):34–70, 1996.

[74] L. Ge. Prime factors. *Proc. Natl. Acad. Sci. USA*, 93(23):12762–12763, 1996.

[75] L. Ge. Applications of free entropy to finite von Neumann algebras. *Amer. J. Math.*, 119(2):467–485, 1997.

[76] L. Ge. Applications of free entropy to finite von Neumann algebras. II. *Ann. Math. (2)*, 147(1):143–157, 1998.

[77] L. Ge. On "Problems on von Neumann algebras by R. Kadison, 1967". *Acta Math. Sin. (Engl. Ser.)*, 19(3):619–624, 2003. With a previously unpublished manuscript by Kadison, International Workshop on Operator Algebra and Operator Theory (Linfen, 2001).

[78] L. Ge and D. Hadwin. Ultraproducts of C^*-algebras. In *Recent advances in operator theory and related topics (Szeged, 1999)*, volume 127 of *Oper. Theory Adv. Appl.*, pages 305–326. Birkhäuser, Basel, 2001.

[79] L. Ge and R. Kadison. On tensor products for von Neumann algebras. *Invent. Math.*, 123(3):453–466, 1996.

[80] L. Ge and S. Popa. On some decomposition properties for factors of type II_1. *Duke Math. J.*, 94(1):79–101, 1998.

[81] L. Ge and J. Shen. Generator problem for certain property T factors. *Proc. Natl. Acad. Sci. USA*, 99(2):565–567 (electronic), 2002.

[82] J. G. Glimm. On a certain class of operator algebras. *Trans. Amer. Math. Soc.*, 95:318–340, 1960.

[83] F. M. Goodman, P. de la Harpe, and V. F. R. Jones. *Coxeter graphs and towers of algebras*, volume 14 of *Mathematical Sciences Research Institute Publications*. Springer-Verlag, New York, 1989.

[84] G. R. Goodson, J. Kwiatkowski, M. Lemańczyk, and P. Liardet. On the multiplicity function of ergodic group extensions of rotations. *Studia Math.*, 102(2):157–174, 1992.

[85] U. Haagerup. Normal weights on W^*-algebras. *J. Funct. Anal.*, 19:302–317, 1975.

[86] U. Haagerup. An example of a nonnuclear C^*-algebra, which has the metric approximation property. *Invent. Math.*, 50(3):279–293, 1978/79.

[87] U. Haagerup. A new proof of the equivalence of injectivity and hyperfiniteness for factors on a separable Hilbert space. *J. Funct. Anal.*, 62(2):160–201, 1985.

[88] U. Haagerup. Orthogonal maximal abelian ∗-subalgebras of the $n \times n$ matrices and cyclic n-roots. In *Operator algebras and quantum field theory (Rome, 1996)*, pages 296–322. International Press, Cambridge, MA, 1997.

[89] C. Houdayer. Construction of type II$_1$ factors with prescribed countable fundamental group. Mathematics ArXiv math.OA/0704.3502.

[90] K. Hrbacek and T. Jech. *Introduction to set theory*, volume 220 of *Monographs and Textbooks in Pure and Applied Mathematics*. Marcel Dekker Inc., New York, third edition, 1999.

[91] A. Ioana, J. Peterson, and S. Popa. Amalgamated free products of w-rigid factors and calculation of their symmetry groups. *Acta. Math.*, to appear. Mathematics ArXiv math.OA/0505589.

[92] P. Jolisaint and Y. Stalder. Singular and strongly mixing MASA's in finite von Neumann algebras. Mathematics ArXiv math.OA/0602158.

[93] P. Jolissaint. Operator algebras related to Thompson's group F. *J. Aust. Math. Soc.*, 79(2):231–241, 2005.

[94] V. F. R. Jones. Index for subfactors. *Invent. Math.*, 72(1):1–25, 1983.

[95] V. F. R. Jones. *Subfactors and knots*, volume 80 of *CBMS Regional Conference Series in Mathematics*. Published for the Conference Board of the Mathematical Sciences, Washington, DC, 1991.

[96] V. F. R. Jones and S. Popa. Some properties of MASAs in factors. In *Invariant subspaces and other topics (Timişoara/Herculane, 1981)*, volume 6 of *Operator Theory: Adv. Appl.*, pages 89–102. Birkhäuser, Basel, 1982.

[97] V. R. Jones and V. S. Sunder. *Introduction to subfactors*, volume 234 of *London Mathematical Society Lecture Note Series*. Cambridge University Press, Cambridge, 1997.

[98] K. Jung. Strongly 1-bounded von Neumann algebras. Mathematics arXiv:math.OA/0510576.

[99] R. V. Kadison. Diagonalizing matrices over operator algebras. *Bull. Amer. Math. Soc. (N.S.)*, 8(1):84–86, 1983.

[100] R. V. Kadison. Diagonalizing matrices. *Amer. J. Math.*, 106(6):1451–1468, 1984.

[101] R. V. Kadison. Which Singer is that? In *Surveys in differential geometry*, Surv. Differ. Geom., VII, pages 347–373. International Press, Somerville, MA, 2000.

[102] R. V. Kadison. Non-commutative conditional expectations and their applications. In *Operator algebras, quantization, and noncommutative geometry*, volume 365 of *Contemp. Math.*, pages 143–179. American Mathematical Society, Providence, RI, 2004.

[103] R. V. Kadison and D. Kastler. Perturbations of von Neumann algebras. I. Stability of type. *Amer. J. Math.*, 94:38–54, 1972.

[104] R. V. Kadison and J. R. Ringrose. *Fundamentals of the theory of operator algebras. Vol. I*, volume 100 of *Pure and Applied Mathematics*. Academic Press Inc. [Harcourt Brace Jovanovich Publishers], New York, 1983. Elementary theory.

[105] R. V. Kadison and J. R. Ringrose. *Fundamentals of the theory of operator algebras. Vol. II*, volume 100 of *Pure and Applied Mathematics*. Academic Press Inc., Orlando, FL, 1986. Advanced theory.

[106] I. Kaplansky. A theorem on rings of operators. *Pacific J. Math.*, 1:227–232, 1951.

[107] J. L. Kelley. *General topology*. D. Van Nostrand Company, Inc., Toronto–New York–London, 1955.

[108] H. Kesten. Symmetric random walks on groups. *Trans. Amer. Math. Soc.*, 92:336–354, 1959.

[109] H. Kosaki. Applications of uniform convexity of noncommutative L^p-spaces. *Trans. Amer. Math. Soc.*, 283(1):265–282, 1984.

[110] J. Kwiatkowski and M. Lemańczyk. On the multiplicity function of ergodic group extensions. II. *Studia Math.*, 116(3):207–215, 1995.

[111] C. Lance. On nuclear C^*-algebras. *J. Funct. Anal.*, 12:157–176, 1973.

[112] D. McDuff. A countable infinity of II_1 factors. *Ann. Math. (2)*, 90:361–371, 1969.

[113] D. McDuff. Uncountably many II_1 factors. *Ann. Math. (2)*, 90:372–377, 1969.

[114] D. McDuff. Central sequences and the hyperfinite factor. *Proc. London Math. Soc. (3)*, 21:443–461, 1970.

[115] R. Mercer. Convergence of Fourier series in discrete crossed products of von Neumann algebras. *Proc. Amer. Math. Soc.*, 94(2):254–258, 1985.

[116] F. J. Murray and J. Von Neumann. On rings of operators. *Ann. Math. (2)*, 37(1):116–229, 1936.

[117] F. J. Murray and J. von Neumann. On rings of operators. II. *Trans. Amer. Math. Soc.*, 41(2):208–248, 1937.

[118] F. J. Murray and J. von Neumann. On rings of operators. IV. *Ann. Math.* *(2)*, 44:716–808, 1943.

[119] S. Neshveyev and E. Størmer. Ergodic theory and maximal abelian subalgebras of the hyperfinite factor. *J. Funct. Anal.*, 195(2):239–261, 2002.

[120] O. A. Nielsen. Maximal abelian subalgebras in hyperfinite factors. *Bull. Amer. Math. Soc.*, 75:579–581, 1969.

[121] O. A. Nielsen. Maximal abelian subalgebras of hyperfinite factors. *Trans. Amer. Math. Soc.*, 146:259–272, 1969.

[122] O. A. Nielsen. Maximal Abelian subalgebras of hyperfinite factors. II. *J. Funct. Anal.*, 6:192–202, 1970.

[123] V. Niţică and A. Török. Maximal abelian and singular subalgebras in $L(F_N)$. *J. Operator Theory*, 30(1):3–19, 1993.

[124] C. L. Olsen and W. R. Zame. Some C^*-algebras with a single generator. *Trans. Amer. Math. Soc.*, 215:205–217, 1976.

[125] N. Ozawa. Solid von Neumann algebras. *Acta Math.*, 192(1):111–117, 2004.

[126] N. Ozawa and S. Popa. Some prime factorization results for type II_1 factors. *Invent. Math.*, 156(2):223–234, 2004.

[127] A. L. T. Paterson. *Amenability*, volume 29 of *Mathematical Surveys and Monographs*. American Mathematical Society, Providence, RI, 1988.

[128] V. Paulsen. *Completely bounded maps and operator algebras*, volume 78 of *Cambridge Studies in Advanced Mathematics*. Cambridge University Press, Cambridge, 2002.

[129] C. Pearcy. W^*-algebras with a single generator. *Proc. Amer. Math. Soc.*, 13:831–832, 1962.

[130] C. Pearcy. On certain von Neumann algebras which are generated by partial isometries. *Proc. Amer. Math. Soc.*, 15:393–395, 1964.

[131] G. K. Pedersen. *C^*-algebras and their automorphism groups*, volume 14 of *London Mathematical Society Monographs*. Academic Press Inc. [Harcourt Brace Jovanovich Publishers], London, 1979.

[132] J. Peterson and S. Popa. On the notion of relative property (T) for inclusions of von Neumann algebras. *J. Funct. Anal.*, 219(2):469–483, 2005.

[133] M. Pimsner and S. Popa. Entropy and index for subfactors. *Ann. Sci. École Norm. Sup. (4)*, 19(1):57–106, 1986.

[134] G. Pisier. *Similarity problems and completely bounded maps*, volume 1618 of *Lecture Notes in Mathematics*. Springer-Verlag, Berlin, expanded edition, 2001. Includes the solution to "The Halmos problem".

[135] G. Pisier. *Introduction to operator space theory*, volume 294 of *London Mathematical Society Lecture Note Series*. Cambridge University Press, Cambridge, 2003.

[136] S. Popa. On a problem of R. V. Kadison on maximal abelian *-subalgebras in factors. *Invent. Math.*, 65(2):269–281, 1981/82.

[137] S. Popa. Maximal injective subalgebras in factors associated with free groups. *Adv. Math.*, 50(1):27–48, 1983.

[138] S. Popa. Orthogonal pairs of *-subalgebras in finite von Neumann algebras. *J. Operator Theory*, 9(2):253–268, 1983.

[139] S. Popa. Singular maximal abelian *-subalgebras in continuous von Neumann algebras. *J. Funct. Anal.*, 50(2):151–166, 1983.

[140] S. Popa. Semiregular maximal abelian *-subalgebras and the solution to the factor state Stone-Weierstrass problem. *Invent. Math.*, 76(1):157–161, 1984.

[141] S. Popa. Notes on Cartan subalgebras in type II_1 factors. *Math. Scand.*, 57(1):171–188, 1985.

[142] S. Popa. A short proof of "injectivity implies hyperfiniteness" for finite von Neumann algebras. *J. Operator Theory*, 16(2):261–272, 1986.

[143] S. Popa. Classification of amenable subfactors of type II. *Acta Math.*, 172(2):163–255, 1994.

[144] S. Popa. *Classification of subfactors and their endomorphisms*, volume 86 of *CBMS Regional Conference Series in Mathematics*. Published for the Conference Board of the Mathematical Sciences, Washington, DC, 1995.

[145] S. Popa. On the distance between MASA's in type II_1 factors. In *Mathematical physics in mathematics and physics (Siena, 2000)*, volume 30 of *Fields Inst. Commun.*, pages 321–324. American Mathematical Society, Providence, RI, 2001.

[146] S. Popa. On the fundamental group of type II_1 factors. *Proc. Natl. Acad. Sci. USA*, 101(3):723–726 (electronic), 2004.

[147] S. Popa. On a class of type II_1 factors with Betti numbers invariants. *Ann. Math. (2)*, 163(3):809–899, 2006.

[148] S. Popa. Some rigidity results for non-commutative Bernoulli shifts. *J. Funct. Anal.*, 230(2):273–328, 2006.

[149] S. Popa. Strong rigidity of II$_1$ factors arising from malleable actions of w-rigid groups. I. *Invent. Math.*, 165(2):369–408, 2006.

[150] S. Popa. Strong rigidity of II$_1$ factors arising from malleable actions of w-rigid groups. II. *Invent. Math.*, 165(2):409–451, 2006.

[151] S. Popa and D. Shlyakhtenko. Cartan subalgebras and bimodule decompositions of II$_1$ factors. *Math. Scand.*, 92(1):93–102, 2003.

[152] S. Popa, A. M. Sinclair, and R. R. Smith. Perturbations of subalgebras of type II$_1$ factors. *J. Funct. Anal.*, 213(2):346–379, 2004.

[153] S. Popa, A. M. Sinclair, and R. R. Smith. Corrigendum to: "Perturbations of subalgebras of type II$_1$ factors" [J. Funct. Anal. **213** (2004), no. 2, 346–379]. *J. Funct. Anal.*, 235(1):355–356, 2006.

[154] L. Pukánszky. On maximal albelian subrings of factors of type II$_1$. *Canad. J. Math.*, 12:289–296, 1960.

[155] T. Pytlik. Radial functions on free groups and a decomposition of the regular representation into irreducible components. *J. Reine Angew. Math.*, 326:124–135, 1981.

[156] F. Rădulescu. Singularity of the radial subalgebra of $\mathcal{L}(F_N)$ and the Pukánszky invariant. *Pacific J. Math.*, 151(2):297–306, 1991.

[157] F. Rădulescu. Random matrices, amalgamated free products and subfactors of the von Neumann algebra of a free group, of noninteger index. *Invent. Math.*, 115(2):347–389, 1994.

[158] M. Rédei. *John Von Neumann: Selected Letters*, volume 27 of *History of Mathematics*. American Mathematical Society and London Mathematical Society, London, 2005.

[159] J. R. Ringrose. Lectures on the trace in a finite von Neumann algebra. In *Lectures on operator algebras (Tulane Univ. Ring and Operator Theory Year, 1970–1971, Vol. II; dedicated to the memory of David M. Topping)*, pages 309–354. Lecture Notes in Math., Vol. 247. Springer, Berlin, 1972.

[160] G. Robertson. Singular MASAS of von Neumann algebras: examples from the geometry of spaces of nonpositive curvature. In *Travaux mathématiques. Fasc. XIV*, Trav. Math., XIV, pages 153–165. Univ. Luxemb., Luxembourg, 2003.

[161] G. Robertson, A. M. Sinclair, and R. R. Smith. Strong singularity for subalgebras of finite factors. *Internat. J. Math.*, 14(3):235–258, 2003.

[162] G. Robertson and T. Steger. Maximal abelian subalgebras of the group factor of an $\tilde{A}_2$ group. *J. Operator Theory*, 36(2):317–334, 1996.

[163] M. Rørdam, F. Larsen, and N. Laustsen. *An introduction to K-theory for C*-algebras*, volume 49 of *London Mathematical Society Student Texts*. Cambridge University Press, Cambridge, 2000.

[164] H. L. Royden. *Real analysis*. Macmillan Publishing Company, New York, third edition, 1988.

[165] S. Sakai. The theory of *W**-algebras. Lecture notes, Yale University, 1962, unpublished.

[166] S. Sakai. An uncountable number of II_1 and II_∞ factors. *J. Functional Analysis*, 5:236–246, 1970.

[167] S. Sakai. *C*-algebras and W*-algebras*. Springer-Verlag, New York, 1971. Ergebnisse der Mathematik und ihrer Grenzgebiete, Band 60.

[168] J. Schwartz. Two finite, non-hyperfinite, non-isomorphic factors. *Comm. Pure Appl. Math.*, 16:19–26, 1963.

[169] J. T. Schwartz. *W*-algebras*. Gordon and Breach Science Publishers, New York, 1967.

[170] J. Shen. Maximal injective subalgebras of tensor products of free group factors. Mathematics ArXiv math.OA/0508305.

[171] J. Shen. Singly generated II_1 factors. Mathematics ArXiv math.OA/0511327.

[172] A. M. Sinclair and R. R. Smith. *Hochschild cohomology of von Neumann algebras*, volume 203 of *London Mathematical Society Lecture Note Series*. Cambridge University Press, Cambridge, 1995.

[173] A. M. Sinclair and R. R. Smith. Cartan subalgebras of finite von Neumann algebras. *Math. Scand.*, 85(1):105–120, 1999.

[174] A. M. Sinclair and R. R. Smith. Strongly singular masas in type II_1 factors. *Geom. Funct. Anal.*, 12(1):199–216, 2002.

[175] A. M. Sinclair and R. R. Smith. The Laplacian MASA in a free group factor. *Trans. Amer. Math. Soc.*, 355(2):465–475 (electronic), 2003.

[176] A. M. Sinclair and R. R. Smith. The Pukánszky invariant for masas in group von Neumann factors. *Illinois J. Math.*, 49(2):325–343 (electronic), 2005.

[177] A. M. Sinclair, R. R. Smith, S. A. White, and A. Wiggins. Strong singularity of singular masas in II_1 factors. *Illinois J. Math.*, to appear.

[178] A. M. Sinclair and S. A. White. A continuous path of singular masas in the hyperfinite II_1 factor. *J. London Math. Soc. (2)*, 75:243–254, 2007.

[179] C. F. Skau. Finite subalgebras of a von Neumann algebra. *J. Funct. Anal.*, 25(3):211–235, 1977.

[180] M. B. Ştefan. Infinite multiplicity of abelian subalgebras in free group subfactors. Mathematics ArXiv math.OA/0402107.

[181] M. B. Ştefan. The primality of subfactors of finite index in the interpolated free group factors. *Proc. Amer. Math. Soc.*, 126(8):2299–2307, 1998.

[182] M. B. Ştefan. Indecomposability of free group factors over nonprime subfactors and abelian subalgebras. *Pacific J. Math.*, 219(2):365–390, 2005.

[183] W. F. Stinespring. Positive functions on C^*-algebras. *Proc. Amer. Math. Soc.*, 6:211–216, 1955.

[184] Ş. Strătilă and L. Zsidó. *Lectures on von Neumann algebras*. Editura Academiei, Bucharest, 1979. Revision of the 1975 original, Translated from the Romanian by Silviu Teleman.

[185] N. Suzuki and T. Saitô. On the operators which generate continuous von Neumann algebras. *Tôhoku Math. J. (2)*, 15:277–280, 1963.

[186] M. Takesaki. On the unitary equivalence among the components of decompositions of representations of involutive Banach algebras and the associated diagonal algebras. *Tôhoku Math. J. (2)*, 15:365–393, 1963.

[187] M. Takesaki. *Theory of operator algebras. I*. Springer-Verlag, New York, 1979.

[188] M. Takesaki. *Theory of operator algebras. II*, volume 125 of *Encyclopaedia of Mathematical Sciences*. Springer-Verlag, Berlin, 2003. Operator Algebras and Non-commutative Geometry, 6.

[189] M. Takesaki. *Theory of operator algebras. III*, volume 127 of *Encyclopaedia of Mathematical Sciences*. Springer-Verlag, Berlin, 2003. Operator Algebras and Non-commutative Geometry, 8.

[190] R. J. Tauer. Maximal abelian subalgebras in finite factors of type II. *Trans. Amer. Math. Soc.*, 114:281–308, 1965.

[191] R. J. Tauer. Semi-regular maximal abelian subalgebras in hyperfinite factors. *Bull. Amer. Math. Soc.*, 71:606–608, 1965.

[192] R. J. Tauer. M-semiregular subalgebras in hyperfinite factors. *Trans. Amer. Math. Soc.*, 129:530–541, 1967.

[193] J. Tomiyama. On the projection of norm one in W^*-algebras. *Proc. Japan Acad.*, 33:608–612, 1957.

[194] J. Tomiyama. On the projection of norm one in W^*-algebras. II. *Tôhoku Math. J. (2)*, 10:204–209, 1958.

[195] J. Tomiyama. On the projection of norm one in W^*-algebras. III. *Tôhoku Math. J. (2)*, 11:125–129, 1959.

[196] D. M. Topping. *Lectures on von Neumann algebras*. Van Nostrand Reinhold Co., London, 1971.

[197] A. R. Trenholme. Maximal abelian subalgebras of function algebras associated with free products. *J. Funct. Anal.*, 79(2):342–350, 1988.

[198] A. van Daele. *Continuous crossed products and type III von Neumann algebras*, volume 31 of *London Mathematical Society Lecture Note Series*. Cambridge University Press, Cambridge, 1978.

[199] D. Voiculescu. The analogues of entropy and of Fisher's information measure in free probability theory. III. The absence of Cartan subalgebras. *Geom. Funct. Anal.*, 6(1):172–199, 1996.

[200] D. V. Voiculescu, K. J. Dykema, and A. Nica. *Free random variables*, volume 1 of *CRM Monograph Series*. American Mathematical Society, Providence, RI, 1992. A noncommutative probability approach to free products with applications to random matrices, operator algebras and harmonic analysis on free groups.

[201] J. von Neumann. Zur Algebra der Funktionaloperationen und Theorie der normalen Operatoren. *Math. Ann.*, 102:370–427, 1930.

[202] J. von Neumann. On rings of operators. III. *Ann. Math. (2)*, 41(2):94–161, 1940.

[203] J. von Neumann. On rings of operators. Reduction theory. *Ann. Math. (2)*, 50:401–485, 1949.

[204] R. C. Walker. *The Stone-Čech compactification*. Springer-Verlag, New York, 1974. Ergebnisse der Mathematik und ihrer Grenzgebiete, Band 83.

[205] S. Wassermann. Injective W^*-algebras. *Math. Proc. Cambridge Philos. Soc.*, 82(1):39–47, 1977.

[206] H. Wenzl. Hecke algebras of type A_n and subfactors. *Invent. Math.*, 92(2):349–383, 1988.

[207] S. A. White. Values of the Pukánszky invariant in McDuff factors. *J. Funct. Anal.*, to appear. Mathematics ArXiv math.OA/0609269.

[208] S. A. White. Tauer masas in the hyperfinite II$_1$ factor. *Quart. J. Math.*, 57:377–393, 2006.

[209] S. A. White and A. Wiggins. Semi-regular masas of transfinite length. *Internat. J. Math.*, to appear. Mathematics ArXiv math.OA/0611615.

[210] W. Wogen. On generators for von Neumann algebras. *Bull. Amer. Math. Soc.*, 75:95–99, 1969.

[211] F. B. Wright. A reduction for algebras of finite type. *Ann. Math. (2)*, 60:560—570, 1954.

[212] G. Zeller-Meier. Deux nouveaux facteurs de type II$_1$. *Invent. Math.*, 7:235–242, 1969.

Index

Index of symbols

Printed in the United States
by Baker & Taylor Publisher Services